T0182148

Texts in Computer Science

Series editors

David Gries, Department of Computer Science, Cornell University, Ithaca, NY, USA

Orit Hazzan, Faculty of Education in Science and Technology, Technion—Israel Institute of Technology, Haifa, Israel

More information about this series at http://www.springer.com/series/3191

Sergei Kurgalin • Sergei Borzunov

The Discrete Math Workbook

A Companion Manual for Practical Study

 Springer

Sergei Kurgalin
Voronezh State University
Voronezh, Russia

Sergei Borzunov
Voronezh State University
Voronezh, Russia

ISSN 1868-0941   ISSN 1868-095X   (electronic)
Texts in Computer Science
ISBN 978-3-030-06479-2   ISBN 978-3-319-92645-2   (eBook)
https://doi.org/10.1007/978-3-319-92645-2

Printed on acid-free paper

This Springer imprint is published by the registered company Springer Nature Switzerland AG
The registered company address is: Gewerbestrasse 11, 6330 Cham, Switzerland

Preface

Rapid development of information and communication technologies is a characteristic feature of the current time period. Training specialists who meet the requirements of the time in the field of such technologies leads to the necessity of paying special attention to their mathematical education. Herewith it is desirable to shape knowledge presented in the form of practical part of mathematical courses in such a way that after performing these tasks, students get the required competencies. One of the mathematical courses which is taught, as a rule, in early semesters is the course of discrete mathematics. A number of general courses and specialist disciplines included in the cycle of specialist training in the field of information technologies are based on the present course. Thus, mathematical logic and theory of algorithms form the theoretical foundation of informatics; Boolean algebra serves as a basis for methods of electronic circuit development; theory of graphs is used when constructing multiprocessor computing systems and computer networks, and in programming. To teach students, the competent practical application of their theoretical knowledge is one of the methodological objectives the given study guide can help with.

The study guide was created over a number of years and is backed up by the teaching experience of the course at the Department of Computer Science of Voronezh State University. Its contents correspond to Federal State Educational Standards on the following training programs: "Information Systems and Technologies," "Program Engineering," and "Radiophysics." The existing problem books on discrete mathematics and mathematical logic do not fully cover the necessary volume and information scope upon the above-mentioned programs to graduate fully skilled subject matter experts; therefore, this study guide provides a large number of problems of a wide range of difficulty for these training programs.

The present study guide is intended for practical training, laboratory practicals, and self-study. It contains basic theoretical concepts and methods of solving the most fundamental problems; it forms the perceptions of sets, relations on sets, and behavior of different types of relations, as well as functions, general concepts of combinatorics, theory of graphs, Boolean algebra, basic principles of the algorithms theory, etc. Some chapters of the study book, such as Chaps. 10 and 11 go beyond the traditional course of discrete mathematics. Nevertheless, we believe it is necessary to include them into the structure of the study guide, as they promote understanding of the methods of construction and analysis of algorithms. According to the famous Church–Turing thesis, the Turing machine is able to imitate all means of step-by-step computation and is considered to be the model of any computing system existing today. The theory of algorithms' asymptotic analysis studies the methods of getting asymptotic estimations of algorithms' computational complexity, which has the paramount importance for the estimation of the need in resource requirement at the specific implementation of algorithm.

It is only natural that we analyzed certain textbooks and problem books while preparing the study guide with a view to the positive experience contained in them and to produce the most efficient method of presenting the material. Most of them are enumerated in the reference list; in particular, these are [4, 15, 25].

The study guide covers two semesters of the study course; some chapters might be used in one-semester courses.

Each chapter begins with a theoretical part, which albeit occupies little space, is of crucial importance as it reviews the basic concepts of the course and sets the utilized terminology. Alongside with the basic concepts, theoretical provisions, and formulae in the text, there are also instructions on their practical use. Then the detailed solution of several most common problems is shown, and the problems for solving in a classroom (a computer laboratory) are distributed. They can be used for self-study. You will find the answers and solutions to all the problems except for the simplest ones. Exercises differ in complexity level. The most difficult ones are marked with an asterisk (*), placed next to the exercise number. If theoretical part is provided with an example, its end is designated by the symbol □.

The differential characteristics of the study guide, in our opinion, are: detailed problem solution containing typical methods and computational schemes; the presence of a big amount of exercises and examples of different complexity level; reflecting the subject field of information and communication technologies in problem definitions where we thought it relevant. Moreover, taking into account the specificity of bachelor's training program in the field of such technologies, it is needed to implement a developed algorithm in a particular programming language.

In Chaps. 9 and 12, fragments of code in the Pascal language are set in order to demonstrate the applicability of studied methods in programming practice. Headings of some problems in Chap. 12 are in semi-bold; the above-mentioned problems are recommended to do in a computer laboratory.

A separate chapter is dedicated to parallel algorithms. Multiprocessor computing systems are extensively used at present for solving tasks on mathematical and computer modeling, management of database and complex software packages, etc. Furthermore, the majority of modern computing systems back up parallel computations on a hardware level. From this perspective, questions connected with construction and analysis of parallel algorithms are becoming increasingly relevant. It should be pointed out that in many cases, developing an effective parallel algorithm of solution of some task requires attraction of new ideas and methods comparing to creating a sequential algorithm version. These are, for instance, practically important problems of searching a target element in data structures, evaluation of an algebraic expression, etc. Theoretical part of this chapter takes a relatively bigger place in comparison with other chapters, as it deals with more complex issues of applied aspects of parallel programming, rather difficult when first studied. This chapter requires good command of methods of problem solution described in the previous chapters and has, in the authors' opinion, increased complexity.

There is a sufficient amount of illustrations in the book, which help students visualize all objects under consideration and connectivity between them. In the final section, you will find reference data which include the Greek alphabet, basic trigonometric formulae, brief summary on differential and integral calculi, and the most important finite summas, which will mitigate the need for resorting to specialist reference literature. This edition is provided with the reference list, as well as the name and the subject indices. After the first mentioning of scientists' surnames, consult the references with their short biographies taken from [82].

Below you can see the scheme of the chapter information dependence in the form of an oriented graph reflecting the preferable order of covering the academic material. For instance, after having studied Chaps. 1, 2, and 3, you can move to one of the three chapters: 4, 5, or 6, in which contents are relatively independent. The dashed border marks the chapters which will be suitable for readers who want to reach a high level of the subject knowledge; the present sections contain more difficult academic material. This way, after studying Chap. 6 you can either come to Chap. 9, or study the material of Chaps. 7 and 8 more thoroughly for better digestion, and only then switch to Chap. 9.

Voronezh, Russia   Sergei Kurgalin
                                                                                                              Sergei Borzunov

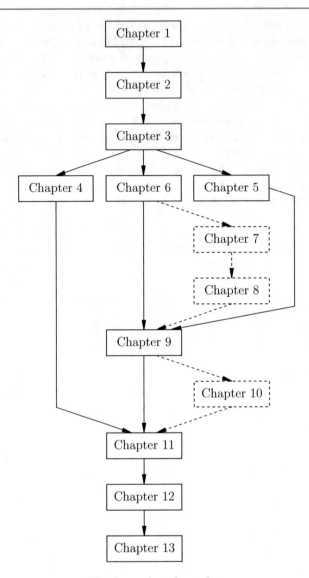

The chapter dependency chart

Contents

Notations

A **and** B	The conjunction of propositions A, B
A **or** B	The disjunction of propositions A, B
$A \Rightarrow B$	The conditional statement, or implication
not A	The negation of the proposition A
$\forall x \, (P(x))$	For all x, $P(x)$ is true
$\exists x \, (P(x))$	There exists an element x such that $P(x)$
F_n	The nth Fibonacci number
L_n	The nth Lucas number
H_n	The nth harmonic number
C_n	The nth Catalan number
$\mathbb{N} = \{0, 1, 2, \ldots\}$	The set of natural numbers
$\mathbb{Z} = \{0, \pm 1, \pm 2, \ldots\}$	The set of integers
$\mathbb{R} = (-\infty, \infty)$	The set of real numbers
$\mathbb{C} = \mathbb{R} \times \mathbb{R}$	The set of complex numbers
$[a, b]$	The closed interval from a to b: $\{x : a \leqslant x \leqslant b\}$
(a, b)	The open interval from a to b: $\{x : a < x < b\}$
$A \subseteq B$	The set A is a subset of the B
$A \subset B$	The set A is a proper subset of the set B
U	The universal set
$\varnothing$	The empty set
$A \cup B$	The union of the sets A and B
$A \cap B$	The intersection of the sets A and B
$A \backslash B$	The complement of the set B with respect to the set A, or the difference of the sets A and B
$\overline{A}$	The complement of the set A
$A \, \Delta \, B$	The symmetric difference of sets A and B
$|A|$	The cardinality of the set A
$\mathcal{P}(A)$	The power set of the set A—the set of all subsets of A
$\aleph_0$	Aleph-null, the cardinality of a countable set
$A \times B$	The Cartesian, or cross, or direct, product of the sets A and B
$(a_1, a_2, \ldots, a_n)$	The vector—the element of a set A^n
$x \, R \, y$	The element x is related to the element y

R^*	The closure of the relation R
$x \prec y$	x is the ancestor y, or the element x is less than the element y
$x \prec\!\!\cdot\, y$	The element y covers the element x
R^{-1}	The inverse relation of the relation R
$S \circ R$	The composite relation for R and S
$g \circ f$	The composition of the functions f and g
$\lfloor x \rfloor$	The floor function of x—the greatest integer less or equal to the real x
$\lceil x \rceil$	The ceiling function of x—the smallest integer greater or equal to the real x
$P(n,k)$	The number of (n,k)-permutations without repetition allowed
$\widetilde{P}(n,k)$	The number of (n,k)-permutations with repetition allowed
$C(n,k)$	The number of (n,k)-combinations without repetition allowed
$\widetilde{C}(n,k)$	The number of (n,k)-combinations with repetition allowed
$G(V,E)$	G is a graph with vertex set V and edge set E
$d(v)$	The degree of vertex d of a graph
$M(i,j)$	The elements of the adjacency matrix
$\widetilde{M}(i,j)$	The elements of the incidence matrix
$G_1 \sim G_2$	The graphs G_1 and G_2 are isomorphic
$\chi(G)$	The chromatic number of the graph G
$\chi'(G)$	The edge chromatic number of the graph G
$d^+(v)$	The out-degree of the vertex v in a digraph
$d^-(v)$	The in-degree of the vertex v in a digraph
$x_1 \wedge x_2$	The conjunction of x_1, x_2
$x_1 \vee x_2$	The disjunction of x_1, x_2
$x_1 \rightarrow x_2$	The implication of x_2 by x_1
$x_1 \leftrightarrow x_2$	The equivalence of x_1 and x_2
$x_1 \oplus x_2$	The Boolean sum of x_1 and x_2
$x_1 \mid x_2$	The Sheffer stroke
$x_1 \downarrow x_2$	The Peirce arrow
$O(g(n))$	The class of functions growing not faster than $g(n)$
$\Omega(g(n))$	The class of functions growing at least as fast as $g(n)$
$\Theta(g(n))$	The class of functions of the same order as $g(n)$
$B(n)$	The best-case time complexity of an algorithm to solve a problem of size n
$A(n)$	The average-case time complexity to solve a problem of size n
$W(n)$	The worst-case time complexity to solve a problem of size n

RAM	Random Access Machine
PRAM	Parallel Random Access Machine
p	The number of nodes in a computing system
S_p	The performance speedup using p processors
E_p	The efficiency of parallel computation using p processors
C_p	The cost of a parallel algorithm for p processors

Fundamentals of Mathematical Logic

Mathematical logic is a science that studies mathematical proofs. Subjects of mathematical logic are mathematical proofs, methods and means for their construction [70].[1]

The simplest division of mathematical logic is the **propositional logic. Proposition** is a statement that has a value of truth; i.e., it can be *true* or *false*. The respective values of truth will be denoted by T or F.

Compound proposition can be built out of atomic propositions using logical operations and brackets. The most common logical operations are: **and** (*conjunction* or *logical multiplication*), **or** (*disjunction* or *logical addition*), **if ... then** (*logical consequence* or *implication*, this operation is also denoted as "⇒"), **not** (*negation*). Conjunction, disjunction, and implication are assigned to **binary** operations, since they use two operands; negation is assigned to **unary** operation, since it requires one operand.

Two compound propositions are called **logically equivalent**, if they take the same values of truth for any set of values of the component parts. A compound proposition that takes true values for any of its components is referred to as **tautology**. A proposition (**not** Q) ⇒ (**not** P) is called **opposite** or **contrapositive** to proposition $P ⇒ Q$.

Notation $P ⇒ Q$ is read as follows: "P implies Q," or "from P follows Q," or "Q is necessary for P," or "P is sufficient for Q."

For justification of the logical equivalence, **truth table** is used, which enumerates the true values of logical expressions for all possible sets of truth values of components (Table 1.1). Equivalence of two propositions can be established by comparing their truth tables. They coincide for the logically equivalent propositions.

---

[1]Numbers in square brackets are references to the literature from the list at the end of the book.

© Springer International Publishing AG, part of Springer Nature 2018
S. Kurgalin and S. Borzunov, *The Discrete Math Workbook*,
Texts in Computer Science, https://doi.org/10.1007/978-3-319-92645-2_1

Table 1.1 Truth table

P	Q	not P	P and Q	P or Q	$P \Rightarrow Q$
T	T	F	T	T	T
T	F	F	F	T	F
F	T	T	F	T	T
F	F	T	F	F	T

If propositions A and B are equivalent, then we write $A \Leftrightarrow B$. The last proposition can be written through the operations of conjunction and implication as $(A \Rightarrow B)$ **and** $(B \Rightarrow A)$.

The basic laws of algebra of logic are listed in Table 1.2. The validity of the mentioned laws is easily proved by constructing the respective truth tables. Some laws of the algebra of logic have direct analogues in the algebra of real numbers, including the commutative, associative, and distributive laws. However, there are such laws as, for example, De Morgan's[2] laws, which have no such analogues.

Statements about properties of variable x are called **predicates** and denoted: $P(x)$, $Q(x)$, ... **Truth domain** of a predicate is a collection of all x, for which the given predicate becomes a true proposition. The predicate properties are studied by **predicate logic**.

For construction of compound logical expressions, we use quantifiers: $\forall$ (**for all**)—universal quantifier—and $\exists$ (**exists**)—existential quantifier. **Quantifier**—logical operation that by predicate $P(x)$ constructs a proposition that characterizes the truth domain $P(x)$ [29].

Logical expression $\forall x \, (P(x))$ (read as "for all x $P(x)$ is true") means that for all possible values x proposition $P(x)$ takes the true value. Expression $\exists x \, (P(x))$ (read "exists x such that $P(x)$ is true") means that for some value x $P(x)$ takes the true value. Brackets after $\forall x$ and $\exists x$ limit the quantifier's range of operation. Often the brackets that define the range of operation are omitted.

The following logical equivalences are known for negation of logical expressions with quantifiers:

$$\textbf{not}\,\forall x \, (P(x)) \;\Leftrightarrow\; \exists x \,(\textbf{not}\, P(x));$$

$$\textbf{not}\,\exists x \, (P(x)) \;\Leftrightarrow\; \forall x \,(\textbf{not}\, P(x)).$$

Note. The names of logical operations derive from Latin words *conjungere* — conjunct, *disjungere* — disjunct, *implicāre* — implicate, *aequālis* — equal, *contrāpositum* — contraposition. Terms "predicate" and "quantifier" derive from Latin *praedicātio* — statement, assertion and *quantum* — how many (much). The names of binary operations derive from Latin *commūtāre* — communicate, *associāre* — associate, *distributere* — distribute.

---

[2]Augustus de Morgan (1806–1871)—Scottish mathematician and logician.

Table 1.2 Laws of algebra of logic

	Idempotent laws	
A or A ⇔ A		A and A ⇔ A
	Properties of constants "T" and "F"	
A or F ⇔ A		A and T ⇔ A
A or T ⇔ T		A and F ⇔ F
	Properties of negation	
A or (not A) ⇔ A	not (not A) ⇔ A	A and (not A) ⇔ F
	Commutative laws	
A or B ⇔ B or A		A and B ⇔ B and A
	Associative laws	
A or (B or C) ⇔ (A or B) or C		A and (B and C) ⇔ (A and B) and C
	Distributive laws	
A or (B and C) ⇔ (A or B) and (A or C)		A and (B or C) ⇔ (A and B) or (A and C)
	Absorption laws	
A or (A and B) ⇔ A		A and (A or B) ⇔ A
	De Morgan's laws	
not (A or B) ⇔ (not A) and (not B)		not (A and B) ⇔ (not A) or (not B)

There are several basic methods for proving the truth of propositions of the form $P \Rightarrow Q$ [27]:

(1) *Forward reasoning*: truth P is suggested and truth Q is derived.
(2) *Backward reasoning*: forward reasoning proves the truth of proposition (**not** Q) $\Rightarrow$ (**not** P) as logically equivalent to $P \Rightarrow Q$.
(3) Method *"by contradiction"*: there are suggested truth of P and false of Q, and on the basis of substantiated reasoning a contradiction is obtained.

A powerful means of proving the truth of propositions with respect to all natural numbers (which include 1, 2, 3, ...) is the mathematical induction principle [4, 14, 69] (from Latin *inductio*).

Mathematical induction principle. *Let $P(n)$ be a predicate defined for all natural numbers n, and let the following conditions be satisfied:*
(1) *$P(1)$ is true.*
(2) *$\forall k \geqslant 1$ implication $P(k) \Rightarrow P(k+1)$ is true.*
Then $P(n)$ is true for any natural n.

Proposition *1* is usually called **basis step**, and proposition *2* is called **inductive step**.

For proof of identity by the mathematical induction method, the following is done. Let predicate $P(k)$ take the true value; then, the considered identity is true for some natural number k. Then, two statements are proved:

(1) induction basis, i.e., $P(1)$;
(2) inductive step, i.e., $P(k) \Rightarrow P(k+1)$ for arbitrary $k \geqslant 1$.

According to the mathematical induction method, a conclusion is made about the truth of the considered identity for all natural values n.

Let us consider the examples of mathematical induction method application for proving identities and inequalities.

Example 1.1 Let us prove that for all natural n the following inequality is fulfilled

$$1^2 + 3^2 + 5^2 + \cdots + (2n-1)^2 = \frac{n(4n^2 - 1)}{3}.$$

Proof Denote predicate "$1^2 + 3^2 + 5^2 + \cdots + (2n-1)^2 = \frac{n(4n^2 - 1)}{3}$" by $P(n)$.

Basis step

Consider the case $n = 1$. Equality takes the form $1^2 = \frac{1(4 \cdot 1^2 - 1)}{3}$, which is a true statement.

Inductive step

Suppose that $P(k)$ for some $k = 1, 2, \ldots$ takes the true value, i.e., $1^2 + 3^2 + 5^2 + \cdots + (2k-1)^2 = \frac{k(4k^2 - 1)}{3}$. Prove that $P(k+1)$ is true.

Write predicate $P(k+1)$ as:

$$1^2 + 3^2 + 5^2 + \cdots + (2k-1)^2 + (2(k+1)-1)^2 = \frac{1}{3}(k+1)(4(k+1)^2 - 1).$$

Use inductive supposition and rewrite the sum $1^2 + 3^2 + 5^2 + \cdots + (2k-1)^2$ as fraction $\frac{1}{3}k(4k^2 - 1)$:

$$\underbrace{1^2 + 3^2 + 5^2 + \cdots + (2k-1)^2}_{k(4k^2-1)/3} + (2(k+1)-1)^2 = \frac{1}{3}(k+1)(4(k+1)^2 - 1);$$

$$\frac{1}{3}k(4k^2 - 1) + (2k+1)^2 = \frac{1}{3}(k+1)(4(k+1)^2 - 1).$$

Transform the left side of the equality:

$$\frac{1}{3}(4k^3 - k + 3(4k^2 + 4k + 1)) = \frac{1}{3}(k+1)(4(k+1)^2 - 1);$$

$$\frac{1}{3}(k+1)(4(k+1)^2 - 1) = \frac{1}{3}(k+1)(4(k+1)^2 - 1).$$

We obtained a true equality. It means that for any $k = 1, 2, \ldots$ implication $P(k) \Rightarrow P(k+1)$ is valid, and the mathematical induction method proved that $1^2 + 3^2 + 5^2 + \cdots + (2n-1)^2 = \dfrac{n(4n^2 - 1)}{3}$ for all natural n. $\qquad\square$

In the next example, as predicate $P(n)$ an inequality is used.

Example 1.2 Prove **Bernoulli's**[3] **inequality** [86]:

$$(1+a)^n \geqslant 1 + na \text{ for } n = 1, 2, \ldots \text{ and } a > -1.$$

Proof Denote predicate "$(1+a)^n \geqslant 1 + na$ for $a > -1$" by $P(n)$.

B a s i s s t e p

For $n = 1$, the inequality takes the form $(1+a)^1 \geqslant 1 + a$, and this is a true statement.

I n d u c t i v e s t e p

Let $P(k)$ be true, i.e., $(1+a)^k \geqslant 1 + ka$ for some natural k and $a > -1$. Check the validity of statement $P(k+1)$: $(1+a)^{k+1} \geqslant 1 + (k+1)a$.

---

[3] Jacob Bernoulli (1654–1705)—Swiss mathematician.

For this, multiply both sides of the inequality that compose the inductive supposition by the positive number $1 + a$:

$$(1 + a)^k (1 + a) \geqslant (1 + ka)(1 + a).$$

Open the brackets in the right side of the obtained inequality:

$$(1 + a)^k (1 + a) \geqslant 1 + (k + 1)a + ka^2.$$

Considering that $ka^2 \geqslant 0$, we obtain:

$$(1 + a)^{k+1} \geqslant 1 + (k + 1)a,$$

which makes the statement $P(k + 1)$. This means that, by the mathematical induction principle, Bernoulli's inequality is valid for all natural n.   □

1.1 Problems

1.1.   Find which of the following sentences are propositions:

(1)   Pascal language belongs to high-level programming languages.
(2)   Drink carrot juice!
(3)   Is there life of Mars?
(4)   The first computers appeared in the eighteenth century.

1.2.   Find which of the following sentences are propositions:

(1)   Any square equation has real roots.
(2)   Triangle ABC is similar to triangle $A'B'C'$.
(3)   It is not true that there is no water on the Moon.
(4)   All hail the Olympic Games!

*1.3.   A sentence is written on a sheet of paper: "This sentence is false." Is it a proposition?

1.4.   Show that the following logical equivalences are fulfilled, known as De Morgan's laws (Table 1.2):

(1)   **not** $(A$ **and** $B)$ $\Leftrightarrow$ **(not** $A)$ **or (not** $B)$;
(2)   **not** $(A$ **or** $B)$ $\Leftrightarrow$ **(not** $A)$ **and (not** $B)$.

∗1.5. Generalize the logical equivalences from the Exercise **1.4** in the case of an arbitrary number of atomic propositions $A_1, A_2, \ldots, A_n$, where $n = 2, 3, \ldots$

1.6. With the help of logical operations, write the following proposition P: "out of three propositions A, B, and C, the true value is taken by exactly one."

1.7. With the help of logical operations, write the following proposition P: "out of three propositions A, B, and C, no less than one of them has the true value."

1.8. With the help of logical operations, write the following proposition P: "out of four propositions A, B, C, and D, three have the same truth value."

1.9. Show that proposition $A \Rightarrow (B \Rightarrow C)$ is logically equivalent to proposition $(A \text{ and } B) \Rightarrow C$.

1.10. Show that proposition $A \Rightarrow (B \Rightarrow C)$ is logically equivalent to proposition $(\text{not } C) \Rightarrow ((\text{not } A) \text{ or } (\text{not } B))$.

1.11. Show that proposition $A \text{ or } (B \text{ and } C)$ is logically equivalent to proposition $(A \text{ or } B) \text{ and } (A \text{ or } C)$.

1.12. Show that proposition $A \text{ and } (B \text{ or } C)$ is logically equivalent to proposition $(A \text{ and } B) \text{ or } (A \text{ and } C)$.

1.13. Prove that proposition

$$((A \text{ or } B) \text{ and } (\text{not } (A \text{ and } B)))) \Leftrightarrow$$
$$((A \text{ and } (\text{not } B)) \text{ or } ((\text{not } A) \text{ and } B))$$

is a tautology.

1.14. Is the proposition $(A \Rightarrow (B \Rightarrow C)) \Rightarrow ((A \text{ and } B) \Rightarrow C)$ a tautology?

1.15. Are the propositions

(1) $(A \text{ and } (B \text{ or } C)) \Rightarrow (A \text{ or } (B \text{ and } C))$,
(2) $(A \text{ and } (B \Rightarrow C)) \text{ and } (B \text{ or } (B \Rightarrow C))$
tautologies?

1.16. Are the propositions

(1) $(A \Rightarrow (B \text{ or } C)) \Rightarrow ((\text{not } B) \Rightarrow A)$,
(2) $((A \text{ and } B) \Rightarrow C) \Rightarrow ((\text{not } C) \Rightarrow A)$
tautologies?

1.17. Denote by A the proposition: "I am hungry," in B "it is three o'clock now," and in C "it is time to have dinner." Write the following propositions as compound propositions, including A, B, and C:

    (1)   If it is three o'clock now or I am hungry, then it is time to have dinner.
    (2)   If it is not time to have dinner, then I am not hungry.
    (3)   If I am hungry, then it is time to have dinner. But I am not hungry. So, either it is not three o'clock now, or it is not time to dinner.

1.18. Denote by A the proposition: "the sun is shining," by B "it is hot outside," and by C "it is snowing." Write the following propositions as compound propositions, including A, B, and C:

    (1)   If it is snowing, then, if the sun is shining, then it is not hot outside.
    (2)   If the sun is shining, then either it is hot outside, or it is snowing (but not simultaneously).

1.19. Formulate a proposition contrapositive to the following:
    "If the Cardinal's guards are nearby, then d'Artagnan is ready to fight."

1.20. Formulate a proposition contrapositive to the following:
    "If Aramis becomes a bishop and Portos obtains a barony, then d'Artagnan will be presented with a marshal's baton."

1.21. Formulate a proposition contrapositive to the following:
    "If Atos intercepts the letter and Portos shakes off the pursuers, then Aramis will disclose the Cardinal's plot."

1.22. Some brackets in compound propositions may be omitted if we agree on the sequence of logical operations in accordance with their binding strength. The strength of binding of operands by logical operations is determined in accordance with the following scheme:

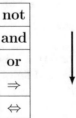

Unary negation operation has the greatest binding strength; then follow binary operations in the following order: conjunction, disjunction, implication, and equivalence operation.
Let A, B, C, D be atomic propositions. Find whether the following propositions are logically equivalent P_1 and P_2:

(1) $P_1 = A$ **or** B **and** C **or** D, $P_2 = (A$ **or** $B)$ **and** $(C$ **or** $D)$;

(2) $P_1 = A \Rightarrow B$ **or** C, $P_2 = A \Rightarrow (B$ **or** $C)$.

1.23. Check by compilation of truth tables whether the following propositions are equivalent:

(1) $X = A$ **or** $(B \Rightarrow C)$, $Y = (A$ **or** $B) \Rightarrow (A$ **or** $C)$;

(2) $X = A \Rightarrow ($**not** $(B$ **and** $C))$,
$Y = $**not** $((A \Rightarrow B)$ **and** $(A \Rightarrow C))$;

(3) $X = A \Rightarrow ((B \Rightarrow C)$ **and** $(C \Rightarrow B))$,
$Y = ((A \Rightarrow B) \Rightarrow (A \Rightarrow C))$ **and** $((A \Rightarrow C) \Rightarrow (A \Rightarrow B))$;

(4) $X = A$ **or** $((B \Rightarrow C)$ **and** $(C \Rightarrow B))$,

$Y = ((A$ **or** $B) \Rightarrow (A$ **or** $C))$ **and** $((A$ **or** $C) \Rightarrow (A$ **or** $B))$.

1.24. Find which of the following propositions are identically false:

(1) $A \Rightarrow (A$ **or** $B)$;

(2) $(A \Rightarrow B) \Rightarrow ($**not** $B \Rightarrow$ **not** $A)$;

(3) $(A$ **and** $B)$ **and** $($**not** A **or not** $B)$;

(4) A **and** $((A \Rightarrow$ **not** $A)$ **and** $($**not** $A \Rightarrow A))$.

1.25. Musketeers noticed a suspicious masked person in the palace park. D'Artagnan believes that it was the Cardinal. Portos states that the unknown is either the Cardinal or Milady. Aramis proves that if Portos is not wrong, then d'Artagnan is right. Atos is sure that Portos and Aramis cannot be wrong. Further investigation proved that Atos's conclusions were right. Who was hiding behind the mask?

1.26. Captain of the King's musketeers, De Treville, received three letters from his subordinates.
The first informed that if Atos had broken a sword in a battle, then Portos and Aramis had broken theirs.
The second stated that Atos and Aramis either both had broken their swords or both had preserved them.
Finally, the third informed: In order for Aramis to break his sword, it is necessary that Portos should break his.
De Treville also knows that out of the three letters sent by the musketeers, one was intercepted and contains false information. Who of the musketeers had broken his sword?

1.27. Three people are suspected of robbery of a jewelry store: Alexandrov, Bykov, and Semenov. Preliminary investigation has lead to the following conclusions:

(1) If Semenov is involved in the crime, then Bykov is also involved.

(2) If Alexandrov is guilty, then Bykov is also guilty.

The first conclusion of the preliminary investigation turned out to be true, and the second turned out to be false. Who committed the robbery?

∗**1.28.**   The workflow provides for the following scheme of operation of four machines S_1–S_4. If the first machine is working, then the second and the third are working too. The third machine is working when and only when the fourth is working. Apart from this, if the second one is working, the fourth must be stopped. Find which machines are working at the moment if it is known that now either the first or the second machine is working (but not simultaneously).

∗**1.29.**   The workflow provides for the following scheme of operation of four machines S_1–S_4. If the first machine is working, then the second is working too. The second machine is working when and only when the third is working, and if the fourth is working, the third must be stopped. Find which machines are working at the moment if it is known that now either the first or the second machine is working (but not simultaneously).

1.30.   A high-performance computing cluster is controlled by two servers: compute and file, and the cluster cooling system consists of three units. The cluster administrator received three diagnostic messages by e-mail:

(1)   Bad file server, and the first cooling unit;
(2)   Bad compute server, and the second cooling unit;
(3)   File server is normal, but bad third cooling unit.

The administrator knows for sure that only one server and no more than one cooling unit are bad. After repair, it turned out that each diagnostic message contained valid information either about the servers' operation or about the cooling units' condition, but not simultaneously.
Which server and which cooling units did the administrator have to repair?

1.31.   A high-performance computing cluster is controlled by two servers: main and standby, and the cluster cooling system consists of three units. The cluster administrator received three diagnostic messages by e-mail:

(1)   Bad main server, and the third cooling unit;
(2)   Bad standby server, and the second cooling unit;
(3)   Main server is normal, but bad first cooling unit.

The administrator knows for sure that only one server and at least two cooling units are bad. After repair, it turned out that each diagnostic message contained valid information either about the servers' operation or about the cooling units' condition, but not simultaneously.
Which server and which cooling units did the administrator have to repair?

1.32. Air service over some territory is provided by four airports: to main (A and B) and two standby (A_1 for A and B_1 for B). The standby airport operates when and only when the respective main airport is closed. The necessary condition for airport B_1 to receive planes is operation of airport A_1. The aircrafts' pilots were instructed that airport B is unable to receive flights due to weather conditions. Find which airports are operating.

1.33. The musketeers of the King's troop go to England to fulfill the Queen's important mission. It is known that Atos and Portos are going (perhaps together), and Portos or Aramis are going (also, perhaps together). If Portos is going, then Aramis is going as well, and if Atos is going, then Aramis is staying in Paris.

Which musketeers are going to fulfill the Queen's mission?

1.34. The musketeers of the King's troop go to England to fulfill the Queen's important mission. It is known that either Atos or Aramis is going (and only one of them). If Portos is not going, then Aramis is staying in Paris. Moreover, if Atos is going, then Aramis is going.

Which musketeers are going to fulfill the Queen's mission?

∗1.35. Four students are on holiday in a recreation camp: Alice, Bogdan, Valeria, and George. Some vacationers are taking a trip, and others are sunbathing at the beach. Either Bogdan or Valeria is taking a trip (perhaps both Bogdan and Valeria have signed up). If Valeria is at the beach and George is at the beach, then Alice is also bathing at the beach. It is not true that if Alice is at the beach then Valeria and George are taking a trip. Which students are taking a trip and who is at the beach?

∗1.36. Four students are on holiday in a recreation camp: Alice, Bogdan, Valeria, and George. Some vacationers are playing at the sports ground, and others are preparing for an amateur show. It is known that either Alice or Bogdan is playing at the sports ground (perhaps for the same team). If Bogdan is preparing for the show and Valeria is preparing for the show, then George is also preparing for the show. Moreover, it is not true that if George is preparing for the show, then Bogdan and Valeria are at the sports ground. Who of the students is playing at the sports ground?

1.37. Consider another two logical operations [4]. The **Sheffer**[4] **stroke** of two propositions A and B is defined as proposition $A \mid B$, which is false when both propositions are true, and is true in other cases:

$$A \mid B = (\textbf{not } A) \textbf{ or } (\textbf{not } B).$$

---

[4]Henry Maurice Sheffer (1882–1964)—American logician.

The **Peirce**[5] **arrow** of two propositions A and B is defined as proposition $A \downarrow B$, which is true when both propositions are false, and is false in other cases:

$$A \downarrow B = (\textbf{not } A) \textbf{ and } (\textbf{not } B).$$

Compile the truth table for propositions $A \mid B$ and $A \downarrow B$.

1.38. Express the basic logical operations (negation, disjunction, conjunction, implication) by:

    (1)   The Sheffer stroke;

    (2)   The Peirce arrow.

***1.39.** Express:

    (1)   Operation "the Peirce arrow" by operation "the Sheffer stroke";

    (2)   Operation "the Sheffer stroke" by operation "the Peirce arrow".

1.40. Which of the following sentences are predicates:

    (1)   Arbitrary even number s can be presented as a sum of two odd numbers.

    (2)   Rational number q is no greater than $-\dfrac{7}{8}$.

    (3)   Triangle ABC is similar to triangle $A'B'C'$.

    (4)   Variables x and y take the same values?

1.41. Write negation of predicate P, if

    (1) $P(n) = \{$natural number n is even$\}$;

    (2) $P(x, y) = \{x < y\}$.

1.42. Present the negation of proposition T so that the negation operations refer to predicates P and Q:

    (1) $T = \forall x\,(\exists y\,(P(x, y)))$;

    (2) $T = \exists x\,(\exists y\,(P(x, y)) \Rightarrow Q(x, y))$.

1.43. Let x and y be real numbers. Find the truth value of the propositions

    (1) $\forall x\,(\exists y\,(x > y))$;

    (2) $\forall x\,(x \neq 0 \Rightarrow \exists y\,(xy = 1))$.

1.44. Let x and y be real numbers. Find the truth value of the propositions

---

[5]Charles Sanders Peirce (1839–1914)—American logician.

(1) $\exists x\,(\forall y\,(xy = y))$;

(2) $\exists x\,(\forall y\,(x + y = 1))$.

1.45. As is known from the course in mathematical analysis [86], the number a is a limit of the numerical sequence $\{a_n\}$ if and only if for any positive number ε there exists such number n_0, that for all natural numbers n, greater or equal to n_0, the inequality $|a_n - a| < \varepsilon$ is met.

With the help of quantifiers, the following statement is written as follows:

$$\lim_{n \to \infty} a_n = a \;\Leftrightarrow\; \forall \varepsilon > 0 \,\exists n_0 \in \mathbb{N}\, \forall n \in \mathbb{N}\, (n \geqslant n_0 \Rightarrow |a_n - a| < \varepsilon).$$

With the help of quantifiers, write the statement: "the limit of the numerical sequence $\{a_n\}$ is not equal to a".

1.46. The number A is a limit of the function $f(x)$ at the point x_0 if and only if $\forall \varepsilon > 0 \,\exists \delta > 0\, \forall x\, (|x - x_0| < \delta \Rightarrow |f(x) - A| < \varepsilon)$.

With the help of quantifiers, write the statement: "the limit of the function $f(x)$ at the point x_0 is not equal to A".

1.47. The function $f(x)$ is continuous at the point x_0 if and only if $\forall \varepsilon > 0 \,\exists \delta > 0\, \forall x\, (|x - x_0| < \delta \Rightarrow |f(x) - f(x_0)| < \varepsilon)$.

With the help of quantifiers, write the statement: "The function $f(x)$ is not continuous at the point x_0".

1.48. With the help of quantifiers, write the statement: "there exists the only value of the variable x, for which the predicate $P(x)$ takes the true value."

1.49. By the mathematical induction method, prove the proposition:
$1 + 3 + 5 + \ldots + (2n - 1) = n^2$ for all natural numbers n.

1.50. By the mathematical induction method, prove the proposition:

$$1 + 4 + 7 + \ldots + (3n - 2) = \frac{1}{2}n(3n - 1)$$

for all natural numbers n.

1.51. By the mathematical induction method, prove the propositions:

(1) $2 + 2^2 + 2^3 + \ldots + 2^n = 2^{n+1} - 2$ for all natural numbers n;

(2) $\dfrac{1}{2} + \dfrac{1}{2^2} + \dfrac{1}{2^3} + \ldots + \dfrac{1}{2^n} = 1 - \dfrac{1}{2^n}$ for all natural numbers n.

1.52. By the mathematical induction method, prove the sum of cubes of the first n natural numbers is equal to the square of their sum:

$$1^3 + 2^3 + \cdots + n^3 = (1 + 2 + \cdots + n)^2, \quad n = 1, 2, 3, \ldots$$

1.53. Prove the formula for the sum of cubes n of the first odd natural numbers

$$1^3 + 3^3 + 5^3 + \cdots + (2n-1)^3 = n^2(2n^2 - 1), \quad n = 1, 2, 3, \ldots$$

1.54. Terms of **arithmetic progression** are defined as $a_n = a_1 + (n-1)d$ for $n = 1, 2, 3, \ldots$, where a_1, d—const. By the mathematical induction method, prove that the sum S_n of the first n terms of arithmetic progression is found by formula $S_n = \dfrac{a_1 + a_n}{2}n$.

1.55. The terms of **geometric progression** are defined as $b_n = b_1 q^{n-1}$ for $n = 1, 2, 3, \ldots$, where b_1, q—const, $q \neq 1$. By the mathematical induction method, prove that the sum S_n of the first n terms of geometric progression is calculated by the formula $S_n = \dfrac{b_1(q^n - 1)}{q - 1}$.

1.56. By the mathematical induction method, prove that $n^2 - n$ is even for all natural n.

1.57. By the mathematical induction method, prove that for all natural numbers n

  (1) $4n^3 + 14n$ is divisible by 3.
  (2) $n^5 - n - 10$ is divisible by 5.
  (3) $6^{2n-1} - 6$ is divisible by 7.
  (4) $n(n-1)(2n-1)$ is divisible by 6.

1.58. By the mathematical induction method, prove that for all natural numbers a and n the number $a^n - 1$ is divisible by $a - 1$.

1.59. By the mathematical induction method, prove that the following numbers are divisible by 10 for all values $n = 1, 2, \ldots$

  (1) $4^n + 4(-1)^n$;
  (2) $3 \cdot 4^n - 8(-1)^n$.

1.60. Prove that the number $\dfrac{1}{10}(79 \cdot 4^n - 4(-1)^n)$ is integer for all values $n = 1, 2, \ldots$

1.61. Prove that the number $\dfrac{79}{30} \cdot 4^n + \dfrac{(-1)^n}{5} - \dfrac{1}{3}$ is integer for all values $n = 1, 2, \ldots$

1.62. By the mathematical induction method, prove that for all natural values n

  (1) $7^{n+1} - 2^{n+1}$ is divisible by 5.
  (2) $13^{n+1} + (-1)^n 12^{n+1}$ is divisible by 25.
  (3) $11 + (-1)^n(2n + 5)$ is divisible by 4.
  (4) $10 - 2^{2n} - 6n$ is divisible by 18.

1.63. Prove that the following expressions take integer values for all natural n:

(1) $\dfrac{1}{10}(2 \cdot 7^n - 7 \cdot 2^n)$;

(2) $\dfrac{1}{60}(7 \cdot 4^n - 42(-1)^n - 10)$;

(3) $\dfrac{1}{36}(9 - (-3)^n(8n - 11))$;

(4) $\dfrac{(-1)^n(2n - 3) + 3}{4}$.

1.64. Prove that the following expressions take integer values for all natural n:

(1) $\dfrac{1}{15}(13 \cdot 5^n - 5 \cdot 2^{n+1})$;

(2) $\dfrac{1}{12}(11 \cdot 5^n - 2^{n+3} - 3)$;

(3) $\dfrac{5}{84}(47 \cdot 7^n - 7 \cdot 3^{n+2} + 28)$;

(4) $\dfrac{7}{60}(23 \cdot 2^{2n} + 12(-1)^n - 20)$.

1.65. Prove the identity

$$\frac{1}{3 \cdot 7} + \frac{1}{7 \cdot 11} + \cdots + \frac{1}{(4n - 1)(4n + 3)} = \frac{n}{3(4n + 3)}$$

for all $n = 1, 2, \ldots$

1.66. Prove the identity

$$\frac{1}{4 \cdot 9} + \frac{1}{9 \cdot 14} + \cdots + \frac{1}{(5n - 1)(5n + 4)} = \frac{n}{4(5n + 4)}$$

for all $n = 1, 2, \ldots$

*1.67.** Suggest the generalization of the identities formulated in Exercises **1.65** and **1.66**.

1.68. Formulate the principle of mathematical induction with the help of quantifiers.

1.69. By forward reasoning, prove the truth of the proposition:

$$n \text{ and } m - even\, numbers \Rightarrow n \cdot m - even\, number.$$

1.70. Prove by backward reasoning:

$$n^2 - odd\, number \Rightarrow n - odd\, number.$$

1.71. Using the method "by contradiction," prove:

$$n \cdot m - odd\,number \implies$$
$$\implies both\,cofactors\,are\,odd\,numbers.$$

1.72. Prove that if a square of some integer p is divisible by 17, then the number p itself is divisible by 17.

1.73. **Irrational number** t can be presented in the form of a fraction with integer numerator and denominator; i.e., there are no such integer p and q, such that $t = \dfrac{p}{q}$ [86].

Using the method "by contradiction," prove that $\sqrt{17}$ is an irrational number.

1.74. Prove irrationality of the number $\sqrt{3} - \sqrt{2}$.

1.75. Prove irrationality of the following numbers:

(1) $\log_2 5$;
(2) $\log_5 16$.

1.76. Let $a = \dfrac{3}{2} - 2\sqrt{2}, b = \sqrt{8} + 1$. At the discrete mathematics and mathematical logic exam, the student was asked to find whether the numbers $a + b$ and $a \cdot b$ are irrational. The student reasons as follows:
"The number a is irrational, the number b is also irrational. Hence, their sum $a + b$ and product $a \cdot b$ are irrational numbers".
Explain what the student's mistake is.

1.77. **Prime number** is such integer $p > 1$ that only has two divisors: unity and the number p itself. The integer $s > 1$, which is not prime, is called **composite** [4, 77].

(1)   Write the first ten prime numbers.
(2)   Using the method "by contradiction," prove that there exist infinitely many prime numbers.

1.78. Write the negation of the predicate

$$P(n) = \{\text{natural number } n \text{ is prime}\}.$$

1.79. Prove that there exist infinitely many prime numbers of the form $8n + 7$, $n = 1, 2, \ldots$

1.80. Prove that there exist infinitely many prime numbers of the form $10n + 9$, $n = 1, 2, \ldots$

1.81. **Factorial** of a positive integer n is the product of all natural numbers up to n inclusive [23,35]:

$$n! = \prod_{i=1}^{n} i$$

(by definition, it is suggested that $0! = 1$ and $1! = 1$). Prove that

$$\sum_{i=1}^{n} \frac{i}{(i+1)!} = 1 - \frac{1}{(n+1)!}.$$

1.82. **Harmonic number** H_n is defined as the sum of the reciprocal values of the first n consecutive natural numbers [24,35]:

$$H_n = \sum_{i=1}^{n} \frac{1}{i}.$$

Using the mathematical induction method, prove the following identities with harmonic numbers for all natural values of the variable n:

(1) $\sum_{i=1}^{n} H_i = (n+1)H_n - n$;

(2) $\sum_{i=1}^{n} i H_i = \frac{n(n+1)}{4}(2H_{n+1} - 1)$.

∗**1.83.** Prove the identities provided in the previous exercise, not applying the mathematical induction method.

1.84. Prove the following inequalities for harmonic numbers [35]:

$$1 + \frac{n}{2} \leqslant H_{2^n} \leqslant 1 + n \quad \text{for all } n = 1, 2, \ldots$$

1.85. Find all values of n, for which the harmonic number H_n is integer.

1.86. **Fibonacci[6] sequence**

$$F_n = 1, 1, 2, 3, 5, 8, 13, 21, 34, 55, \ldots \quad \text{for } n = 1, 2, \ldots$$

is defined by the recurrence relation: $F_{n+2} = F_n + F_{n+1}$ with the initial conditions $F_1 = F_2 = 1$ [23,79]. (For more detail about recurrent relations, see chapter "Recurrence Relations" on p. 273.) Using the mathematical induction method, prove the following properties of the Fibonacci numbers for all natural n:

---

[6]Under the name Fibonacci is known Middle Age mathematician Leonardo Pisano (1170–1250).

(1) $\displaystyle\sum_{i=1}^{n} F_i = F_{n+2} - 1$;

(2) $\displaystyle\sum_{i=1}^{n} F_{2i-1} = F_{2n}$;

(3) $\displaystyle\sum_{i=1}^{n} F_{2i} = F_{2n+1} - 1$;

(4) F_{3n} are even, and F_{3n+1} and F_{3n+2} are odd.

1.87. Prove the validity of the relation $\displaystyle\sum_{i=1}^{n} F_i^2 = F_n F_{n+1}$ for all natural numbers n.

1.88. Prove that for all natural n the identities are fulfilled

(1) $\displaystyle\sum_{i=1}^{n} i F_i = n F_{n+2} - F_{n+3} + 2$;

(2) $\displaystyle\sum_{i=1}^{n} (n - i + 1) F_i = F_{n+4} - n - 3$.

1.89. Calculate the sum $F_1 - F_2 + F_3 - F_4 + \cdots + (-1)^{n+1} F_n$.

1.90. **The Lucas[7] numbers** are defined as $L_0 = 2, L_1 = 1$ and $L_n = L_{n-1} + L_{n-2}$ for integers $n \geqslant 2$ [23]. Write the first ten Lucas numbers.

1.91. Prove the formula for the sum of the Lucas numbers:

$$\sum_{k=0}^{n} L_k = L_{n+2} - 1, \quad n \geqslant 1.$$

1.92. It is often more convenient to use an equivalent formula of the mathematical induction principle [4, 26].
The second form of the mathematical induction principle (the complete induction principle). *Let $P(n)$ be the predicate, determined for all natural numbers n, and let the following conditions be fulfilled:*
(1) $P(1)$ is true.

(2) For arbitrary k from the truth of $P(i)$ for all $i \leqslant k$ follows the truth of $P(k + 1)$.

Then $P(n)$ is true for any natural n.

The complete induction principle is used when for the proof of the predicate $P(k + 1)$ it is required to apply not only $P(k)$, but also some previous predicates.
Formulate the complete induction principle with the help of quantifiers.

---

[7]François Édouard Anatole Lucas (1842–1891)—French mathematician.

*1.93. Prove the equivalence of the first and the second forms of the mathematical induction principle.

1.94. A real number x is set, such that $x + \dfrac{1}{x}$ is integer. Prove that $x^n + \dfrac{1}{x^n}$ is also integer for all natural n.

1.95. A computer game character fights many-headed dragons. At his disposal are two swords, the first cutting exactly three monster's heads in one move and the second exactly five heads. Prove that this character can defeat any dragon with a number of heads no less than eight. A dragon is defeated when no heads are left after several moves.

1.96. A computer game character fights many-headed dragons. At his disposal are two swords, the first cutting exactly four monster's heads in one move and the second exactly 11 heads. What should be the maximum number of a dragon's heads in order for it to be invincible with the help of the available weapon?

1.97. Prove the Fibonacci series terms addition formula

$$F_{n+m} = F_{n-1} F_m + F_n F_{m+1}$$

for all natural n and m, $n > 1$.

1.98. Prove that the expression $F_n^2 + F_{n+1}^2$ is also a Fibonacci number.

*1.99. Prove that the explicit expression for the Fibonacci numbers F_n, where $n = 1, 2, 3, \ldots$, is given by the formula [23]

$$F_n = \frac{1}{\sqrt{5}} \left(\frac{1+\sqrt{5}}{2} \right)^n - \frac{1}{\sqrt{5}} \left(\frac{1-\sqrt{5}}{2} \right)^n,$$

known as the **Binet[8]formula**.

1.100. Prove that for all positive integers $n = 2, 3, \ldots$, the following equality is fulfilled $\varphi^n = F_n \varphi + F_{n-1}$, where $\varphi = \dfrac{1+\sqrt{5}}{2}$ is the golden ratio.

1.101. Prove that the relation $F_{2n-1} F_{2n+1} - F_{2n}^2 = 1$ is fulfilled for $n = 1, 2, 3, \ldots$

1.102. Prove that the relation $F_{2n+1} F_{2n+2} - F_{2n} F_{2n+3} = 1$ is fulfilled for $n = 1, 2, 3, \ldots$

---

[8]Jacques Philippe Marie Binet (1786–1856)—French mathematician and mechanician.

∗1.103. Prove the identity $\arctan \dfrac{1}{F_{2n}} = \arctan \dfrac{1}{F_{2n+1}} + \arctan \dfrac{1}{F_{2n+2}}, \; n \geqslant 1.$

1.104. Prove that the Lucas numbers can be expressed by the Fibonacci numbers

$$L_n = F_{n-1} + F_{n+1}, \quad n \geqslant 2.$$

1.105. Having presented the recurrent relation for the Fibonacci numbers in the form $F_n = F_{n+2} - F_{n+1}$, we can generalize F_n for all integer values of n. Find the expression for the Fibonacci numbers with negative indices.

1.106. Having presented the recurrent relation for the Lucas numbers in the form $L_n = L_{n+2} - L_{n+1}$, we can generalize L_n for all integer values of n. Find the expression for the Lucas numbers with negative indices.

∗1.107. Prove that the explicit expression for the Lucas numbers L_n, where $n = 1, 2, 3, \ldots$, is given by the formula

$$L_n = \left(\frac{1 + \sqrt{5}}{2}\right)^n + \left(\frac{1 - \sqrt{5}}{2}\right)^n.$$

1.108. Prove the following identities for $n = 1, 2, 3, \ldots$:

    (1) $L_{4n} = L_{2n}^2 - 2;$

    (2) $L_{4n+2} = L_{2n+1}^2 + 2.$

1.109. Prove the following identities valid for $n = 1, 2, 3, \ldots$:

    (1) $L_n^2 = L_{2n} + 2(-1)^n;$

    (2) $L_n^3 = L_{3n} + 3(-1)^n L_n.$

1.110. Prove the following identities valid for $n = 1, 2, 3, \ldots$:

    (1) $L_n^4 = L_{4n} + 4(-1)^n L_{2n} + 6;$

    (2) $L_n^5 = L_{5n} + 5(-1)^n L_{3n} + 10L_n.$

1.111. Simplify the expression $L_{n-1}L_m + L_n L_{m+1}$ for the natural values n and m.

1.112. Prove the identity for nonnegative integers n and k, $0 \leqslant k \leqslant n$:

$$L_{n-k}L_{n+k} = L_{2n} + (-1)^{n+k} L_{2k}.$$

1.113. Prove the identities for integer i:

    (1) $L_{6i} = L_{2i}(L_{4i} - 1);$

(2)　$L_{8i} = L_{2i}L_{6i} - L_{4i}$.

1.114. Prove the following relation that combines the Fibonacci numbers and the Lucas numbers :

$$F_{n+k} + (-1)^k F_{n-k} = F_n L_k, \quad n, k = 1, 2, 3, \ldots, n > k.$$

1.115. Prove the identities establishing the connection between the Fibonacci numbers and the Lucas numbers:

(1)　$F_n^2 = \dfrac{L_n^2 - 4(-1)^n}{5}, \quad n = 1, 2, 3, \ldots;$

(2)　$F_{n-1}F_{n+1} = \dfrac{L_n^2 + (-1)^n}{5}, \quad n = 2, 3, \ldots$

1.116. Prove the identities for all integer $n > 3$:

(1)　$F_{n-2}F_{n+2} = \dfrac{1}{5}(L_n^2 - 9(-1)^n);$

(2)　$F_{n-3}F_{n+3} = \dfrac{1}{5}(L_n^2 + 16(-1)^n).$

1.117. Having generalized the identities from Exercises **1.115** and **1.116**, express the product of the Fibonacci numbers $F_{n-k}F_{n+k}$ for nonnegative integer n and k, with $k \leqslant n - 1$, by the Lucas numbers.

1.118. Prove the identity that is valid for the natural values of n:

$$L_{n-1}L_{n+1} - L_n^2 = 5(-1)^{n-1}.$$

1.119. An arbitrary natural number N can decimally be presented only in the form [4,77,78]

$$N = a_k \cdot 10^k + a_{k-1} \cdot 10^{k-1} + \cdots + a_1 \cdot 10 + a_0,$$

where each of $a_i, i = 0, 1, \ldots, k$, is equal to one of the numbers $0, 1, \ldots, 9$. Such a number is written as $N = \overline{a_k a_{k-1} \ldots a_1 a_0}$.
Prove the **criterion of divisibility by** 3: *For the number N to be divisible by 3, it is necessary and sufficient that the sum of the digits of its decimal notation is divisible by 3.*

1.120. Prove the **criterion of divisibility by** 7: *For the number N to be divisible by 7, it is necessary and sufficient that the sum of the tripled number of tens and number of units of its decimal notation is divisible by 7.*

1.121. Prove the **criterion of divisibility by** 9: *For the number N to be divisible by 9, it is necessary and sufficient that the sum of the digits of its decimal notation is divisible by 9.*

1.122. Prove the **criterion of divisibility by** 11: *For the number N to be divisible by 11, it is necessary and sufficient that the difference between the sums of the digits standing in its decimal notation on even and odd places is divisible by 11.*

∗**1.123**. **Fundamental theorem of arithmetic** states that *each natural number greater than unity is either a prime number or can be presented in the form of the product of prime numbers. Such a presentation is unique to an accuracy of the arrangement of the cofactors* [29,77].
Using the mathematical induction principle in the second form, prove the fundamental theorem of arithmetic.

1.124. Prove that for an arbitrary number n in the natural series $1, 2, 3, \ldots$, there exist n of consecutive composite numbers.

1.125. Find ten consecutive composite numbers.

1.2 Answers, Hints, Solutions

1.1. *Solution.*
As follows from the definition, proposition is a sentence about which it is possible to say that it is true or false. Statement "Pascal language belongs to high-level programming languages" is true, while "The first computers appeared in the eighteenth century" is false. Interrogative and exclamatory sentences are not assigned to propositions. As a result, we obtain that only sentences (1) and (4) are propositions.

1.2. *Answer*: Propositions are sentences (1) and (3).

1.3. *Answer*: No, since a contradiction follows from the supposition about truth or falsity of the considered sentence.

1.4. *Proof.*
Compile a truth table for the statements provided:

A	B	**not** $(A$ **and** $B)$	$($ **not** $A)$ **or** $($ **not** $B)$	**not** $(A$ **or** $B)$	$($ **not** $A)$ **and** $($ **not** $B)$
T	T	F	F	F	F
T	F	T	T	F	F
F	T	T	T	F	F
F	F	T	T	T	T

The columns corresponding to the right and left sides of the equality coincide. This proves the truth of the logical equivalences **not** $(A$ **and** $B) \Leftrightarrow ($ **not** $A)$ **or** $($ **not** $B)$ and **not** $(A$ **or** $B) \Leftrightarrow ($ **not** $A)$ **and** $($ **not** $B)$.

1.5. *Answer*:

(1) **not** $(A_1$ **and** A_2 **and** ... **and** $A_n)$ $\Leftrightarrow$ (**not** A_1) **or** (**not** A_2) **or** ... **or** (**not** A_n);

(2) **not** $(A_1$ **or** A_2 **or** ... **or** $A_n)$ $\Leftrightarrow$ (**not** A_1) **and** (**not** A_2) **and** ... **and** (**not** A_n).

1.6. *Solution*.

Proposition P may be presented as disjunction of three propositions: $P = P_A$ **or** P_B **or** P_C, where by P_A is denoted proposition "A—true, B—false, and C—false," and P_B and P_C for other two variants of the truth values of propositions A, B, and C are written similarly. Finally, we obtain:

$$P = (A \text{ and not } B \text{ and not } C) \text{ or } (\text{not } A \text{ and } B \text{ and not } C) \text{ or}$$
$$(\text{ not } A \text{ and not } B \text{ and } C).$$

1.7. *Answer*: $P = A$ **or** B **or** C.

1.8. *Answer*:

$$P = \textbf{not} \left((A \text{ and } B \text{ and not } C \text{ and not } D) \text{ or } (A \text{ and not } B \text{ and } C \text{ and not } D) \text{ or} \right.$$
$$(A \text{ and not } B \text{ and not } C \text{ and } D) \text{ or } (\text{not } A \text{ and } B \text{ and } C \text{ and not } D) \text{ or}$$
$$\left. (\text{not } A \text{ and } B \text{ and not } C \text{ and } D) \text{ or } (\text{not } A \text{ and not } B \text{ and } C \text{ and } D) \right).$$

1.9. *Solution*.

Compile a truth table for the statements provided:

A	B	C	A and B	$B \Rightarrow C$	$A \Rightarrow (B \Rightarrow C)$	$(A$ and $B) \Rightarrow C$
T	T	T	T	T	T	T
T	T	F	T	F	F	F
T	F	T	F	T	T	T
T	F	F	F	T	T	T
F	T	T	F	T	T	T
F	T	F	F	F	T	T
F	F	T	F	T	T	T
F	F	F	F	T	T	T

The two last columns coincide, this is why the respective expressions are logically equivalent.

1.10. *Solution*.

Denote: $P_1 = A \Rightarrow (B \Rightarrow C)$, $P_2 = (\textbf{not } C) \Rightarrow ((\textbf{not } A) \textbf{ or } (\textbf{not } B))$. Truth table for propositions P_1 and P_2 is presented below.

From comparison of the two last columns follows the logical equivalence P_1 and P_2.

A	B	C	B ⇒ C	(not A) or (not B)	P_1	P_2
T	T	T	T	F	T	T
T	T	F	F	F	F	F
T	F	T	T	T	T	T
T	F	F	T	T	T	T
F	T	T	T	T	T	T
F	T	F	F	T	T	T
F	F	T	T	T	T	T
F	F	F	T	T	T	T

1.11. *Solution.*

The problem is solved by compiling a truth table for propositions $P_1 = A$ **or** $(B$ **and** $C)$, $P_2 = (A$ **or** $B)$ **and** $(A$ **or** $C)$.

A	B	C	A or B	A or C	B and C	P_1	P_2
T	T	T	T	T	T	T	T
T	T	F	T	T	F	T	T
T	F	T	T	T	F	T	T
T	F	F	T	T	F	T	T
F	T	T	T	T	T	T	T
F	T	F	T	F	F	F	F
F	F	T	F	T	F	F	F
F	F	F	F	F	F	F	F

1.12. *Solution.*

The problem is solved by compiling a truth table for propositions $P_1 = A$ **and** $(B$ **or** $C)$, $P_2 = (A$ **and** $B)$ **or** $(A$ **and** $C)$.

A	B	C	A and B	A and C	B or C	P_1	P_2
T	T	T	T	T	T	T	T
T	T	F	T	F	T	T	T
T	F	T	F	T	T	T	T
T	F	F	F	F	F	F	F
F	T	T	F	F	T	F	F
F	T	F	F	F	T	F	F
F	F	T	F	F	T	F	F
F	F	F	F	F	F	F	F

1.13. *Proof.*

Let us introduce the following notations:

$P_1 = (A$ **or** $B)$ **and** $($**not** $(A$ **and** $B))$,

$P_2 = (A$ **and** $($**not** $B)) $ **or** $((\text{\textbf{not}}\ A)$ **and** $B)$.

For all possible sets of truth values of propositions A and B proposition $P_1 \Leftrightarrow P_2$ is true, this is why the provided proposition is a tautology.

A	B	A **and not** B	**not** A **and** B	P_1	P_2	$P_1 \Leftrightarrow P_2$
T	T	F	F	F	F	T
T	F	T	F	T	T	T
F	T	F	T	T	T	T
F	F	F	F	F	F	T

1.14. *Solution.*

Compile a truth table for the compound proposition $P = (A \Rightarrow (B \Rightarrow C)) \Rightarrow ((A \text{ and } B) \Rightarrow C)$:

A	B	C	$B \Rightarrow C$	$A \Rightarrow (B \Rightarrow C)$	A **and** B	$(A \text{ and } B) \Rightarrow C$	P
T	T	T	T	T	T	T	T
T	T	F	F	F	T	F	T
T	F	T	T	T	F	T	T
T	F	F	T	T	F	T	T
F	T	T	T	T	F	T	T
F	T	F	F	T	F	T	T
F	F	T	T	T	F	T	T
F	F	F	T	T	F	T	T

Analysis of the table shows that for all possible sets of truth values of atomic propositions A, B, and C proposition P is true. Hence, the provided proposition is a tautology.

1.15. *Solution.*

(1) Compile a truth table for the provided proposition $P = (A \text{ and } (B \text{ or } C)) \Rightarrow (A \text{ or } (B \text{ and } C))$. For all sets of values of propositions, A, B, and C, being parts of P, P take the true value.

A	B	C	B **or** C	A **and** $(B$ **or** $C)$	B **and** C	A **or** $(B$ **and** $C)$	P
T	T	T	T	T	T	T	T
T	T	F	T	T	F	T	T
T	F	T	T	T	F	T	T
T	F	F	F	F	F	T	T
F	T	T	T	F	T	T	T
F	T	F	T	F	F	F	T
F	F	T	T	F	F	F	T
F	F	F	F	F	F	F	T

Thus, proposition P is a tautology.

(2) Let $P = (A \text{ and } (B \Rightarrow C))$ **and** $(B \text{ or } (A \Rightarrow C))$. From the analysis of the truth table for proposition P follows that for some sets of values of propositions A, B, and C, being a part of P (e.g., for $A = B = $ T, $C = $ F), P takes a false value.

Hence, proposition P is not a tautology.

A	B	C	$B \Rightarrow C$	A and $(B \Rightarrow C)$	$A \Rightarrow C$	B or $(A \Rightarrow C)$	P
T	T	T	T	T	T	T	T
T	T	F	F	F	F	T	F
T	F	T	T	T	T	T	T
T	F	F	T	T	F	F	F
F	T	T	T	F	T	T	F
F	T	F	F	F	T	T	F
F	F	T	T	F	T	T	F
F	F	F	T	F	T	T	F

1.16. *Answer*:

(1) No, it is not.
(2) No, it is not.

1.17. *Solution*.

Note that the sequence of propositions written in the form of separate sentences with the help of logical operations should be written in the form of conjunction of atomic propositions. Thereby, and using the basic logical operations, we obtain the following symbol form of the compound propositions:

(1) $(B$ **or** $A) \Rightarrow C$;
(2) $(\textbf{not } C) \Rightarrow (\textbf{not } A)$;
(3) $((A \Rightarrow C)$ **and** $(\textbf{not } A)) \Rightarrow ((\textbf{not } B)$ **or** $(\textbf{not } C))$.

1.18. *Answer*:

(1) $C \Rightarrow (A \Rightarrow (\textbf{not } B))$;
(2) $A \Rightarrow ((B$ **and** $(\textbf{not } C))$ **or** $((\textbf{not } B)$ **and** $C))$.

1.19. *Solution*.

We consider the provided proposition as a compound one. Let us denote the propositions that are parts of it by

$A = $ "the Cardinal's guards are nearby";
$B = $ "d'Artagnan is ready to fight".

Contrapositive to proposition $A \Rightarrow B$ by definition is **not** $B \Rightarrow$ **not** A. Hence, the answer to this exercise is **not** $B \Rightarrow$ **not** $A = $ "if d'Artagnan is not ready to fight, then there are no Cardinal's guards nearby".

1.20. *Solution*.

Let us denote the propositions that are parts of the provided compound proposition as

$A = $ "Aramis becomes a bishop";
$B = $ "Portos obtains a barony";
$C = $ "d'Artagnan will be presented with a marshal's baton."

Contrapositive to proposition $(A$ **and** $B) \Rightarrow C$ is $(\textbf{not } C) \Rightarrow (\textbf{not } (A$ **and** $B))$, which is logically equivalent to the following: $(\textbf{not } C) \Rightarrow ((\textbf{not } A)$ **or** $(\textbf{not } B))$. Thus, the answer to this exercise is $(\textbf{not } C) \Rightarrow ((\textbf{not } A)$ **or** $(\textbf{not } B)) = $ "if d'Artagnan is not presented with a marshal's baton, then either Aramis will not become a bishop or Portos will not obtain a barony."

1.21. *Answer*:

"If Aramis does not disclose the Cardinal's plot, then either Atos will not intercept the letter, or Portos will not shake off the pursuers."

1.22. *Answer*:

(1) P_1 and P_2 are not logically equivalent.
(2) Equivalent propositions.

1.23. *Answer*:

(1) Equivalent propositions;
(2) X and Y are not logically equivalent;
(3) Equivalent propositions;
(4) Equivalent propositions.

1.24. *Answer*:

Identically false are propositions of items (3) and (4).

1.25. *Solution*.

Let us introduce the notations:

C = "incognito—the Cardinal";
M = "incognito—Milady".

Let us write each musketeer's proposition with the help of logical operations:

(1) d'Artagnan: C is true, $C = \text{T}$;
(2) Portos: $P = (C \text{ **or** } M) \text{ **and not** } (C \text{ **and** } M)$.

By compilation of a truth table, it is easy to show that Portos's proposition is logically equivalent to

$$P = (C \text{ **and not** } M) \text{ **or** } (M \text{ **and not** } C).$$

(3) Aramis: $P \Rightarrow C$.
(4) Atos: $P \text{ **and** } (P \Rightarrow C)$.

According to the condition, Atos's proposition is true, $P \text{ **and** } (P \Rightarrow C) = \text{T}$. Conjunction of two propositions is true if and only if each of the propositions takes the true value, this is why

$$\begin{cases} P = \text{T}; \\ P \Rightarrow C = \text{T}. \end{cases}$$

From the second equation follows $\text{T} \Rightarrow C = \text{T}$, hence, proposition C is true, and the Cardinal was hiding behind the mask.

1.26. *Answer*: Portos and Aramis had broken their swords.

1.27. *Solution*.

Let us introduce the notations:

A —"Alexandrov is guilty";

B —"Bykov is guilty";

C —"Semenov is guilty."

Let $P = (C \Rightarrow B)$ **and not** $(A \Rightarrow B)$. The problem statement implies that $P = $ T. Let us write the truth table for the propositions in the left side of the obtained equality.

A	B	C	$C \Rightarrow B$	$A \Rightarrow B$	**not** $(A \Rightarrow B)$	P
T	T	T	T	T	F	F
T	T	F	T	T	F	F
T	F	T	F	F	T	F
T	F	F	T	F	T	T
F	T	T	T	T	F	F
F	T	F	T	T	F	F
F	F	T	F	T	F	F
F	F	F	T	T	F	F

The compound proposition $(C \Rightarrow B)$ **and not** $(A \Rightarrow B)$ takes the true value only in one case, namely when propositions B and C are false and proposition A is true. The conclusion follows from here that Alexandrov is guilty of the robbery.

1.28. *Solution.*

First method

Let $S_i = $ T, if ith machine is working, where $i = 1, 2, 3, 4$, and $S_i = $ F otherwise. Let us compile a truth table for the logical expressions corresponding to machine's operating conditions (Table 1.3). From the problem statement, we conclude that the following compound propositions have the true value:

$$P_1 = S_1 \Rightarrow (S_2 \text{ and } S_3), \qquad P_2 = S_3 \Leftrightarrow S_4,$$
$$P_3 = S_2 \Rightarrow \text{not } S_4, \qquad P_4 = (S_1 \text{ and not } S_2) \text{ or } (\text{not } S_1 \text{ and } S_2).$$

From the truth table, it follows that

$$S_1 = S_3 = S_4 = \text{F}, \quad S_2 = \text{T}.$$

We obtain that only the second machine is working at the moment.

Second method

Since the number of rows in the truth table is 2^n, where n is the number of variables in the logical expression [4], then as n grows, the size of the truth table grows rapidly, and it becomes difficult to use the truth table. Thereby, the preferred method is the method of simplification of logical expressions using equivalence transformations.

Consider logical expression

$$P = P_1 \text{ and } P_2 \text{ and } P_3 \text{ and } P_4.$$

Table 1.3 To Exercise **1.28**

S_1	S_2	S_3	S_4	P_1	P_2	P_3	P_4	P
T	T	T	T	T	T	F	F	F
T	T	T	F	T	F	T	F	F
T	T	F	T	F	F	F	F	F
T	T	F	F	F	T	T	F	F
T	F	T	T	F	T	T	T	F
T	F	T	F	F	F	T	T	F
T	F	F	T	F	F	T	T	F
T	F	F	F	F	T	T	T	F
F	T	T	T	T	T	F	T	F
F	T	T	F	T	F	T	T	F
F	T	F	T	T	F	F	T	F
F	T	F	F	T	T	T	T	T
F	F	T	T	T	T	T	F	F
F	F	T	F	T	F	T	F	F
F	F	F	T	T	F	T	F	F
F	F	F	F	T	T	T	F	F

According to the problem conditions, the given expression takes the true value. Simplify P, using the laws of the algebra of logic (Table 1.2).

Since the change of the order of the expressions bound by the conjunction operation takes place in a logically equivalent formula (commutative law), then let us present expression P in the form

$$P = P_1 \text{ and } P_2 \text{ and } P_3 \text{ and } P_4 = ((P_1 \text{ and } P_4) \text{ and } P_2) \text{ and } P_3.$$

Now, we sequentially simplify the obtained expression:

$P_1 \text{ and } P_4 = (S_1 \Rightarrow (S_2 \text{ and } S_3)) \text{ and } [(S_1 \text{ and not } S_2) \text{ or } (\text{not } S_1 \text{ and } S_2)] \Leftrightarrow$
$\Leftrightarrow (\text{not } S_1 \text{ or } (S_2 \text{ and } S_3)) \text{ and } [(S_1 \text{ and not } S_2) \text{ or } (\text{not } S_1 \text{ and } S_2)] \Leftrightarrow$
$\Leftrightarrow [(\text{not } S_1 \text{ or } (S_2 \text{ and } S_3) \text{ and } (S_1 \text{ and not } S_2)] \text{ or}$
$\qquad \text{or } [(\text{not } S_1 \text{ or } (S_2 \text{ and } S_3)) \text{ and } (\text{not } S_1 \text{ and } S_2)] \Leftrightarrow$
$\Leftrightarrow [(\text{not } S_1 \text{ and } S_1 \text{ and not } S_2) \text{ or } (S_2 \text{ and } S_3 \text{ and } S_1 \text{ and not } S_2)] \text{ or}$
$\qquad \text{or } [(\text{not } S_1 \text{ and not } S_1 \text{ and } S_2) \text{ or } (S_2 \text{ and } S_3 \text{ and not } S_1 \text{ and } S_2)] \Leftrightarrow$
$\Leftrightarrow (\text{F or F}) \text{ or } (\text{not } S_1 \text{ and } S_2) \text{ or } (\text{not } S_1 \text{ and } S_2 \text{ and } S_3) = \text{not } S_1 \text{ and } S_2.$

Then, we calculate $(P_1$ and $P_4)$ and P_3:

$$(P_1 \text{ and } P_4) \text{ and } P_3 = (\text{not } S_1 \text{ and } S_2) \text{ and } (S_2 \Rightarrow \text{not } S_4) \Leftrightarrow$$
$$\Leftrightarrow (\text{not } S_1 \text{ and } S_2) \text{ and } (\text{not } S_2 \text{ or not } S_4) \Leftrightarrow$$
$$\Leftrightarrow (\text{not } S_1 \text{ and } S_2 \text{ and not } S_2) \text{ or } (\text{not } S_1 \text{ and } S_2 \text{ and not } S_4) \Leftrightarrow$$
$$\Leftrightarrow \text{not } S_1 \text{ and } S_2 \text{ and not } S_4.$$

Let us perform the final logical multiplication:

$$P = ((P_1 \text{ and } P_4) \text{ and } P_3) \text{ and } P_2 =$$
$$= (\text{not } S_1 \text{ and } S_2 \text{ and not } S_4) \text{ and } [(S_3 \text{ and } S_4) \text{ or } (\text{not } S_3 \text{ and not } S_4)] \Leftrightarrow$$
$$\Leftrightarrow (\text{not } S_1 \text{ and } S_2 \text{ and not } S_4 \text{ and } S_3 \text{ and } S_4) \text{ or }$$
$$\text{or } (\text{not } S_1 \text{ and } S_2 \text{ and not } S_4 \text{ and not } S_3 \text{ and not } S_4) \Leftrightarrow$$
$$\Leftrightarrow \text{F or } (\text{not } S_1 \text{ and } S_2 \text{ and not } S_4 \text{ and not } S_3 \text{ and not } S_4) \Leftrightarrow$$
$$\Leftrightarrow \text{not } S_1 \text{ and } S_2 \text{ and not } S_3 \text{ and not } S_4.$$

The obtained equivalent formula

$$P = P_1 \text{ and } P_2 \text{ and } P_3 \text{ and } P_4 = \text{not } S_1 \text{ and } S_2 \text{ and not } S_3 \text{ and not } S_4$$

allows easily finding the values of logical expressions S_1–S_4:

$$S_1 = S_3 = S_4 = \text{F}, \quad S_2 = \text{T}.$$

Hence, only the second machine is working at the moment.

1.29. *Solution.*
First method
Let $S_i = \text{T}$, if ith machine is working, where $i = 1, 2, 3, 4$, and $S_i = \text{F}$ otherwise. The answer to the problem's question can be obtained by compiling a truth table for the logical expressions corresponding to the machine's operating conditions (Table 1.4). From the problem statement, we conclude that the following compound propositions have the true value:

$$P_1 = S_1 \Rightarrow S_2, \qquad P_2 = S_2 \Leftrightarrow S_3,$$
$$P_3 = S_4 \Rightarrow \text{not } S_3, \qquad P_4 = (S_1 \text{ and } S_2) \text{ or } (\text{not } S_1 \text{ and not } S_2).$$

From the truth table, it follows that

$$S_1 = S_4 = \text{F}, \quad S_2 = S_3 = \text{T}.$$

We finally obtain that the second and the third machines are working at the moment.

Table 1.4 To Exercise **1.29**

S_1	S_2	S_3	S_4	P_1	P_2	P_3	P_4	P
T	T	T	T	T	T	F	F	F
T	T	T	F	T	T	T	F	F
T	T	F	T	T	F	T	F	F
T	T	F	F	T	F	T	F	F
T	F	T	T	F	F	F	T	F
T	F	T	F	F	F	T	T	F
T	F	F	T	F	T	T	T	F
T	F	F	F	F	T	T	T	F
F	T	T	T	T	T	F	T	F
F	T	T	F	T	T	T	T	T
F	T	F	T	T	F	T	T	F
F	T	F	F	T	F	T	T	F
F	F	T	T	T	F	F	F	F
F	F	T	F	T	F	T	F	F
F	F	F	T	T	T	T	F	F
F	F	F	F	T	T	T	F	F

Second method
Consider logical expression

$$P = P_1 \text{ and } P_2 \text{ and } P_3 \text{ and } P_4.$$

According to the problem statement, the following expression takes a true value.
Let us try to simplify P.

Applying the commutative law, we present expression P in the form

$$P = P_1 \text{ and } P_2 \text{ and } P_3 \text{ and } P_4 = ((P_1 \text{ and } P_4) \text{ and } P_2) \text{ and } P_3.$$

Now, we sequentially simplify the obtained expression:

$P_1 \text{ and } P_4 = (S_1 \Rightarrow S_2) \text{ and } [(S_1 \text{ and not } S_2) \text{ or } (\text{not } S_1 \text{ and } S_2)] \Leftrightarrow$
$\Leftrightarrow (\text{not } S_1 \text{ or } S_2) \text{ and } [(S_1 \text{ and not } S_2) \text{ or } (\text{not } S_1 \text{ and } S_2)] \Leftrightarrow$
$\Leftrightarrow [(\text{not } S_1 \text{ or } S_2) \text{ and } (S_1 \text{ and not } S_2)] \text{ or } [(\text{not } S_1 \text{ or } S_2) \text{ and } (\text{not } S_1 \text{ and } S_2)] \Leftrightarrow$
$\Leftrightarrow [(\text{not } S_1 \text{ and } S_1 \text{ and not } S_2) \text{ or } (S_2 \text{ and } S_1 \text{ and not } S_2)] \text{ or }$
$\quad \text{or } [(\text{not } S_1 \text{ and not } S_1 \text{ and } S_2) \text{ or } (S_2 \text{ and not } S_1 \text{ and } S_2)] \Leftrightarrow$
$\Leftrightarrow (\text{F or F}) \text{ or } (\text{not } S_1 \text{ and } S_2) \text{ or } (S_2 \text{ and not } S_1) \Leftrightarrow \text{not } S_1 \text{ and } S_2.$

Then, we calculate $(P_1$ and $P_4)$ and P_2:

$(P_1$ and $P_4)$ and $P_2 = ($not S_1 and $S_2)$ and $[S_2$ and S_3 or (not S_2 and not $S_3)] \Leftrightarrow$
$$\Leftrightarrow (\text{not } S_1 \text{ and } S_2 \text{ and } S_2 \text{ and } S_3) \text{ or}$$
$$\text{or (not } S_1 \text{ and } S_2 \text{ and not } S_2 \text{ and not } S_3) \Leftrightarrow$$
$$\Leftrightarrow (\text{not } S_1 \text{ and } S_2 \text{ and } S_3) \text{ or F} \Leftrightarrow \text{not } S_1 \text{ and } S_2 \text{ and } S_3.$$

Now, we only need to perform the last logical multiplication:

$P = ((P_1$ and $P_4)$ and $P_2)$ and $P_3 = ($not S_1 and S_2 and $S_3)$ and $(S_4 \Rightarrow$ not $S_3) \Leftrightarrow$
$$\Leftrightarrow (\text{not } S_1 \text{ and } S_2 \text{ and } S_3) \text{ and (not } S_3 \text{ or not } S_4) \Leftrightarrow$$
$$\Leftrightarrow (\text{not } S_1 \text{ and } S_2 \text{ and } S_3 \text{ and not } S_3) \text{ or (not } S_1 \text{ and } S_2 \text{ and } S_3 \text{ and not } S_4) \Leftrightarrow$$
$$\Leftrightarrow \text{F or (not } S_1 \text{ and } S_2 \text{ and } S_3 \text{ and not } S_4) \Leftrightarrow \text{not } S_1 \text{ and } S_2 \text{ and } S_3 \text{ and not } S_4.$$

Of course, we may choose any other order of logical multiplication of expressions P_1–P_4, if this will provide for convenience of calculations.

The obtained equivalent formula

$$P = P_1 \text{ and } P_2 \text{ and } P_3 \text{ and } P_4 = \text{not } S_1 \text{ and } S_2 \text{ and } S_3 \text{ and not } S_4$$

allows easily finding the values of logical expressions S_1–S_4:

$$S_1 = S_4 = \text{F}, \quad S_2 = S_3 = \text{T}.$$

Hence, the second and the third machines are working at the moment.

1.30. *Answer*:
Bad were the compute server and the first cooling unit.

1.31. *Answer*:
The administrator had to repair the main server and the first and the second cooling units.

1.32. *Solution*.
The problem can be solved by compiling a truth table or by the method of equivalence transformations of logical expressions. However, it is easier to note that since the standby airport operates when and only when the respective main airport is closed, it is easy to conclude: operates B_1, $B_1 = \text{T}$. Out of the remaining airports A and A_1, one and only one can receive planes. Further, operation of A_1 is necessary for B_1, this is why $A_1 = \text{T}$. Thus, the operating airports are A_1 and B_1.

1.33. *Solution*.
Let us introduce logical variables:
      A — "Atos is going";
      B — "Portos is going";
      C — "Aramis is going."

From the problem statement, we conclude that
$(A$ or $B)$ and $(B$ or $C)$ and $(B \Rightarrow C)$ and $(A \Rightarrow$ not $C) = $ T.
Let us simplify the expression in the left side of the obtained equality:

$$(A \text{ or } B) \text{ and } (B \text{ or } C) \text{ and } (B \Rightarrow C) \text{ and } (A \Rightarrow \text{ not } C) \Leftrightarrow$$
$$\Leftrightarrow (A \text{ or } B) \text{ and } (B \text{ or } C) \text{ and } (\text{not } B \text{ or } C) \text{ and } (\text{not } A \text{ or } \text{ not } C) \Leftrightarrow$$
$$\Leftrightarrow (A \text{ or } B) \text{ and } [(B \text{ and not } B) \text{ or } C] \text{ and } (\text{not } A \text{ or } \text{ not } C) \Leftrightarrow$$
$$\Leftrightarrow (A \text{ or } B) \text{ and } C \text{ and } (\text{not } A \text{ or } \text{ not } C) \Leftrightarrow$$
$$\Leftrightarrow (A \text{ or } B) \text{ and } [(C \text{ and not } A) \text{ or } (C \text{ and } \text{ not } C)] \Leftrightarrow$$
$$\Leftrightarrow (A \text{ or } B) \text{ and } (\text{not } A \text{ and } C) \Leftrightarrow$$
$$\Leftrightarrow (A \text{ and } (\text{not } A \text{ and } C)) \text{ or } (\text{not } A \text{ and } B \text{ and } C) \Leftrightarrow$$
$$\Leftrightarrow \text{ not } A \text{ and } B \text{ and } C.$$

Hence, Portos and Aramis are going.

1.34. *Answer*: Aramis and Portos.

1.35. *Solution*.
Since each vacationer is either taking a trip or bathing at the beach, we may introduce logical variables A, B, V, and G, whose value determines the occupation of each student. Denote $A = $ T, if Alice is taking a trip, and $A = $ F, if Alice is at the beach. By a similar rule, determine the values of variables B, V, G.

According to the problem statement, expression

$$(B \text{ or } V) \text{ and } ((\text{not } V \text{ and not } G) \Rightarrow \text{ not } A) \text{ and } (\text{not } (\text{not } A \Rightarrow (V \text{ or } G)))$$

takes the true value:

$$(B \text{ or } V) \text{ and } ((\text{not } V \text{ and not } G) \Rightarrow \text{ not } A) \text{ and } (\text{not } (\text{not } A \Rightarrow (V \text{ or } G))) \Leftrightarrow$$
$$\Leftrightarrow (B \text{ or } V) \text{ and } (V \text{ or } G \text{ or not } A) \text{ and } (\text{not } A \text{ and not } V \text{ and } \text{ not } G) \Leftrightarrow$$
$$\Leftrightarrow ((B \text{ and not } A \text{ and not } V \text{ and not } G) \text{ or}$$
$$\text{or } (V \text{ and not } A \text{ and not } V \text{ and not } G)) \text{ and}$$
$$\text{and } (V \text{ or } G \text{ or not } A) \Leftrightarrow$$
$$\Leftrightarrow ((\text{not } A \text{ and } B \text{ and not } V \text{ and not } G) \text{ or } F) \text{ and } (V \text{ or } G \text{ or not } A) \Leftrightarrow$$
$$\Leftrightarrow (\text{not } A \text{ and } B \text{ and not } V \text{ and not } G) \text{ and } (\text{not } A \text{ or } V \text{ or } G) \Leftrightarrow$$
$$\Leftrightarrow (\text{not } A \text{ and } B \text{ and not } V \text{ and not } G \text{ and not } A) \text{ or}$$
$$\text{or } (\text{not } A \text{ and } B \text{ and not } V \text{ and not } G \text{ and } V) \text{ or}$$
$$\text{or } (\text{not } A \text{ and } B \text{ and not } V \text{ and not } G \text{ and } G) \Leftrightarrow$$
$$\Leftrightarrow (\text{not } A \text{ and } B \text{ and not } V \text{ and not } G) \text{ or } F \text{ or } F \Leftrightarrow$$
$$\Leftrightarrow \text{ not } A \text{ and } B \text{ and not } V \text{ and not } G.$$

Since **not** A **and** B **and not** V **and not** $G = $ T, then we obtain the following result: Bogdan is taking a trip, and Alice, Valeria, and George are at the beach.

1.36. *Answer*: Alice.

1.37. *Answer*:

A	B	$A \mid B$	$A \downarrow B$
T	T	F	F
T	F	T	F
F	T	T	F
F	F	T	T

1.38. *Solution*.

(1) Using transformations for logically equivalent expressions for operations of negation, disjunction, conjunction, and implication, we obtain:

> **not** $A \Leftrightarrow ($ **not** $A)$ **or** $($ **not** $A) = A \mid A$;
>
> A **or** $B \Leftrightarrow$ **not** $($ **not** $A)$ **or not** $($ **not** $B) = (A \mid A) \mid (B \mid B)$;
>
> A **and** $B \Leftrightarrow$ **not** $($ **not** A **or not** $B) = (A \mid B) \mid (A \mid B)$;
>
> $A \Rightarrow B \Leftrightarrow ($ **not** $A)$ **or** $B \Leftrightarrow ($ **not** $A)$ **or not** $($ **not** $B) = A \mid (B \mid B)$.

(2) The task is fulfilled similarly to item 1):

> **not** $A \Leftrightarrow ($ **not** $A)$ **and** $($ **not** $A) = A \downarrow A$;
>
> A **or** $B \Leftrightarrow$ **not** $($ **not** A **and not** $B) = (A \downarrow B) \downarrow (A \downarrow B)$;
>
> A **and** $B \Leftrightarrow$ **not** $($ **not** $A)$ **and not** $($ **not** $B) \Leftrightarrow$
>
>   **not** $A \downarrow$ **not** $B = (A \downarrow A) \downarrow (B \downarrow B)$;
>
> $A \Rightarrow B \Leftrightarrow ($ **not** $A)$ **or** $B \Leftrightarrow$ **not** $(A$ **and not** $B) \Leftrightarrow$ **not** $($ **not** $A \downarrow B) \Leftrightarrow$
>
> $\Leftrightarrow$ **not** $((A \downarrow A) \downarrow B) = ((A \downarrow A) \downarrow B) \downarrow ((A \downarrow A) \downarrow B)$.

1.39. *Answer*:

> (1) $A \downarrow B \Leftrightarrow ((A \mid A) \mid (B \mid B)) \mid ((A \mid A) \mid (B \mid B))$;
> (2) $A \mid B \Leftrightarrow ((A \downarrow A) \downarrow (B \downarrow B)) \downarrow ((A \downarrow A) \downarrow (B \downarrow B))$.

1.40. *Answer*: Predicates are sentences (2)–(4).

1.41. *Answer*:

> (1) **not** $P(n) = \{$natural number n is odd$\}$;
> (2) **not** $P(x, y) = \{x \geqslant y\}$.

1.42. *Solution*.

Let us write the negation of the expression T and use logical equivalences
not $\forall x\,(P(x)) \Leftrightarrow \exists x\,\textbf{not}\,(P(x))$ and **not** $\exists x\,(P(x)) \Leftrightarrow \Leftrightarrow \forall x\,\textbf{not}\,(P(x))$:

$$
\begin{aligned}
\text{(1)} \quad \textbf{not}\,T \;&=\; \textbf{not}\,\forall x\,(\exists y\,(P(x,\,y))) \Leftrightarrow \exists x\,(\textbf{not}\,\exists y\,(P(x,\,y))) \Leftrightarrow \\
&\Leftrightarrow \exists x\,(\forall y\,(\textbf{not}\,P(x,\,y)));
\end{aligned}
$$

$$
\begin{aligned}
\text{(2)} \quad \textbf{not}\,T \;&=\; \textbf{not}\,\exists x\,(\exists y\,(P(x,\,y)) \Rightarrow Q(x,\,y)) \Leftrightarrow \\
&\Leftrightarrow \forall x\,(\textbf{not}\,(\exists y(P(x,\,y))) \Rightarrow Q(x,\,y)) \Leftrightarrow \\
&\Leftrightarrow \forall x\,(\exists y\,(P(x,\,y))\,\textbf{and not}\,Q(x,\,y)).
\end{aligned}
$$

1.43. *Solution*.

(1) The proposition $\forall x\,(\exists y\,(x > y))$ means "for each real number x there exists a certain number y, that is less than x". This is a true proposition.

(2) The proposition $\forall x\,(x \neq 0 \Rightarrow \exists y\,(xy = 1))$ means "for each nonzero value x there exists a reciprocal number y, such that $xy = 1$". This is a true proposition.

1.44. *Answer*:

(1) True;
(2) False.

1.45. *Solution*.
Using the definition of a limit of numerical sequence, we obtain

$$
\lim_{n\to\infty} a_n \neq a \Leftrightarrow \textbf{not}\ \forall \varepsilon > 0\,\exists n_0 \in \mathbb{N}\,\forall n \in \mathbb{N}\ (n \geqslant n_0 \Rightarrow |a_n - a| < \varepsilon).
$$

We sequentially introduce the negation operation under the signs of the quantifiers:

$$
\begin{aligned}
\lim_{n\to\infty} a_n \neq a \;&\Leftrightarrow\; \textbf{not}\ \forall \varepsilon > 0\,\exists n_0 \in \mathbb{N}\,\forall n \in \mathbb{N}\ (n \geqslant n_0 \Rightarrow |a_n - a| < \varepsilon) \Leftrightarrow \\
&\Leftrightarrow \exists \varepsilon > 0\ \textbf{not}\ \exists n_0 \in \mathbb{N}\,\forall n \in \mathbb{N}\ (n \geqslant n_0 \Rightarrow |a_n - a| < \varepsilon) \Leftrightarrow \\
&\Leftrightarrow \exists \varepsilon > 0\,\forall n_0 \in \mathbb{N}\ \textbf{not}\ \forall n \in \mathbb{N}\ (n \geqslant n_0 \Rightarrow |a_n - a| < \varepsilon) \Leftrightarrow \\
&\Leftrightarrow \exists \varepsilon > 0\,\forall n_0 \in \mathbb{N}\,\exists n \in \mathbb{N}\ \textbf{not}\ (n \geqslant n_0 \Rightarrow |a_n - a| < \varepsilon) \Leftrightarrow \\
&\Leftrightarrow \exists \varepsilon > 0\,\forall n_0 \in \mathbb{N}\,\exists n \in \mathbb{N}\ (n \geqslant n_0\ \textbf{and}\ |a_n - a| \geqslant \varepsilon).
\end{aligned}
$$

When calculating the negation of implication, we used equivalence
$P \Rightarrow Q \Leftrightarrow \textbf{not}\ P\ \textbf{or}\ Q$.

1.46. *Answer*: $\exists \varepsilon > 0\,\forall \delta > 0\,\exists x\ (|x - x_0| < \delta\ \textbf{and}\ |f(x) - A| \geqslant \varepsilon)$.

1.47. *Answer*: $\exists \varepsilon > 0\,\forall \delta > 0\,\exists x\ (|x - x_0| < \delta\ \textbf{and}\ |f(x) - f(x_0)| \geqslant \varepsilon)$.

1.48. *Answer*: $\exists x\ (P(x)\ \textbf{and}\ (\forall y\,((P(y)) \Rightarrow (y = x))))$.

1.49. *Proof*.
Let $P(n)$ be the predicate "$1 + 3 + 5 + \ldots + (2n - 1) = n^2$".

Basis step

For $n = 1$ we obtain $1 = 1^2$, i.e., $P(1)$ is true.

Inductive step

Let for $n = k$, the proposition $1 + 3 + 5 + \ldots + (2k - 1) = k^2$ be true. Let us prove the truth of $P(k + 1)$:

$$1 + 3 + 5 + \ldots + (2k - 1) + (2(k + 1) - 1) =$$
$$= k^2 + (2(k + 1) - 1) = k^2 + 2k + 1 = (k + 1)^2.$$

Thus, for any natural k the following implication is valid $P(k) \Rightarrow P(k + 1)$. Hence, by the mathematical induction principle the predicate $P(n)$ has the true value for all natural n.

1.50. *Proof.*

Denote predicate "$1 + 4 + 7 + \ldots + (3n - 2) = \dfrac{1}{2}n\,(3n - 1)$" by $P(n)$.

Basis step

For $n = 1$ we obtain $1 = \dfrac{1}{2} \cdot 1 \cdot (3 - 1)$; therefore $P(1)$ is true.

Induction step

Let for $n = k$, the proposition $1 + 4 + 7 + \ldots + (3k - 2) = \dfrac{1}{2}k(3k - 1)$ be true. Then

$$1 + 4 + 7 + \ldots + (3k - 2) + (3(k + 1) - 2) =$$
$$= \frac{1}{2}k(3k - 1) + (3(k + 1) - 2) = \frac{1}{2}(3k^2 + 5k + 2) =$$
$$= \frac{1}{2}(k + 1)(3(k + 1) - 1).$$

Thus, for any $k = 1, 2, \ldots$ the implication $P(k) \Rightarrow P(k + 1)$ is valid. Hence, according to the mathematical induction principle the predicate $P(n)$ as the true value for all natural n.

1.52. *Solution.*

It is required to prove the identity

$$1^3 + 2^3 + \cdots + n^3 = (1 + 2 + \cdots + n)^2$$

for all natural n. Denote by $P(n)$ the predicate $1^3 + 2^3 + \cdots + n^3 = (1 + 2 + \cdots + n)^2$.

Basis step
For $n = 1$ we obtain $1 = 1$; therefore, $P(1)$ is true.

Inductive step

Let for $n = k$, the proposition $P(k)$ be true. Then

$$1^3 + 2^3 + \cdots + k^3 + (k+1)^3 = (1 + 2 + \cdots + k)^2 + (k+1)^3.$$

As is known, $1 + 2 + \cdots + k = \dfrac{1}{2}k(k+1)$; hence

$$1^3 + 2^3 + \cdots + k^3 + (k+1)^3 = \left(\dfrac{1}{2}k(k+1)\right)^2 + (k+1)^3 =$$

$$= (k+1)^2 \left(\dfrac{1}{4}k^2 + (k+1)\right) = \left(\dfrac{(k+1)(k+2)}{2}\right)^2.$$

Taking into account that $1 + 2 + \cdots + k + (k+1) = \dfrac{1}{2}(k+1)(k+2)$, we finally obtain:

$$1^3 + 2^3 + \cdots + k^3 + (k+1)^3 = (1 + 2 + \cdots + k + (k+1))^2.$$

Hence, according to the mathematical induction principle the predicate $P(n)$ has the true value for all natural n and the sum of cubes of the first n natural numbers is equal to the square of their sum.

1.54. *Proof.*
Let us use the mathematical induction method. Denote by $P(n)$ the predicate $S_n = \dfrac{a_1 + a_n}{2}n$.

Basis step

If we assume that $n = 1$, then the predicate $P(n)$ takes the form $S_1 = \dfrac{a_1 + a_1}{2}$. $1 = a_1$, which is the true proposition.

Inductive step

Let $P(k)$, $k = 1, 2, 3, \ldots$ be true. Let us consider the sum of the first $k + 1$ terms of arithmetic progression:

$$S_{k+1} = S_k + a_{k+1}.$$

According to inductive supposition $S_k = \dfrac{a_1 + a_k}{2}k$. Then

$$S_{k+1} = \dfrac{a_1 + a_k}{2}k + a_{k+1} = \dfrac{1}{2}\left[(a_1 + a_k)k + 2a_{k+1}\right] =$$

$$= \dfrac{1}{2}[(a_1 + a_1 + (k-1)d)k + 2(a_1 + kd)] = \dfrac{1}{2}[2a_1(k+1) + k(k+1)d] =$$

$$= \dfrac{1}{2}(a_1 + a_1 + kd)(k+1) = \dfrac{a_1 + a_{k+1}}{2}(k+1).$$

By the mathematical induction method, it is proved that $P(n)$ takes the true value for all natural n.

1.55. *Hint.* $S_{k+1} = S_k + b_1 q^k$.

1.56. *Proof.*
Denote $f(n) = n^2 - n$ and $P(n)$—the predicate "$f(n)$ is divisible by 2".

B a s i s   s t e p

For $n = 1$, we obtain $f(1) = 1^2 - 1 = 0$—even number; therefore $P(1)$ is true.

I n d u c t i v e   s t e p

Suppose that $f(k)$ is even for natural $k \geqslant 1$. Let us prove that this implies evenness of $f(k + 1)$:

$$f(k+1) = (k+1)^2 - (k+1) = k^2 + 2k + 1 - k - 1 = (k^2 - k) + 2k = f(k) + 2k.$$

Since in the right side of the obtained relation stands a sum of two even numbers, then $f(k+1)$ is divisible by 2.

Note. The statement of the problem becomes apparent if we present $k^2 - k$ in the form $k^2 - k = k(k - 1)$. Out of two consecutive natural numbers, one is necessarily even, and their product is divisible by 2.

1.57. *Proof.*
(1) Denote the predicate "$4n^3 + 14n$ is divisible by 3" by $P(n)$.

B a s i s   s t e p

For $n = 1$ we obtain: 18 is divisible by 3, which is the true proposition.

I n d u c t i v e   s t e p

Let $n = k$ and $4k^3 + 14k$ be divisible by 3. Show that from the last statement follows the truth of $P(k + 1)$:

$$4(k + 1)^3 + 14(k + 1) = 4(k^3 + 3k^2 + 3k + 1) + 14(k + 1) =$$
$$= (4k^3 + 14k) + 3(4k^2 + 4k + 6).$$

The first summand in the sum is divisible by 3 by the inductive supposition, and the second includes number 3 as a cofactor.

Thus, according to the mathematical induction principle the predicate $P(n)$ has the true value for all natural n.

(2) Denote the predicate "$n^5 - n - 10$ is divisible by 5" as $P(n)$. Since $10 = 2 \cdot 5$, then it is sufficient to prove that $f(n) = n^5 - n$ is divisible by 5 for all natural values n.

B a s i s   s t e p

For $n = 1$ we obtain: $f(1) = 0$ is divisible by 5.

Inductive step

Suppose that $f(k) = k^5 - k = 5l$ for some integer l. From this supposition, it follows that $f(k+1)$ is divisible by 5. Let us prove this statement.

Actually,

$$
\begin{aligned}
f(k+1) &= (k+1)^5 - (k+1) = (k+1)\left((k+1)^4 - 1\right) = \\
&= (k+1)\left((k+1)^2(k+1)^2 - 1\right) = \\
&= (k+1)\left((k^2 + 2k + 1)(k^2 + 2k + 1) - 1\right) = \\
&= (k+1)\left((k^4 + 4k^3 + 6k^2 + 4k + 1) - 1\right) = \\
&= (k+1)(k^4 + 4k^3 + 6k^2 + 4k) = \\
&= k^5 + 5k^4 + 10k^3 + 10k^2 + 4k = (k^5 - k) + 5(k^4 + 2k^3 + 2k^2 + k).
\end{aligned}
$$

Using the inductive supposition, we obtain that $f(k+1)$ is divisible by 5. Therefore, $n^5 - n + 10$ is divisible by 5 for all natural n.

(3) Basis step

For $n = 1$ we obtain: $6^{2n-1} - 6 = 0$ is divisible by 7.

Inductive step

Suppose that $6^{2k-1} - 6$ is divisible by 7, $k = 1, 2, \ldots$

Therefore, $6^{2k-1} - 6 = 7l$, and $6^{2k-1} = 7l + 6$ for some integer l. Let us prove that $6^{2(k+1)-1} - 6 = 6^{2k+1} - 6$ is divisible by 7:

$$
\begin{aligned}
6^{2k+1} - 6 &= 6^{2k-1+2} - 6 = 36 \cdot 6^{2k-1} - 6 = \\
&= 36(7l + 6) - 6 = 36 \cdot 7l + 35 \cdot 6.
\end{aligned}
$$

Note that both summands in the obtained sum are divisible by 7; therefore, the expression $6^{2k+1} - 6$ is divisible by 7. The inductive transition is proved. Therefore, $6^{2n-1} - 6$ is divisible by 7 for all natural numbers n.

(4) Denote $f(n) = n(n-1)(2n-1)$.

Basis step

For $n = 1$ we obtain: $f(1) = 0$ is divisible by 6.

Inductive step

Let $f(k) = k(k-1)(2k-1)$ be divisible by 6. Let us prove that from thus supposition follows divisibility by 6 of the value $f(k+1)$:

$$
\begin{aligned}
f(k+1) &= (k+1)k\left(2(k+1) - 1\right) = k(k+1)(2k+1) = \\
&= (k^2 + k)(2k+1) = 2k^3 + 3k^2 + k.
\end{aligned}
$$

Calculate the difference $f(k + 1) - f(k)$:

$$f(k+1) - f(k) = 2k^3 + 3k^2 + k - k(k-1)(2k-1) =$$
$$= 2k^3 + 3k^2 + k - (2k^3 - 3k^2 + k) = 6k^2.$$

In the equality $f(k + 1) = f(k) + 6k^2$, the right side is divisible by 6 according to the inductive supposition. Thereby, $f(k + 1)$ is divisible by 6, and according to the mathematical induction method, $f(n) = n(n - 1)(2n - 1)$ is divisible by 6 for all natural n.

1.58. *Proof.*

Basis step

For $n = 1$ we obtain: $a - 1$ is divisible by $a - 1$—the true proposition.

Inductive step

Let $a^k - 1$ be divisible by $a - 1$ for some natural k. Therefore, there exists such polynomial $p(a)$, the equality $a^k - 1 = (a - 1)p(a)$ is valid, or $a^k = (a - 1)p(a) + 1$. Let us prove that $a^{k+1} - 1$ is divisible by $a - 1$:

$$a^{k+1} - 1 = a \cdot a^k - 1 = a\,[(a-1)p(a) + 1] - 1 = a(a-1)p(a) + (a-1).$$

From the last equality follows

$$a^{k+1} - 1 = (a-1)(a\,p(a) + 1) = (a-1)\widetilde{p}(a),$$

where $\widetilde{p}(a)$ is some polynomial.

This proves that the polynomial $a^n - 1$ is divisible by $(a - 1)$ for all natural n.

Note. The statement of this point is easy to prove by directly dividing the polynomial $a^n - 1$ by $(a - 1)$ by the method known from the algebra course [41]:

$$
\begin{array}{rl|l}
a^n \quad\quad\quad\quad\quad -1 & a - 1 \\
\cline{2-2}
\underline{a^n - a^{n-1}} & a^{n-1} + a^{n-2} + \cdots + a + 1 \\
\quad a^{n-1} \\
\quad \underline{a^{n-1} - a^{n-2}} \\
\quad\quad\quad \vdots \\
\quad\quad\quad a^2 \\
\quad\quad\quad \underline{a^2 - a} \\
\quad\quad\quad\quad a - 1 \\
\quad\quad\quad\quad \underline{a - 1} \\
\quad\quad\quad\quad\quad 0
\end{array}
$$

Note that the formula is obtained for the sum of geometric progression (see Exercise **1.55**) in the particular case $b_1 = 1$.

1.59. *Hint*. Use the mathematical induction method.

1.60. *Hint*. $79 \cdot 4^n - 4(-1)^n = 80 \cdot 4^n - 4^n - 4(-1)^n$.

1.61. *Solution*.

For the proof, let us use the mathematical induction principle. Denote $a_n = \dfrac{79}{30} \times 4^n + \dfrac{(-1)^n}{5} - \dfrac{1}{3}$.

B a s i s   s t e p

For $n = 1$ we obtain: $a_1 = 10$ is integer.

I n d u c t i v e   s t e p

Suppose that a_k is integer for some integer value k. Let us prove that a_{k+1} is also integer. For this, calculate the difference $a_{k+1} - a_k$:

$$a_{k+1} - a_k = \left(\frac{79}{30} \cdot 4^{k+1} + \frac{(-1)^{k+1}}{5} - \frac{1}{3}\right) - \left(\frac{79}{30} \cdot 4^k + \frac{(-1)^k}{5} - \frac{1}{3}\right) =$$
$$= \frac{79}{10} \cdot 4^k - 2\frac{(-1)^k}{5} = \frac{1}{10}(79 \cdot 4^k - 4(-1)^k).$$

The number $\dfrac{1}{10}(79 \cdot 4^k - 4(-1)^k)$ is integer (see the previous exercise). Therefore, $a_n = \dfrac{1}{10}(79 \cdot 4^n - 4(-1)^n)$ is integer for all values of $n = 1, 2, \ldots$

1.65. *Proof*.

First method

Denote by $P(n)$ the predicate

$$\frac{1}{3 \cdot 7} + \frac{1}{7 \cdot 11} + \cdots + \frac{1}{(4n-1)(4n+3)} = \frac{n}{3(4n+3)}$$

for some n.

B a s i s   s t e p

The inductive base consists in the statement $P(1)$. For $n = 1$ the sum on the left side consist of the only summand: $\dfrac{1}{3 \cdot 7} = \dfrac{1}{3 \cdot (4+3)}$; therefore $P(1)$ is true.

Inductive step

Let us prove the truth of the predicate $P(k+1)$:

$$\underbrace{\frac{1}{3\cdot7}+\frac{1}{7\cdot11}+\cdots+\frac{1}{(4k-1)(4k+3)}}_{k/(3(4k+3))}+\frac{1}{(4(k+1)-1)(4(k+1)+3)}=$$

$$=\frac{k}{3(4k+3)}+\frac{1}{(4k+3)(4k+7)}=\frac{1}{3(4k+3)(4k+7)}\left(k(4k+7)+3\right)=$$

$$=\frac{k+1}{3(4(k+1)+3)}.$$

Therefore, the initial identity is true for all $n=1,2,\ldots$

Second method

This method is only based in algebraic transformations. Note that each summand in the sum is presented in the form of the difference

$$\frac{1}{(4k-1)(4k+3)}=\frac{1}{4}\left(\frac{1}{4k-1}-\frac{1}{4k+3}\right).$$

Rewrite the sum in the following form:

$$\frac{1}{3\cdot7}+\cdots+\frac{1}{(4n-1)(4n+3)}=$$

$$=\frac{1}{4}\left[\left(\frac{1}{3}-\frac{1}{7}\right)+\left(\frac{1}{7}-\frac{1}{11}\right)+\cdots+\right.$$

$$+\left(\frac{1}{4k-1}-\frac{1}{4k+3}\right)+\left(\frac{1}{4k+3}-\frac{1}{4k+7}\right)+\cdots+$$

$$\left.+\left(\frac{1}{4n-1}-\frac{1}{4n+3}\right)\right].$$

Regroup the summands in the obtained sum so that to unite the summands opposite in sign:

$$\frac{1}{3\cdot7}+\cdots+\frac{1}{(4n-1)(4n+3)}=$$

$$=\frac{1}{4}\left[\frac{1}{3}+\left(-\frac{1}{7}+\frac{1}{7}\right)+\cdots+\left(-\frac{1}{4k+3}+\frac{1}{4k+3}\right)+\right.$$

$$+\left(-\frac{1}{4k+7}+\frac{1}{4k+7}\right)+\cdots-\left.\frac{1}{4n+3}\right]=\frac{1}{4}\left(\frac{1}{3}-\frac{1}{4n+3}\right)=\frac{n}{3(4n+3)}.$$

1.66. *Hint.* See the solution of the previous exercise.

1.67. *Answer*:

$$\sum_{k=1}^{n}\frac{1}{(ak+b)(ak+a+b)}=\frac{n}{(a+b)(an+a+b)}.$$

The results of Exercises **1.65** and **1.66** are obtained for the values of the parameters $a = 4, b = -1$ and $a = 5, b = -1$, respectively.

1.68. *Solution.*

The basis step is the proposition $P(1)$, and the inductive step is $(\forall k\ P(k) \Rightarrow P(k + 1))$. From the conjunction of these proposition follows the truth of $P(n)$ for an arbitrary natural number n. Using the logical operations and the universal quantifier, we obtain the statement of the mathematical induction principle with the help of quantifiers:

$$(P(1) \textbf{ and } (\forall k\ P(k) \Rightarrow P(k + 1))) \Rightarrow \forall n\ P(n).$$

1.69. *Proof.*

Any even number a can be presented in the form $a = 2k$, where k is some integer number; therefore, let us write n and m in the form $n = 2k_1, m = 2k_2$, where k_1, k_2 are integer. The product is $m \cdot n = (2k_1) \cdot (2k_2) = 2 \cdot (2k_1 k_2)$. Since k_1, k_2 are integer numbers, the twice the product of them will also be some integer number $k_3 = 2k_1 k_2$. Thus, $m \cdot n = 2k_3$, where k_3 is integer; therefore, the product of the integers n and m is even.

1.70. *Proof.*

The problem statement is logically equivalent to the following: "if n is not odd, then n^2 is not odd," or "if n is even, then n^2 is even."

Let us prove the last statement by forward reasoning. Present the integer n in the form $n = 2k$, where k is integer. Then $n^2 = 4k^2 = 2(2k^2) = 2k'$, where k' is integer. From this, we conclude that n^2 is even. Thus, the statement is proved "if n^2 is odd, then n is odd."

1.71. *Proof.*

Suppose that at least one cofactor, for example, n, is an even number. As any even number, n can be presented in the form $n = 2k$, where k is integer. Then, we obtain that the product $n \cdot m = 2k \cdot m$ is divisible by 2 and is an even number. We come to a contradiction with the statement. Therefore, our supposition was false, and both cofactors are odd.

1.72. *Proof.*

Let $p = 17l + m$, where l, m are integer, and $0 \leqslant m < 17$ is a remainder of dividing p by 17. Then

$$p^2 = (17l + m)^2 = 17^2 l^2 + 34lm + m^2.$$

We obtain that p^2 is divisible by 17 if and only if m^2 is divisible by 17. And the last is only possible if $m = 0$; other possible values of m lead to the fractional number $\dfrac{m^2}{17}$, which is checked by direct brute force. Therefore, if p^2 is divisible by 17, then p is divisible by 17.

1.73. *Proof.*

Suppose that $\sqrt{17}$ is rational, in other words, the value $\sqrt{17}$ can be presented in the form of the irreducible fraction $\sqrt{17} = \dfrac{p}{q}$, where p and q are some integers having no common divisor. Both sides of the equality are positive; therefore, squaring both sides results in the equivalent equality

$$17 = \frac{p^2}{q^2} \text{ or } p^2 = 17q^2.$$

From the obtained relation follows that p^2 is divisible by 17; therefore, the number p itself is divisible by 17 (see the previous exercise). Hence, there exists such integer s, such that $p = 17s$. Substituting the obtained expression into the relation $p^2 = 17q^2$, we obtain

$$(17s)^2 = 17q^2 \quad \Rightarrow \quad 17s^2 = q^2.$$

From the last equality follows that q is divisible by 17, i.e., there exists such integer s', such that $q = 17s'$. But it contradicts the initial supposition about irreducibility of the fraction $\dfrac{p}{q} = \dfrac{17s}{17s'}$. The obtained contradiction proves irrationality of the number $\sqrt{17}$.

1.74. *Hint.*

Use the method "by contradiction" and suppose that the difference $\sqrt{3} - \sqrt{2}$ can be presented in the form $\sqrt{3} - \sqrt{2} = \dfrac{p}{q}$, where p and q are integers, having no common divisor.

1.76. *Hint.*

The sum and the product of two irrational numbers are not necessarily irrational.

1.77. *Solution.*

(1) Using the definition of a prime number, we directly find the first ten primes $p_1, p_2, \ldots, p_{10}$:

p_1	p_2	p_3	p_4	p_5	p_6	p_7	p_8	p_9	p_{10}
2	3	5	7	11	13	17	19	23	29

(2) Suppose that the number of primes is finite, the greatest of them is denoted by p_N. Consider the product of all the prime numbers $P = p_1 \cdot p_2 \ldots p_N$ and add the unity to them. The number $P' = P + 1$ when divided by p_1 gives the remainder 1, when divided by p_2 the remainder is also 1, and so on. We obtain that P' is divisible by neither of the numbers $p_1, p_2, \ldots, p_N$.

According to the fundamental theorem of arithmetic, any number $n = 2, 3, \ldots$ can uniquely be presented in the form of a product of prime cofactors (see Exercise **1.123**); therefore we conclude that either the number P' is prime or it can be presented in the form $P' = p_a \cdot p_b \ldots$, and $p_a, p_b, \ldots$ are prime numbers, each not being equal to any of $p_1, p_2, \ldots$

The obtained contradiction proves that there exist infinitely many prime numbers.

Note. This proof of the fact of the infinity of the primes was suggested by Euclid[9] [2].

1.78. *Answer*:

not $P(n) = \{$natural number n is composite or equal to $1\}$.

1.79. *Proof.*

Suppose that there only exists a finite number of primes of the form $8n + 7$, denote them by $p_1, p_2, \ldots, p_N$. The number $8p_1 p_2 \ldots p_N - 1$ is not divisible by any of the numbers $p_i, i = 1, 2, \ldots, n$ and can be presented in the form $8n + 7$. The obtained contradiction means that there exist infinitely many prime numbers of the form $8n + 7$.

1.80. *Hint.* See the previous exercise.

1.81. *Hint.* Proof can be performed by the methods considered in the Exercise **1.65**.

1.82. *Proof.*
First of all, let us write the first several terms of the sequence of harmonic numbers $\{H_n\}$:

$$H_1 = 1, \quad H_2 = 1 + \frac{1}{2} = \frac{3}{2}, \quad H_3 = 1 + \frac{1}{2} + \frac{1}{3} = \frac{11}{6}, \quad \ldots$$

(1) Denote by $P(n)$ the predicate $\sum_{i=1}^{n} H_i = (n+1)H_n - n$.

B a s i s s t e p

If $n = 1$, then we have $H_1 = (1+1) \cdot 1 - 1$; since $H_1 = 1$, we have obtained the true equality.

I n d u c t i v e s t e p

Let for some natural k the predicate $P(k)$ take the true value; i.e., the equality $\sum_{i=1}^{k} H_i = (k+1)H_k - k$ is true.

Calculate the sum $\sum_{i=1}^{k+1} H_i$:

$$\sum_{i=1}^{k+1} H_i = \sum_{i=1}^{k} H_i + H_{k+1} = (k+1)H_k - k + H_{k+1}.$$

---

[9] Euclid (Εὐχλείδης) (about 325 BC–before 265 BC)—Ancient Greek mathematician.

From the definition of harmonic numbers follows that $H_{k+1} = H_k + \dfrac{1}{k+1}$ or $H_k = H_{k+1} - \dfrac{1}{k+1}$. Further,

$$\sum_{i=1}^{k+1} H_i = (k+1)\left(H_{k+1} - \frac{1}{k+1}\right) - k + H_{k+1} = (k+2)H_{k+1} - (k+1).$$

Therefore, the implication $P(k) \Rightarrow P(k+1)$ is true, and $P(n)$ is fulfilled for all natural n.

(2) Denote by $P(n)$ the predicate $\displaystyle\sum_{i=1}^{n} i H_i = \frac{n(n+1)}{4}(2H_{n+1} - 1)$.

B a s i s   s t e p

For $n = 1$, we obtain $H_1 = \dfrac{1 \cdot 2}{4}(2H_2 - 1)$ is the true equality.

I n d u c t i v e   s t e p

Let for some natural k the predicate $P(k)$ take the true value; i.e., the equality $\displaystyle\sum_{i=1}^{k} i H_i = \frac{k(k+1)}{4}(2H_{k+1} - 1)$ is true. Check the truth of $P(k+1)$:

$$\sum_{i=1}^{k+1} i H_i = \sum_{i=1}^{k} i H_i + (k+1)H_{k+1} = \frac{k(k+1)}{4}(2H_{k+1} - 1) + (k+1)H_{k+1} =$$

$$= \frac{(k+1)(k+2)}{4}(2H_{k+2} - 1).$$

We obtained that $P(n)$ takes the true value for all natural n.

1.83. *Proof.*

The suggested identities can be proved by algebraic transformations [35].

(1) Consider the left side of the relation $\displaystyle\sum_{i=1}^{n} H_i = (n+1)H_n - n$ and use the definition of a harmonic number $H_i = \displaystyle\sum_{j=1}^{i} \frac{1}{j}$:

$$\sum_{i=1}^{n} H_i = \sum_{i=1}^{n}\sum_{j=1}^{i} \frac{1}{j} =$$

$$= 1 + \left(1 + \frac{1}{2}\right) + \left(1 + \frac{1}{2} + \frac{1}{3}\right) + \cdots + \left(1 + \frac{1}{2} + \frac{1}{3} + \cdots + \frac{1}{n}\right).$$

In the given sum, change the sequence of the summands:

$$\sum_{i=1}^{n} H_i = \underbrace{(1 + 1 + \cdots + 1)}_{n \text{ summands}} + \underbrace{\left(\frac{1}{2} + \frac{1}{2} + \cdots + \frac{1}{2}\right)}_{n-1 \text{ summand}} + \cdots +$$

$$+ \left(\frac{1}{n-1} + \frac{1}{n-1}\right) + \frac{1}{n} = \sum_{j=1}^{n} \frac{1}{j} \sum_{i=j}^{n} 1.$$

Transform the obtained expression using the formulae (A.55) and (A.56) from Appendix "Reference Data":

$$\sum_{i=1}^{n} H_i = \sum_{j=1}^{n} \frac{n-j+1}{j} = (n+1) \sum_{j=1}^{n} \frac{1}{j} - \sum_{j=1}^{n} 1 = (n+1)H_n - n.$$

Thus, we obtained the required identity.

(2) Transform the left side of the suggested relation
$\sum_{i=1}^{n} i H_i = \dfrac{n(n+1)}{4}(2H_{n+1} - 1)$ as follows:

$$\sum_{i=1}^{n} i H_i = \sum_{i=1}^{n} i \sum_{j=1}^{i} \frac{1}{j} =$$

$$= 1 \cdot 1 + 2 \left(1 + \frac{1}{2}\right) + 3 \left(1 + \frac{1}{2} + \frac{1}{3}\right) + \cdots + n \left(1 + \frac{1}{2} + \frac{1}{3} + \cdots + \frac{1}{n}\right) =$$

$$= \underbrace{(1 + 2 + 3 + \cdots + n)}_{n \text{ summands}} + \frac{1}{2} \cdot \underbrace{(2 + 3 + \cdots + n)}_{n-1 \text{ summand}} + \cdots + \frac{1}{n} \cdot n = \sum_{j=1}^{n} \frac{1}{j} \sum_{i=j}^{n} i.$$

Using the relations (A.55) and (A.56), we obtain

$$\sum_{i=1}^{n} i H_i = \sum_{j=1}^{n} \frac{1}{j} \cdot \frac{(n+j)(n-j+1)}{2} = \frac{1}{2} \sum_{j=1}^{n} \frac{n^2 + n - j^2 + j}{j} =$$

$$= \frac{n^2 + n}{2} H_n - \frac{n(n-1)}{4} = \frac{n(n+1)}{4}(2H_{n+1} - 1).$$

1.85. *Solution.*
Analysis of the first several terms of the sequence $\{H_n\}$ shows that
$$H_1 = 1 \in \mathbb{Z}, \; H_2 = \frac{3}{2} \notin \mathbb{Z}, \; H_3 = \frac{11}{6} \notin \mathbb{Z}, \; H_4 = \frac{25}{12} \notin \mathbb{Z}, \; \ldots,$$
whereby $\mathbb{Z}$ is denoted a set of integers (for more details, see chapter "Set Theory" on p. 63). The supposition arises that none of the harmonic numbers H_n, except H_1, is integer. Let us prove it.

For any n can be found such an integer k that $2^k \leqslant n < 2^{k+1}$. Reduce the sum $H_n = 1 + \dfrac{1}{2} + \dfrac{1}{3} + \cdots + \dfrac{1}{n}$ to a common denominator. The numerator of the obtained fraction will be odd, and the denominator is divisible by 2^k. Therefore, $H_n \notin \mathbb{Z}$. As a result, we obtain that there exists the only harmonic integer, namely $H_1 = 1$.

1.86. *Proof.*

(1) Let $P(n)$ be the predicate $\displaystyle\sum_{i=1}^{n} F_i = F_{n+2} - 1$.

B a s i s   s t e p

Let $n = 1$. Then we obtain $F_1 = F_3 - 1$, or $1 = 2 - 1$ is the true equality.

I n d u c t i v e   s t e p

Now, let the statement $P(k)$ be true for all natural values k, i.e., $\displaystyle\sum_{i=1}^{k} F_i = F_{k+2} - 1$.

Prove the truth of $P(k + 1)$, or the equality $\displaystyle\sum_{i=1}^{k+1} F_i = F_{k+3} - 1$. Actually,

$$\sum_{i=1}^{k+1} F_i = \sum_{i=1}^{k} F_i + F_{k+1} = (F_{k+2} - 1) + F_{k+1} = (F_{k+1} + F_{k+2}) - 1 = F_{k+3} - 1.$$

Hence, the predicate $P(n)$ for all $n = 1, 2, 3, \ldots$ takes the true value.

(2) Denote by $P(n)$ the predicate $\displaystyle\sum_{i=1}^{n} F_{2i-1} = F_{2n}$.

B a s i s   s t e p

For $n = 1$, the inequality takes the form $F_1 = F_2$, or $1 = 1$, i.e., is true.

I n d u c t i v e   s t e p

In the supposition of validity of the relation

$$\sum_{i=1}^{k} F_{2i-1} = F_{2k}$$

prove that

$$\sum_{i=1}^{k+1} F_{2i-1} = F_{2k+2}.$$

Present the sum on the left side of the last equality in the form

$$\sum_{i=1}^{k+1} F_{2i-1} = \sum_{i=1}^{k} F_{2i-1} + F_{2(k+1)-1} = \sum_{i=1}^{k} F_{2i-1} + F_{2k+1}.$$

Now, let us use the inductive supposition

$$\sum_{i=1}^{k+1} F_{2i-1} = F_{2k} + F_{2k+1}.$$

The sum of two consecutive terms of the Fibonacci series F_{2k} and F_{2k+1} is equal, according to the definition property, to F_{2k+2}, which proves the inductive supposition. Hence, the analysed identity is true for all natural values n.

(3) Denote by $P(n)$ the predicate $\sum_{i=1}^{n} F_{2i} = F_{2n+1} - 1$, $n = 1, 2, 3, \ldots$

B a s i s   s t e p

Let $n = 1$. Then, the equality takes the form $F_2 = F_3 - 1$ is the true equality.

I n d u c t i v e   s t e p

Suppose the validity of the relation

$$\sum_{i=1}^{k} F_{2i} = F_{2k+1} - 1$$

for some k. Let us prove that

$$\sum_{i=1}^{k+1} F_{2i} = F_{2k+3} - 1.$$

Actually,

$$\sum_{i=1}^{k+1} F_{2i} = \sum_{i=1}^{k} F_{2i} + F_{2k+2} =$$

$$= (F_{2k+1} - 1) + F_{2k+2} = (F_{2k+1} + F_{2k+2}) - 1 = F_{2k+3} - 1.$$

according to the mathematical induction principle $\sum_{i=1}^{n} F_{2i} = F_{2n+1} - 1$ for $n \geqslant 1$.

(4) Denote by $P(n)$ the statement of the given item.

B a s i s   s t e p

Check the validity of the statement for $n = 1$.
We obtain that $F_3 = 2$ even, and $F_4 = 3$ and $F_5 = 5$ are odd.

Inductive step

In the supposition that there exist such integers l_0, l_1, l_2, that $F_{3k} = 2l_0$, $F_{3k+1} = 2l_1 + 1$, $F_{3k+2} = 2l_2 + 1$, prove that: $F_{3(k+1)} = F_{3k+3}$ is even, and $F_{3(k+1)+1} = F_{3k+4}$ and $F_{3(k+1)+2} = F_{3k+5}$ are odd. For this, note that

$$
\begin{aligned}
F_{3k+3} &= F_{3k+1} + F_{3k+2} = (2l_1 + 1) + (2l_2 + 1) = \\
&= 2(l_1 + l_2 + 1) - \text{even}, \\
F_{3k+4} &= F_{3k+2} + F_{3k+3} = (2l_2 + 1) + 2(l_1 + l_2 + 1) = \\
&= 2(l_1 + 2l_2 + 1) + 1 \text{ is odd}, \\
F_{3k+5} &= F_{3k+3} + F_{3k+4} = 2(l_1 + l_2 + 1) + 2(l_1 + 2l_2 + 1) + 1 = \\
&= 2(2l_1 + 3l_2 + 2) + 1 \text{ is odd}.
\end{aligned}
$$

Therefore, for all natural n we have: F_{3n} are even, and F_{3n+1} and F_{3n+2} are odd.

1.87. *Proof.*

Denote predicate $\sum_{i=1}^{n} F_i^2 = F_n F_{n+1}$ by $P(n)$.

Basis step

Let $n = 1$. We have: $F_1^2 = F_1 F_2$, or $1 = 1 \cdot 1$ is the true equality.

Inductive step

Suppose the validity of the relation $\sum_{i=1}^{k} F_i^2 = F_k F_{k+1}$ for $k = 1, 2, \ldots$ Let us prove that $\sum_{i=1}^{k+1} F_i^2 = F_{k+1} F_{k+2}$. For this, write the sum on the left side of the equality in the form

$$
\sum_{i=1}^{k+1} F_i^2 = \sum_{i=1}^{k} F_i^2 + F_{k+1}^2.
$$

Using the inductive supposition, we obtain

$$
\sum_{i=1}^{k+1} F_i^2 = F_k F_{k+1} + F_{k+1}^2 = F_{k+1}(F_k + F_{k+1}).
$$

According to the fundamental recurrence relation for the Fibonacci numbers $F_k + F_{k+1} = F_{k+2}$, therefore

$$
\sum_{i=1}^{k+1} F_i^2 = F_{k+1} F_{k+2}.
$$

So, according to the mathematical induction principle $\sum_{i=1}^{n} F_i^2 = F_n F_{n+1}$ for all natural numbers n.

1.88. *Hint.*

For proving the second identity, use the formula for the sum of the Fibonacci numbers from Exercise **1.86**, (1).

1.89. *Solution.*

As is known from Exercise **1.86**, (2) and (3), for $k \geqslant 1$ the equalities are fulfilled

$$\sum_{i=1}^{k} F_{2i-1} = F_{2k}, \quad \sum_{i=1}^{k} F_{2i} = F_{2k+1} - 1.$$

Deducting the second equality from the first one term by term, we obtain

$$F_1 - F_2 + F_3 - F_4 + F_5 - F_6 + \cdots + F_{2k-1} - F_{2k} = F_{2k} - F_{2k+1} + 1.$$

The obtained equality is equivalent to the following:

$$F_1 - F_2 + F_3 - F_4 + \cdots + (-1)^{n+1} F_n = (-1)^{n+1} F_{n-1} + 1, \ n \geqslant 1,$$

which can easily be checked having considered the cases of even and odd values of n.

1.90. *Answer*:

n	0	1	2	3	4	5	6	7	8	9
L_n	2	1	3	4	7	11	18	29	47	76

1.91. *Proof.*

For the proof, let us use the mathematical induction method. Denote by $P(n)$ the predicate $\sum_{k=0}^{n} L_k = L_{n+2} - 1$.

Basis step

The inductive base consists in the statement $P(1)$. For $n = 1$ we obtain: $\sum_{k=0}^{1} L_k = L_0 + L_1 = 3 = L_3 - 1$; therefore $P(1)$ is true.

Inductive step

Assume validity of the equality $\sum_{i=0}^{k} L_i = L_{k+2} - 1$. Let us prove that the implication $P(k) \Rightarrow P(k+1)$ is true. Actually, since

$$\sum_{i=0}^{k+1} L_i = \sum_{i=0}^{k} L_i + L_{k+1} = (L_{k+2} - 1) + L_{k+1} = (L_{k+1} + L_{k+2}) - 1,$$

then by the definition property of the Lucas numbers $L_{k+3} = L_{k+1} + L_{k+2}$. We obtain the relation:

$$\sum_{i=0}^{k+1} L_i = L_{k+3} - 1.$$

Therefore, the equality $\sum_{k=0}^{n} L_k = L_{n+2} - 1$ is true for all $n \geqslant 1$.

1.92. *Solution.*

The basis step is the proposition $P(1)$, and the inductive step is $(\forall k, (\forall m < k) P(m) \Rightarrow P(k))$. From the conjunction of these propositions follows the truth of $P(n)$ for all natural n. Using the logical operations and the universal quantifier, we obtain the following symbolic form of the complete induction principle:

$$(P(1) \text{ and } (\forall k, (\forall m < k) P(m) \Rightarrow P(k))) \Rightarrow \forall n \, P(n).$$

1.93. *Proof.*

It is required to prove that the second form of the principle follows from the first one, and the first one follows from the second one.

The generalized mathematical induction principle for the series of propositions $P(k)$ coincides with the mathematical induction principle (in the first form) for the series of statements.

$\widetilde{P}(k) = $ "statements $P(1), P(2), \ldots, P(k)$ are true".

The first form of the mathematical induction principle follows directly from the second one, since

$$\forall k \ (P(1) \text{ and } P(2) \text{ and } \ldots \text{ and } P(k)) \Rightarrow P(k+1))$$

implies the validity of the statement

$$\forall k \, P(k) \Rightarrow P(k+1).$$

1.94. *Hint.*

The basis step consists in the statements "$x + \dfrac{1}{x}$ is integer" and "$x^2 + \dfrac{1}{x^2}$ is integer." In order to prove the inductive transition, we should consider the product $\left(x^n + \dfrac{1}{x^n}\right)\left(x + \dfrac{1}{x}\right)$.

1.95. *Proof.*

The problem is reduced to proving the statement: "For any natural number $n \geqslant 8$ there exists such a pair of nonnegative integers (c_1, c_2), that the equality $n = 3c_1 + 5c_2$ is fulfilled."

For the sake of illustration, let us insert several values of n for $c_1, c_2 = 0, 1, 2, \ldots$ into the table

Values of the sum $3c_1 + 5c_2$

c_2/c_1	0	1	2	3	4	5	...
0	0	3	6	9	12	15	
1	5	8	11	14	17	20	
2	10	13	16	19	22	25	
3	15	18	21	24	27	30	...
4	20	23	26	29	32	35	
5	25	28	31	34	37	40	
...			...				

From the provided table, we conclude that natural numbers 1, 2, 4, and 7 cannot are presented in the required form. For example, in order to obtain the value $n = 7$, the coefficients c_1 and c_2 shall satisfy the inequalities $0 \leqslant c_1 < 3$ and $0 \leqslant c_2 < 2$. All such pairs of (c_1, c_2) are listed in the table, and none of them results in the value $n = 7$.

If we continue filling in the table for the increasingly greater values c_1 and c_2, then each of the values $n \geqslant 8$ will be encountered at least once. Let us prove it.

Let us consider the predicate

$$P(n) = \text{"for some nonnegative integers } c_1 \text{ and } c_2$$
$$\text{the equality } n = 3c_1 + 5c_2 \text{ is fulfilled"}.$$

Using the complete mathematical induction method, let us prove the truth of $P(n)$ for all $n \geqslant 8$.

B a s i s   s t e p

The propositions $P(8)$, $P(9)$, and $P(10)$ take the true values and form the basis step:

$$8 = 3 \cdot 1 + 5 \cdot 1;$$
$$9 = 3 \cdot 3 + 5 \cdot 0;$$
$$10 = 3 \cdot 0 + 5 \cdot 2.$$

I n d u c t i v e   s t e p

Suppose that proposition $P(k)$ is true for some $k > 10$. Note that the truth $P(k)$ for $k = 7, 8, 9$ is already proved. Let us check the fulfillment of the implication $P(k - 2) \Rightarrow P(k + 1)$.

In view of the inductive supposition, the expression $k - 2$ for an arbitrary $k > 10$ can be presented in the form $k - 2 = 3c_1 + 5c_2$. Then

$$k + 1 = (k - 2) + 3 = (3c_1 + 5c_2) + 3 = 3(c_1 + 1) + 5c_2.$$

Thus, $P(k + 1)$ takes the true value, and according to the complete mathematical induction method, the statement $P(n)$ is proved for all $n \geqslant 8$.

1.96. *Answer*: 29 heads.

1.97. *Proof*.
We will perform the proof by the method of mathematical induction by a variable m.

Denote predicate "$F_{n+m} = F_{n-1}F_m + F_n F_{m+1}$ for all natural numbers n" by $P(m)$.

B a s i s   s t e p

For $m = 1$ $F_{n+1} = F_{n-1}F_1 + F_n F_2$. Hence, since $F_1 = F_2 = 1$, we come to the true identity $F_{n+1} = F_{n-1} + F_n$.

Now let $m = 2$. Then $F_{n+2} = F_{n-1}F_2 + F_n F_3$. Taking into account the value of the third term of the Fibonacci series $F_3 = 2$, let us rewrite the last equality in the form

$$F_{n+2} = F_{n-1} + 2F_n = (F_{n-1} + F_n) + F_n = F_{n+1} + F_n.$$

Thus, the statements $P(1)$ and $P(2)$ are proved.

I n d u c t i v e   s t e p

Let $P(k - 1)$ and $P(k)$ be true, $k > 2$. It means that the equalities $F_{n+k-1} = F_{n-1}F_{k-1} + F_n F_k$ and $F_{n+k} = F_{n-1}F_k + F_n F_{k+1}$ are true for all natural n. Let us show that the two last equalities entail $P(k + 1)$, or $F_{n+k+1} = F_{n-1}F_{k+1} + F_n F_{k+2}$ for all $n = 1, 2, \ldots$ Actually,

$$F_{n+k+1} = F_{n+k-1} + F_{n+k} = F_{n-1}F_{k-1} + F_n F_k + F_{n-1}F_k + F_n F_{k+1} =$$
$$= F_{n-1}(F_{k-1} + F_k) + F_n(F_k + F_{k+1}).$$

Now, we only have to use the fundamental recurrent relation. As a result, we obtain

$$F_{n+k+1} = F_{n-1}F_{k+1} + F_n F_{k+2}.$$

Thereby, $P(m)$ is true for all natural m.

1.98. *Proof*.
Denote $a_n = F_n^2 + F_{n+1}^2$. Let us compile a table containing the first several terms of the series a_n:

n	1	2	3	4	5	6	7	8
F_n	1	1	2	3	5	8	13	21
a_n	2	5	13	34	89	233	610	1597

The table analysis shows that $a_n = F_{2n+1}$.

The identity $F_{2k+1} = F_k^2 + F_{k+1}^2$ can be proved by way of substitution of $n = k + 1$, $m = k$ into the addition formula for the Fibonacci series terms studied in

Exercise **1.97**. We obtain:

$$F_{n+m} = F_{n-1}F_m + F_nF_{m+1} \Rightarrow \left\{ \begin{array}{l} n = k+1 \\ m = k \end{array} \right\} \Rightarrow$$

$$\Rightarrow F_{2k+1} = F_{(k+1)-1}F_k + F_{k+1}F_{k+1} = F_k^2 + F_{k+1}^2.$$

1.99. *Proof.*

Let us use the complete mathematical induction method. Let $P(n)$ denote the predicate $F_n = \frac{1}{\sqrt{5}}\left(\frac{1+\sqrt{5}}{2}\right)^n - \frac{1}{\sqrt{5}}\left(\frac{1-\sqrt{5}}{2}\right)^n$.

Let us introduce the following notations: $\varphi = \frac{1+\sqrt{5}}{2}$, $\widetilde{\varphi} = \frac{1-\sqrt{5}}{2}$. Thus, it is required to prove that $F_n = \frac{1}{\sqrt{5}}\varphi^n - \frac{1}{\sqrt{5}}\widetilde{\varphi}^n$ for $n = 1, 2, 3, \ldots$

Preparatory to proving this statement, let us write the identities satisfied by the values φ and $\widetilde{\varphi}$.

From the definitions of φ and $\widetilde{\varphi}$ directly follows:

$$\begin{cases} \varphi + \widetilde{\varphi} = 1; \\ \varphi - \widetilde{\varphi} = \sqrt{5}. \end{cases}$$

Moreover, since

$$\varphi^2 = \left(\frac{1+\sqrt{5}}{2}\right)^2 = \frac{1+2\sqrt{5}+5}{4} = \frac{3+\sqrt{5}}{2} = 1 + \varphi;$$

$$\widetilde{\varphi}^2 = \left(\frac{1-\sqrt{5}}{2}\right)^2 = \frac{1-2\sqrt{5}+5}{4} = \frac{3-\sqrt{5}}{2} = 1 + \widetilde{\varphi},$$

then for all integer k it is easy to check the fulfillment of the following pair of equalities:

$$\begin{cases} \varphi^{k+1} = \varphi^k + \varphi^{k-1}; \\ \widetilde{\varphi}^{k+1} = \widetilde{\varphi}^k + \widetilde{\varphi}^{k-1}. \end{cases}$$

Now, we can proceed to proving $P(n)$ for all $n = 1, 2, 3, \ldots$

B a s i s   s t e p

The basis step consists in the statements $P(1)$ and $P(2)$.

For $n = 1$, we obtain the equality: $F_1 = \frac{1}{\sqrt{5}}(\varphi - \widetilde{\varphi})$, or $1 = \frac{1}{\sqrt{5}} \cdot \sqrt{5}$ is the true proposition.

Further, let us check the truth of $P(2)$. For $n = 2$, we have: $F_2 = \frac{1}{\sqrt{5}}\varphi^2 - \frac{1}{\sqrt{5}}\widetilde{\varphi}^2$, or $1 = \frac{1}{\sqrt{5}}(\varphi - \widetilde{\varphi})(\varphi + \widetilde{\varphi})$ is the true equality.

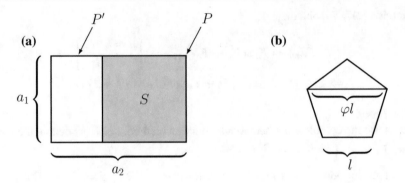

Fig. 1.1 Golden ratio

Inductive step

Suppose that for $i \leqslant k$ the equality $F_i = \frac{1}{\sqrt{5}}\left(\varphi^i - \widetilde{\varphi}^i\right)$ is fulfilled. Let us prove
that $F_{k+1} = \frac{1}{\sqrt{5}}\left(\varphi^{k+1} - \widetilde{\varphi}^{k+1}\right)$. For this, let us present φ^{k+1} and $\widetilde{\varphi}^{k+1}$ in the form
$\varphi^{k+1} = \varphi^k + \varphi^{k-1}, \widetilde{\varphi}^{k+1} = \widetilde{\varphi}^k + \widetilde{\varphi}^{k-1}$ and substitute it into the expression for F_{k+1}:

$$F_{k+1} = \frac{1}{\sqrt{5}}\left(\varphi^{k+1} - \widetilde{\varphi}^{k+1}\right) = \frac{1}{\sqrt{5}}\left((\varphi^k + \varphi^{k-1}) - (\widetilde{\varphi}^k + \widetilde{\varphi}^{k-1})\right).$$

Let us group the summands in brackets:

$$F_{k+1} = \frac{1}{\sqrt{5}}\left((\varphi^k - \widetilde{\varphi}^k) + (\varphi^{k-1} - \widetilde{\varphi}^{k-1})\right) = \frac{1}{\sqrt{5}}\left(\varphi^{k-1} - \widetilde{\varphi}^{k-1}\right) + \frac{1}{\sqrt{5}}\left(\varphi^k - \widetilde{\varphi}^k\right).$$

Using the inductive supposition, we finally obtain the true equality

$$F_{k+1} = F_{k-1} + F_k.$$

As a result, we conclude that for all natural numbers n the following relation is
fulfilled $F_n = \frac{1}{\sqrt{5}}\left(\frac{1+\sqrt{5}}{2}\right)^n - \frac{1}{\sqrt{5}}\left(\frac{1-\sqrt{5}}{2}\right)^n$.

Note. The value φ is referred to as the **golden ratio** or the **Phidias**[10] **number**.
This value appears in many geometrical constructions. It is possible to show [2]
that if the side ratio of a rectangle P is equal to $\dfrac{a_2}{a_1} = \varphi$, then after cutting from P
the square of S the remaining rectangle P' will be identical to the initial one (see
Fig. 1.1, a). Moreover, in a regular pentagon, the diagonal length is φ times greater
that the length of the side (see Fig. 1.1, b) [2].

1.100. *Hint.* Use the property $\varphi^2 = \varphi + 1$.

---

[10]Phidias (Φειδίας) (about 490 BC–about 430 BC)—Ancient Greek sculptor and architect.

1.101. *Proof.*
Let us substitute into the left side of the relation the explicit expressions for the Fibonacci numbers and perform the algebraic transformations:

$$F_{2n-1}F_{2n+1} - F_{2n}^2 = \frac{\varphi^{2n-1} - \widetilde{\varphi}^{2n-1}}{\sqrt{5}} \times \frac{\varphi^{2n+1} - \widetilde{\varphi}^{2n+1}}{\sqrt{5}} - \left(\frac{\varphi^{2n} - \widetilde{\varphi}^{2n}}{\sqrt{5}}\right)^2 =$$

$$= \frac{1}{5}\left[\varphi^{4n} - \varphi^{2n-1}\widetilde{\varphi}^{2n+1} - \varphi^{2n+1}\widetilde{\varphi}^{2n-1} + \widetilde{\varphi}^{4n} - \left(\varphi^{4n} - 2\varphi^{2n}\widetilde{\varphi}^{2n} + \widetilde{\varphi}^{4n}\right)\right] =$$

$$= \frac{1}{5}\left[-(\varphi\widetilde{\varphi})^{2n-1}(\varphi^2 + \widetilde{\varphi}^2) + 2(\varphi\widetilde{\varphi})^{2n}\right].$$

Then, let us take advantage of the fact that $\varphi\widetilde{\varphi} = -1$, $\varphi^2 + \widetilde{\varphi}^2 = 3$. Hence,

$$F_{2n-1}F_{2n+1} - F_{2n}^2 = 1 \text{ for } n = 1, 2, 3, \ldots$$

1.102. *Proof.*
Transform the left side of the equality using the definition property of the Fibonacci numbers:

$$F_{2n+2} = F_{2n} + F_{2n+1};$$
$$F_{2n+3} = F_{2n+1} + F_{2n+2} = F_{2n} + 2F_{2n+1}.$$

Substitute the obtained equalities into the initial relation:

$$F_{2n+1}F_{2n+2} - F_{2n}F_{2n+3} = F_{2n+1}(F_{2n} + F_{2n+1}) - F_{2n}(F_{2n} + 2F_{2n+1}) =$$
$$= F_{2n+1}^2 - F_{2n}^2 - F_{2n}F_{2n+1} = F_{2n+1}(F_{2n+1} - F_{2n}) - F_{2n}^2 = F_{2n-1}F_{2n+1} - F_{2n}^2.$$

According to Exercise **1.101** $F_{2n-1}F_{2n+1} - F_{2n}^2 = 1$. Therefore, $F_{2n+1}F_{2n+2} - F_{2n}F_{2n+3} = 1$ for all $n = 1, 2, 3, \ldots$

1.103. *Proof.*
Transform the right side of the identity using the known trigonometric relation

$$\arctan x + \arctan y = \arctan\frac{x + y}{1 - xy}, \quad xy < 1:$$

$$\arctan\frac{1}{F_{2n+1}} + \arctan\frac{1}{F_{2n+2}} = \arctan\left(\frac{F_{2n+1}^{-1} + F_{2n+2}^{-1}}{1 - F_{2n+1}^{-1}F_{2n+2}^{-1}}\right) =$$

$$= \arctan\left(\frac{F_{2n+1} + F_{2n+2}}{F_{2n+1}F_{2n+2} - 1}\right).$$

Simplify the obtained expression. The fraction denominator under the arctangent sign we transform to the form $F_{2n+1}F_{2n+2} - 1 = F_{2n}F_{2n+3}$ (see Exercise **1.102**), and the numerator is equal to $F_{2n+1} + F_{2n+2} = F_{2n+3}$ by the definition property of the series $\{F_n\}$.

As a result, we come the sought relation

$$\arctan \frac{1}{F_{2n+1}} + \arctan \frac{1}{F_{2n+2}} = \arctan \frac{1}{F_{2n}}, \quad n \geq 1.$$

Note. Consecutive application of the proved formula, staring from $n = 1$, taking into account $\lim_{n \to \infty} \arctan(1/F_{2n+2}) = 0$, results in the sum

$$\frac{\pi}{4} = \sum_{k=1}^{\infty} \arctan \frac{1}{F_{2k+1}}.$$

1.104. *Proof.*

Let $P(n-1)$ be the predicate $L_n = F_{n-1} + F_{n+1}$ for some integer $n = 2, 3, 4, \ldots$

B a s i s   s t e p

Check the truth of $P(1)$ and $P(2)$. For $n = 2$ we obtain: $L_2 = F_1 + F_3 = 3$; therefore $P(1)$ is true. For $n = 3$ we obtain: $L_3 = F_2 + F_4 = 4$, therefore $P(2)$ also takes the true value.

I n d u c t i v e   s t e p

Let us prove that from $L_{k-1} = F_{k-2} + F_k$ and $L_k = F_{k-1} + F_{k+1}$ follows the equality $L_{k+1} = F_k + F_{k+2}$. Actually,

$$L_{k+2} = L_k + L_{k+1} = (F_{k-1} + F_{k+1}) + (F_k + F_{k+2}) =$$
$$= F_{k-1} + F_k + F_{k+1} + F_{k+2} = F_{k+1} + F_{k+3}.$$

Hence, $L_n = F_{n-1} + F_{n+1}$, for all $n = 2, 3, \ldots$

1.105. *Solution.*

Using the recurrent relation $F_n = F_{n+2} - F_{n+1}$ with the known initial conditions $F_1 = F_2 = 1$, we obtain

$$
\begin{array}{llllll}
n = 0: & F_0 & = F_2 - F_1 & = & 1 - 1 = & 0, \\
n = -1: & F_{-1} & = F_1 - F_0 & = & 1 - 0 = & 1, \\
n = -2: & F_{-2} & = F_0 - F_{-1} & = & 0 - 1 = & -1, \\
n = -3: & F_{-3} & = F_{-1} - F_{-2} & = & 1 - (-1) = & 2, \\
n = -4: & F_{-4} & = F_{-2} - F_{-3} & = & -1 - 2 = & -3, \\
n = -5: & F_{-5} & = F_{-3} - F_{-4} & = & 2 - (-3) = & 5, \\
n = -6: & F_{-6} & = F_{-4} - F_{-5} & = & -3 - 5 = & -8, \\
n = -7: & F_{-7} & = F_{-5} - F_{-6} & = & 5 - (-8) = & 13, \\
n = -8: & F_{-8} & = F_{-6} - F_{-7} & = & (-8) - 13 = & -21
\end{array}
$$

and so on. Let us present the obtained values in the form of a table.

n	0	1	2	3	4	5	6	7	8
F_n	0	1	1	2	3	5	8	13	21
F_{-n}	0	1	-1	2	-3	5	-8	13	-21

Suppose that for any integer n the numbers F_n and F_{-n} are equal in the absolute value and differ in sign, if $|n|$ is even. In other words, suppose that the following equality is fulfilled

$$F_{-n} = (-1)^{n-1} F_n, \quad n = 0, 1, 2, \ldots$$

For the proof, let us use the complete mathematical induction method (see Exercise **1.92**).

Let $P(n)$ be the predicate $F_{-n} = (-1)^{n-1} F_n$.

B a s i s   s t e p

The basis step consists in the statements $P(1)$ and $P(2)$. For $n = 1$ we obtain: $F_{-1} = F_1 = 1$, and for $n = 2$: $F_{-2} = -F_2 = -1$ in accordance with the previously written table of values.

I n d u c t i v e   s t e p

Now let $P(i)$ be true for all $i \leqslant k$. Let us prove the truth of the predicate $P(k+1)$ by the recurrent relation for the Fibonacci numbers. According to the inductive supposition, $F_{-k} = (-1)^{k-1} F_k$, $F_{-k+1} = (-1)^{k-2} F_{k-1}$. Hence,

$$F_{-(k+1)} = (-1)^{k-2} F_{k-1} - (-1)^{k-1} F_k.$$

In the right side of the obtained equality, introduce the common factor $(-1)^{k-2}$ and apply the fundamental recurrent relation:

$$F_{-(k+1)} = (-1)^{k-2} (F_{k-1} - (-F_k)) = (-1)^{k-2} (F_{k-1} + F_k) = (-1)^{k-2} F_{k+1}.$$

Since $(-1)^{k-2} = (-1)^k = (-1)^{(k+1)-1}$, then we finally obtain

$$F_{-(k+1)} = (-1)^{(k+1)-1} F_{k+1},$$

which proves the truth of the predicate $P(k+1)$.

Therefore, by the mathematical induction method, we have proved the formula for calculating the Fibonacci numbers with negative values of the index $F_{-n} = (-1)^{n-1} F_n$ for integer $n = 1, 2, 3, \ldots$

1.106. *Answer*: $L_{-n} = (-1)^n L_n$, $n = 1, 2, \ldots$

1.107. *Hint*. See solution of Exercise **1.99**.

1.108. *Hint*. Having used the results of Exercise **1.107**, write the Lucas number in the form $L_n = \varphi^n + \widetilde{\varphi}^n$, where φ is the golden ratio, $\varphi\widetilde{\varphi} = -1$.

1.110. *Hint*. Present the Lucas numbers in the form $L_n = \varphi^n + \widetilde{\varphi}^n$.

Note. The identities of Exercises **1.107–1.110** are valid for all integer n.

1.111. *Answer:* $L_{n-1}L_m + L_n L_{m+1} = L_{n+m-1} + L_{n+m+1}$ $\forall n, m \geqslant 1$.

1.113. *Solution.*
The suggested equalities for the positive $i > 0$ follow directly from the identity of the previous exercise, if $n = 3i, k = i$ and $n = 4i, k = 2i$. The case $i \leqslant 0$ is considered taking into account the result of the Exercise **1.106**.

1.114. *Hint.* Express the terms of the Fibonacci and Lucas series by constant φ and $\widetilde{\varphi}$.

1.117. *Solution.*
From the analysis of the identities formulated in Exercises **1.115** and **1.116**, we may offer a hypothesis that the following equality is valid

$$F_{n-k}F_{n+k} = \frac{1}{5}\left(L_n^2 - (-1)^{n+k}L_k^2\right), \quad n \in \mathbb{N}, \ k = 1, 2, \ldots, n.$$

Let us write the Lucas numbers in the form $L_k = \varphi^k + \widetilde{\varphi}^k$ and the Fibonacci numbers in the form $F_k = \frac{1}{\sqrt{5}}(\varphi^k - \widetilde{\varphi}^k)$, where φ is the golden ratio, $\varphi\widetilde{\varphi} = -1$:

$$F_{n-k}F_{n+k} = \frac{1}{\sqrt{5}}(\varphi^{n-k} - \widetilde{\varphi}^{n-k}) \cdot \frac{1}{\sqrt{5}}(\varphi^{n+k} - \widetilde{\varphi}^{n+k}) =$$

$$= \frac{1}{5}(\varphi^{2n} - \varphi^{n-k}\widetilde{\varphi}^{n+k} - \varphi^{n+k}\widetilde{\varphi}^{n-k} + \widetilde{\varphi}^{2n}) =$$

$$= \frac{1}{5}\left(\varphi^{2n} + \widetilde{\varphi}^{2n} - (\varphi\widetilde{\varphi})^{n-k}(\varphi^{2k} + \widetilde{\varphi}^{2k})\right) =$$

$$= \frac{1}{5}(L_{2n} - (-1)^{n-k}L_{2k}).$$

Then, we only have to use the identities $L_{2k} = L_k^2 - 2(-1)^k$ (see Exercise **1.109**) and $(-1)^{n-k} = (-1)^{n+k}$:

$$F_{n-k}F_{n+k} = \frac{1}{5}(L_{2n} - (-1)^{n-k}L_{2k}) =$$

$$= \frac{1}{5}\left((L_n^2 - 2(-1)^n) - (-1)^{n-k}(L_k^2 - 2(-1)^k)\right) = \frac{1}{5}(L_n^2 - (-1)^{n+k}L_k^2),$$

i.e., the required equality.

1.119. *Proof.*
Let $N = \overline{a_k a_{k-1} \ldots a_0}$.   Denote $P = \{N \text{ is divisible by 3}\}$ and $Q = \{\sum_{i=0}^{k} a_i \text{ is divisible by 3}\}$. It is required to prove that $P \Leftrightarrow Q$. For this, consider the

difference $N - \sum_{i=0}^{k} a_i$. Using the decimal notation of N, we obtain

$$N - \sum_{i=0}^{k} a_i = a_k(10^k - 1) + a_{k-1}(10^{k-1} - 1) + \cdots + a_2(10^2 - 1) + a_1(10 - 1).$$

For any $k = 1, 2, 3, \ldots$, the value $10^k - 1$ is divisible by 3 according to the algebraic identity

$$d^k - 1 = (d - 1)(d^{k-1} + \cdots + d + 1).$$

Therefore, the difference $N - \sum_{i=0}^{k} a_i$ is divisible by 3, and then both numbers N and $\sum_{i=0}^{k} a_i$ are either divisible by 3 or not simultaneously; hence, $P \Leftrightarrow Q$.

1.122. *Hint.*
Use the easy-to-check algebraic identities

$$d^{2n+1} + 1 = (d + 1)(d^{2n} - d^{2n-1} + \cdots - d + 1);$$
$$d^{2n} - 1 = (d + 1)(d - 1)(d^{2n-2} + d^{2n-4} + \cdots + d^2 + 1).$$

1.123. *Proof.*
Denote the predicate "natural number n, greater than unity is either a prime number or can be presented in the form of the product of prime numbers" by $P(n)$. For the proof, let us use the complete induction principle.

Basis step

The basis step consists in an infinite number of cases of prime values of n. It is clear that in such cases the predicate $P(n)$ takes the true value.

Inductive step

Let us prove that $P(n)$ is valid for all composite numbers $n > 2$. For this, consider some number $k + 1$, which is not prime, and prove the validity of the implication $(\forall i < k \, P(k)) \Rightarrow P(k + 1)$.

Since $k + 1$ is composite, then it has at least one divisor, for example, l. Then $k + 1 = l \cdot m$, and the natural numbers l and m are not equal to unity: $l \neq 1, m \neq 1$. According to the inductive supposition, both l and m can be presented in the form of a product of several prime numbers; let their quantity be s and s', respectively:

$$l = p_1 p_2 \ldots p_s, \quad m = p_1' p_2' \ldots p_{s'}', \quad s, s' \text{ are integers.}$$

Hence,

$$k + 1 = p_1 p_2 \ldots p_s \cdot p_1' p_2' \ldots p_{s'}'.$$

Thus, it is proved that any natural number greater than unity is either a prime number or can be presented in the form of the product of prime numbers.

1.124. *Proof.*
Consider n of consecutive natural numbers

$$(n + 1)! + 2, \ (n + 1)! + 3, \ \ldots, \ (n + 1)! + n + 1, \quad n \geqslant 2.$$

In the constructed sequence, the first term $(n + 1)! + 2$ is divisible by 2, the second $(n + 1)! + 3$ is divisible by 3, and so on [44]. The sequence consists of the total of n elements, each of them being a composite number. We obtain that for an arbitrary number n in the series of natural numbers there exist n of consecutive composite numbers.

1.125. *Solution.*
Using the idea described in the previous exercise, we obtain ten consecutive composite numbers:

$$11! + 2, \ 11! + 3, \ \ldots, \ 11! + 11.$$

However, we can specify a set of thirteen smaller numbers, for example

$$114, \ 115, \ \ldots, \ 126.$$

Set Theory

One of the fundamental concepts of mathematics is the notion of a set. A **set** is a collection of objects which we conceive as a whole [34, 70]. These objects are called **elements** of a set. The belonging of some element a to the set A can be denoted as follows: $a \in A$. This record reads as follows: "a is an element of the set A" or "element a belongs to the set A." If a does not belong to A, then we write $a \notin A$.

There are several ways to denote which elements belong to a set; the most common are the following [4, 65]:

(1) Enumerating the elements $A = \{a_1, a_2, \ldots, a_n\}$. The elements of the set A are enclosed in braces and separated by commas.

(2) Using the characteristic predicate $A = \{x : P(x)\}$. The **characteristic predicate** is a statement, allowing one to establish the fact that the object x belongs to the set A. If for some x the predicate $P(x)$ takes a true value, then $x \in A$; otherwise $x \notin A$.

Some widely used sets have specific notations [5, 86]:

$\mathbb{N} = \{1, 2, 3, \ldots\}$ is the **set of natural numbers**.

$\mathbb{Z} = \{0, \pm 1, \pm 2, \pm 3, \ldots\}$ is the **set of integers**.

$\mathbb{Q} = \{p/q : p, q$ are integers, $q \neq 0\}$ is the **set of rational numbers**.

A **set of real numbers** is denoted by $\mathbb{R} = (-\infty, +\infty)$ and of **complex numbers** by $\mathbb{C}$.

Consider a set of real numbers satisfying the inequalities $a \leqslant x \leqslant b$, where $a < b$. Such a set will be called a **closed interval** or a **segment** and denoted by $[a, b]$, i.e.,

$$[a, b] = \{x : a \leqslant x \leqslant b\}.$$

The set

$$(a, b) = \{x : a < x < b\}.$$

is called an **open interval**.

© Springer International Publishing AG, part of Springer Nature 2018
S. Kurgalin and S. Borzunov, *The Discrete Math Workbook*,
Texts in Computer Science, https://doi.org/10.1007/978-3-319-92645-2_2

The set A is called a **subset** of the set B (notation: $A \subseteq B$) if $x \in A \Rightarrow x \in B$; i.e., all elements of A belong to the set B. Two sets are **equal** $(A = B)$ if and only if $A \subseteq B$ and $B \subseteq A$. If $A \subseteq B$ and $A \neq B$, then A is called a **proper subset** of the set B $(A \subset B)$.

A **universal set** U contains as a subset any of the sets occurring in the problem under consideration. The **empty set** that does not contain any element is denoted by the symbol $\varnothing$. It is assumed that an empty set is a subset of any set.

Consider basic operations on sets:

$A \cup B = \{x : x \in A \text{ or } x \in B\}$ is a **union** of the sets A and B.

$A \cap B = \{x : x \in A \text{ and } x \in B\}$ is an **intersection** of A and B.

$A \setminus B = \{x : x \in A \text{ and } x \notin B\}$ is a **complement** of the set B to A, or a **set-theoretic difference** of A and B.

$\overline{A} = \{x : x \notin A\}$ is a **complement** of the set A (to the universal set U).

$A \triangle B = \{x : (x \in A \text{ and } x \notin B) \text{ or } (x \in B \text{ and } x \notin A)\}$ is a **symmetric difference of two sets** A and B. The symmetric difference of two sets can be expressed in terms of operations of union, intersection, and complement: $A \triangle B = (A \cap \overline{B}) \cup (\overline{A} \cap B)$.

The union, intersection, set-theoretic difference, and symmetric difference of two sets are **binary operations**; the complement of a set to the universal one is a **unary operation**.

Example 2.1 Let $\mathbb{R}$ be a universal set. Then for the sets $A = [0, 1]$ and $\mathbb{R}^+ = (0, \infty)$ we have:

$$A \cup \mathbb{R}^+ = [0, \infty), \qquad\qquad A \cap \mathbb{R}^+ = (0, 1],$$
$$A \setminus \mathbb{R}^+ = \{0\}, \qquad\qquad \mathbb{R}^+ \setminus A = (1, \infty),$$
$$\overline{A} = (-\infty, 0) \cup (1, \infty), \qquad\qquad \overline{\mathbb{R}^+} = (-\infty, 0],$$
$$A \triangle \mathbb{R}^+ = \mathbb{R}^+ \triangle A = \{0\} \cup (1, \infty). \qquad\qquad \square$$

For a graphic illustration of the relations in set theory, we use **Venn**[1] **diagrams**. The Venn diagram shows a rectangle, whose points represent the elements of the universal set U.

In this rectangle, the subsets of the set U are represented by circles or ellipses. Examples of some common sets are shown in Figs. 2.1 and 2.2.

The basic laws of set algebra are presented in Table 2.1. The comparison of Tables 2.1 and 1.2 shows the correspondence between the laws of set algebra and the laws of algebra of logic.

If we perform a replacement of operations $\cup \rightleftarrows \cap$ and sets $\varnothing \rightleftarrows U$ in an arbitrary identity of the algebra of logic, then we obtain a **dual** identity. The latter statement is called the **duality law**, which is clearly illustrated in Table 2.1: *Each of the identities*

---

[1]John Venn (1834–1923) was the British mathematician and logician.

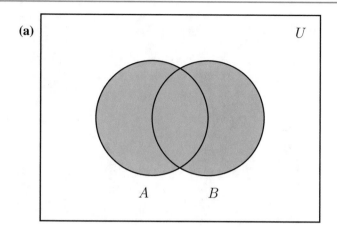

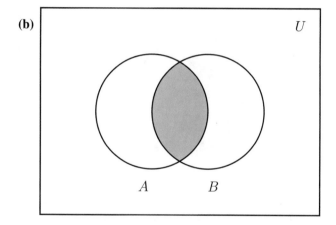

Fig. 2.1 Venn diagrams of the sets **a** $A \cup B$ and **b** $A \cap B$

in the left column is dual to the corresponding identity in the right column and vice versa.

A set is called **finite** if it contains a finite number of elements, and it is called **infinite** in the contrary case. The number of elements of the finite set A is called its **cardinality** (denoted by $|A|$).

To calculate the cardinality of a union of two finite sets, one can use the **inclusion and exclusion formula**:

$$|A \cup B| = |A| + |B| - |A \cap B|.$$

The set whose elements are all subsets of the set A is called a **power set** (or powerset) $\mathcal{P}(A)$, that is, $\mathcal{P}(A) = \{X : X \subseteq A\}$. The cardinality of a power set can be determined by the formula $|\mathcal{P}(A)| = 2^{|A|}$; therefore, one more notation is used for the power set of the set A, namely 2^A [65].

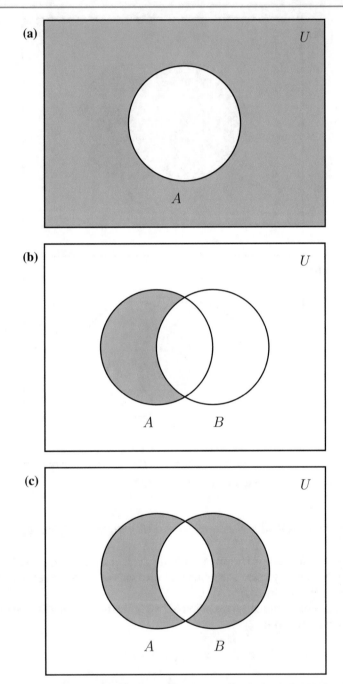

Fig. 2.2 Venn diagrams of the sets **a** $\overline{A}$, **b** $A \setminus B$, and **c** $A \triangle B$

Table 2.1 The laws of set algebra

Idempotent Laws	$A \cup A = A$	$A \cap A = A$
Identity and Dominating Laws	$A \cup \emptyset = A$ $A \cup U = U$	$A \cap U = A$ $A \cap \emptyset = \emptyset$
Complement Laws	$A \cup \overline{A} = U$ $\overline{U} = \emptyset$ $\overline{\overline{A}} = A$	$A \cap \overline{A} = \emptyset$ $\overline{\emptyset} = U$
Commutative Laws	$A \cup B = B \cup A$	$A \cap B = B \cap A$
Associative Laws	$A \cup (B \cup C) = (A \cup B) \cup C$	$A \cap (B \cap C) = (A \cap B) \cap C$
Distributive Laws	$A \cup (B \cap C) = (A \cup B) \cap (A \cup C)$	$A \cap (B \cup C) = (A \cap B) \cup (A \cap C)$
Absorption Laws	$A \cup (A \cap B) = A$	$A \cap (A \cup B) = A$
De Morgan's Laws	$\overline{(A \cup B)} = \overline{A} \cap \overline{B}$	$\overline{(A \cap B)} = \overline{A} \cup \overline{B}$

The concept of cardinality can be generalized for the case of infinite sets. Two sets A and B are said to be **equinumerous** if among their elements there is a one-to-one correspondence, which implies that each element of one set corresponds to one and only one element of the other (the term **equipotence** is sometimes used instead). The sets equinumerous to the set of natural numbers $\mathbb{N}$ are called **countable**; the cardinality of such sets is denoted by $\aleph_0$ (reads as "aleph-zero" or "aleph-null"). The sets $\mathbb{Z}$ and $\mathbb{Q}$ are countable, and the union of a finite (and also countable) number of countable sets is also countable.

The theorem of an uncountable set. *The set of numbers of the interval* $(0, 1)$ *is uncountable.*

We prove this statement by contradiction. Let there exists a way according to which each real number x_k of the interval $(0, 1)$, where $k = 1, 2, \ldots$, corresponds to some natural number. We write each of the numbers x_k in the form of an infinite decimal decomposition:

$$x_1 = 0{,}x_{11}x_{12}\ldots x_{1k}\ldots,$$
$$x_2 = 0{,}x_{21}x_{22}\ldots x_{2k}\ldots,$$
$$\cdots\cdots\cdots\cdots\cdots$$
$$x_k = 0{,}x_{k1}x_{k2}\ldots x_{kk}\ldots,$$
$$\cdots\cdots\cdots\cdots\cdots$$

Let us introduce into consideration the number $\widetilde{x} = 0{,}\widetilde{x}_1\widetilde{x}_2\ldots\widetilde{x}_k\ldots$, defined according to the following rule:

$$\widetilde{x}_k = \begin{cases} 1, & \text{if } x_{kk} \neq 1, \\ 2, & \text{if } x_{kk} = 1. \end{cases}$$

The number $\widetilde{x}$ does not contain zeros and nines in its decimal notation, which makes it possible to exclude from consideration a sequence of the form $0{,}4999\ldots = 0{,}5000\ldots$ Note that $\widetilde{x}$ differs from x_1 in the first digit after the comma, from x_2 in the second digit, and so on. Because of this, for any $k \in \mathbb{N}$ the inequality $\widetilde{x} \neq x_k$ holds. Therefore, one cannot number all numbers in the interval $(0, 1)$, and the set of numbers in the interval $(0, 1)$ is not countable. The theorem on an uncountable set is proved.

Note that the considered method for constructing an auxiliary number $\widetilde{x}$ is known as the **Cantor**[2] **diagonal procedure**.

As shown above, the interval $(0, 1)$ is an example of an uncountable set. Its cardinality is called the **cardinality of the continuum** and is denoted by $\mathfrak{c}$. The set of all irrational numbers $\mathbb{R} \setminus \mathbb{Q}$ and the set of real numbers $\mathbb{R}$ have the cardinality of the continuum. The cardinality of the set $2^{\mathbb{N}}$ is equal to the cardinality of the continuum.

---

[2]Georg Ferdinand Ludwig Philipp Cantor (1845–1918) was a German mathematician.

The set $A \times B = \{(a, b) : a \in A \text{ and } b \in B\}$ is called the **Cartesian**[3] or a **direct product** of the sets A and B.

The elements (a, b) of the set $A \times B$ are called **ordered pairs**. The cardinality of a direct product of finite sets A and B is equal to $|A \times B| = |A| \times |B|$. The Cartesian product of an arbitrary number of sets is defined as follows:

$$A_1 \times A_2 \times \ldots \times A_n = \{(a_1, a_2, \ldots, a_n) : a_i \in A_i, \; i = 1, 2, \ldots, n\}.$$

If all the sets A_i are the same, $A_i = A$, then to denote a direct product $\underbrace{A \times A \times \ldots \times A}_{n \text{ times}}$ we write A^n. An arbitrary element of the set A^n is called a **vector**.

Example 2.2 Let $A = \{\varnothing\}$, $B = \{1, 2\}$. Let us list the elements of the sets $A \times B$ and $(2^A)^2$.
Solution.
According to the definition of the Cartesian product of sets, the elements of $A \times B$ are all possible ordered pairs (a, b), where $a \in A$, $b \in B$. Thus, $A \times B = \{(\varnothing, 1), (\varnothing, 2)\}$. The power set of the set A consists of all subsets of the set $A = \{\varnothing\}$; therefore, $2^A = \{\varnothing, \{\varnothing\}\}$. We only need to write out the elements of the Cartesian product $(2^A)^2 = 2^A \times 2^A$:

$$(2^A)^2 = \{(\varnothing, \varnothing), (\varnothing, \{\varnothing\}), (\{\varnothing\}, \varnothing), (\{\varnothing\}, \{\varnothing\})\}. \qquad \square$$

A **bit string** (of length n) is an element of the set $\mathbb{B}^n$, where $\mathbb{B} = \{0, 1\}$. If $A \subseteq U$, where $U = \{u_1, u_2, \ldots, u_n\}$ is a universal set, then a **characteristic vector a** of the set A is a bit string of length n of the form $(a_1, a_2, \ldots, a_n)$, where $a_i = 1$, if $u_i \in A$, and $a_i = 0$ in the contrary case. Characteristic vectors are used for modeling operations on finite sets [27].

Example 2.3 Let $U = \{1, 2, \ldots, 10\}$ be a universal set. We write the characteristic vectors of the sets A, B, $\overline{A}$, $A \cup B$, $A \cap B$ if $A = \{1, 2, 3\}$, $B = \{2, 4, 6, 8, 10\}$.
Solution.
Denote the characteristic vector of the set A by **a** and the characteristic vector of the set B by **b**. By the definition of the characteristic vector, we have:

$$\mathbf{a} = (1, 1, 1, 0, 0, 0, 0, 0, 0, 0), \quad \mathbf{b} = (0, 1, 0, 1, 0, 1, 0, 1, 0, 1).$$

The logical operations **not, or, and** over the values $1 \equiv T$, $0 \equiv F$ correspond to the operations of negation, union, and intersection of sets.

Therefore, the set $\overline{A}$ is represented by a bit string

$$\mathbf{not\ a} = (0, 0, 0, 1, 1, 1, 1, 1, 1, 1),$$

---

[3]René Descartes (1596–1650) was a French philosopher, mathematician, physicist, and physiologist.

the set $A \cup B$ by a bit string

$$\mathbf{a \ or \ b} = (1, 1, 1, 1, 0, 1, 0, 1, 0, 1),$$

the set $A \cap B$ by a bit string

$$\mathbf{a \ and \ b} = (0, 1, 0, 0, 0, 0, 0, 0, 0, 0).$$

By the well-known characteristic vector of the set, it is possible to write out the elements of this set, in particular, $\overline{A} = \{4, 5, 6, 7, 8, 9, 10\}$, $A \cup B = \{1, 2, 3, 4, 6, 8, 10\}$, $A \cap B = \{2\}$ □

We emphasize once again that $\{u_1, u_2, \ldots, u_n\}$ is an unordered collection or set, and $(u_1, u_2, \ldots, u_n)$ is an ordered set or vector.

Note. Strictly speaking, the concept of a set used in this study guide is not mathematically correct. The inner inconsistency of set theory, constructed on this definition, follows from the analysis of **Russell's**[4] **paradox** and other paradoxes [14]. At present, the axiomatic systems are used for the construction of a consistent set theory, among them the most famous **Zermelo**[5]**–Fraenkel**[6] **system** of axioms [65].

2.1 Exercises

2.1.   List the elements of the following sets:

  (1) A_1 is a set of all prime numbers not exceeding the value of 11.
  (2) A_2 is a set of Galilean[7] satellites of Jupiter.
  (3) A_3 is a set of prime even numbers.
  (4) A_4 is a set of students who have visited Mercury.

2.2.   Let A_1–A_4 be the sets defined in the previous exercise. Establish the truth value of the given statements:

  (1) $A_3 \subset A_1$.
  (2) $A_3 \in A_1$.
  (3) $A_4 \subset A_2$.
  (4) $A_4 \in A_2$.

---

[4]Bertrand Arthur William Russell (1872–1970) was a British mathematician and philosopher.
[5]Ernst Friedrich Ferdinand Zermelo (1871–1953) was a German mathematician.
[6]Abraham Halevi Fraenkel (1891–1965) was an Israeli mathematician.
[7]Galileo Galilei (1564–1642) was an Italian mathematician, physicist, and astronomer.

2.3.   Let $A = \{2, 3\}$, $B = \{\varnothing, 1, \{2, 3\}\}$. Determine which of the statements given below are true and which are false:

(1) $\{1\} \subset A$.
(2) $\{1\} \in A$.
(3) $\varnothing \subset B$.
(4) $\varnothing \in B$.

2.4.   Let $A = \{\varnothing, a, \{b, c\}\}$, $B = \{b, c\}$. Determine which of the statements given below are true and which are false:

(1) $\{a\} \subset A$.
(2) $\{a\} \in A$.
(3) $B \subset A$.
(4) $B \in A$.

2.5.   Write down the following sets, using the characteristic predicates:

(1) $S_1 = \{1, 4, 7, 10, \ldots\}$.
(2) $S_2 = \{10, 2, -6, -14, \ldots\}$.
(3) $S_3 = \{1, 3, 6, 10, \ldots\}$.
(4) $S_4 = \{\pm 3, \pm 9, \pm 27, \pm 81, \ldots\}$.

2.6.   Write down the following sets, using the characteristic predicates:

(1) $S_1 = \left\{1, \dfrac{1}{2}, \dfrac{1}{4}, \dfrac{1}{8}, \ldots\right\}$.

(2) $S_2 = \left\{\dfrac{2}{3}, \dfrac{8}{9}, \dfrac{26}{27}, \dfrac{80}{81}, \ldots\right\}$.

(3) $S_3 = \left\{\dfrac{1}{2}, -\dfrac{1}{3}, \dfrac{1}{4}, -\dfrac{1}{5}, \ldots\right\}$.

(4) $S_4 = \left\{0, \dfrac{1}{3}, \dfrac{1}{2}, \dfrac{3}{5}, \ldots\right\}$.

2.7.   List the elements of the following sets: A, $B \cup C$, $(B \cap C) \triangle A$, where $A = \{x : x \in \mathbb{Z} \text{ and } x^2 \leqslant 15\}$, $B = \{x : x \in \mathbb{Z} \text{ and } x^2 - 4x - 12 = 0\}$, $C = \{x : x \in \mathbb{Z} \text{ and } x^2 - 2x - 8 \leqslant 0\}$.

2.8.   List elements of the following sets: $A \cap C$, $(B \cup C) \triangle A$, where $A = \{x : x \in \mathbb{Z} \text{ and } -7 \leqslant x^3 \leqslant 555\}$; $B = \{x : x \in \mathbb{Z} \text{ and } 4x^2 + 3x - 1 \leqslant 0\}$; $C = \{x : x \in \mathbb{R} \text{ and } 4x^2 + 3x - 1 = 0\}$.

2.9. The sets $A = \{2, 3, 4\}$, $B = \{1, 5, 10\}$, $C = \{-1, 1, 2\}$, $D = \{2, 5, 10\}$ are given. Define the elements of sets:

(1) $(A \cap B) \cup (C \cap D)$.
(2) $(A \cup (B \cap C)) \cap ((A \cup B) \cap C)$.
(3) $A \setminus (B \setminus (C \setminus D))$.
(4) $(A \setminus (B \cup C)) \setminus (A \setminus (B \cup D))$.

2.10. The sets $A = \{0, 1, 7\}$, $B = \{3, 4, 5\}$, $C = \{0, 1, 2\}$, $D = \{-1, 1, 10\}$ are given. Define the elements of the sets:

(1) $(A \cap B) \cup (A \cap C) \cup (B \cap C)$.
(2) $(A \cap (B \cup C)) \cap (A \cup B \cup D)$.
(3) $(A \setminus C) \setminus (B \setminus D)$.
(4) $(A \setminus (B \cap C)) \cap (D \setminus A)$.

2.11. We assume that the universal set of the given problem has the form $U = \{a, b, c, e, f, g, h, k, m\}$. Let $A = \{a, b, c\}$, $B = \{a, c, f, h, m\}$, $C = \{c, e, g, k, m\}$. Find the elements of the following sets: $\overline{A} \cap B, (A \bigtriangleup B) \cup C, (A \cup C) \setminus B$.

2.12. We assume that the universal set of the given problem has the form $U = \{a, b, c, e, f, g, h, k, m\}$. Let $A = \{a, b, c\}$, $B = \{a, c, f, h, m\}$, $C = \{c, e, g, k, m\}$. Find the elements of the following sets: $A \cup \overline{C}, (A \bigtriangleup C) \cup B, (B \cap \overline{C}) \setminus A$.

2.13. We assume that the universal set of this problem has the form $U = \{a, b, c, e, f, g, h, k, m\}$. Let $A = \{a, b, c\}$, $B = \{a, c, f, h, m\}$, $C = \{c, e, g, k, m\}$. Find elements of the following sets: $B \bigtriangleup C, (B \setminus \overline{C}) \cup A, (B \bigtriangleup C) \cap \overline{A}$.

2.14. Consider the sets $A = \{1, 2, \{3, 4\}\}$, $B = \{2, 3\}$. Write out elements of the sets $A \cup B, A \cap B, A \bigtriangleup B$.

2.15. Consider the sets $A = \{\varnothing, \{2\}\}$, $B = \{\{1\}, \{2\}\}$. Write out elements of the sets $A \cup B, A \cap B, A \bigtriangleup B$.

2.16. Write out all subsets of the set $A = \{a, b, c\}$.

2.17. Write out all subsets of the set $A = \{\{a, b, c\}\}$.

2.18. Write out all subsets of the set $A = \{a, \{b, c\}\}$.

2.19. Find the power set:

(1) Of the empty set $\varnothing$;
(2) Of the power set of the empty set $\mathcal{P}(\varnothing)$.

2.20. Express the operation of set-theoretic difference of sets in terms of the operation of complement to a universal set and intersection of sets.

2.21. Prove the following identity with the help of set algebra: $(A \setminus B) \setminus C = A \setminus (B \cup C)$.

∗**2.22.** Determine whether binary operations of complement and symmetric difference have the following properties:

(1) Commutative;
(2) Associative.

2.23. Using laws of set algebra, prove the following identity: $A \vartriangle A \vartriangle A = A$.

2.24. Using laws of set algebra, prove the following identity: $(A \cup B) \vartriangle (A \cup C) = (B \vartriangle C) \setminus A$.

2.25. Let A, B, C be arbitrary sets. Using laws of set algebra, prove the identities:

(1) $\overline{(A \cup \overline{B})} = \overline{A} \cap B$.
(2) $(A \cup (A \cap \overline{B})) \cap (A \cap (A \cup B)) = A$.
(3) $(A \cup B) \setminus C = (A \setminus C) \cup (B \setminus C)$.
(4) $(A \setminus (B \setminus C)) \setminus ((A \setminus B) \setminus C) = A \cap C$.
(5) $(A \setminus B) \vartriangle (B \setminus A) = A \vartriangle B$.
(6) $(A \cup B \cup C) \vartriangle (A \cap B \cap C) = (A \cup B \cup C) \setminus (A \cap B \cap C)$.

Represent the respective Venn diagrams.

2.26. For the universal set $U = \{-5, -4, \ldots, 5\}$ and the set $A = \{x : f(x) = 0$ **and** $x \in \mathbb{Z}\}$, write elements of the set $\overline{A}$, $A \vartriangle U$, $A \vartriangle A$. What is the cardinality of the power set $\mathcal{P}(A)$? Consider the following cases:

(1) $f(x) = x^4 - 5x^2 + 4$.
(2) $f(x) = (x^2 - 2)(x^2 - 3x + 2)$.
(3) $f(x) = x^4 + 2x^3 - 25x^2 - 26x + 120$.
(4) $f(x) = x^4 - x^3 - 19x^2 - 11x + 30$.

2.27. Represent the set $(A \setminus B) \cap C$ in the Cartesian coordinate system if the sets A, B, $C \subset \mathbb{R} \times \mathbb{R}$ are determined as follows: $A = \{(x, y) : x^2 + y^2 \leqslant 4\}$, $B = \{(x, y) : x^2 - 2x + y^2 \leqslant 0\}$, $C = \{(x, y) : y > 1\}$.

2.28. Represent the set $(A \cup B) \vartriangle C$ in the Cartesian coordinate system if the sets A, B, $C \subset \mathbb{R} \times \mathbb{R}$ are as follows:

(1) $A = \{(x, y) : x^2 + y^2 - 5x < 0\}$, $B = \{(x, y) : x^2 \leqslant 4\}$,
$C = \{(x, y) : -3 \leqslant y \leqslant 0\}$.

(2) $A = \{(x, y): y - x^2 - 1 \leqslant 0\}$, $B = \{(x, y): y - x^2 - 1 \geqslant 0\}$,
      $C = \{(x, y): x^2 + y^2 < 7\}$.

(3) $A = \{(x, y): x^2 + y^2 - y \leqslant 0\}$, $B = \{(x, y): x^2 + y^2 + y \leqslant 0\}$,
      $C = \{(x, y): y > |x| - \frac{1}{2}\}$.

(4) $A = \{(x, y): |x + y| < 10\}$, $B = \{(x, y): -5 \leqslant x \leqslant 5\}$,
      $C = \{(x, y): x + y > 0\}$.

2.29. Represent the set $(A \triangle B) \cap C$ in the Cartesian coordinate system if the sets $A, B, C \subset \mathbb{R} \times \mathbb{R}$ are as follows:

(1) $A = \{(x, y): x - y^2 - 3 \geqslant 0\}$, $B = \{(x, y): x + y^2 - 3 \leqslant 0\}$,
      $C = \{(x, y): x^2 - 6x + y^2 < 0\}$.

(2) $A = \{(x, y): x^2 + y^2 \leqslant 2\}$, $B = \{(x, y): |x| + |y| \leqslant 2\}$,
      $C = \{(x, y): y - x > 0\}$.

(3) $A = \{(x, y): x^2 + 6x + y^2 < 0\}$, $B = \{(x, y): |x| \leqslant 3, |y| \leqslant 3\}$,
      $C = \{(x, y): x + 3y \leqslant 0\}$.

(4) $A = \{(x, y): y \geqslant |x|\}$, $B = \{(x, y): y \leqslant 2 - |x|\}$,
      $C = \{(x, y): x^2 + (y - 1)^2 \leqslant 1\}$.

2.30. Prove that for arbitrary finite sets A and B

$$|A \triangle B| = |A| + |B| - 2|A \cap B|.$$

2.31. Prove that for arbitrary finite sets A, B, and C

$$|A \triangle B \triangle C| = |A| + |B| + |C| - 2(|A \cap B| + |A \cap C| + |B \cap C|) + 4|A \cap B \cap C|.$$

2.32. Let A, B, and C be finite sets. Prove the **inclusion and exclusion formula for the three sets** A, B, and C:

$$|A \cup B \cup C| = |A| + |B| + |C| - |A \cap B| - |A \cap C| - |B \cap C| + |A \cap B \cap C|.$$

2.33. The class consists of 29 pupils. Out of them, 13 go to the sports club, 6 to the aeromodeling club, and 19 to the extra classes in math. Two go in for aeromodeling and sports, 7 sports and math, and 4 aeromodeling and math. None of the pupils goes in for all the out-of-school activities. How many pupils visit only one elective and how many pupils are not interested in them at all?

2.34. A class in a girl's high school offers circles for knitting, sewing, and embroidery. It is known that 9 pupils go in for embroidery, 10 knitting, 15 sewing, 2 embroidery and knitting, 3 sewing and knitting, 5 embroidery and sewing, and 1 girl visits all the circles. How many girls visit circles and how many of them go in for sewing only?

2.35. The group monitor presented the achievement sheet to the curator. The midterm attestation results showed that 12 student received grades "excellent," 15 "good," and 8 "satisfactory." Ten students received grades "excellent" and "good," two "good" and "satisfactory," and one student "excellent" and "satisfactory." None received a set of all the three grades. How many students are there in the group?

2.36. It is known that $|A| = |B| = 16, |C| = 25, |A \cap B| = 10, |A \cap C| = |B \cap C| = 9, |A \cap B \cap C| = 3$. Calculate $|(A \triangle B) \cup C|$.

2.37. It is known that $|A| = |C| = 29, |B| = 19, |A \cap B| = 11, |A \cap C| = |B \cap C| = 9, |A \cap B \cap C| = 5$. Calculate $|A \cup (B \triangle C)|$.

2.38. It is known that $|A| = 18, |B| = |C| = 23, |A \cap B| = |A \cap C| = |B \cap C| = 7, |A \cap B \cap C| = 5$. Calculate $|(A \triangle C) \cup B|$.

∗2.39. Let $A_1, A_2, \ldots, A_n$ be finite sets, $n = 1, 2, 3, \ldots$ Prove **the inclusion and exclusion formula for n sets**:

$$|A_1 \cup A_2 \cup \ldots \cup A_n| = \sum_{i=1}^{n} |A_i| - \sum_{\substack{i,j=1 \\ i<j}}^{n} |A_i \cap A_j| +$$

$$+ \sum_{\substack{i,j,k=1 \\ i<j<k}}^{n} |A_i \cap A_j \cap A_k| - \cdots + (-1)^{n-1} |A_1 \cap A_2 \cap \ldots \cap A_n|.$$

2.40. Write the inclusion and exclusion formula for four sets.

2.41. The Cardinal's guards have been interrogated. The only question was: "Which musketeer bothers you most of all?" Each respondent named at least one musketeer out of the four: d'Artagnan, Atos, Portos, or Aramis. The interrogation results are shown in Table 2.2.
How many guards are there at the Cardinal's service?

2.42. An interrogation has been conducted among 100 visitors of the electronic library. The only question was: "Poems of what poets do you know by heart?" The interrogation results are shown in Table 2.3.
How many people could not answer or failed to name any author?

2.43. Determine which of the examples below have cardinality $\aleph_0$:

    (1) $A_1 = \{n : n > 10, \ n \in \mathbb{Z}\}$.
    (2) $A_2 = \{x : \sin x = 0\}$.
    (3) $A_3 = \{n : \sqrt{n} < 10^{10}, \ n \in \mathbb{N}\}$.
    (4) $A_4 = \{n : n - \text{a prime number}\}$.

Table 2.2 To Exercise **2.41**

N	Musketeers	Number of respondents
1	d'Artagnan	25
2	Atos	12
3	Portos	19
4	Aramis	16
5	d'Artagnan, Atos	7
6	d'Artagnan, Portos	7
7	d'Artagnan, Aramis	9
8	Atos, Portos	5
9	Atos, Aramis	4
10	Portos, Aramis	5
11	d'Artagnan, Atos, Portos	2
12	d'Artagnan, Atos, Aramis	2
13	d'Artagnan, Portos, Aramis	2
14	Atos, Portos, Aramis	2
15	d'Artagnan, Atos, Portos, Aramis	1

2.44. Prove that the set of all even nonnegative numbers $P = \{n : n = 2(k - 1)$, where $k \in \mathbb{N}\}$, is countable.

2.45. Prove that the set of integers $\mathbb{Z}$ is countable.

2.46. Prove that the union of the countable set A and the finite set B is countable.

∗**2.47.** Prove that Cartesian product of the two countable sets A and B is countable.

∗**2.48.** Find the number received by each element of the set $A \times B$ in the enumeration in accordance with the scheme from the previous exercise's solution.

2.49. Prove that the kth power of the set of the natural numbers $\mathbb{N}^k$ for all $k \in \mathbb{N}$ is a countable set.

2.50. Prove that the union of a countable number of countable sets $A_1, A_2, \ldots$ is countable.

2.51. Prove that any two intervals (a, b) and (c, d), where $a, b, c, d \in \mathbb{R}, a < b$, $c < d$, are equinumerous. What is the cardinality of these sets?

2.52. The following sets are specified $A = \{a, b\}, B = \{1, 2\}$. Find elements of the sets $(A \times B) \cup (B \times A)$ and $(A \times B) \cap (B \times A)$.

2.53. The following sets are specified $A = \{1, 2\}, B = \{2, 3\}$. Find elements of the sets $(A \times B) \cup (B \times A)$ and $(A \times B) \cap (B \times A)$.

Table 2.3 To Exercise 2.42

N	Poets named by the visitors	Number of respondents
1	A. S. Pushkin	33
2	M. Yu. Lermontov	27
3	S. A. Yesenin	27
4	M. I. Tsvetayeva	25
5	A. S. Pushkin, M. Yu. Lermontov	9
6	A. S. Pushkin, S. A. Yesenin	10
7	A. S. Pushkin, M. I. Tsvetayeva	8
8	M. Yu. Lermontov, S. A. Yesenin	11
9	M. Yu. Lermontov, M. I. Tsvetayeva	9
10	S. A. Yesenin, M. I. Tsvetayeva	10
11	A. S. Pushkin, M. Yu. Lermontov, S. A. Yesenin	5
12	A. S. Pushkin, M. Yu. Lermontov, M. I. Tsvetayeva	2
13	A. S. Pushkin, S. A. Yesenin, M. I. Tsvetayeva	3
14	M. Yu. Lermontov, S. A. Yesenin, M. I. Tsvetayeva	3
15	A. S. Pushkin, M. Yu. Lermontov, S. A. Yesenin, M. I. Tsvetayeva	1

2.54. Find $|A \times B|$ if $A = \{1, 2, \ldots, 10\}$, $B = \{1, 3, 5\}$.

2.55. Find $|A \times B|$ if $A = \{a, b, c, d\}$, $B = \{-10, -9, \ldots, 9, 10\}$.

2.56. About some sets A, B, and C, the following is known: $|\mathcal{P}(A)| = |\mathcal{P}(B)| = 4$, $|\mathcal{P}(C)| = 16$. How many elements are contained in the set $A \times B \times C$?

2.57. About some sets A, B, and C, it is known that $|\mathcal{P}(A)| = 32$, $|\mathcal{P}(B)| = 16$, $|\mathcal{P}(C)| = 1024$. How many elements are contained in the set $A \times B \times C$?

2.58. Cardinality of the power set A is 512. What is the cardinality of the set A^3?

2.59. Cardinality of the power set A is 1024. What is the cardinality of the set A^3?

2.60. Find whether the following equalities are fulfilled

    (1) $(A \times B) \times C = A \times (B \times C)$.
    (2) $(A \times B) \times (C \times D) = A \times (B \times C) \times D$

for arbitrary sets A, B, C, and D.

2.61. We assume that the universal set has the form $U = \{1, 3, 5, 7, 9, 11, 13\}$. Write out the characteristic vectors of the subsets $A = \{1, 5, 7, 11\}$ and $B = \{3, 5, 7, 9, 13\}$. Find the characteristic vectors of the subsets $\overline{A} \cap B$ and $U \setminus (A \bigtriangleup B)$, following which, list their elements.

2.62. We assume that the universal set has the form $U = \{1, 2, 3, 4, 5, 6, 7, 8\}$. Write out the characteristic vectors of the subsets $A = \{1, 3, 5, 7, 8\}$ and $B = \{3, 4, 6, 7\}$. Find the characteristic vectors of the subsets $A \cup \overline{B}$ and $U \bigtriangleup (A \setminus \overline{B})$, following which, list their elements.

2.63. We assume that the universal set has the form $U = \{2, 4, 6, 8, 10, 12, 14, 16\}$. Write out the characteristic vectors of the subsets $A = \{2, 8, 10, 12, 14\}$ and $B = \{4, 8, 10, 16\}$. Find the characteristic vectors of the subsets $A \setminus \overline{B}$ and $U \bigtriangleup (A \cap \overline{B})$, following which, list their elements.

2.2 Answers, Hints, Solutions

2.1. *Answer:*

    (1) $A_1 = \{2, 3, 5, 7, 11\}$.
    (2) $A_2 = \{\text{Io, Europe, Ganymede, Callisto}\}$.
    (3) $A_3 = \{2\}$.
    (4) $A_4 = \varnothing$.

2.2. *Answer:*

    (1) True;
    (2) False;
    (3) True;
    (4) False.

2.3. *Answer:*

    (1) False;
    (2) False;
    (3) True;
    (4) True.

2.4. *Answer:*

(1) True;
(2) False;
(3) False;
(4) True.

2.5. *Answer:*

(1) $S_1 = \{n : n = 3k - 2, \ k \in \mathbb{N}\}$.
(2) $S_2 = \{n : n = 18 - 8k, \ k \in \mathbb{N}\}$.
(3) $S_3 = \{n : n = k(k + 1)/2, \ k \in \mathbb{N}\}$.
(4) $S_4 = \{n : n = \pm 3^k, \ k \in \mathbb{N}\}$.

2.6. *Answer:*

(1) $S_1 = \{x : x = 2^{1-k}, \ k \in \mathbb{N}\}$.
(2) $S_2 = \{x : x = 1 - 3^{-k}, \ k \in \mathbb{N}\}$.
(3) $S_3 = \left\{x : x = \dfrac{(-1)^{k+1}}{k + 1}, \ k \in \mathbb{N}\right\}$.
(4) $S_4 = \left\{x : x = \dfrac{k - 1}{k + 1}, \ k \in \mathbb{N}\right\}$.

2.7. *Solution.*
Since all integers satisfying the inequality $x^2 \leqslant 15$ belong to the set A, then, consequently, $A = \{0, \pm 1, \pm 2, \pm 3\}$. To define the elements of the set B, we solve the quadratic equation $x^2 - 4x - 12 = 0$. As is known, the quadratic equation $ax^2 + bx + c = 0$ has two roots that are calculated by the formula

$$x_{1,2} = \frac{-b \pm \sqrt{b^2 - 4ac}}{2a}.$$

The value $D = b^2 - 4ac$ is called the **discriminant** of the quadratic equation (from Latin *discrīmināre*). In our case, the discriminant $D = (-4)^2 - 4 \times (-12) = 64$, the roots of the equation are $x_1 = 6$, $x_2 = -2$. Hence, $B = \{-2, 6\}$.

Next, consider the set C. To define its elements, we solve the quadratic equation $x^2 - 2x - 8 = 0$: $x_1 = 4$, $x_2 = -2$. Using the substitution method, we will define the constant sign intervals of the function $f(x) = x^2 - 2x - 8$: $f(-3) = 7$, $f(0) = -8$, $f(5) = 7$. Obviously, $f(x) \leqslant 0$ in the interval $[-2, 4]$. Hence, $C = \{-2, -1, 0, 1, 2, 3, 4\}$.

Now, we easily find the elements of the remaining sets:

$$B \cup C = \{-2, 6\} \cup \{-2, -1, 0, 1, 2, 3, 4\} = \{-2, -1, 0, 1, 2, 3, 4, 6\},$$
$$(B \cap C) \, \triangle \, A =$$
$$= (\{-2, 6\} \cap \{-2, -1, 0, 1, 2, 3, 4\}) \, \triangle \, \{-3, -2, -1, 0, 1, 2, 3\} =$$
$$= \{-2\} \, \triangle \, \{-3, -2, -1, 0, 1, 2, 3\} = \{-3, -1, 0, 1, 2, 3\}.$$

2.8. *Answer:*

$A = \{-1, 0, 1, 2, 3, 4, 5, 6, 7, 8\}$, $B = \{-1, 0\}$, $C = \{-1, \frac{1}{4}\}$, $A \cap C = \{-1\}$,

$(B \cup C) \vartriangle A = \{\frac{1}{4}, 1, 2, 3, 4, 5, 6, 7, 8\}$.

2.9. *Answer:*

(1) $(A \cap B) \cup (C \cap D) = \{2\}$.
(2) $(A \cup (B \cap C)) \cap ((A \cup B) \cap C) = \{1, 2\}$.
(3) $A \setminus (B \setminus (C \setminus D)) = \{2, 3, 4\}$.
(4) $(A \setminus (B \cup C)) \setminus (A \setminus (B \cup D)) = \varnothing$.

2.10. *Answer:*

(1) $(A \cap B) \cup (A \cap C) \cup (B \cap C) = \{0, 1\}$.
(2) $(A \cap (B \cup C)) \cap (A \cup B \cup D) = \{0, 1\}$.
(3) $(A \setminus C) \setminus (B \setminus D) = \{7\}$.
(4) $(A \setminus (B \cap C)) \cap (D \setminus A) = \varnothing$.

2.11. *Answer:*

$\overline{A} \cap B = \{f, h, m\}$, $(A \vartriangle B) \cup C = \{b, c, e, f, g, h, k, m\}$, $(A \cup C) \setminus B = \{b, e, g, k\}$.

2.12. *Answer:*

$A \cup \overline{C} = \{a, b, c, f, h\}$, $(A \vartriangle C) \cup B = \{a, b, c, e, f, g, h, k, m\} = U$, $(B \cap \overline{C}) \setminus A = \{f, h\}$.

2.13. *Answer:*

$B \vartriangle C = \{a, e, f, g, h, k\}$, $(B \setminus \overline{C}) \cup A = \{a, b, c, m\}$, $(B \vartriangle C) \cap \overline{A} = \{e, f, g, h, k\}$.

2.14. *Answer:*

$A \cup B = \{1, 2, 3, \{3, 4\}\}$, $A \cap B = \{2\}$, $A \vartriangle B = \{1, 3, \{3, 4\}\}$.

2.15. *Answer:*

$A \cup B = \{\varnothing, \{1\}, \{2\}\}$, $A \cap B = \{\{2\}\}$, $A \vartriangle B = \{\varnothing, \{1\}\}$.

2.16. *Answer:* $\mathcal{P}(A) = \{\varnothing, \{a\}, \{b\}, \{c\}, \{a, b\}, \{a, c\}, \{b, c\}, \{a, b, c\}\}$.

2.17. *Solution.*
The set A contains only one element, namely $\{a, b, c\}$. This is why A has only two subsets: empty and containing the only element $\{a, b, c\}$. As a result, $\mathcal{P}(A) = \{\varnothing, \{\{a, b, c\}\}\}$.

2.18. *Answer:* $\mathcal{P}(A) = \{\varnothing, \{a\}, \{\{b, c\}\}, \{a, \{b, c\}\}\}$.

2.19. *Answer:*

(1) $P(\varnothing) = \{\varnothing\}$.
(2) $P(P(\varnothing)) = \{\varnothing, \{\varnothing\}\}$.

2.20. *Solution.*
According to the definition, the set-theoretic difference of the sets A and B is

$$A \setminus B = \{x : x \in A \text{ and } x \notin B\}.$$

Show that the set coincides with $A \cap \overline{B}$ for any sets $A, B \in U$. Actually,

$$A \setminus B = \{x : x \in A \text{ and } x \notin B\} = A \cap \overline{B}.$$

2.21. *Solution.*
Using the identity $A \setminus B = A \cap \overline{B}$ proved in the previous exercise obtains:

$$(A \setminus B) \setminus C = (A \setminus B) \cap \overline{C} =$$
$$= (A \cap \overline{B}) \cap \overline{C} = \qquad \text{(associative law)}$$
$$= A \cap (\overline{B} \cap \overline{C}) = \qquad \text{(de Morgan's law)}$$
$$= A \cap \overline{(B \cup C)} = A \setminus (B \cup C) \qquad \text{(identity from Exercise \textbf{2.20}).}$$

2.22. *Solution.*

(1) Commutativity of some operation "$\otimes$" of set algebra means that $\forall A, B \in U$ $A \otimes B = B \otimes A$. Check this relation for complement of the set B to the set A and symmetric difference of the two sets:

$$A \setminus B = A \cap \overline{B} \neq B \setminus A,$$
$$A \bigtriangleup B = (A \cap \overline{B}) \cup (\overline{A} \cap B) = (\overline{A} \cap B) \cup (A \cap \overline{B}) = B \bigtriangleup A.$$

Hence, the operation of symmetric difference $B \bigtriangleup A$ is commutative, and the complement operation $A \setminus B$ is not commutative.

(2) Associativity of some operation "$\otimes$" of set algebra means that $\forall A, B, C \in U$ $(A \otimes B) \otimes C = A \otimes (B \otimes C)$. Check this relation for complement of the set B to the set A:

$$(A \setminus B) \setminus C = (A \cap \overline{B}) \cap \overline{C} = A \setminus (B \cup C) \neq A \setminus (B \setminus C).$$

Consider the operation of symmetric difference of some sets A and B. Represent the expression $(A \bigtriangleup B) \bigtriangleup C$, consecutively using the definition of symmetric difference, as well as distributive and de Morgan's laws:

$$(A \bigtriangleup B) \bigtriangleup C = ((A \cap \overline{B}) \cup (\overline{A} \cap B)) \bigtriangleup C =$$
$$= (((A \cap \overline{B}) \cup (\overline{A} \cap B)) \cap \overline{C}) \cup (\overline{((A \cap \overline{B}) \cup (\overline{A} \cap B))} \cap C) =$$
$$= (A \cap \overline{B} \cap \overline{C}) \cup (\overline{A} \cap B \cap \overline{C}) \cup (((\overline{A} \cup B) \cap (A \cup \overline{B})) \cap C).$$

Note that the expression $(\overline{A} \cup B) \cap (A \cup \overline{B})$ can be written as

$$(\overline{A} \cup B) \cap (A \cup \overline{B}) = ((\overline{A} \cup B) \cap A) \cup ((\overline{A} \cup B) \cap \overline{B}) =$$
$$= (\overline{A} \cap A) \cup (A \cap B) \cup (\overline{A} \cap \overline{B}) \cup (B \cap \overline{B}) = (A \cap B) \cup (\overline{A} \cap \overline{B}).$$

Using the obtained relation, we obtain

$$(A \,\triangle B) \,\triangle\, C = (A \cap \overline{B} \cap \overline{C}) \cup (\overline{A} \cap B \cap \overline{C}) \cup (((A \cap B) \cup (\overline{A} \cap \overline{B})) \cap C) =$$
$$= (A \cap \overline{B} \cap \overline{C}) \cup (\overline{A} \cap B \cap \overline{C}) \cup (A \cap B \cap C) \cup (\overline{A} \cap \overline{B} \cap C).$$

Then, consider the set $A \,\triangle\, (B \,\triangle\, C)$. Perform transformations similar to those specified above:

$$A \,\triangle\, (B \,\triangle\, C) = A \,\triangle\, ((B \cap \overline{C}) \cup (\overline{B} \cap C)) =$$
$$= (A \cap \overline{((B \cap \overline{C}) \cup (\overline{B} \cap C))}) \cup (\overline{A} \cap ((B \cap \overline{C}) \cup (\overline{B} \cap C))) =$$
$$= (A \cap ((\overline{B} \cup C) \cap (B \cup \overline{C}))) \cup (\overline{A} \cap B \cap \overline{C}) \cup (\overline{A} \cap \overline{B} \cap C).$$

Since $(\overline{B} \cup C) \cap (B \cup \overline{C}) = (B \cap C) \cup (\overline{B} \cap \overline{C})$, then

$$A \,\triangle\, (B \,\triangle\, C) = (A \cap ((B \cap C) \cup (\overline{B} \cap \overline{C}))) \cup (\overline{A} \cap B \cap \overline{C}) \cup (\overline{A} \cap \overline{B} \cap C) =$$
$$= (A \cap B \cap C) \cup (A \cap \overline{B} \cap \overline{C}) \cup (\overline{A} \cap B \cap \overline{C}) \cup (\overline{A} \cap \overline{B} \cap C).$$

Comparing the obtained expression with the expression for $(A \,\triangle\, B) \,\triangle\, C$, we obtain that $(A \,\triangle\, B) \,\triangle\, C = A \,\triangle\, (B \,\triangle\, C) = A \,\triangle\, B \,\triangle\, C$. The respective Venn diagram is shown in Fig. 2.3.

Hence, the operation of symmetric difference $A \,\triangle\, B$ is associative, and the complement operation $A \setminus B$ is not associative.

Fig. 2.3 Venn diagram of the set $A \,\triangle\, B \,\triangle\, C$ to Exercise **2.22**

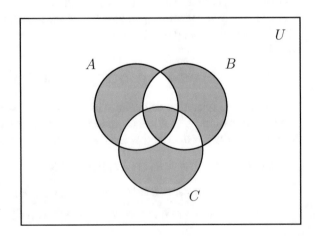

2.23. *Proof.*

Considering the definition of symmetric difference and the well-known laws of set algebra, we obtain:

$$A \bigtriangleup A \bigtriangleup A =$$

$$= (A \bigtriangleup A) \bigtriangleup A = \qquad \text{(group the summands)}$$

$$= \left((A \cap \overline{A}) \cup (\overline{A} \cap A)\right) \bigtriangleup A = \qquad \text{(definition of symmetric difference)}$$

$$= (\varnothing \cup \varnothing) \bigtriangleup A = \varnothing \bigtriangleup A = \qquad \text{(complement and idempotent properties)}$$

$$= (U \cap A) \cup (\overline{A} \cap \varnothing) = A \qquad \text{(properties of } \varnothing \text{ and } U).$$

2.25. *Solution.*

(1) Use de Morgan's laws and complement properties:

$$\overline{(A \cup \overline{B})} = \overline{A} \cap \overline{\overline{B}} = \overline{A} \cap B.$$

(2) According to absorption laws, $A \cup (A \cap \overline{B}) = A$ and $A \cap (A \cup B) = A$. It remains to apply idempotent law:

$$(A \cup (A \cap \overline{B})) \cap (A \cap (A \cup B)) = A \cap A = A.$$

(3) Having expressed complement of the set C to the set $(A \cup B)$, we obtain:

$$(A \cup B) \setminus C = (A \cup B) \cap \overline{C} = (A \cap \overline{C}) \cup (B \cap \overline{C}) = (A \setminus C) \cup (B \setminus C).$$

(4) The identity can be proved by the following transformations.

$$(A \setminus (B \setminus C)) \setminus ((A \setminus B) \setminus C) =$$

$$= (A \cap \overline{(B \setminus C)}) \setminus ((A \setminus B) \cap \overline{C}) = (A \cap \overline{(B \cap \overline{C})}) \cap \overline{((A \cap \overline{B}) \cap \overline{C})} =$$

$$= (A \cap (\overline{B} \cup C)) \cap \overline{((A \cap \overline{B}) \cup C)} = (A \cap (\overline{B} \cup C)) \cap ((\overline{A} \cup B) \cup C) =$$

$$= ((A \cap \overline{B}) \cup (A \cap C)) \cap (\overline{A} \cup B \cup C) =$$

$$= ((A \cap \overline{B}) \cap (\overline{A} \cup B \cup C)) \cup ((A \cap C) \cap (\overline{A} \cup B \cup C)) =$$

$$= (A \cap \overline{B} \cap \overline{A}) \cup (A \cap \overline{B} \cap B) \cup (A \cap \overline{B} \cap C) \cup$$

$$\cup (A \cap C \cap \overline{A}) \cup (A \cap C \cap B) \cup (A \cap C \cap C) =$$

$$= \varnothing \cup \varnothing \cup (A \cap \overline{B} \cap C) \cup \varnothing \cup (A \cap B \cap C) \cup (A \cap C) =$$

$$= ((A \cap C) \cap (\overline{B} \cup B)) \cup (A \cap C) = (A \cap C) \cup (A \cap C) = A \cap C.$$

(5) Using the definition of symmetric difference, we obtain:

$$(A \setminus B) \vartriangle (B \setminus A) = (A \cap \overline{B}) \vartriangle (B \cap \overline{A}) =$$
$$= ((A \cap \overline{B}) \cap \overline{(B \cap \overline{A})}) \cup (\overline{(A \cap \overline{B})} \cap (B \cap \overline{A})) =$$
$$= ((A \cap \overline{B}) \cap (\overline{B} \cup A)) \cup ((\overline{A} \cup B) \cap (B \cap \overline{A})) =$$
$$= (A \cap \overline{B} \cap \overline{B}) \cup (A \cap \overline{B} \cap A) \cup (\overline{A} \cap B \cap \overline{A}) \cup (B \cap B \cap \overline{A}) =$$
$$= (A \cap \overline{B}) \cup (B \cap \overline{A}) = A \vartriangle B.$$

(6) According to the definition of symmetric difference of sets $(A \cup B \cup C)$ and $(A \cap B \cap C)$, we obtain:

$$(A \cup B \cup C) \vartriangle (A \cap B \cap C) =$$
$$= \left((A \cup B \cup C) \cap \overline{(A \cap B \cap C)} \right) \cup \left(\overline{(A \cup B \cup C)} \cap (A \cap B \cap C) \right).$$

Transform the second summand:

$$\overline{(A \cup B \cup C)} \cap (A \cap B \cap C) = (\overline{A} \cap \overline{B} \cap \overline{C}) \cap (A \cap B \cap C) =$$
$$= \overline{A} \cap \overline{B} \cap \overline{C} \cap A \cap B \cap C = \varnothing.$$

Therefore,

$$(A \cup B \cup C) \vartriangle (A \cap B \cap C) = \left((A \cup B \cup C) \cap \overline{(A \cap B \cap C)} \right) \cup \varnothing =$$
$$= (A \cup B \cup C) \cap \overline{(A \cap B \cap C)} = (A \cup B \cup C) \setminus (A \cap B \cap C).$$

The respective Venn diagrams are shown in Figs. 2.4 and 2.5.

2.26. *Solution.*

(1) Determine elements of the set A. The equation $f(x) = 0$ is biquadratic, and by replacement $t = x^2$ it is reduced to. We obtain that $f(x) = 0$ has four roots: $x = \pm 1, x = \pm 2$. (According to the fundamental theorem of algebra, the theorem of the form $f(x) = 0$, where $f(x)$ is a polynomial, has exactly n roots, where n is the polynomial order; for more details, see chapter "Complex Numbers" on p. 245.) Therefore,
$A = \{-2, -1, 1, 2\}$,
$\overline{A} = \{-5, -4, -3, 0, 3, 4, 5\}$.
Using the definition of the symmetric difference, we obtain:
$A \vartriangle U = \{-5, -4, -3, 0, 3, 4, 5\}$,
$A \vartriangle A = \varnothing$.
Since the number of elements of the set A is $|A| = 4$, then the cardinality of the power set $|\mathcal{P}(A)| = 2^4 = 16$.

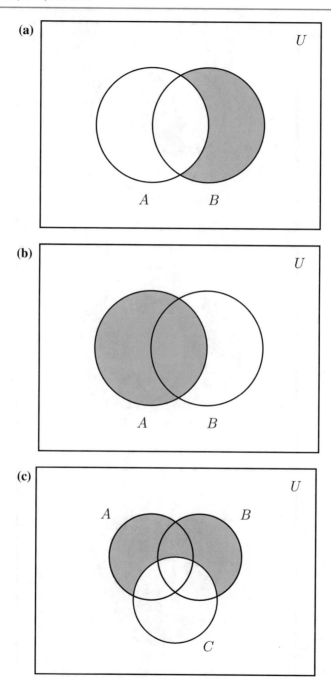

Fig. 2.4 Venn diagrams of sets **a** $\overline{(A \cup \overline{B})}$, **b** $(A \cup (A \cap \overline{B})) \cap (A \cap (A \cup B))$, and **c** $(A \cup B) \setminus C$ to Exercise **2.25**

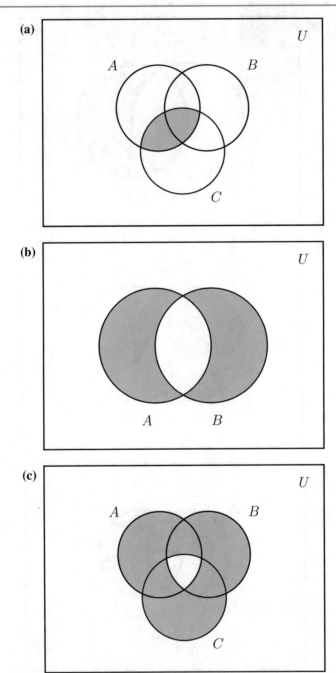

Fig. 2.5 Venn diagrams of sets **a** $(A \setminus (B \setminus C)) \setminus ((A \setminus B) \setminus C)$, **b** $(A \setminus B) \triangle (B \setminus A)$, and **c** $(A \cup B \cup C) \triangle (A \cap B \cap C)$ to Exercise 2.25

(2) The equation $f(x) = 0$ has four roots: $x_{1,2} = \pm\sqrt{2}$, $x_3 = 1$, and $x_4 = 2$. Therefore, $A = \{1, 2\}$, $\overline{A} = \{-5, -4, -3, -2, -1, 0, 3, 4, 5\}$, $A \bigtriangleup U = \overline{A}$, $A \bigtriangleup A = \varnothing$.

Since the number of elements of the set A is $|A| = 2$, then the cardinality of the power set $|\mathcal{P}(A)| = 4$.

(3) By brute force, we find the root $x = 2$. In order to find the remaining three roots, divide the polynomial $x^4 + 2x^3 - 25x^2 - 26x + 120$ by $x - 2$:

$$
\begin{array}{r|l}
\underline{x^4 + 2x^3 - 25x^2 - 26x + 120} & \,x - 2 \\
\underline{x^4 - 2x^3} & \overline{x^3 + 4x^2 - 17x - 60} \\
\quad 4x^3 - 25x^2 \\
\underline{4x^3 - 8x^2} \\
\qquad -17x^2 - 26x \\
\underline{-17x^2 + 34x} \\
\qquad\quad -60x + 120 \\
\underline{-60x + 120} \\
\qquad\qquad 0
\end{array}
$$

The equation $x^3 + 4x^2 - 17x - 60 = 0$ has the root $x = 4$:

$$
\begin{array}{r|l}
\underline{x^3 + 4x^2 - 17x - 60} & \,x - 4 \\
\underline{x^3 - 4x^2} & \overline{x^2 + 8x + 15} \\
\quad 8x^2 - 17x \\
\underline{8x^2 - 32x} \\
\qquad 15x - 60 \\
\underline{15x - 60} \\
\qquad\quad 0
\end{array}
$$

And, finally, $x^2 + 8x + 15 = (x + 3)(x + 5)$.

As a result, we obtain $f(x) = (x + 5)(x + 3)(x - 2)(x - 4)$, which allows writing out the elements of the sought sets:

$A = \{-5, -3, 2, 4\}$, $A \bigtriangleup U = \overline{A} = \{-4, -2, -1, 0, 1, 3, 5\}$, $A \bigtriangleup A = \varnothing$.

(4) By brute force, we find the root $x = 1$. In order to find the remaining three roots, divide the polynomial $x^4 - x^3 - 19x^2 - 11x + 30$ by $x - 1$:

$$
\begin{array}{r|l}
\underline{x^4 - x^3 - 19x^2 - 11x + 30} & \,x - 1 \\
\underline{x^4 - x^3} & \overline{x^3 - 19x - 30} \\
\quad -19x^2 - 11x \\
\underline{-19x^2 + 19x} \\
\qquad -30x + 30 \\
\underline{-30x + 30} \\
\qquad\quad 0
\end{array}
$$

The equation $x^3 - 19x - 30 = 0$ has the root $x = -2$:

$$
\begin{array}{r|l}
\underline{x^3 } - 19x - 30 & x + 2 \\
x^3 + 2x^2 & x^2 - 2x - 15 \\
\underline{-2x^2 - 19x} & \\
-2x^2 - 4x & \\
\underline{-15x - 30} & \\
-15x - 30 & \\
0 &
\end{array}
$$

And, finally, $x^2 - 2x - 15 = (x + 3)(x - 5)$. As a result, we obtain $A = \{-3, -2, 1, 5\}$, $A \bigtriangleup U = \overline{A} = \{-5, -4, -1, 0, 2, 3, 4\}$, $A \bigtriangleup A = \varnothing$.

2.27. *Solution.*
As is known from the analytic geometry course, the equation

$$(x - a)^2 + (y - b)^2 = R^2$$

determines in the Cartesian plane the circle of radius R with the center in a point with coordinates $x = a$, $y = b$. Therefore, the set A includes all points lying inside the circle of radius $R_1 = 2$ with the center in the point $(0, 0)$. Representing the definition property of the set B in the form $x^2 - 2x + y^2 \leqslant 0 \Leftrightarrow (x - 1)^2 + y^2 \leqslant 1$, we conclude that the set B includes all points of the plane lying inside the circle of radius $R_2 = 1$ with the center in the point $(1, 0)$. The set C is formed by a half-plane located above the line $y = 1$.

The sought set $(A \setminus B) \cap C$ is formed by elements belonging to A and not belonging to B, and, apart from this, belonging to the set C. In Fig. 2.6, the respective area is filled.

2.28. *Answer:*
The set $(A \cup B) \bigtriangleup C$ in the Cartesian coordinate system is shown in Figs. 2.7 and 2.8.

2.30. *Proof.*
Since $A \bigtriangleup B = (A \setminus B) \cup (B \setminus A)$ and $(A \setminus B) \cap (B \setminus A) = \varnothing$,

$$
|A \bigtriangleup B| = |A \setminus B| + |B \setminus A| = |A| - |A \cap B| + |B| - |B \cap A| =
$$
$$
= |A| + |B| - 2|A \cap B|.
$$

2.31. *Hint.*
Use the identity from the previous exercise and distributivity of the set intersection operation with respect to the symmetric difference operation $(A \bigtriangleup B) \cap C = (A \cap C) \bigtriangleup (B \cap C)$ [65].

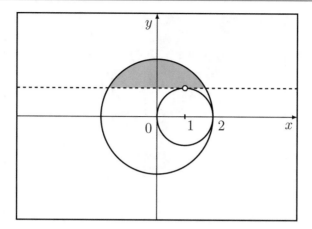

Fig. 2.6 Set $(A \setminus B) \cap C$ to Exercise **2.27**

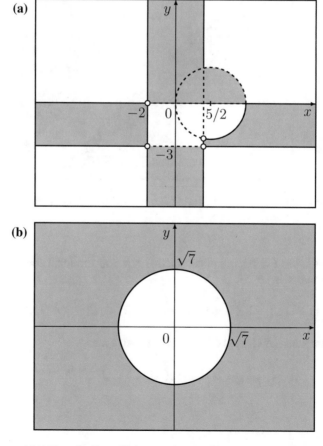

Fig. 2.7 The set $(A \cup B) \bigtriangleup C$ of item (1) is a panel **a** and of item (2) is a panel **b** to Exercise **2.28**

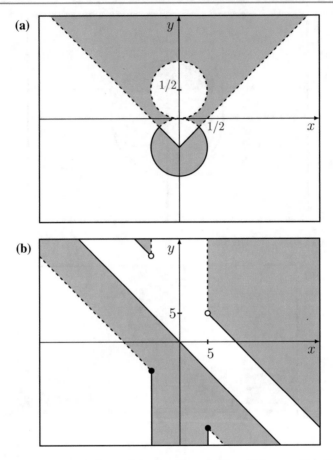

Fig. 2.8 The set $(A \cup B) \bigtriangleup C$ of item (3) is a panel **a** and of item (4) is a panel **b** to Exercise 2.28

2.32. *Proof.*
For the proof, use the set union associative law and the inclusion and exclusion
formula for two sets:

$$|A \cup B \cup C| = |(A \cup B) \cup C| = |A \cup B| + |C| - |(A \cup B) \cap C|.$$

Expand summand $|(A \cup B) \cap C|$, using distributive law:

$$|(A \cup B) \cap C| = |(A \cap C) \cup (B \cap C)|.$$

Applying the inclusion and exclusion formula for two sets once again, we obtain:

$$|A \cup B \cup C| = |(A \cup B) \cup C| = |A \cup B| + |C| - |(A \cup B) \cap C| =$$
$$= |A| + |B| - |A \cap B| + |C| - |(A \cap C) \cup (B \cap C)| =$$
$$= |A| + |B| - |A \cap B| + |C| -$$
$$- (|A \cap C| + |B \cap C| - |(A \cap C) \cap (B \cap C)|) =$$
$$= |A| + |B| + |C| - |A \cap B| - |A \cap C| - |B \cap C| + |A \cap B \cap C|.$$

The inclusion and exclusion formula for three sets is proved.

2.33. *Solution.*
Denote by A the set of pupils who go to the sports club, by B the set of pupils who go in for aeromodeling, and by C the set of pupils who are keen on math. Apply the inclusion and exclusion formula for three sets:

$$|A \cup B \cup C| =$$
$$= |A| + |B| + |C| - |A \cap B| - |A \cap C| - |B \cap C| + |A \cap B \cap C| =$$
$$= 13 + 6 + 19 - 2 - 7 - 4 = 25.$$

So, 25 pupils go to at least one elective; hence, $29 - 25 = 4$ are not interested in them at all.

Calculate the number of pupils who visit only one out-of-school activity:

$$|A \cup B \cup C| - |A \cap B| - |A \cap C| - |B \cap C| + 2|A \cap B \cap C| = 25 - 2 - 7 - 4 = 12.$$

For solving such problems, it is convenient to use Venn diagram for the sets A, B, and C. On each segment of the diagram, one can specify the number of elements of the respective set. The diagram should be filled in starting from the central part $A \cap B \cap C$. Then, we can place on the diagram the cardinalities of the sets $(A \cup B) \setminus C$, $(A \cup C) \setminus B$, and $(B \cup C) \setminus A$. Then, specify the number of elements in the sets $A \setminus (B \cup C)$, $B \setminus (A \cup C)$, and $C \setminus (A \cup B)$, and, finally, $U \setminus (A \cup B \cup C)$ (Fig. 2.9).

According to the problem statement, $|A \cap B \cap C| = 0$ and $|A \cap B| = 2$. Therefore, since $(A \cap B) \setminus C = (A \cap B) \setminus (A \cap B \cap C)$, we obtain $|(A \cap B) \setminus C| = |(A \cap B) \setminus (A \cap B \cap C)| = 2$. Similarly, we have: $|(A \cap C) \setminus B| = 7$, $|(B \cap C) \setminus A| = 4$, $|A \setminus (B \cup C)| = 4$, $|B \setminus (A \cup C)| = 0$, $|C \setminus (A \cup B)| = 8$.

The final form of Venn diagram is shown in Fig. 2.10.

As a result, we obtain the answer: Twelve pupils go to only one elective; four are not interested in them at all.

2.34. *Solution.*
Denote by A the set of pupils who visit circles for knitting, by B the set of pupils who go in for sewing, and by C the number of pupil who is keen on embroidery.

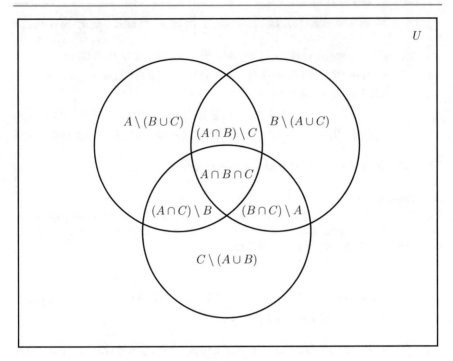

Fig. 2.9 Venn diagram for the sets A, B, and C to Exercise **2.33**

Fig. 2.10 To Exercise **2.33**

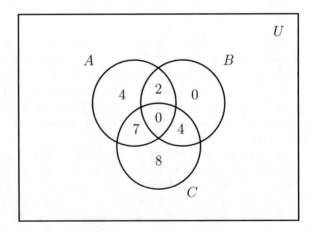

Fig. 2.11 To Exercise **2.34**

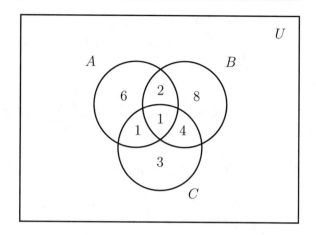

Represent a respective Venn diagram (Fig. 2.11).
We obtain that 25 pupils go to the circles, where eight girls do in for sewing only.

2.35. *Answer:* There are 22 students in the group.

2.36. *Solution.*
For calculation of $|(A \triangle B) \cup C|$, use the inclusion and exclusion formula for two sets:

$$|(A \triangle B) \cup C| = |A \triangle B| + |C| - |(A \triangle B) \cap C| =$$
$$= |A \triangle B| + |C| - |(A \cap C) \triangle (B \cap C)|.$$

Considering the relation from Exercise **2.30** (see also hints to Exercise **2.31**), we have:

$$|(A \triangle B) \cup C| = |A| + |B| - 2|A \cap B| + |C| - (|A \cap C| + |B \cap C| - 2|A \cap B \cap C|) =$$
$$= |A| + |B| + |C| - 2|A \cap B| - |A \cap C| - |B \cap C| + 2|A \cap B \cap C|.$$

Substituting numerical values, we obtain $|(A \triangle B) \cup C| = 25$.

2.37. *Answer:* $|A \cup (B \triangle C)| = 49$.

2.38. *Answer:* $|(A \triangle C) \cup B| = 46$.

2.39. *Proof.*
Use the mathematical induction method.
   Let $P(n)$ be a statement in the condition of the problem for $n = 1, 2, 3, \ldots$

   B a s i s   s t e p
   For $n = 1$, we obtain: $|A_1| = \sum_{i=1}^{1} |A_i| = |A_1|$.

Inductive step

As an inductive supposition, we have $P(m)$, where
$m = 1, 2, 3, \ldots$, i.e.,

$$|A_1 \cup A_2 \cup \ldots \cup A_m| = \sum_{i=1}^{m} |A_i| - \sum_{\substack{i,j=1 \\ i<j}}^{m} |A_i \cap A_j| +$$

$$+ \sum_{\substack{i,j,k=1 \\ i<j<k}}^{m} |A_i \cap A_j \cap A_k| - \cdots + (-1)^{m-1}|A_1 \cap A_2 \cap \ldots \cap A_m|.$$

Prove the validity of $P(m + 1)$:

$$|A_1 \cup A_2 \cup \ldots \cup A_{m+1}| = |A_1 \cup A_2 \cup \ldots \cup A_m| + |A_{m+1}| -$$
$$- |(A_1 \cup A_2 \cup \ldots \cup A_m) \cap A_{m+1}| = |A_1 \cup A_2 \cup \ldots \cup A_m| + |A_{m+1}| -$$
$$- \underbrace{|(A_1 \cap A_{m+1}) \cup (A_2 \cap A_{m+1}) \cup \ldots \cup (A_m \cap A_{m+1})|}_{m \text{ terms}}.$$

The last summand can be rewritten using the inductive supposition. Then, we
obtain

$$|A_1 \cup A_2 \cup \ldots \cup A_{m+1}| = |A_1 \cup A_2 \cup \ldots \cup A_m| + |A_{m+1}| -$$

$$- \left(\sum_{i=1}^{m} |A_i \cap A_{m+1}| - \sum_{\substack{i,j=1 \\ i<j}}^{m} |(A_i \cap A_{m+1}) \cap (A_j \cap A_{m+1})| + \right.$$

$$+ \sum_{\substack{i,j,k=1 \\ i<j<k}}^{m} |(A_i \cap A_{m+1}) \cap (A_j \cap A_{m+1}) \cap (A_k \cap A_{m+1})| - \cdots +$$

$$\left. + (-1)^{m-1}|(A_1 \cap A_{m+1}) \cap (A_2 \cap A_{m+1}) \cap \ldots \cap (A_m \cap A_{m+1})| \right).$$

Open the brackets under the summation sign:

$$|A_1 \cup A_2 \cup \ldots \cup A_{m+1}| = |A_1 \cup A_2 \cup \ldots \cup A_m| + |A_{m+1}| -$$

$$- \sum_{\substack{i,j=1 \\ i<j}}^{m} |A_i \cap A_{m+1}| + \sum_{\substack{i,j,k=1 \\ i<j<k}}^{m} |A_i \cap A_j \cap A_{m+1}| - \cdots +$$

$$+ (-1)^{m} |A_1 \cap \ldots \cap A_{m+1}|.$$

Apply the inductive supposition, and express the first summand $|A_1 \cup A_2 \cup \ldots \cup A_m|$ by $|A_i|$:

$$|A_1 \cup A_2 \cup \ldots \cup A_{m+1}| = \left(\sum_i^m |A_i| - \sum_{\substack{i,j=1 \\ i<j}}^m |A_i \cap A_j| + \right.$$

$$\left. + \sum_{\substack{i,j,k=1 \\ i<j<k}}^m |A_i \cap A_j \cap A_k| - \cdots + (-1)^{m-1}|A_1 \cap A_2 \cap \ldots \cap A_m| \right) +$$

$$+ |A_{m+1}| - \sum_{i=1}^m |A_i \cap A_{m+1}| + \sum_{\substack{i,j=1 \\ i<j}}^m |A_i \cap A_j \cap A_{m+1}| - \cdots +$$

$$+ (-1)^m |A_1 \cap \ldots \cap A_{m+1}|.$$

Uniting similar sums in the last equality, we obtain the final result:

$$|A_1 \cup A_2 \cup \ldots \cup A_{m+1}| = \sum_{i=1}^{m+1} |A_i| - \sum_{\substack{i,j=1 \\ i<j}}^{m+1} |A_i \cap A_j| +$$

$$+ \sum_{\substack{i,j,k=1 \\ i<j<k}}^{m+1} |A_i \cap A_j \cap A_k| - \cdots + (-1)^m |A_1 \cap A_2 \cap \ldots \cap A_{m+1}|.$$

This is why, according to the mathematical induction method, we obtain the validity of the inclusion and exclusion formula for an arbitrary number of sets.

Note.

An alternative method of proving is shown in the solution of Exercise 4.59.

2.40. *Solution.*

Assuming in the inclusion and exclusion formula for an arbitrary number of set (see Exercise 2.39) value $N = 4$, we obtain the answer:

$$|A_1 \cup A_2 \cup A_3 \cup A_4| = \sum_{i=1}^4 |A_i| - \sum_{\substack{i,j=1 \\ i<j}}^4 |A_i \cap A_j| +$$

$$+ \sum_{\substack{i,j,k=1 \\ i<j<k}}^4 |A_i \cap A_j \cap A_k| - |A_1 \cap A_2 \cap A_3 \cap A_4| =$$

$$= |A_1| + |A_2| + |A_3| + |A_4| -$$

$$- (|A_1 \cap A_2| + |A_1 \cap A_3| + |A_1 \cap A_4| + |A_2 \cap A_3| + |A_2 \cap A_4| + |A_3 \cap A_4|) +$$

$$+ (|A_1 \cap A_2 \cap A_3| + |A_1 \cap A_2 \cap A_4| + |A_1 \cap A_3 \cap A_4| + |A_2 \cap A_3 \cap A_4|) -$$

$$- |A_1 \cap A_2 \cap A_3 \cap A_4|.$$

2.41. *Solution.*

Introduce the notations:

A_1—the set of respondents who named d'Artagnan;

A_2—the set of respondents who named Atos;

A_3—the set of respondents who named Portos;

A_4—the set of respondents who named Aramis.

The number of interrogated guards who named at least one musketeer is equal to the cardinality of the union of the sets $A_1 \cup A_2 \cup A_3 \cup A_4$. Use the inclusion and exclusion formula for four sets (see Exercise **2.40**):

$$|A_1 \cup A_2 \cup A_3 \cup A_4| =$$
$$= |A_1| + |A_2| + |A_3| + |A_4| -$$
$$- (|A_1 \cap A_2| + |A_1 \cap A_3| + |A_1 \cap A_4| + |A_2 \cap A_3| + |A_2 \cap A_4| + |A_3 \cap A_4|) +$$
$$+ (|A_1 \cap A_2 \cap A_3| + |A_1 \cap A_2 \cap A_4| + |A_1 \cap A_3 \cap A_4| + |A_2 \cap A_3 \cap A_4|) -$$
$$- |A_1 \cap A_2 \cap A_3 \cap A_4|.$$

Finding the values $A_1 \cap A_2$, $A_1 \cap A_3$, etc., from Table 2.2, we obtain

$$|A_1 \cup A_2 \cup A_3 \cup A_4| =$$
$$= 25 + 12 + 19 + 16 - (7 + 7 + 9 + 5 + 4 + 5) + (2 + 2 + 2 + 2) - 1 = 42.$$

Venn diagram is shown in Fig. 2.12.

Fig. 2.12 To Exercise **2.41**

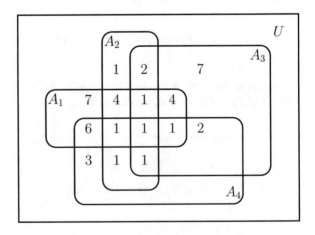

2.42. *Answer:* 33 visitors of the library named none of the poets.

2.43. *Answer:*
The sets A_1, A_2, and A_4 have cardinality equal to $\aleph_0$. The set A_3 is finite.

2.44. *Proof.*
Determine the function $f(n)$, which associates each element of the set of even non-negative numbers with some natural number $f(n) = n/2 + 1$. This association is shown below:

$$
\begin{aligned}
0 &\longleftrightarrow 1, \\
2 &\longleftrightarrow 2, \\
4 &\longleftrightarrow 3, \\
6 &\longleftrightarrow 4, \\
8 &\longleftrightarrow 5, \\
&\cdots
\end{aligned}
$$

Each natural number $n \in \mathbb{N}$ is associated with one and only one element of the set P; the constructed one-to-one correspondence proves that $|P| = \aleph_0$.

2.45. *Hint.*
Arrange the element of the set $\mathbb{Z}$ in the form of a table, and enumerate the elements from top to bottom and from left to right, as shown in the following diagram.

$$
\begin{array}{ccccccc}
0 & 1 & 2 & 3 & 4 & 5 & \cdots \\
\downarrow \;\; \nearrow & \downarrow \;\; \nearrow & \downarrow \;\; \nearrow & \downarrow \;\; \nearrow & \downarrow \;\; \nearrow & \downarrow \;\; \nearrow & \\
-1 & -2 & -3 & -4 & -5 & -6 & \cdots
\end{array}
$$

2.46. *Proof.*
Enumerate elements of the set B by numbers $1, 2, \ldots, |B|$. Place the remaining natural numbers $|B|, |B| + 1, \ldots$ into one-to-one correspondence with elements of the countable set by the law

$$
f(n) = n + 1 - |B|, \quad n \geqslant |B|.
$$

2.47. *Hint.*
Arrange elements of the set $A \times B$ in the form of the matrix

$$(a_1, b_1) \;\rightarrow\; (a_1, b_2) \qquad (a_1, b_3) \qquad (a_1, b_4) \qquad \cdots$$

$$(a_2, b_1) \qquad (a_2, b_2) \qquad (a_2, b_3) \qquad (a_2, b_4) \qquad \cdots$$

$$(a_3, b_1) \qquad (a_3, b_2) \qquad (a_3, b_3) \qquad (a_3, b_4) \qquad \cdots$$

$$\vdots \qquad\qquad \vdots \qquad\qquad \vdots \qquad\qquad \vdots \qquad\qquad \ddots$$

and enumerate in accordance with the suggested scheme.

2.48. *Solution.*
We obtain a formula for the number assigned to the pair (a_i, b_j).

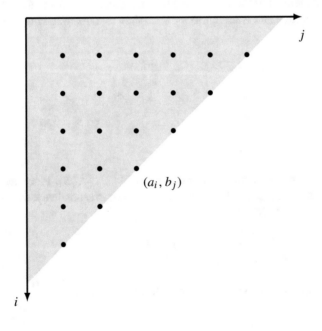

The lengths of legs of the filled right triangle, on whose hypotenuse the considered pair (a_i, b_j) is situated, are equal to $i + j$. On hypotenuse, there are $i + j - 2$ points, and the total number of points inside the triangle is

$$N(i, j) = 1 + 2 + 3 + \cdots + (i + j - 2) = \frac{(i + j - 2)(i + j - 1)}{2}.$$

Then, the number of the matrix element standing at the intersection of the ith row and the jth column is determined by the formula

$$N(i, j) = \frac{1}{2}[(i + j - 2)(i + j - 1)] + i.$$

2.49. *Hint.* Use the mathematical induction method.

2.51. *Solution.*
The one-to-one correspondence between the elements of the sets (a, b) and (c, d) is specified by the linear function $f(x) = \dfrac{d - c}{b - a}(x - a) + c$. A graphic representation of $f(x)$ is shown below.

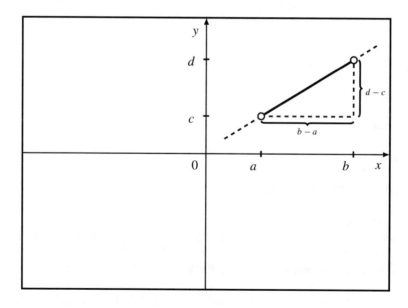

If we assume that $c = 0, d = 1$, then we obtain that

$$|(a, b)| = |(c, d)| = |(0, 1)| = \mathfrak{c}.$$

2.52. *Solution.*
From the definition of a direct product of sets, we have:

$$A \times B = \{(a, 1), (a, 2), (b, 1), (b, 2)\},$$
$$B \times A = \{(1, a), (2, a), (1, b), (2, b)\}.$$

Perform operations of union and intersection of the obtained sets:

$$(A \times B) \cup (B \times A) = \{(a, 1), (a, 2), (b, 1), (b, 2), (1, a), (2, a), (1, b), (2, b)\},$$
$$(A \times B) \cap (B \times A) = \emptyset.$$

2.53. *Answer:*
$$(A \times B) \cup (B \times A) = \{(1, 2), (1, 3), (2, 1), (2, 2), (2, 3), (3, 1), (3, 2)\},$$
$$(A \times B) \cap (B \times A) = \{(2, 2)\}.$$

2.54. *Solution.*
Cardinality of a direct product of finite sets is equal to the product of cardinalities of the initial sets:

$$|A \times B| = |A| \times |B| = 10 \cdot 3 = 30.$$

2.55. *Answer:* $|A \times B| = 4 \cdot 21 = 84.$

2.56. *Answer:* $|A \times B \times C| = 16.$

2.57. *Answer:* $|A \times B \times C| = 200.$

2.58. *Answer:* $|A^3| = 729.$

2.59. *Answer:* $|A^3| = 1000.$

2.60. *Answer:* They are not fulfilled.

2.61. *Solution.*
The characteristic vector of the set A is the vector $\mathbf{a} = (1, 0, 1, 1, 0, 1, 0)$. The characteristic vector of the set B is $\mathbf{b} = (0, 1, 1, 1, 1, 0, 1)$.
  Calculate the characteristic vector of the set $\overline{A} \cap B$. It is

(not a) and b $= (0, 1, 0, 0, 1, 0, 1)$ **and** $(0, 1, 1, 1, 1, 0, 1) = (0, 1, 0, 0, 1, 0, 1)$.

Hence, $\overline{A} \cap B = \{3, 9, 13\}$.
  In order to determine the set $U \setminus (A \bigtriangleup B)$, perform the following transformations:

$$U \setminus (A \bigtriangleup B) = \overline{(A \bigtriangleup B)} = \overline{(A \cap \overline{B}) \cup (B \cap \overline{A})} =$$
$$= \overline{(A \cap \overline{B})} \cap \overline{(B \cap \overline{A})} = (\overline{A} \cup \overline{\overline{B}}) \cap (\overline{B} \cup \overline{\overline{A}}) = (\overline{A} \cup B) \cap (\overline{B} \cup A).$$

The characteristic vector of the set $U \setminus (A \bigtriangleup B)$ is

$$\textbf{((not a) or b) and ((not b) or a)} =$$
$$= \textbf{((not} (1, 0, 1, 1, 0, 1, 0)) \textbf{ or } (0, 1, 1, 1, 1, 0, 1)) \textbf{ and}$$
$$\textbf{and ((not} (0, 1, 1, 1, 1, 0, 1)) \textbf{ or } (1, 0, 1, 1, 0, 1, 0)) =$$
$$= (0, 1, 1, 1, 1, 0, 1) \textbf{ and } (1, 0, 1, 1, 0, 1, 0) =$$
$$= (0, 0, 1, 1, 0, 0, 0).$$

Thus, $U \setminus (A \bigtriangleup B) = \{5, 7\}$.

2.62. *Answer:*
$\textbf{a} = (1, 0, 1, 0, 1, 0, 1, 1)$, $\textbf{b} = (0, 0, 1, 1, 0, 1, 1, 0)$;
$\textbf{a or (not b)} = (1, 1, 1, 0, 1, 0, 1, 1)$, $A \cup \overline{B} = \{1, 2, 3, 5, 7, 8\}$;
$\textbf{(not a) or (not b)} = (1, 1, 0, 1, 1, 1, 0, 1)$,
$U \bigtriangleup (A \setminus \overline{B}) = \{1, 2, 4, 5, 6, 8\}$.

2.63. *Answer:*
$\textbf{a} = (1, 0, 0, 1, 1, 1, 1, 0)$, $\textbf{b} = (0, 1, 0, 1, 1, 0, 0, 1)$;
$\textbf{a and b} = (0, 0, 0, 1, 1, 0, 0, 0)$, $A \setminus \overline{B} = \{8, 10\}$;
$\textbf{(not a) or b} = (0, 1, 1, 1, 1, 0, 0, 1)$, $U \bigtriangleup (A \cap \overline{B}) = \{4, 6, 8, 10, 16\}$.

Relations and Functions

3

Binary relation between the elements of the sets A and B is the subset R of the direct product $A \times B$. It is said that R is the **relation on** A, if $A = B$. If R is some binary relation, then instead of notation $(x, y) \in R$ designation $x \ R \ y$ is often used [4,27].

Let $A = \{x_1, x_2, \ldots, x_n\}$, $B = \{y_1, y_2, \ldots, y_m\}$. **Logical matrix** of the relation R on $A \times B$ is defined as matrix M of size $n \cdot m$, whose elements $M_{ij} = T$, if $(x_i, y_j) \in R$, and $M_{ij} = F$, if $(x_i, y_j) \notin R$.

A binary relationship between elements of finite sets can be represented by a listing of ordered pairs, using suitable predicates, as well as in the form of a logical matrix or a directed graph (digraph, see chapter "Graphs" on p. 176). If at least one of the sets A or B is not finite, the relation $R \subseteq A \times B$ can be specified with the help of predicates.

Example 3.1 Let the relation R between elements of the sets $A = \{1, 2, 3\}$ and $B = \{a, b, c, d\}$ be specified by a listing of the ordered pairs:

$$R = \{(1, b), (2, a), (2, d), (3, b)\}.$$

In this case, the logical matrix of the relation R has the form:

$$M = \begin{array}{c} \\ 1 \\ 2 \\ 3 \end{array} \begin{array}{cccc} a & b & c & d \\ \left[\begin{array}{cccc} F & T & F & F \\ T & F & F & T \\ F & T & F & F \end{array}\right] \end{array}.$$

The graphic representation of the relation R in the form of a digraph is shown in Fig. 3.1. Vertices of the digraph correspond to the elements of each of the sets A and B; edges of the form uv connect the vertices representing components of the pair $(u, v) \in R$. □

© Springer International Publishing AG, part of Springer Nature 2018
S. Kurgalin and S. Borzunov, *The Discrete Math Workbook*,
Texts in Computer Science, https://doi.org/10.1007/978-3-319-92645-2_3

Fig. 3.1 Digraph of the
relation R from Example 3.1

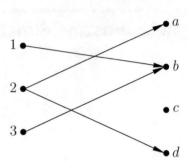

Note that in case $A = B$, the digraph vertices of the relation $R \subseteq A^2$ only show the elements of the set A; i.e., the number of vertices of the digraph of the relation on the set A is $|A|$.

A relation R on the set A is called

reflexive, if $\forall x \in A$ is valid $x\,R\,x$;
symmetric, if $\forall x, y \in A$ $x\,R\,y \Rightarrow y\,R\,x$;
antisymmetric, if $\forall x, y \in A$ $(x\,R\,y$ and $y\,R\,x) \Rightarrow (x = y)$;
transitive, if $\forall x, y, z \in A$ $(x\,R\,y$ and $y\,R\,z) \Rightarrow x\,R\,z$.

Closure of the relation R with respect to the property S is commonly such a relation R^* that [60]:

(1) R^* has the property S.
(2) $R \subset R^*$.
(3) R^* is a subset of any other relation containing R and having the property S.

Equivalence relation on the set A is a reflexive, symmetric, or transitive relation E. **Equivalence class** of the element $x \in A$ is the subset $E_x \subseteq E$:

$$E_x = \{z \in A : z\,E\,x\}.$$

In case the E is the equivalence relation on A, then various equivalence classes form a **partition** A. **Partition of set** A is the collection of nonempty subsets $A_1, A_2, \ldots, A_n$ in A, satisfying the following requirements:

$$A = A_1 \cup A_2 \cup \cdots \cup A_n \text{ and } A_i \cap A_j = \varnothing \text{ for } i \neq j.$$

Also, the subsets A_i, where $i = 1, \ldots, n$, are called **blocks** of a partition.

Partial order on the set A is a reflexive, antisymmetric, or transitive relation of P. **Partially ordered set** or **poset** is a set on which the relation of partial order is determined.

Linear order on the set A is a partial order L, for which any pair of elements can be compared; in other words, $\forall x, y \in A$ $(x\,L\,y)$ **or** $(y\,L\,x)$.

If P is the partial order relation on the set A and $x\,P\,y$, where $x \neq y$, then x is called an **ancestor** y and the notation $x \prec y$ is used; in this case, y is the **successor** for x. If there is no element $z \in A$ such that $x\,P\,z$ **and** $z\,P\,y$, then it is said that x is the **direct ancestor** y (denoted as $x \rightarrow y$). In any finite ordered set, an element can be singled out that has no ancestors (it is called **minimal**) and an element that has no successors (it is called **maximal**).

By definition, the **least** element of some partially ordered set P is an ancestor of all other elements of this set. The **greatest** element of the set P is the successor for all other elements of P.

Hasse[1] **diagram** is a directed graph whose vertices represent the elements of a partially ordered set. The vertex x is located one level below the vertex y and is connected with it by an edge if and only if $x \rightarrow y$. No arrows are shown on the edges of Hasse diagram.

Consider some set A with a partial order introduced on it. The set $C \subseteq A$ is called a **chain** if all elements of C are pairwise comparable. From this, definition follows that the elements of the set C can be linearly ordered. In any finite chain, there exists both the least and the greatest element.

The set $\widetilde{C} \subseteq A$ is called an **antichain** if all elements of $\widetilde{C}$ are pairwise incomparable. **Width** of a finite partially ordered set is a cardinality of the maximal antichain in this set.

Dilworth's[2] **theorem**. *The minimal number of chains into which the finite partially ordered set can be partitioned is equal to the width of this set.*

Example 3.2 Find the minimal number of chains partitioning the set $A = \{a_1, a_2, \ldots, a_6\}$ with the partial order relation P introduced into it if the logical matrix of the relation P has the form:

$$
M_P = \begin{array}{c} \\ a_1 \\ a_2 \\ a_3 \\ a_4 \\ a_5 \\ a_6 \end{array}
\begin{array}{c} a_1\, a_2\, a_3\, a_4\, a_5\, a_6 \\
\begin{bmatrix}
T & T & T & T & T & F \\
F & T & T & T & T & F \\
F & F & T & T & T & F \\
F & F & F & T & F & F \\
F & F & F & F & T & F \\
F & T & T & T & T & T
\end{bmatrix}
\end{array} .
$$

Solution.

Compile a table with information about ancestors and direct ancestors of elements of the set A (see Table 3.1).

With the help of Table 3.1, construct Hasse diagram of the partially ordered set A. Note that the elements a_1 and a_6 have no ancestors and are the minimal elements of the set A. They should be placed at the lower level of Hasse diagram. The next level will be occupied by the element a_2, above it we place the element a_3. Finally, the upper level is comprised of the elements a_4 and a_5, which are the maximal elements of the set A.

---

[1]Helmut Hasse (1898–1979)—German mathematician.
[2]Robert Palmer Dilworth (1914–1993)—American mathematician.

Table 3.1 Ancestors and direct ancestors of the relation $P \subset A^2$

Element	Ancestor	Direct ancestor
a_1	Not	Not
a_2	a_1, a_6	a_1, a_6
a_3	a_1, a_2, a_6	a_2
a_4	a_1, a_2, a_3, a_6	a_3
a_5	a_1, a_2, a_3, a_6	a_3
a_6	Not	Not

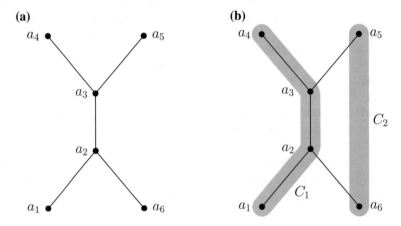

Fig. 3.2 Hasse diagram of the partially ordered set A is a panel **a**, and partition of A into chains C_1 and C_2 is a panel **b**

The final Hasse diagram is shown in Fig. 3.2a.

From the figure, it is clear that each antichain of the set A contains no more than two elements; hence, the width of A is two. According to Dilworth's theorem, the minimal number of the chains partitioning the considered set is two. It is easy to see that the chains $C_1 = \{a_1, a_2, a_3, a_4\}$ and $C_2 = \{a_5, a_6\}$ contain in total all the elements of the set $A = C_1 \cup C_2$ and $C_1 \cap C_2 = \varnothing$. Partition of A into chains is schematically shown in Fig. 3.2b.   □

Inverse relation for the relation R between the elements of the sets A and B is the relation R^{-1} between the elements of the sets B and A, which is specified as follows: $R^{-1} = \{(b, a) \colon (a, b) \in R\}$.

Let R be a relation between the elements of the sets A and B, and S a relation between the elements of the set B and some third set C. Then, the **composition of relations** R and S (from Latin *componēre*—to compose) is a binary relation $S \circ R$ between the elements of the sets A and C, calculated by the formula:

$$S \circ R = \{(a, c) \colon a \in A, \ c \in C \ \text{and} \ a R b, \ b S c \ \text{for} \ b \in B\}.$$

If the relations R and S are specified in the form of the logical matrices M_R and M_S, respectively, then the logical matrix of composition of the relations R and S is equal to the **logical product** of the matrices $M_R M_S$ [27].

Functions

Function from a set A to a set B is a binary relation, for which each element from A is associated with the only element from the set B. This definition can be written in a different form: For any $a \in A$, there exists the only pair from the relation of the form (a, b), where $b \in B$ [4,27].

A function from the set A to the set B is denoted as $f : A \to B$, where the set A is called the **domain** of the function f and B is the **range**. Another denotation of the function: $y = f(x)$, in this notation $y \in B$ is the **value** of the function f, corresponding to the argument x or the **image** x for mapping of f.

Set of values of the function f is the set of values of arguments $x \in A$ in B: $f(A) = \{f(x) : x \in A\}$.

Consider an arbitrary set $A' \subseteq A$. The **image of set** A' by definition contains those and only those elements $y \in B$, to which the elements $x \in A'$ are mapped:

$$f(A') = \{y \in B : \exists x ((x \in A') \textbf{and} (y = f(x)))\}.$$

Thus, the set of values of a function is the image of the domain of this function.

Let $f : A \to B$ be a function acting from the set A to the set B. For the two sets A_1 and A_2 such that $A_1, A_2 \subseteq A$, the following relations are valid:

(1) $f(A_1 \cup A_2) = f(A_1) \cup f(A_2)$.
(2) $f(A_1 \cap A_2) \subseteq f(A_1) \cap f(A_2)$.
(3) $(A_1 \subseteq A_2) \Rightarrow (f(A_1) \subseteq f(A_2))$.

Example 3.3 Let us show that the image of intersection of two sets does not necessarily coincide with the intersection of images of these sets.

Solution.
Consider a function $f : A \to B$, determined on a set $A = \{-1, 0, 1\}$ and acting by the rule $f(x) = x^2$.
Then, for the sets $A_1 = \{-1, 0\} \subset A$ and $A_2 = \{0, 1\} \subset A$ we have:

$$f(A_1 \cap A_2) = f(\{0\}) = \{0\},$$
$$f(A_1) \cap f(A_2) = f(\{-1, 0\}) \cap f(\{0, 1\}) = \{0, 1\} \cap \{0, 1\} = \{0, 1\}.$$

It means that $f(A_1 \cap A_2) \subseteq f(A_1) \cap f(A_2)$, but $f(A_1 \cap A_2) \neq f(A_1) \cap f(A_2)$. Hence, the image of intersection of two sets, generally speaking, does not coincide with the intersection of their images. $\square$

A function $f: A \to B$ is called **injective**, or **injection**, if from the equality of images of arbitrary elements a_1 and a_2 of the set A follows the equality of these elements to each other, or, in a symbolic notation, $\forall a_1, a_2 \in A \ (f(a_1) = f(a_2)) \Rightarrow (a_1 = a_2)$.

A function $f: A \to B$ is called **surjective**, or **surjection**, if $f(A) = B$; in other words, if for any element $b \in B$, there exists some element $a \in A$ such that $f(a) = b$.

A function that is both injection and surjection is called **bijective** or **bijection**. A bijective function realizes a one-to-one correspondence between the elements of the domain and the range.

Concepts of injection and surjection are applied not only to functions, but also to arbitrary relations. The introduced definitions are schematically illustrated in Fig. 3.3.

If the inverse relation to the function f is also a function, then it is said that the function is f **invertible**. A function $f: A \to B$ is invertible if and only if it is bijective. A function inverse to f is commonly denoted by $f^{-1}: B \to A$. If $f(a) = b$, then $f^{-1}(b) = a$.

Composition of functions $f: A \to B$ and $g: B \to C$ is defined as

$$g \circ f = \{(a, c) \in A \times C: \ \exists b \in B, \ b = f(a) \ \text{and} \ c = g(b)\}.$$

A composition of two functions is a function acting by the rule

$$(g \circ f)(x) = g(f(x)).$$

Among the **basic elementary function** are the following [29]:

$$y = x^a, \qquad y = b^x, \qquad y = \log_c x,$$
$$y = \sin x, \qquad y = \cos x, \qquad y = \tan x, \qquad y = \cot x,$$
$$y = \arcsin x, \qquad y = \arccos x, \qquad y = \arctan x, \qquad y = \operatorname{arccot} x,$$

where a, b, c—const, $b > 0, 0 < c \neq 1$.

Elementary functions are the functions obtained by a finite number of arithmetical operations with the basic elementary functions, as well as by composition of these functions [29].

For visual representation of functional dependences, graphs of functions are plotted on a coordinate plane. **Graph** of a function $y = f(x)$ is a set of points on a coordinate plane (x, y) consisting of points with coordinates $(x, f(x))$. When plotting a graph of the function $f: A \to B$, the domain A is placed on the X-axis, and the range B is placed on the Y-axis.

Suppose we have a graph of some function $y = f(x)$, denote it by Γ. Then:

(1) Graph $y = f(x - c)$, where $c > 0$, can be obtained by shift of Γ along the X-axis by c units right.
(2) Graph $y = f(x) + d$, where $d > 0$, can be obtained by shift of Γ along the Y-axis by d units up.

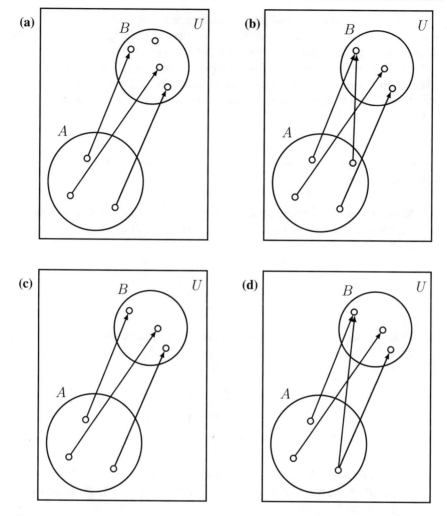

Fig. 3.3 Properties of binary relations between the sets A and B: **a** injective but not surjective; **b** surjective but not injective; **c** injective and surjective (bijective); **d** relation but not function

(3) Graph $y = hf(x)$, where $h > 1$, can be obtained by extension of Γ along the Y-axis by h times.

(4) Graph $y = f(kx)$, where $k > 1$, can be obtained by contraction of Γ along the X-axis by k times.

(5) Graphs $y = -f(x)$ and $y = f(-x)$ can be obtained by reflection of Γ with respect to the X-axis and with respect to the Y-axis, respectively.

Pigeonhole principle (Dirichlet's[3] **drawer principle**): *Let* $f : A \rightarrow B$, *where A and B are finite sets; if* $|A| > k|B|$ *for some natural k, then a function f will take the same values at least for* $k + 1$ *different arguments* [4,26,27].

Note. The names of the relations' properties are derived from Latin words *reflectere* and *trānsīre*. The word "function" is derived from Latin *fūnctio*—execution, performance. The terms *injection*, *surjection*, and *bijection* were suggested by Bourbaki.[4]

3.1 Problems

3.1. Write out the set of ordered pairs and draw the directed graph of the relation between the elements of the sets $\{a, b, c\}$ and $\{1, 2, 3, 4\}$, specified by the logical matrix:

$$
\begin{array}{c}
\begin{array}{cccc} 1 & 2 & 3 & 4 \end{array} \\
\begin{array}{c} a \\ b \\ c \end{array}
\begin{bmatrix}
T & F & T & F \\
F & T & F & F \\
F & F & T & T
\end{bmatrix}.
\end{array}
$$

3.2. Write out the set of ordered pairs and draw the directed graph of the relation between the elements of the sets $\{1, 2, 3, 4\}$ and $\{a, b, c\}$, specified by the logical matrix:

$$
\begin{array}{c}
\begin{array}{ccc} a & b & c \end{array} \\
\begin{array}{c} 1 \\ 2 \\ 3 \\ 4 \end{array}
\begin{bmatrix}
T & F & T \\
F & T & T \\
F & F & T \\
T & F & F
\end{bmatrix}.
\end{array}
$$

3.3. Write out the set of ordered pairs and draw the directed graph of the relation between the elements of the sets $\{a, b, c, d\}$ and $\{1, 2, 3\}$, specified by the logical matrix:

$$
\begin{array}{c}
\begin{array}{ccc} 1 & 2 & 3 \end{array} \\
\begin{array}{c} a \\ b \\ c \\ d \end{array}
\begin{bmatrix}
F & T & T \\
F & T & F \\
T & F & T \\
F & F & T
\end{bmatrix}.
\end{array}
$$

3.4. Write out the set of ordered pairs and draw the directed graph of the relation between the elements of the sets $\{1, 2, 3\}$ and $\{a, b, c, d\}$, specified by the

---

[3] Johann Peter Gustav Lejeune Dirichlet (1805–1859)—German mathematician.
[4] Nicolas Bourbaki—a collective pseudonym of a group of mathematicians.

logical matrix:

$$
\begin{array}{c} \quad a \ \ b \ \ c \ \ d \\ \begin{array}{c} 1 \\ 2 \\ 3 \end{array} \begin{bmatrix} F & F & T & T \\ T & T & F & F \\ T & F & F & T \end{bmatrix} \end{array}.
$$

3.5. For each of the following relations on the set of natural numbers $\mathbb{N}$, describe the ordered pairs belonging to the relations:

(1) $R = \{(n, m): 3n + m = 16\}$;
(2) $S = \{(n, m): n + 2m < 7\}$.

3.6. For each of the following relations on the set of natural numbers $\mathbb{N}$, describe the ordered pairs belonging to the relations:

(1) $R = \{(n, m): n + 2m = 10\}$;
(2) $S = \{(n, m): 3n + m < 9\}$.

3.7. For each of the following relations on the set of nonnegative integers $\mathbb{Z}_0 = \{0\} \cup \mathbb{N}$, describe the ordered pairs belonging to the relations:

(1) $R = \{(n, m): 3n + 2m = 30\}$;
(2) $S = \{(n, m): n + 3m < 5\}$.

3.8. Let R be a relation on the set $\{1, 2, 3, 4\}$, defined by the condition: $n \, R \, m$ if and only if $2n + m$ is an even number. Represent R by each of the following methods:

(1) As a set of ordered pairs;
(2) In a graphic form;
(3) In the form of a logical matrix.

3.9. Let R be a relation on the set $\{1, 2, 3, 4\}$, defined by the condition: $n \, R \, m$ if and only if $2n + m < 8$. Represent R by each of the following methods:

(1) As a set of ordered pairs;
(2) In a graphic form;
(3) In the form of a logical matrix.

3.10. Find which of the following relations on the set of people are reflexive, symmetric, or transitive:

(1) "...lives in the same settlement as...";
(2) "...is heavier or lighter that...";
(3) "...is no older than...";
(4) "...is a daughter-in-law to...".

3.11. Find which of the following relations on the set of people are reflexive, symmetric, or transitive:

(1) "...is taller or shorter than ...";
(2) "...is a relative of...";
(3) "...is a sister of...";
(4) "...lives in the same house as...".

3.12. Find which of the following relations on $\mathbb{Z}$ are reflexive, symmetric, and transitive:

(1) "$n + 2m$ is an odd number."
(2) "$n \cdot m$ is an even number."

3.13. Find which of the following relations on $\mathbb{Z}$ are reflexive, symmetric, and transitive:

(1) "$n + 2m$ is an even number."
(2) "$n^2 + m$ is an even number."

3.14. Is the relation S reflexive, symmetric, antisymmetric, transitive if:

(1) $S = \{(x, y): xy > 0, \ x, y \in \mathbb{R}\}$.
(2) $S = \{(x, y): x + y = 0, \ x, y \in \mathbb{R}\}$.
(3) $S = \{(n, m): n + m = 2, \ n, m \in \mathbb{Z}\}$.
(4) $S = \{(n, m): n \neq m + 1, \ n, m \in \mathbb{Z}\}$?

*3.15. Consider the set $A = \{a, b, c, d, e\}$. Give an example of a binary relation on A, if it is possible, which is:

(1) Reflexive, but is neither symmetric nor transitive;
(2) Symmetric, but is neither reflexive nor transitive;
(3) Transitive, but is neither reflexive nor symmetric;
(4) Reflexive and symmetric, but is not transitive;
(5) Symmetric and transitive, but is not reflexive;
(6) Reflexive and transitive, but is not symmetric;
(7) Neither symmetric nor antisymmetric;
(8) Symmetric and antisymmetric.

*3.16. At the discrete mathematics and mathematical logic exam, the student states that no antisymmetric relation has a symmetric property. Is he right?

*3.17. At the discrete mathematics and mathematical logic exam, the student states that any symmetric and transitive relation is reflexive, and provides the following "proof" [13]: "Suppose that R is a symmetric and transitive relation on X. Let us show that $a \, R \, a$ for $a \in X$. Take an arbitrary b, such that $a \, R \, b$.

From the symmetry, we obtain $b\,R\,a$. Then, from the transitivity, we obtain $a\,R\,a$."

Explain what the error of this reasoning is.

3.18. Find closures with respect to reflexive, symmetric, and transitive properties of the relation

$$\{(a, a), (a, c), (a, g), (c, e), (e, a), (e, g), (g, c), (g, g)\},$$

specified on the set $A = \{a, c, e, g\}$.

3.19. Find closures with respect to reflexive, symmetric, and transitive properties of the relation

$$\{(k, n), (l, k), (l, m), (m, l), (m, m), (n, k), (n, m), (n, n)\},$$

specified on the set $A = \{k, l, m, n\}$.

3.20. List the ordered pairs belonging to the relations specified on the set $\{n: \ n \in \mathbb{Z}$ **and** $1 \leqslant n \leqslant 10\}$:

(1) $R = \{(n, m): nm = 30\}$.
(2) $S = \{(n, m): 2n + m = 10\}$.
(3) Closure R with respect to symmetry.
(4) Closure S with respect to transitivity.

3.21. List the ordered pairs belonging to the relations specified on the set $\{n: \ n \in \mathbb{Z}$ **and** $1 \leqslant n \leqslant 12\}$:

(1) $R = \{(n, m): nm = 12\}$.
(2) $S = \{(n, m): n + 2m = 10\}$.
(3) Closure R with respect to transitivity.
(4) Closure S with respect to symmetry.

3.22. For each of the following equivalence relations on the given set A, describe the blocks into which the set A is partitioned:

(1) A are months of the year, and R is defined by the condition: $x\,R\,y$ if and only if the season x coincides with the season y.
(2) A is the set of people, and R is defined by the condition: $x\,R\,y$ if and only if x has the same zodiac sign as y.

3.23. For each of the following equivalence relations on the given set A, describe the blocks into which the set A is partitioned:

(1) $A = \mathbb{Z}$, R is specified by the condition: $x\,R\,y$ if and only if $x + y$ is an even number.

(2) $A = \mathbb{R}^3$, R is specified by the rule: $(x, y, z) R (\tilde{x}, \tilde{y}, \tilde{z})$ if and only if $x^2 + y^2 + z^2 = \tilde{x}^2 + \tilde{y}^2 + \tilde{z}^2$.

3.24. Draw Hasse diagram for the following partially ordered set:

$\{n: 1 \leqslant n \leqslant 20$ **and** n is divisible by $2\}$ with the relation "n divides m."

3.25. Draw Hasse diagram for the following partially ordered set:

$\{n: 1 \leqslant n \leqslant 30$ **and** n is divisible by $3\}$ with the relation "n divides m."

3.26. Draw Hasse diagram for the following partially ordered sets:

(1) A set of all subsets in $\{a, b, c\}$ with the relation "X is the subset of Y";
(2) A set of all subsets in $\{a, b, c, d\}$ with the relation "X is the subset of Y."

3.27. Draw Hasse diagram for the partially ordered set $A = \{a, b, c, d\}$ with the partial order relation $P = \{(a, a), (b, b), (c, c), (d, d)\}$. Are there maximal and minimal elements in this set? Is it possible to indicate the greatest and the least element in A?

3.28. Draw Hasse diagram for the partially ordered set $A = \{a, b, c, d\}$ with the partial order relation

$$P = \{(a, a), (a, b), (a, c), (b, b), (b, c), (b, d), (c, c), (c, d), (d, d)\}.$$

Indicate the maximal and the minimal element. Is it possible to indicate the greatest and the least element in this set?

3.29. Let R be a relation between the sets $\{1, 2\}$ and $\{1, 2, 3, 4, 5\}$, specified by listing of the pairs: $R = \{(1, 2), (1, 4), (1, 5), (2, 3), (2, 5)\}$, and S be a relation between the sets $\{1, 2, 3, 4, 5\}$ and $\{1, 2, 3\}$, consisting of the pairs: $S = \{(1, 1), (2, 1), (2, 3), (3, 3), (5, 2)\}$. Calculate $S \circ R$ and $(S \circ R)^{-1}$.

3.30. Let R be a relation between the sets $\{1, 2, 3, 4, 5\}$ and $\{1, 2\}$, specified by listing of the pairs: $R = \{(2, 1), (3, 2), (4, 1), (5, 1), (5, 2)\}$, and S be a relation between the sets $\{1, 2, 3, 4, 5\}$ and $\{1, 2, 3\}$, consisting of the pairs: $S = \{(1, 1), (2, 1), (2, 3), (3, 3), (5, 2)\}$. Calculate R^{-1} and $S \circ R^{-1}$.

3.31. Let R be a relation between the sets $\{1, 2, 3\}$ and $\{1, 2, 3, 4\}$, specified by listing of the pairs: $R = \{(1, 2), (1, 3), (2, 1), (2, 4), (3, 4)\}$, and S be a relation between the sets $\{1, 2, 3, 4\}$ and $\{1, 2\}$, consisting of the pairs: $S = \{(1, 1), (2, 2), (3, 2), (4, 1), (4, 2)\}$. Calculate $S \circ R$ and R^{-1}.

3.32. Let R be a relation between the sets $\{1, 2, 3\}$ and $\{1, 2, 3, 4\}$, specified by listing of the pairs: $R = \{(1, 2), (1, 3), (2, 1), (2, 4), (3, 4)\}$, and S be a relation between the sets $\{1, 2\}$ and $\{1, 2, 3, 4\}$, consisting of the pairs: $S = \{(1, 1), (1, 4), (2, 2), (2, 3), (2, 4)\}$. Calculate R^{-1} and $R^{-1} \circ S$.

3.33. Let R be a relation between the sets $\{1, 2, 3\}$ and $\{1, 2, 3, 4, 5\}$, specified by listing of the pairs: $R = \{(1, 1), (2, 3), (2, 4), (3, 2), (3, 5)\}$, and S be a relation between the sets $\{1, 2, 3, 4, 5\}$ and $\{1, 2, 3, 4\}$, consisting of the pairs: $S = \{(1, 2), (1, 3), (2, 2), (4, 1), (4, 4)\}$. Calculate S^{-1} and $S \circ R$.

3.34. Let R be a relation between the sets $\{1, 2, 3\}$ and $\{1, 2, 3, 4, 5\}$, specified by listing of the pairs: $R = \{(1, 1), (2, 3), (2, 4), (3, 2), (3, 5)\}$, and S be a relation between the sets $\{1, 2, 3, 4\}$ and $\{1, 2, 3, 4, 5\}$, consisting of the pairs: $S = \{(1, 4), (2, 1), (2, 2), (3, 1), (4, 4)\}$. Calculate S^{-1} and $S^{-1} \circ R$.

3.35. Let R be the relation "...brother..." and S be the relation "...husband..." on the set of all people. Give a brief verbal description of the relations R^{-1}, S^{-1}, and $S^{-1} \circ R$.

3.36. Let R be the relation "...brother...", and S be the relation "...husband..." on the set of all people. Give a brief verbal description of the relations $R \circ S^{-1}$ and $S \circ R$.

3.37. Let R be the relation "...sister..." and S be the relation "...husband..." on the set of all people. Give a brief verbal description of the relations R^{-1} and $S^{-1} \circ R$.

3.38. Denote by R the relation "...parent..." and by S the relation "...husband..." on the set of all people. Give a brief verbal description of the relations: R^{-1}, S^{-1}, $S \circ R$, and $S^{-1} \circ R$.

3.39. Let $A = \{1, 2, 3\}$, $B = \{x, y, z\}$. Find $T_2 \circ T_2$, $(T_2 \circ T_1)^{-1}$, $(T_1^{-1} \circ T_2^{-1})$ if:

(1) $T_1 = \{(1, x), (1, y), (2, x), (3, z)\}$, $T_2 = \{(x, x), (y, x), (z, z)\}$.
(2) $T_1 = \{(1, x), (1, y), (2, x), (3, z)\}$,
$T_2 = \{(x, y), (x, z), (y, x), (y, z), (z, x), (z, y)\}$.

3.40. Relations R and S are specified by logical matrices M_R and M_S, respectively, where

$$M_R = \begin{bmatrix} F & F & T \\ T & T & T \end{bmatrix} \quad \text{and} \quad M_S = \begin{bmatrix} F & T & T & T \\ F & F & T & F \\ F & T & F & T \end{bmatrix}.$$

Write the logical matrix of the relation $S \circ R$.

3.41. Relations R and S are specified by logical matrices M_R and M_S, respectively, where

$$M_R = \begin{bmatrix} T & T & F \\ F & T & T \end{bmatrix} \quad \text{and} \quad M_S = \begin{bmatrix} T & F & T & F \\ F & T & F & F \\ T & T & F & T \end{bmatrix}.$$

Write the logical matrix of the relation $S \circ R$.

3.42. Relations R and S are specified by logical matrices M_R and M_S, respectively, where

$$M_R = \begin{bmatrix} F\ T\ F \\ T\ F\ T \end{bmatrix} \quad \text{and} \quad M_S = \begin{bmatrix} T\ F\ F\ T \\ T\ T\ F\ F \\ F\ T\ T\ F \end{bmatrix}.$$

Write the logical matrix of the relation $S \circ R$.

3.43. Let a binary relation R be specified by the logical matrix M_R. Express the logical matrix of the relation R^{-1} by M_R.

3.44. Binary relations R, S, and T are specified by logical matrices M_R, M_S, and M_T, respectively, where

$$M_R = \begin{bmatrix} F\ T\ F\ F \\ T\ F\ F\ F \end{bmatrix}, \quad M_S = \begin{bmatrix} T\ T\ T\ T \\ F\ T\ T\ T \\ F\ F\ T\ T \\ F\ F\ F\ T \end{bmatrix} \quad \text{and} \quad M_T = \begin{bmatrix} F\ F\ F \\ T\ T\ F \\ T\ F\ T \\ F\ T\ T \end{bmatrix}.$$

Prove that the logical matrices of the relations $R^{-1} \circ S \circ T^{-1}$ and $(T \circ S^{-1} \circ R)^{-1}$ coincide.

3.45. Binary relations R, S, and T are specified by logical matrices M_R, M_S and M_T respectively, where

$$M_R = \begin{bmatrix} F\ T\ F \\ F\ F\ T \\ T\ F\ F \end{bmatrix}, \quad M_S = \begin{bmatrix} F\ F\ T \\ F\ T\ F \\ T\ F\ F \end{bmatrix} \quad \text{and} \quad M_T = \begin{bmatrix} F\ T\ F \\ T\ F\ F \\ F\ F\ T \end{bmatrix}.$$

Write the logical matrix of the following relation:

$$Q = S \circ T \circ R^{-1} \circ S^{-1} \circ T^{-1}.$$

3.46. Check the fulfillment of Dilworth's theorem for each of the following partially ordered sets:

(1) $A = \{a_1, a_2, a_3, a_4\}$ with the relation

$$R = \{(a_1, a_1), (a_2, a_2), (a_3, a_3), (a_4, a_4)\};$$

(2) $A = \{a_1, a_2, a_3, a_4\}$ with the relation

$$R = \{(a_1, a_1), (a_1, a_2), (a_1, a_3), (a_2, a_2), (a_2, a_3), (a_2, a_4), (a_3, a_3),$$
$$(a_3, a_4), (a_4, a_4)\};$$

(3) $A = \{a_1, a_2, \ldots, a_6\}$ with the relation

$$R = \{(a_1, a_1), (a_1, a_3), (a_1, a_5), (a_1, a_6), (a_2, a_2), (a_2, a_3), (a_3, a_3),$$
$$(a_4, a_3), (a_4, a_4), (a_5, a_3), (a_5, a_5), (a_6, a_3), (a_6, a_6)\};$$

(4) $A = \{a_1, a_2, \ldots, a_5\}$ with the relation

$$R = \{(a_1, a_1), (a_1, a_2), (a_1, a_3), (a_1, a_4), (a_2, a_2), (a_2, a_3), (a_2, a_4),$$
$$(a_3, a_3), (a_4, a_4), (a_5, a_2), (a_5, a_3), (a_5, a_4), (a_5, a_5)\}.$$

3.47. Check the fulfillment of Dilworth's theorem for each of the following partially ordered sets:

(1) A set of all subsets in $\{a_1, a_2, a_3\}$ with the relation "X is the subset of Y";
(2) A set of all subsets in $\{a_1, a_2, a_3, a_4\}$ with the relation "X is the subset of Y."

3.48. Find the minimal number of pairwise disjoint chains containing all elements of the partially ordered set $A = \{a_1, a_2, \ldots, a_7\}$ with the relation R, whose logical matrix is

$$M_R = \begin{array}{c} \\ a_1 \\ a_2 \\ a_3 \\ a_4 \\ a_5 \\ a_6 \\ a_7 \end{array} \begin{array}{c} a_1\, a_2\, a_3\, a_4\, a_5\, a_6\, a_7 \\ \left[\begin{array}{ccccccc} T & T & T & T & F & F & T \\ F & T & T & F & F & F & F \\ F & F & T & F & F & F & F \\ F & F & F & T & F & F & F \\ F & F & F & T & T & F & F \\ F & F & T & T & T & T & T \\ F & F & T & T & F & F & T \end{array} \right] \end{array}.$$

*3.49. Prove the statement dual to Dilworth's theorem: The minimal number of antichains partitioning a finite partially ordered set is equal to the cardinality of the maximal chain in this set.

*3.50. A computer firm's staff consists of 17 programmers. For each of them, we know the record of service in this firm (s) and the total record of service in the information technologies sphere (σ). Show that there exists such a sequence of five employees, for whom either s grows with the growth of σ or s lowers with the growth of σ.

3.51. Let $A = \{a, b, c, d\}$ and $B = \{1, 2, 3, 4\}$. Which of the following relations between the sets A and B are functions defined on A with values in B:

(1) $\{(a, 4), (b, 2), (b, 4), (c, 3), (d, 1)\}$;
(2) $\{(a, 2), (b, 1), (c, 3), (d, 4)\}$;
(3) $\{(a, 3), (b, 1), (d, 2)\}$;
(4) $\{(a, 1), (b, 3), (c, 4), (d, 1)\}$?

Which of these functions are injective, and which are surjective?

3.52. Let $A = \{1, 2, 3, 4\}$ and $B = \{a, b, c, d\}$. Which of the following relations between the sets A and B are functions defined on A with values in B:

(1) $\{(4, b), (1, c), (2, a), (3, d)\}$;
(2) $\{(2, d), (4, a), (3, c), (1, a)\}$;
(3) $\{(3, a), (2, b), (1, c), (2, a), (4, d)\}$;
(4) $\{(1, d), (4, a), (2, c)\}$?

Which of these functions are injective, and which are surjective?

3.53. Let $A = \{a, b, c, d\}$ and $B = \{1, 2, 3, 4\}$. Which of the following relations between the sets A and B are functions defined on A with values in B:

(1) $\{(a, 1), (b, 3), (c, 1), (d, 4)\}$;
(2) $\{(a, 3), (b, 4), (d, 1)\}$;
(3) $\{(a, 3), (b, 2), (c, 4), (d, 1)\}$;
(4) $\{(a, 1), (b, 2), (c, 1), (c, 4), (d, 3)\}$?

Which of these functions are injective, and which are surjective?

3.54. For each of the following functions acting from $\mathbb{Z}$ to $\mathbb{Z}$, establish whether they are injections, surjections, or bijections:

(1) $f(n) = -n$.
(2) $\text{sign}(n) = \begin{cases} 1, & \text{if } n > 0, \\ 0, & \text{if } n = 0, \\ -1, & \text{if } n < 0. \end{cases}$

3.55. For each of the following functions acting from $\mathbb{Z}$ to $\mathbb{Z}$, indicate whether that are injections, surjections, or bijections:

(1) $f(n) = 3n - 2$.
(2) $g(n) = \begin{cases} \dfrac{1}{2}n, & \text{if } n \text{ is even,} \\ 3n, & \text{if } n \text{ is odd.} \end{cases}$
(3) $h(n) = 1 - n^2$.

3.56. For each of the following functions acting from $\mathbb{Z}$ in $\mathbb{Z}$, indicate whether they are injections, surjections, or bijections:

(1) $h_1(n) = n - 6$.
(2) $h_2(n) = \begin{cases} \dfrac{1}{3}n, & \text{if } n \text{ is divisible by 3,} \\ 3n, & \text{if } n \text{ is not divisible by 3.} \end{cases}$

(3) $h_3(n) = 1 - n^3$.

3.57. Draw graphs of the functions:

(1) $f : \mathbb{R} \to \mathbb{R}, \quad f(x) = \sin x$.

(2) $g : \mathbb{R} \to \mathbb{R}, \quad g(x) = \begin{cases} 6 - 2x, & \text{if } x \geqslant 1, \\ \dfrac{3}{2} - \dfrac{1}{2}x, & \text{if } x < 1. \end{cases}$

Find the sets of values and indicate which of the functions are injective, and which are surjective.

3.58. Modulus (or **absolute value**) of a real number $|x|$ is defined as the number x itself if $x \geqslant 0$, and $-x$ for $x < 0$. Prove that for real x, y

(1) $\sqrt{x^2} = |x|$.
(2) $|x + y| \leqslant |x| + |y|$.

3.59. Is the function $|x|$ elementary?

3.60. Prove the inequality valid for all real numbers a and b:

$$|a - b| \geqslant |a| - |b|.$$

3.61. Express the maximum and the minimum of the two real numbers a and b

$$\max(a, b) = \begin{cases} a, & \text{if } a \geqslant b, \\ b, & \text{if } a < b; \end{cases}$$

$$\min(a, b) = \begin{cases} a, & \text{if } a \leqslant b, \\ b, & \text{if } a > b; \end{cases}$$

by the modulus of their difference.

3.62. Prove that for arbitrary real numbers a and b the correlations are fulfilled:

(1) $\max(a, b) + \min(a, b) = a + b$.
(2) $\max(a, b) - \min(a, b) = |a - b|$.

3.63. Express the modulus of the real number a by:

(1) Operation max;
(2) Operation min.

3.64. Draw graphs of the functions:

(1) $f : \mathbb{R} \to \mathbb{R}, \quad f(x) = 2^x$.
(2) $g : \mathbb{R} \to \mathbb{R}, \quad g(x) = 3x - |x|$.

Indicate the set of values of each of them and tell which of these functions are injective, and which are surjective.

3.65. Draw graphs of the functions:

(1) $f : \mathbb{R} \to \mathbb{R}, \quad f(x) = x^3$;

(2) $g : \mathbb{R} \to \mathbb{R}, \quad g(x) = \begin{cases} x + 2, & \text{if } x \geqslant 3, \\ 2x - 3, & \text{if } x < 3. \end{cases}$

Indicate the set of values of each of them and tell which of these functions are injective, and which are surjective.

3.66. Construct the graphs of the basic trigonometric functions: $\sin x$, $\cos x$, $\tan x$, $\cot x$.

3.67. Construct the graphs of the following functions:

(1) $f_1(x) = 2 \sin x + 1$.

(2) $f_2(x) = \dfrac{1}{2} \cos x - \dfrac{1}{2}$.

(3) $f_3(x) = \sin(2x)$.

(4) $f_4(x) = \tan(x/2 - \pi/2) + 1$.

∗3.68. Construct the graphs of the functions

(1) $f_1(x) = \sin(3 \arcsin(x))$.

(2) $f_2(x) = \cos(3 \arccos(x))$.

∗3.69. Construct the graphs of the functions

(1) $f_1(x) = \cos(4 \arccos(x))$.

(2) $f_2(x) = \tan(2 \arctan(x))$.

3.70. Give several examples (if possible) of the functions $f : [0, 1] \to [0, 1]$, which are:

(1) Injective and surjective;

(2) Injective, but not surjective;

(3) Surjective, but not injective;

(4) Neither injective nor surjective.

3.71. The function $f : \mathbb{R} \to \mathbb{R}$ is specified by the formula: $f(x) = 2x - 5$. Show that f is bijective, and find the function inverse to it.

3.72. The function $f : \mathbb{R} \to \mathbb{R}$ is specified by the formula: $f(x) = \dfrac{x}{3} + 2$.

Show that f is bijective, and find the function inverse to it.

3.73. The function $f: \mathbb{R} \setminus \{0\} \to \mathbb{R} \setminus \{2\}$ is specified by the formula: $f(x) = \dfrac{3}{x} + 2$. Show that f bijective, and find the function inverse to it.

*3.74.** Establish bijection between the sets:

    (1) $[0, 1]$ *and* $[a, b]$, where a, b—const;
    (2) $(0, 1)$ *and* $\mathbb{R}$.

3.75. Functions $f: \mathbb{R} \to \mathbb{R}$ and $g: \mathbb{R} \to \mathbb{R}$ are specified by the following rules:

$$f(x) = (x + 2)^2 \text{ and } g(x) = \begin{cases} 2x - 6, & \text{if } x \geqslant 1, \\ -x - 1, & \text{if } x < 1. \end{cases}$$

Express by formulas the compositions $f \circ g$ and $g \circ g$.

3.76. Functions $f: \mathbb{R} \to \mathbb{R}$ and $g: \mathbb{R} \to \mathbb{R}$ are specified by the following rules:

$$f(x) = (x - 1)^2 \text{ and } g(x) = \begin{cases} 2x - 6, & \text{if } x \geqslant 1, \\ -x - 1, & \text{if } x < 1. \end{cases}$$

Express by formulas the compositions $f \circ f$ and $g \circ f$.

3.77. Functions $f: \mathbb{R} \to \mathbb{R}$ and $g: \mathbb{R} \to \mathbb{R}$ are specified by the following rules:

$$f(x) = x^3 \text{ and } g(x) = \begin{cases} 3x - 1, & \text{if } x \geqslant 0, \\ -x, & \text{if } x < 0. \end{cases}$$

Express by formulas the compositions $f \circ f$ and $g \circ g$.

3.78. Functions $f: \mathbb{R} \to \mathbb{R}$ and $g: \mathbb{R} \to \mathbb{R}$ are specified by the following rules:

$$f(x) = x^3 \text{ and } g(x) = \begin{cases} 3x - 1, & \text{if } x \geqslant 0, \\ -x, & \text{if } x < 0. \end{cases}$$

Express by formulas the compositions $f \circ g$ and $g \circ f$.

3.79. Integer part of a real number x is a function defined as the greatest integer, which is less than or equal to x:

$$\lfloor \cdot \rfloor: \mathbb{R} \to \mathbb{Z}, \quad \lfloor x \rfloor = \max(n \in \mathbb{Z}, \ n \leqslant x).$$

The integer part of a number is also called a "**floor**". Similarly, the function "**ceiling**" is defined, as the least integer, which is greater than or equal to x:

$$\lceil \cdot \rceil: \mathbb{R} \to \mathbb{Z}, \quad \lceil x \rceil = \min(n \in \mathbb{Z}, \ n \geqslant x).$$

In the foreign literature, for $\lfloor x \rfloor$ and $\lceil x \rceil$ the terms **floor function** and **ceiling function**, respectively, are used. Prove the following properties of the considered functions for $x \in \mathbb{R}, n \in \mathbb{Z}$:

(1) $\lfloor x + n \rfloor = \lfloor x \rfloor + n, \ \lceil x + n \rceil = \lceil x \rceil + n.$

(2) $\left\lfloor \dfrac{n}{2} \right\rfloor + \left\lceil \dfrac{n}{2} \right\rceil = n.$

(3) $\lfloor -x \rfloor = -\lceil x \rceil, \ \lceil -x \rceil = -\lfloor x \rfloor.$

(4) $\lceil \log_2 (n + 1) \rceil = \lfloor \log_2 n \rfloor + 1, \ \ n \geqslant 1.$

3.80. Find the values of the expressions:

(1) $\lceil -\lfloor 2 \rfloor \rceil$;

(2) $\lfloor \lceil 5/2 \rceil + 1 \rfloor$;

(3) $\lfloor -\lfloor 5/2 \rfloor \cdot \lceil 5/2 \rceil \rfloor$;

(4) $\lfloor \lceil \lfloor 7/8 \rfloor / 8 \rceil - 1 \rfloor.$

3.81. Find the values of the expressions:

(1) $\lfloor \sqrt{\pi} \rfloor$;

(2) $-\lfloor -\sqrt{\pi} \rfloor$;

(3) $-\lfloor -\pi \rfloor - \lceil -\pi \rceil$;

(4) $\lceil -\lfloor \pi \cdot \lfloor -\pi \rfloor \rfloor \rceil.$

3.82. Draw graphs of the functions "floor" and "ceiling", defined on the set of real numbers (see Exercise **3.79**). Are the functions under consideration injective, surjective, bijective?

3.83. Let natural numbers p and n be specified. Prove that the number of natural numbers divisible by p and not exceeding n is equal to $\left\lfloor \dfrac{n}{p} \right\rfloor$.

3.84. Find the quantity of real numbers not exceeding 100, which are divisible either by 3, or by 5, or by 11.

3.85. Find the quantity of real numbers not exceeding 2^{20}, which are

(1) Divisible either by 3, or by 4, or by 5;

(2) Divisible either by 3 or by 4, but not divisible by 5;

(3) Divisible by 3, but not divisible either by 4 or by 5;

(4) Not divisible by any of numbers 3, 4, 5.

3.86. Find the quantity of positive integers not exceeding 1024 and divisible either by 3, or by 5, or by 7, or by 17.

∗3.87. Find the quantity of positive integers not exceeding 1024, which are not divisible by any of numbers 6, 10, and 15, but are divisible by 28.

∗3.88. Find the number of integers belonging to each of the following sets:

(1) (x_1, x_2);

(2) $[x_1, x_2]$,

where x_1, x_2 are real numbers, $x_1 < x_2.$

3.89. How many applicants should enter the same faculty so that

    (1) Birthdays of at least two of them will coincide.
    (2) Twenty-five of them were born in the same month.

3.90. How many times three dice should be thrown so that one could state for sure: The sum of the points will appear at least twice?

3.91. How many movies should be in the video rental shop so that the names of some four movies will begin with the same letter?

3.92. Let $U = \{1, 2, \ldots, 20\}$. What least quantity of odd numbers should be taken from the set U, so that at least two of them will add up to 20?

3.93. Let $U = \{1, 2, \ldots, 20\}$. What least quantity of even numbers should be taken from the set U, so that one could state for sure that the difference of two of them is equal to 8?

3.2 Answers, Hints, Solutions

3.1. *Answer*:
Set of ordered pairs: $R = \{(a, 1), (a, 3), (b, 2), (c, 3), (c, 4)\}$. The graphic form of the relation is shown in Fig. 3.4.
3.5. *Solution*.

    (1) To the relation R belong such pairs of natural numbers (n, m), for which the condition $3n + m = 16$ or $m = 16 - 3n$ is fulfilled. Assuming $n = 1$, $n = 2$, $\ldots$, $n = 5$, we obtain the respective values of m. For $n > 5$, such natural m does not exist. This is why we write out the answer:
    $R = \{(1, 13), (2, 10), (3, 7), (4, 4), (5, 1)\}$.
    (2) Similarly to item (1), we have:
    $S = \{(1, 1), (1, 2), (2, 1), (2, 2), (3, 1), (4, 1)\}$.
3.6. *Answer*:
    (1) $R = \{(2, 4), (4, 3), (6, 2), (8, 1)\}$.

Fig. 3.4 To Exercise **3.1**

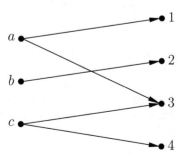

Fig. 3.5 To Exercise **3.8**

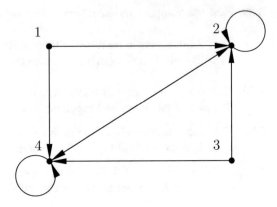

(2) $S = \{(1, 1), (1, 2), (1, 3), (1, 4), (1, 5), (2, 1), (2, 2)\}$.

3.7. *Answer*:
(1) $R = \{(0, 15), (2, 12), (4, 9), (6, 6), (8, 3), (10, 0)\}$.
(2) $S = \{(0, 0), (0, 1), (1, 0), (1, 1), (2, 0), (3, 0)\}$.

3.8. *Answer*:
(1) Set of ordered pairs:
$R = \{(1, 2), (1, 4), (2, 2), (2, 4), (3, 2), (3, 4), (4, 2), (4, 4)\}$.
(2) The graphic form of the relation is shown in Fig. 3.5.
(3) The matrix of the relation R has the form:

$$M_R = \begin{bmatrix} F & T & F & T \\ F & T & F & T \\ F & T & F & T \\ F & T & F & T \end{bmatrix}.$$

3.10. *Answer*:
Use the definitions of reflexive, symmetric, and transitive relations:
(1) Reflexive, symmetric, and transitive;
(2) Not reflexive, symmetric, and not transitive;
(3) Reflexive, not symmetric, and transitive;
(4) Not reflexive, not symmetric, and not transitive.

3.12. *Solution*.

(1) The relation is not reflexive, since the number $n + 2n = 3n$ will be even only if n is an even number. It is not symmetric, since the sum $n + 2m$ is an odd number only if n is odd, and m can be both even or odd, but the sum $m + 2n$ will be odd only if m is odd. Transitivity of this relation follows from the fact that the sums $n + 2m$ and $m + 2l$ will be odd only if n and m are odd, and the sum $n + 2l$ turns out to be odd.

(2) This relation is symmetric due to commutativity of the product: $nm = mn$. It is not reflexive, since n^2 is even only for even n, and not transitive, since for odd n, l and even m, the products nm and ml are even, and nl is odd.

3.13. *Answer*:
(1) Not reflexive, not symmetric, transitive;
(2) Reflexive, symmetric, and transitive.

3.14. *Answer*:
(1) Not reflexive, symmetric, not antisymmetric, transitive;
(2) Not reflexive, symmetric, not antisymmetric, not transitive;
(3) Not reflexive, symmetric, not antisymmetric, not transitive;
(4) Reflexive, not symmetric, not antisymmetric, not transitive.

3.16. *Answer*: No.

3.18. *Solution*.
Let $R = \{(a, a), (a, c), (a, g), (c, e), (e, a), (e, g), (g, c), (g, g)\}$.

The reflexive closure of R is the relation $R \cup \{(c, c), (e, e)\}$.

$R \cup \{(a, e), (c, a), (c, g), (e, c), (g, a), (g, e)\}$ is the closure with respect to symmetry.

The closure with respect to transitivity is constructed as follows: $a\,R\,c$ and $c\,R\,e \Rightarrow a\,R\,e$, $c\,R\,e$ and $e\,R\,a \Rightarrow c\,R\,a$, $c\,R\,e$ and $e\,R\,g \Rightarrow c\,R\,g$, $e\,R\,a$ and $a\,R\,c \Rightarrow e\,R\,c$, $c\,R\,a$ and $a\,R\,c \Rightarrow c\,R\,c$, $e\,R\,a$ and $a\,R\,e \Rightarrow e\,R\,e$, $g\,R\,c$ and $c\,R\,a \Rightarrow g\,R\,a$, $g\,R\,c$ and $c\,R\,e \Rightarrow g\,R\,e$.

So, we have obtained a transitive closure:

$$R \cup \{(a, e), (c, a), (c, c), (c, g), (e, c), (e, e), (g, a), (g, e)\} = A \times A.$$

3.19. *Answer*:
The reflexive closure of R is the relation $R \cup \{(k, k), (l, l)\}$. The symmetric closure is $R \cup \{(k, l), (m, n)\}$. The transitive closure:

$$R \cup \{(k, k), (k, l), (k, m), (l, l), (l, n), (m, k), (m, n), (n, l)\} = A \times A.$$

3.20. *Answer*:
(1) $R = \{(3, 10), (5, 6), (6, 5), (10, 3)\}$.
(2) $S = \{(1, 8), (2, 6), (3, 4), (4, 2)\}$.
(3) $R^* = R$ (the initial relation is symmetric).
(4) $S^* = S \cup \{(3, 2), (3, 6), (4, 6)\}$.

3.21. *Answer*:
(1) $R = \{(1, 12), (2, 6), (3, 4), (4, 3), (6, 2), (12, 1)\}$.
(2) $S = \{(2, 4), (4, 3), (6, 2), (8, 1)\}$.
(3) $R^* = R \cup \{(1, 1), (2, 2), (3, 3), (4, 4), (6, 6), (12, 12)\}$.
(4) $S^* = S \cup \{(1, 8), (2, 6), (3, 4), (4, 2)\}$.

Element	Ancestor	Direct ancestor
2	No item	No item
4	2	2
6	2	2
8	2, 4	4
10	2	2
12	2, 4, 6	4, 6
14	2	2
16	2, 4, 8	8
18	2, 6	6
20	2, 4, 10	4, 10

Table 3.2 Ancestors and direct ancestors of the elements of the relation P

Fig. 3.6 To Exercise **3.24**

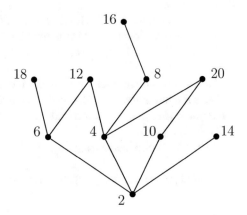

3.22. *Answer*:

(1) Here, we have four equivalence classes that correspond to seasons: {*December, January, February*}, {*March, April, May*}, {*June, July, August*}, and {*September, October, November*}.

(2) Each equivalence class consists of all people whose zodiac signs coincide.

3.24. *Solution*.

Let $A = \{n\colon\ 1 \leqslant n \leqslant 20$ **and** n is divisible by 2$\}$, and P is the relation "n divides m." Write out the elements of the partially ordered set A: $A = \{2, 4, 6, 8, 10, 12, 14, 16, 18, 20\}$. We construct Hasse diagram taking into account the information about the direct ancestors of the elements of the set A, mentioned in Table 3.2. One of the possible forms of Hasse diagram is shown in Fig. 3.6.

3.29. *Solution*.

Using the definitions of relation composition and inverse relation, we obtain the following sets:

$S \circ R = \{(1, 1), (1, 2), (1, 3), (2, 2), (2, 3)\}$,

$(S \circ R)^{-1} = \{(1, 1), (2, 1), (2, 2), (3, 1), (3, 2)\}$.

3.30. *Answer*:
$R^{-1} = \{(1, 2), (1, 4), (1, 5), (2, 3), (2, 5)\}$,
$S \circ R^{-1} = \{(1, 1), (1, 2), (1, 3), (2, 2), (2, 3)\}$.

3.34. *Answer*:
$S^{-1} = \{(1, 2), (1, 3), (2, 2), (4, 1), (4, 4)\}$,
$S^{-1} \circ R = \{(1, 2), (1, 3), (2, 1), (2, 4), (3, 2)\}$.

3.35. *Answer*:
R^{-1}: "...brother or sister...";
S^{-1}: "...wife...";
$S^{-1} \circ R$: "...wife's brother..." or "...brother-in-law...".

3.39. *Answer*:
(1) $T_2 \circ T_2 = \{(x, x), (y, x), (z, z)\}$,
$(T_2 \circ T_1)^{-1} = T_1^{-1} \circ T_2^{-1} = \{(x, 1), (x, 2), (z, 3)\}$.
(2) $T_2 \circ T_2 = \{(x, x), (x, y), (x, z), (y, x), (y, y), (y, z), (z, x), (z, y), (z, z)\} = B \times B$,
$(T_2 \circ T_1)^{-1} = T_1^{-1} \circ T_2^{-1} = \{(x, 1), (x, 3), (y, 1), (y, 2), (y, 3), (z, 1), (z, 2)\}$.

3.40. *Solution*.
The logical matrix of the relation $S \circ R$ is the product of $M_R M_S$:

$$M_R M_S = \begin{bmatrix} F & F & T \\ T & T & T \end{bmatrix} \begin{bmatrix} F & T & T & T \\ F & F & T & F \\ F & T & F & T \end{bmatrix} =$$

$$= \begin{bmatrix} (\text{FandF}) \text{ or } (\text{FandF}) \text{ or } (\text{TandF}) & (\text{FandT}) \text{ or } (\text{FandF}) \text{ or } (\text{TandT}) \ldots \\ (\text{TandF}) \text{ or } (\text{TandF}) \text{ or } (\text{TandF}) & (\text{TandT}) \text{ or } (\text{TandF}) \text{ or } (\text{TandT}) \ldots \end{bmatrix} =$$

$$= \begin{bmatrix} F & T & F & T \\ F & T & T & T \end{bmatrix}.$$

3.41. *Answer*:
$$M_{S \circ R} = \begin{bmatrix} T & T & T & F \\ T & T & F & T \end{bmatrix}.$$

3.42. *Answer*:
$$M_{S \circ R} = \begin{bmatrix} T & T & F & F \\ T & T & T & T \end{bmatrix}.$$

3.43. *Answer*:
The logical matrix $M_{R^{-1}}$ of the relation R^{-1} is connected with M_R by the transposition operation: $M_{R^{-1}} = (M_R)^T$.

3.44. *Hint*. Use the result of Exercise **3.43**.

3.45. *Answer*:
$$M_Q = \begin{bmatrix} F & T & F \\ F & F & T \\ T & F & F \end{bmatrix}.$$

3.48. *Answer*: The minimal number of chains containing all elements of the partially ordered set A with the relation R is 3.

3.49. *Proof.*
Let the cardinality of the maximal chain in the partially ordered set A be M. Associate each element $a_i \in A$, $1 \leqslant i \leqslant |A|$ with $C^{(i)}$—the maximal chain out of the chains ending on this element. Consider the sets

$$A^{(d)} = \{a_i : |C^{(i)}| = d\}, \text{ where } d = 1, 2, \ldots, M,$$

formed by the elements located on the level d of Hasse diagram of the partially ordered set A. The family of sets $A^{(d)}$ forms partitioning of A into antichains. Hence, the minimal number of antichains into which an arbitrary finite set can be partitioned is M, i.e., the cardinality of the maximal chain in this set.

3.50. *Hint.*
Apply the result of the previous exercise to the partially ordered set of the firm's programmers with the relation R of the form

$$a_i \ R \ a_j \ \Leftrightarrow \ (s(a_i) \leqslant s(a_j)) \textbf{ and } (\sigma(a_i) \leqslant \sigma(a_j)).$$

3.51. *Solution.*
Relations of items (2) and (4) are functions.

Relation of item (1) is not a function, since the element $b \in A$ in it is associated with two elements of the set B: 2 and 4.

Relation of item (3) is not a function either, since the element $c \in A$ is not associated with any element of the set B.

Function of item (4) is not injective, since it takes two different elements of the set A, namely a and d, to one and the same element $1 \in B$. This function is not surjective either, since the element $2 \in B$ is not included into its set of values.

Function of item (2) is both injective and surjective; i.e., it is bijective.

3.52. *Answer:*
Relations of items (1) and (2) are functions, and the function from (1) is bijective.

Function of item (2) is not injective and not surjective.

Relations (3) and (4) are not functions.

3.53. *Answer:*
Relations of items (1) and (3) are functions.

Function of item (1) is not injective and not surjective.

Function of item (3) is a bijection.

Relations (2) and (4) are not functions.

3.54. *Solution.*
(1) If $f(a_1) = f(a_2)$, then $-a_1 = -a_2$, whence $a_1 = a_2$. Therefore, f is an injective function. Further, since $f(n) = -n$, where n is an arbitrary number, then the set of values of this function coincides with the entire set $\mathbb{Z}$. This means that the function f is surjective. Since this function is both injection and surjection, f is a bijection.

(2) Function $\text{sign}(n)$ is not injective, since, for example, for any positive n it is equal to one. Moreover, $\text{sign}(A) = \{-1, 0, 1\}$; i.e., the set of values of the function

sign does not coincide with its range. Function sign is not a bijection, since it is not injection and surjection.

Note.

The name of function sign is derived from Latin *sīgnum*.

3.55. *Solution.*

(1) Check whether the function f is injective. An injectivity condition is implication $f(a_1) = f(a_2) \Rightarrow a_1 = a_2$ for any a_1 and a_2 from the domain. We obtain the true proposition $3a_1 - 2 = 3a_2 - 2 \Rightarrow a_1 = a_2$; therefore, f is injective.

Surjectivity condition: $\forall b \in \mathbb{Z}$ there exists integer $a \in \mathbb{Z}$ such that $f(a) = b$. But, for example, for the integer $b = 0$ there is no such $a \in \mathbb{Z}$ that $3a - 2 = 0$. Therefore, f is not surjective.

Function f is not bijective, since the surjectivity condition is not fulfilled.

(2) Check the fulfillment of the injectivity condition:

$g(a_1) = g(a_2) \Rightarrow a_1 = a_2$ for $a_1, a_2 \in \mathbb{Z}$. This implication is false since the values of the function g on various arguments $a_1, a_2 \in \mathbb{Z}$ coincide, for example, $g(6) = g(1) = 3$. Hence, the function g is not injective.

Surjectivity condition $\forall b \in \mathbb{Z}$ $\exists a \in \mathbb{Z}$: $f(a) = b$ is fulfilled in virtue of the equality $g(2b) = b$ $\forall b \in \mathbb{Z}$.

Function g is not bijective, since the injectivity condition is not fulfilled.

(3) Injectivity condition $h(a_1) = h(a_2) \Rightarrow a_1 = a_2$ for $a_1, a_2 \in \mathbb{Z}$ is not fulfilled, since the function is even and $h(a) = h(-a)$.

Surjectivity condition: $\forall b \in \mathbb{Z}$ $\exists a \in \mathbb{Z}$: $1 - a^2 = b$ is not fulfilled either, since, for example, for $b > 1$ there are no such integer values of a.

Finally, h is not bijective, since it is neither injective nor surjective.

3.56. *Answer*:

(1) $h_1(n)$ is a bijective function.
(2) function $h_2(n)$ is surjective, but not injective.
(3) function $h_3(n)$ is injective, but not surjective.

3.57. *Solution.*

Graphs of the functions are shown in Fig. 3.7.

(1) The set of values of the function f is $[-1, 1]$; it does not coincide with the range $f(A) \neq B$; therefore, the function is not surjective. Since the function is periodic, for example, $f(-\pi) = f(\pi) = -1$, then f is not an injection.

(2) The set of values of the function g coincides with the set $\mathbb{R}$, whence its surjectivity follows. Since $g(2) = g(-1) = 2$, it is not injective.

3.58. *Proof.*

(1) Considering the cases of nonnegative and negative values of the argument separately, we obtain

$$\sqrt{x^2} = \begin{cases} x, & \text{if } x \geqslant 0, \\ -x, & \text{if } x < 0 \end{cases} = |x|.$$

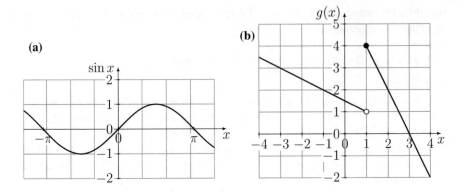

Fig. 3.7 To Exercise **3.57**

(2) Both left and right sides of the inequality take only nonnegative values; therefore, for the proof, we can take the square of both sides of the inequality.

$$|x + y| \leqslant |x| + |y| \Leftrightarrow |x + y|^2 \leqslant (|x| + |y|)^2 \Leftrightarrow x^2 + 2xy + y^2 \leqslant x^2 + 2|x||y| + y^2.$$

Since $x \leqslant |x|$ and $xy \leqslant |x||y|$, then it is proved that for all real x and y the inequality $|x + y| \leqslant |x| + |y|$ is fulfilled.

3.59. *Solution.*
Yes, since $|x| = \sqrt{x^2}$ is a composition of the basic elementary functions x^2 and $\sqrt{x}$.

3.60. *Proof.*
Assume in the inequality $|x + y| \leqslant |x| + |y|$ proved in Exercise **3.58**, $x = a - b$, $y = b$. Then

$$|a - b| + |b| \geqslant |a| \Leftrightarrow |a - b| \geqslant |a| - |b|.$$

3.61. *Solution.*
Show that $\max(a, b)$ can be represented in the form

$$\max(a, b) = \frac{1}{2}(a + b + |a - b|).$$

Actually, expanding the right side of the last equality by the definition of modulus, we come to the definition of the maximum of two numbers.

$$\frac{1}{2}(a + b + |a - b|) = \begin{cases} \frac{1}{2}(a + b + (a - b)), & \text{if } a - b \geqslant 0, \\ \frac{1}{2}(a + b - (a - b)), & \text{if } a - b < 0 \end{cases} =$$

$$= \begin{cases} a, & \text{if } a \geqslant b, \\ b, & \text{if } a < b \end{cases} = \max(a, b).$$

The minimum of two numbers can be written similarly, namely $\frac{1}{2}(a + b - |a - b|)$.

$$\frac{1}{2}(a + b - |a - b|) = \begin{cases} \frac{1}{2}(a + b - (a - b)), & \text{if } a - b \geqslant 0, \\ \frac{1}{2}(a + b + (a - b)), & \text{if } a - b < 0 \end{cases} =$$

$$= \begin{cases} b, & \text{if } a \geqslant b, \\ a, & \text{if } a < b \end{cases} = \min(a, b).$$

Thus, the maximum and the minimum of two numbers can be represented in an explicit form in terms of the elementary functions:

$$\max(a, b) = \frac{1}{2}(a + b + |a - b|),$$

$$\min(a, b) = \frac{1}{2}(a + b - |a - b|).$$

3.63. *Answer*:

(1) $|x| = \max(a, -a)$.

(2) $|x| = -\min(a, -a)$.

3.64. *Solution*.

(1) Since f is an exponential function, then the set of its values $\{x \in \mathbb{R}: x > 0\}$, i.e., all positive real numbers. Further, since $2^{a_1} = 2^{a_2} \Rightarrow a_1 = a_2$ for any $a_1, a_2 \in \mathbb{R}$, then $f(x) = 2^x$ is injective.

The set of values does not coincide with the range, when a conclusion follows that f is not surjective.

(2) When constructing a graph of the function g, it is convenient to turn to presentation of g in the form

$$g(x) = 3x - |x| = \begin{cases} 2x, & \text{if } x \geqslant 0, \\ 4x, & \text{if } x < 0. \end{cases}$$

Each value of $g(x)$ is associated with the only argument x; therefore, g is injective. The set of values $\mathbb{R}$ coincides with the range, and this function is surjective.

Graphs of the functions are shown in Fig. 3.8, panels a and b, respectively.

3.66. *Answer*: The graphs of the basic trigonometric functions are shown in Fig. 3.9.

3.68. *Solution*.

The function $f_1(x)$ is defined for all $x \in [-1, 1]$. Since

$$\sin 3x = 3 \sin x - 4 \sin^3 x,$$

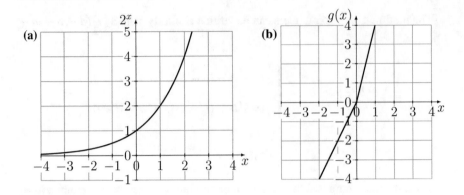

Fig. 3.8 To Exercise **3.64**

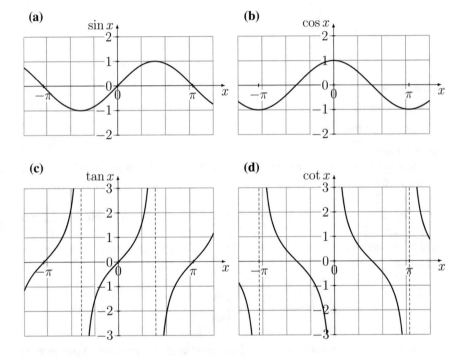

Fig. 3.9 Basic trigonometric functions

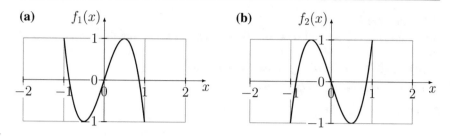

Fig. 3.10 To Exercise **3.68**

then the function $f_1(x)$ on the closed segment $[-1, 1]$ coincides with the polynomial $p_1(x) = 3x - 4x^3$.

(2) The function $f_2(x)$ is defined for all $x \in [-1, 1]$. Since

$$\cos 3x = 4\cos^3 x - 3\cos x,$$

then the function $f_1(x)$ on the closed segment $[-1, 1]$ coincides with the polynomial $p_1(x) = 4x^3 - 3x$.

Graphs of the functions $f_1(x)$, $f_2(x)$ are shown in Fig. 3.10.

3.69. *Answer*:

(1) The function $f_1(x)$ on the closed segment $[-1, 1]$ coincides with the polynomial $p_1(x) = 8x^4 - 8x^2 + 1$.

(2) The function $f_2(x)$ on the set $\mathbb{R} \setminus \{-1, 1\}$ coincides with the function $p_2(x) = \dfrac{2x}{1 - x^2}$.

3.70. *Answer*:

The condition of item (1) is fulfilled, for example, by the functions $f_1(x) = 1 - x$, $f_2(x) = x^2$, $f_3(x) = \sqrt{1 - x^2}$ and other functions invertible on segment $[0, 1]$. Other examples are given similarly.

3.71. *Solution*.

If $f(a) = f(b)$, then $2a - 5 = 2b - 5$, which results in the equality $a = b$. This implies injectivity of the function f. Let us present the equation $f(x) = y$ in the form $2x - 5 = y$, whence $x = \dfrac{y + 5}{2}$. The last expression is defined on the entire set of real numbers; i.e., by any element of the set $B = \mathbb{R}$, it is possible to construct x, which is taken to this element by the function f. Therefore, f is surjective.

We have shown that f is both injective and surjective. Thus, it is a bijection. The inverse to it function $f^{-1} \colon B \to A$ is defined by the formula $f^{-1}(x) = \dfrac{x + 5}{2}$.

3.72. *Answer*: $f^{-1}(x) = 3(x - 2)$.

3.73. *Answer*: $f^{-1}(x) = 3(x - 2)^{-1}$.

3.74. *Answer*:

The following are examples of invertible functions satisfying the conditions of the problem:

(1) $f : [0, 1] \to [a, b], \quad f(x) = (b - a)x + a$.

(2) $f : (0, 1) \to \mathbb{R}, \quad f(x) = \cot(\pi x)$.

3.75. *Solution*.

$$(f \circ g)(x) = f(g(x)) = \begin{cases} (2x - 4)^2, & \text{if } x \geqslant 1, \\ (1 - x)^2, & \text{if } x < 1. \end{cases}$$

In order to calculate the composition $(g \circ g)(x)$, one should consider four possible cases:

(1) $x \geqslant 1, \; g(x) \geqslant 1$;

(2) $x \geqslant 1, \; g(x) < 1$;

(3) $x < 1, \; g(x) \geqslant 1$;

(4) $x < 1, \; g(x) < 1$.

$$(g \circ g)(x) = g(g(x)) = g\left(\begin{cases} 2x - 6, & x \geqslant 1, \\ -x - 1, & x < 1 \end{cases} \right) =$$

$$= \begin{cases} 2(2x - 6) - 6, & x \geqslant 1, \; 2x - 6 \geqslant 1, \\ -(2x - 6) - 1, & x \geqslant 1, \; 2x - 6 < 1, \\ 2(-x - 1) - 6, & x < 1, \; -x - 1 \geqslant 1, \\ -(-x - 1) - 1, & x < 1, \; -x - 1 < 1 \end{cases} = \begin{cases} 4x - 18, & x \geqslant 7/2, \\ -2x + 5, & 1 \leqslant x < 7/2, \\ -2x - 8, & x \leqslant -2, \\ x, & -2 < x < 1. \end{cases}$$

We obtain the final answer:

$$(g \circ g)(x) = \begin{cases} 4x - 18, & \text{if } x \geqslant 7/2, \\ -2x + 5, & \text{if } 1 \leqslant x < 7/2, \\ x, & \text{if } -2 < x < 1, \\ -2x - 8, & \text{if } x \leqslant -2. \end{cases}$$

3.76. *Answer*:

$f \circ f = ((x - 1)^2 - 1)^2$;

$$g \circ f = \begin{cases} 2x^2 - 4x - 4, & \text{if } x \geqslant 1, \\ -x^2 + 2x - 2, & \text{if } x < 1. \end{cases}$$

3.79. *Proof*.

(1) Definition of the integer part of a number can be written in the form

$$\lfloor x \rfloor = a \; \Leftrightarrow \; a \leqslant x < a + 1, \quad a \in \mathbb{Z}.$$

Adding to both sides of the inequality the integer n, we obtain

$$\lfloor x \rfloor + n = a + n \; \Leftrightarrow \; a + n \leqslant x + n < a + n + 1,$$

$$\lfloor x \rfloor + n = a + n \; \Leftrightarrow \; \lfloor x + n \rfloor = a + n.$$

From the last correlation follows the equality of item 1) for the function "floor"; the equality for the function "ceiling" is proved similarly.

(2) Consider the cases of even and odd arguments of n. Let n be even, $n = 2k$, $k \in \mathbb{Z}$. Then, $\left\lfloor \dfrac{n}{2} \right\rfloor = \left\lceil \dfrac{n}{2} \right\rceil = k$,

$$\left\lfloor \frac{n}{2} \right\rfloor + \left\lceil \frac{n}{2} \right\rceil = k + k = 2k = n.$$

In case of off n, we have: $n = 2k + 1$, $k \in \mathbb{Z}$. Therefore, $\left\lfloor \dfrac{n}{2} \right\rfloor = k$, $\left\lceil \dfrac{n}{2} \right\rceil = k + 1$,

$$\left\lfloor \frac{n}{2} \right\rfloor + \left\lceil \frac{n}{2} \right\rceil = k + (k + 1) = 2k + 1 = n.$$

Thus, the equality of item (2) is valid for all integer n.

(3) Prove the equality $\lfloor -x \rfloor = -\lceil x \rceil$. As is known, any real number x can only be represented in the form $x = n + \delta$, where $n \in \mathbb{Z}$, $0 \leqslant \delta < 1$. Hence,

$$\lfloor -x \rfloor = \lfloor -(n + \delta) \rfloor = \lfloor -n - \delta \rfloor = \begin{cases} -n, & \text{if } \delta = 0; \\ -n - 1, & \text{if } \delta \in (0, 1). \end{cases}$$

The right side of the analyzed equality

$$-\lceil x \rceil = -\lceil n + \delta \rceil = \begin{cases} -n, & \text{if } \delta = 0; \\ -n - 1, & \text{if } \delta \in (0, 1). \end{cases}$$

The left and the right sides of the correlation $\lfloor -x \rfloor = -\lceil x \rceil$ coincide for all real $x \in \mathbb{R}$. The equality $\lceil -x \rceil = -\lfloor x \rfloor$ is proved similarly.

(4) Represent n as a sum of $n = 2^k + \delta$, where $0 \leqslant \delta < 2^k$, $k, \delta \in \mathbb{Z}_0$. Transform the left side of the equality:

$$\left\lceil \log_2(n + 1) \right\rceil = \left\lceil \log_2(2^k + \delta + 1) \right\rceil = \left\lceil \log_2 2^k \left(1 + (\delta + 1)/2^k \right) \right\rceil =$$
$$= \left\lceil k + \log_2 \left(1 + (\delta + 1)/2^k \right) \right\rceil = k + \left\lceil \log_2 \left(1 + (\delta + 1)/2^k \right) \right\rceil.$$

Since $0 < (\delta + 1)/2^k \leqslant 1$, then $0 < \log_2 \left(1 + (\delta + 1)/2^k \right) \leqslant 1$, and the summand $\left\lceil \log_2 \left(1 + (\delta + 1)/2^k \right) \right\rceil$ is equal to one. Then,

$$\left\lceil \log_2(n + 1) \right\rceil = k + 1.$$

Now consider the right side of the equality of item 4):

$$\lfloor \log_2 n \rfloor + 1 = \lfloor \log_2 n + 1 \rfloor = \lfloor \log_2(2n) \rfloor = \lfloor \log_2 2(2^k + \delta) \rfloor =$$
$$= \lfloor \log_2 2^{k+1}(1 + \delta/2^k) \rfloor = \lfloor k + 1 + \log_2(1 + \delta/2^k) \rfloor =$$
$$= k + 1 + \lfloor \log_2(1 + \delta/2^k) \rfloor.$$

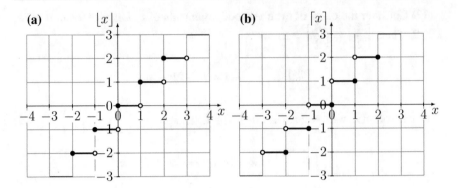

Fig. 3.11 To Exercise **3.82**

Since $0 \leqslant \delta/2^k < 1$, then the inequalities $1 \leqslant 1 + \delta/2^k < 2$ and $0 \leqslant \log_2(1 + \delta/2^k) < 1$ are valid, whence

$$\lfloor \log_2 n \rfloor + 1 = k + 1.$$

We finally obtain: $\lceil \log_2(n + 1) \rceil = \lfloor \log_2 n \rfloor + 1$.

3.80. *Answer*:
   (1) -2;
   (2) 4;
   (3) -6;
   (4) -1.

3.81. *Answer*:
   (1) 1;
   (2) 2;
   (3) 7;
   (4) 13.

3.82. *Solution.*
Graphs of the functions are shown in Fig. 3.11.
   Both functions are surjective, since $\forall n \in \mathbb{Z}$ there always exist real x_1, x_2 such that $\lfloor x_1 \rfloor = n$, $\lceil x_2 \rceil = n$. Functions "floor" and "ceiling" are not injective, since, for example, $\lfloor 1,1 \rfloor = \lfloor 1,2 \rfloor = 1$, $\lceil 1,1 \rceil = \lceil 1,2 \rceil = 2$. Hence, they are not bijective.

3.84. *Solution.*
Let the set A contain all numbers out of the universal set $U = \{1, 2, \ldots, 100\}$, which are divisible by 3. Further, let the set B contain all numbers out of U, which are divisible by 5, and, finally, let the set C contain all numbers out of U, divisible by

Fig. 3.12 To Exercise **3.84**

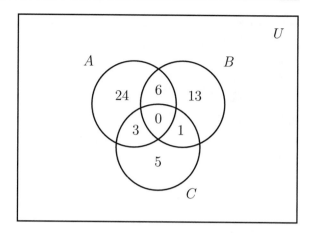

11. It is required to find the cardinality of the set $A \cup B \cup C$. For calculation of $|A \cup B \cup C|$, let us use the inclusion and exclusion formula:

$$|A \cup B \cup C| = |A| + |B| + |C| - |A \cap B| - |A \cap C| - |B \cap C| + |A \cap B \cap C|.$$

We will find the cardinalities of the sets in the right side of the formula applying the result of Exercise **3.83**:

$$|A| = \left\lfloor \frac{100}{3} \right\rfloor = 33, \ |B| = \left\lfloor \frac{100}{5} \right\rfloor = 20, \ |C| = \left\lfloor \frac{100}{11} \right\rfloor = 9.$$

The sets $A \cap B$, $A \cap C$, and $B \cap C$ are formed by numbers from U, divisible by $3 \cdot 5$, $3 \cdot 11$, and $5 \cdot 11$, respectively. Therefore,

$$|A \cap B| = \left\lfloor \frac{100}{15} \right\rfloor = 6, \ |A \cap C| = \left\lfloor \frac{100}{33} \right\rfloor = 3, \ |B \cap C| = \left\lfloor \frac{100}{55} \right\rfloor = 1.$$

The set $A \cap B \cap C$ is empty, since $\left\lfloor \frac{100}{3 \cdot 5 \cdot 11} \right\rfloor = 0$. We obtain the quantity of natural numbers not exceeding 100, which are divisible either by 3, or by 5, or by 11:

$$|A \cup B \cup C| = 33 + 20 + 9 - 6 - 3 - 1 + 0 = 52.$$

Respective Venn diagram is shown in Fig. 3.12.

3.85. *Answer*:

(1) 629 146;

(2) 419 431;

(3) 209 715;

(4) 419 430.

3.86. *Answer*: 582.

3.87. *Answer*: 19.

3.88. *Answer*:

(1) $\lceil x_2 \rceil - \lfloor x_1 \rfloor - 1$;

(2) $\lfloor x_2 \rfloor - \lceil x_1 \rceil + 1$.

3.89. *Answer*:

(1) 367 applicants;

(2) $24 \cdot 12 + 1 = 289$ applicants.

3.90. *Solution*.

The sum of the points will be minimal if all sides show "1" and maximal if they show "6". This is why the set of possible values of the sum of points on the dice $B = \{3, \ldots, 18\}$ contains $|B| = 16$ elements. According to pigeonhole principle, if the dice is thrown 17 times, the sum of the points will be repeated at least twice.

Hence, three dice should be thrown at least 17 times.

3.91. *Solution*.

Let us denote by B the set of English alphabet letters, with which the name of a movie can begin. The cardinality of this set is $|B| = 26$, since any letter can stand at the first position of the movie's name.

Let the function $f : A \to B$ associate the first letter of the movie's name with the movie. The elements of the set A are all movies available at the video rental shop. Then, in accordance with pigeonhole principle, if $|A| > k|B|$, then among all records at least $k + 1$ will begin with the same letter. According to the condition, $k = 3$, therefore $|A| > 3 \cdot 26 = 78$.

Hence, in order to satisfy the condition of the problem, the video rental shop should have no less than 79 movies.

3.92. *Solution*.

Consider the subset A of the set of odd numbers belonging to the set U: $A \subset \{a \in U$ **and** a is odd$\}$. Construct the set B, whose elements are unordered pair of numbers of the form $\{a, b\}$, where $a, b \in \{a \in U$ **and** a is odd$\}$ and $a + b = 20$. The set B contains elements

$$B = \{\{1, 19\}, \{3, 17\}, \{5, 15\}, \{7, 13\}, \{9, 11\}\}.$$

The cardinality of the obtained set is $|B| = 5$.

Let the function $f : A \rightarrow B$ associate the pair $\{a, b\} \in B$ with the odd number $a \in A$, where $b = 20 - a$. In accordance with pigeonhole principle, if $|A| > |B|$, then there exist two numbers $a_1, a_2 \in A$, the sum of which is $a_1 + a_2 = 20$. Therefore, $|A| > 5$, and if we take six numbers out of the set U, then at least two of them will add up to $a_1 + a_2 = 20$.

3.93. *Answer*: Seven numbers.

Combinatorics

4

Combinatorics solves the problems of counting the number of various configurations (e.g., permutations), formed by the elements of finite sets, on which various restrictions can be imposed [4]. The number of configurations formed by the elements of the set is counted in accordance with the sum rule and the product rule [4,27].

The **sum rule** states that *if A_1, A_2,...,A_k are independent events, and there exists n_1 of possible outcomes of the event A_1 **or** A_2 **or** ... **or** A_k possible outcomes of the event A_2,..., n_k possible outcomes of the event A_k, then the possible number of outcomes of the event "A_1 **or** A_2 **or** ... **or** A_k" is equal to the sum $n_1 + n_2 + \cdots + n_k$.*

The **product rule** states that *if we have a sequence of k events with n_1 possible outcomes of the first one, n_2 of the second one, and so on, up to n_k possible outcomes of the last one, then the total number of outcomes of the sequence of k events is equal to the product $n_1 \cdot n_2 \cdot \ldots \cdot n_k$.*

(n, k)-**Sample** or **sample** of volume k **out of** n **elements** is any set of elements $x_1, x_2, \ldots, x_k$ out of the set A of cardinality n. And if the sequence of elements is of the essence, then the sample is called **ordered**, otherwise **unordered**. An ordered (n, k)-sample with repeated elements is called (n, k)-**permutation with repetitions**, and without repeated elements — (n, k)-**permutation without repetitions**.

(n, k)-**Combination with repetitions** is an unordered (n, k)-sample, whose elements can be repeated. Finally, (n, k)-**combination without repetitions** is an unordered (n, k)-sample, whose elements are prohibited to be repeated.

The number of various (n, k)-permutations without repetitions is equal to $P(n, k) = \dfrac{n!}{(n - k)!}$, (n, k)-combinations without repetitions $C(n, k) = \dfrac{n!}{(n - k)!\, k!}$.

The number of various (n, k)-permutations with repetitions is equal to $\widetilde{P}(n, k) = n^k$, (n, k)-combinations with repetitions $\widetilde{C}(n, k) = \dfrac{(n + k - 1)!}{k!\,(n - 1)!}$.

© Springer International Publishing AG, part of Springer Nature 2018
S. Kurgalin and S. Borzunov, *The Discrete Math Workbook*,
Texts in Computer Science, https://doi.org/10.1007/978-3-319-92645-2_4

The total number of all (n, k)-permutations and (n, k)-combinations, both with and without repetitions, is shown in Table 4.1.

Example 4.1 A computer firm employs 25 system administrators, four of which are experienced enough and can perform the functions of the duty shift master. Count the number of options for forming a night shift consisting of two duty system administrators and a shift master. Options when different employees receive the position of the shift master will be considered as different options.
Solution.
Consider the events A_1 "the shift master is chosen" and A_2 "two ordinary system administrators are chosen." The first event can be implemented in four different ways, since one person out of four possible candidates is chosen. After assignment of the shift master, 24 employees are left, two of which should be included in the list of duty administrators. Hence, the event A_2 is a $(24, 2)$-combination without repetitions. The number of ways for implementation of the event A_2 is, according to Table 4.1,

$$C(24, 2) = \frac{24!}{2!\,(24 - 2)!} = \frac{24!}{2!\,22!} = \frac{23 \cdot 24}{2} = 276.$$

According to the product rule, the number of ways for forming the night shift consisting of two duty system administrators and the shift master is $4 \cdot 276 = 1104$. $\square$

Example 4.2 Count the number of bijective functions $f : A \to B$, where $A = \{a_1, a_2, \ldots, a_n\}$, $B = \{b_1, b_2, \ldots, b_m\}$.
Solution.
Consider two functions of the form $f : A \to B$, where $|A| = n$, $|B| = m$. Denote the number of functions f, having the property of bijectivity, by Q.

As is known (see chapter "Relations and Functions" on p. 108), a bijective function performs one-to-one correspondence between the elements of the domain A and the range B. Each element from A must be associated with one and only one element from B.

Let us analyze each of the two cases $m \neq n$ and $m = n$ separately.

(1) Let the following condition be fulfilled $m \neq n$. Cardinalities of the sets A and B are different, and the bijectivity property cannot be fulfilled. Hence, the number of bijective functions in this case is $Q = 0$.

(2) Let $m = n$, and the cardinality of the sets A and B coincides. An arbitrary function $f : A \to B$ is completely defined by the set of its values on the domain. $f(a_1)$ can take one of n possible options of values. Further, for the value $f(a_2)$ there remain $(n - 1)$ options, since the element $b_1 = f(a_1)$ cannot be chosen again. For $f(a_3)$, we have $(n - 2)$ possible options, etc. Such a listing of values of the function f is (n, n)-permutation without repetitions $P(n, n)$. According to Table 4.1, the equality $P(n, n) = n!$ is fulfilled.

Table 4.1 Permutations and combinations

	The elements are repeated		The elements are not repeated	
The sequence is of the essence	Permutations with repetitions	n^k	Permutations without repetitions	$\dfrac{n!}{(n-k)!}$
The order is not of the essence	Combinations with repetitions	$\dfrac{(n+k-1)!}{k!\,(n-1)!}$	Combinations without repetitions	$\dfrac{n!}{(n-k)!\,k!}$

As a result, the number of bijective functions $f : A \to B$, where $|A| = n$, $|B| = m$, can be written in the form

$$Q = \begin{cases} n!, & m = n, \\ 0, & m \neq n. \end{cases}$$

Note that it is convenient to represent the answer in a more compact form with the help of the Kronecker[1] symbol δ_{ij}, depending on two integer arguments i and j: $\delta_{ij} = \begin{cases} 1, & \text{if } i = j, \\ 0, & \text{if } i \neq j. \end{cases}$ Note that the Kronecker symbol is often used for simplification of formulae notations in discrete mathematics, linear algebra, and other divisions of mathematics.

We finally obtain that the number of bijective functions f, acting from the set A of n elements to the set B of m elements, is $Q = n!\,\delta_{nm}$. □

For all natural values n and real a and b, valid is the formula **Newton[2] binomial**:

$$(a+b)^n = \sum_{k=0}^{n} C(n,k)a^{n-k}b^k =$$

$$= C(n,0)a^n + C(n,1)a^{n-1}b + C(n,2)a^{n-2}b^2 + \cdots + C(n,n)b^n.$$

Therefore, the expressions $C(n,k)$ are also referred to as **binomial coefficients**. The binomial coefficients are often denoted as C_n^k or $\dbinom{n}{k}$.

Generalization of the binomial expansion is the identity

$$(a_1 + a_2 + \cdots + a_m)^n = \sum_{\substack{k_1 \geqslant 0,\,\ldots,\,k_m \geqslant 0 \\ k_1 + \cdots + k_m = n}} \frac{n!}{k_1!\,k_2!\,\ldots\,k_m!} a_1^{k_1} a_2^{k_2} \ldots a_m^{k_m},$$

---

[1] Leopold Kronecker (1823–1891)—German mathematician.

[2] Isaac Newton (1643–1727)—English mathematician, physicist, mechanic, astronomer.

valid for all real a_i, $i = 1, \ldots, m$, and natural n. The sum is taken over all nonnegative integers $k_1, k_2, \ldots, k_m$ such that $k_1 + k_2 + \cdots + k_m = n$. Values $\dfrac{n!}{k_1! k_2! \ldots k_m!}$ are called **polynomial coefficients** and are denoted in one of the following ways: $C(n; k_1, k_2, \ldots, k_m)$, $C_n^{k_1, k_2, \ldots, k_m}$ or $\begin{pmatrix} n \\ k_1, k_2, \ldots, k_m \end{pmatrix}$. It should be noted that for $m = 2$ the polynomial expansion is transformed into formula of Newton binomial.

Permutation of elements of the set $A = \{a_1, a_2, \ldots, a_n\}$ is any ordered set of the length n of elements $a_1, a_2, \ldots, a_n$.

Permutation theorem states that *there exist* $\dfrac{n!}{k_1! k_2! \ldots k_m!}$ *of various permutations of n objects, k_1 of which belong to the first type, k_2 to the second type, and so on, up to k_m objects of the type m.*

4.1 Problems

4.1. Let A be a set of four-digit numbers, whose decimal notation includes only even figures.

(1) How many numbers are there in the set A?
(2) How many numbers from A exceed the value of 5000?

4.2. Palindrome is a string of characters that reads the same both from left to right and from right to left. How many numbers have the property: The decimal notation of the number consists of eleven characters, contains only odd figures, and is a palindrome.

4.3. Calculate the number of palindromes among the bit strings of the length n, where $n > 1$.

4.4. A password to a file consists of seven characters. The first two of them are lowercase Latin letters (total of 26 letters), the next two are figures, and the remaining three can be both figures and lowercase letters. How many different passwords can be written?

4.5. There are seven pairs of earrings, three pairs of clips, one chain, and three pendants in a box. How many different sets of decorations can be assembled if it is permissible to wear either earrings or clips, and only one pendant is worn over the chain?

4.6. How many are the options of distributing nine students for work placement at an enterprise if the cooperation contract is concluded with six of them? There is no limitation on the number of students for work placement at the same enterprise.

4.7. Find the number of square matrices of n rows and n columns with elements from the set

  (1) $\mathbb{B} = \{0, 1\}$.
  (2) $\mathbb{T} = \{-1, 0, 1\}$.

4.8. Find the number of square matrices of n rows and n columns with elements from the set $\mathbb{T} = \{-1, 0, 1\}$, which are

  (1) Symmetric;
  (2) Antisymmetric

  with respect to the main diagonal.

4.9. What is the number of various binary relations on some set A, having the property of

  (1) Symmetry;
  (2) Antisymmetry?

4.10. Find the number of three-digit numbers, whose decimal notation contains at least one figure "5".

4.11. Find the number of four-digit numbers, whose decimal notation contains at least one figure "9".

4.12. Having generalized the results of Exercises **4.10** and **4.11**, find the number of n-digit numbers, whose decimal notation contains at least one of the figures "5" or "9".

4.13. Calculate the following values and explain to which samples they correspond:

  (1) $P(7, 3)$, $P(9, 6)$, and $P(5, 2)$;
  (2) $C(11, 6)$, $C(8, 3)$, and $C(10, 8)$;
  (3) $\widetilde{P}(3, 2)$, $\widetilde{P}(2, 5)$, and $\widetilde{P}(5, 5)$;
  (4) $\widetilde{C}(5, 1)$, $\widetilde{C}(3, 5)$, and $\widetilde{C}(4, 14)$.

4.14. How many are the possibilities for awarding five prize-winning places to 20 participants of the computer program contest?

4.15. A group of students includes 22 people. How many are the ways of electing the body of activists of the group, which should consist of a monitor and his/her deputy?

4.16. A theater is preparing to stage play "A Dog in the Manger" by Lope de Vega. The cast of characters includes 18 people, and only four roles are female. Nine women and 17 men take part in the casting. How many are the ways for assembling the actors?

4.17. The King's residence is guarded by a 100 men strong regiment of musketeers.

(1) How many are the ways of selecting a detachment of ten men for a night watch?

(2) How will the number of selection ways change in comparison with the previous case if d'Artagnan must be in this detachment?

4.18. The King of France is guarded by a 100 men strong regiment of musketeers.

(1) How many are the ways of selecting a detachment of 25 men for accompanying the King in a military expedition?

(2) How will the number of selection ways change in comparison with the previous case if d'Artagnan must be in this detachment?

4.19. A basketball team should be assembled (five people) out of ten professionals and six amateurs.

(1) How many different teams can be assembled?

(2) How many teams will consist of professionals or of amateurs only?

(3) How many basketball teams can be assembled so that the number of professionals and amateurs in them differs by no more than one person?

4.20. You are buying copybooks in a shop that can offer seven different types of copybooks.

(1) How many different sets of 14 copybooks can you buy?

(2) How many buying sets can be assembled if you are limited by three types of copybooks only, but buy 14 copybooks anyway?

4.21. A flower vendor sells tulips of five different colors: purple, cream, gold, amber, and violet. How many different bouquets of different colors can be assembled out of fifteen tulips if the violet flowers should not be included in a bouquet with either amber or gold ones?

4.22. You have decided to spend from 180 to 200 rubles on products in the grocery store. The grocery offers you milk: price of one carton—38 rubles, choose out of seven names; eggs: price of one dozen—52 rubles, choose out of four names; and vegetable mix: price of one package—46 rubles, choose out of six names. How many different purchases can you make within the specified price range?

4.23. You are going to a celebration and decide to buy a present for 800 to 1000 rubles. (No matter if the present is one article or a set, its total price should be within the specified interval.) Roses: price per piece—110 rubles, choose out of eight, but their number should always be odd; books: price per piece—252 rubles, choose out of six names; and DVDs: price per package—640 rubles, choose out of four names. How many different presents can you purchase?

4.24. By how many ways can we record 10 Gb of information to disks if we have 5 DVDs of 4.7 Gb, 3 DVDs of 1.4 Gb, and 3 CDs of 700 Mb?

4.25. For a graduation party, we need to assemble an organizing committee of 4 people, and we should choose as follows: out of 6 people—in 11 "A" class, out of 4 people—in 11 "B" class, out of 5 people—in 11 "C" class, and out of 4 people—in 11 "D" class.

(1) How many various compositions of the organizing committee can be assembled in the specified conditions?
(2) How many various compositions of the organizing committee can be assembled if the pupils of 11 "A" and 11 "C" classes should not be its members at the same time?

4.26. For a graduation party, we need to assemble an organizing committee of 4 people, and we should choose as follows: out of 5 people—in 11 "A" class, out of 7 people—in 11 "B" class, out of 9 people—in 11 "C" class, and out of 4 people—in 11 "D" class. How many various compositions of the organizing committee can be assembled if it should include at least two pupils from 11 "A" class and at least one pupil from 11 "D" class?

4.27. A computer firm employs six engineers, ten programmers, and eight software testers. In order to discuss the New Year bonuses, it was decided to establish a commission of six people. How many are the ways to do it if:

(1) All groups of employees should be represented equally?
(2) At least two engineers should be in the commission?

4.28. A computer firm employs five engineers, eight programmers, and seven software testers. In order to discuss the distribution of the New Year bonuses, it was decided to establish a commission of six people. How many are the ways to do it if it is necessary to elect at least three programmers?

4.29. A computer firm employs 11 engineers, 15 programmers, and 9 software testers. In order to discuss the award of the New Year bonuses, it was decided to establish a commission of eight people. How many are the ways to do it if it is necessary to ensure participation of the representatives of all the three groups of employees?

4.30. In the shooting of a movie about the students' life, wish to participate 3 first-year students, 6 second-year students, 5 third-year students, and 4 fourth-year students. Eight actors should be chosen out of them. How many are the ways to do it if:

(1) Students of all years should participate equally?
(2) At least one second-year student should participate?

4.31. In the shooting of a movie about the students' life, wish to participate 3 first-year students, 6 second-year students, 5 third-year students, and 4 fourth-year students. Eight actors should be chosen out of them. How many are the ways to do it if students of all years should participate, except the first?

4.32. In the shooting of a movie about the students' life, wish to participate 6 first-year students, 7 second-year students, 5 third-year students, and 4 fourth-year students. Eight actors should be chosen out of them. How many are the ways to do it if the first- and the fourth-year students cannot act in the movie simultaneously?

4.33. Find how many solutions in natural numbers the equation $x_1 + x_2 + \cdots + x_n = a$ has, where $n, a \in \mathbb{N}$.

4.34. Find how many solutions in nonnegative integers the equation $x_1 + x_2 + \cdots + x_n = a$ has, where $n, a \in \mathbb{N}$.

4.35. How many integer solutions has the equation $x_1 + x_2 + x_3 + x_4 = 10$ provided that $x_1, x_2 \geqslant 0$, $x_3, x_4 \geqslant 4$?

4.36. How many integer solutions has the equation $x_1 + x_2 + x_3 + x_4 + x_5 = 35$ provided that $x_i \geqslant i$ for all $1 \leqslant i \leqslant 5$?

*4.37. How many integer solutions has the equation $x_1 + x_2 + x_3 = a$ provided that $0 \leqslant x_i \leqslant c$ for $i = 1, 2, 3$ and $a, c \in \mathbb{N}$?

*4.38. Find the number of vectors $\mathbf{x} = (x_1, x_2, x_3, x_4)$ with integer components satisfying the equality $x_1 + x_2 + x_3 + x_4 = 90$ if the following constraints are imposed $0 \leqslant x_1 \leqslant 10$, $10 \leqslant x_2 \leqslant 20$, $20 \leqslant x_3 \leqslant \leqslant 30$, $30 \leqslant x_4 \leqslant 40$.

4.39. Prove the symmetry property of the binomial coefficients:

$$C(n, k) = C(n, n - k), \ 0 \leqslant k \leqslant n,$$

for all natural numbers n.

4.40. Prove **Pascal's**[3] **identity**

$$C(n, k) = C(n - 1, k - 1) + C(n - 1, k),$$

where $0 \leqslant k \leqslant n$, using two methods—algebraic and combinatorial.

4.41. Prove Newton binomial formula by mathematical induction method.

*4.42. Prove the following properties of binomial coefficients for all natural numbers n [60]:

(1) $C(n, k) = \dfrac{n}{k} C(n - 1, k - 1), \ 1 \leqslant k \leqslant n$.

(2) $C(n, k) C(k, m) = C(n, m) C(n - m, k - m)$,
     $0 \leqslant m \leqslant k \leqslant n$.

---

[3]Blaise Pascal (1623–1662)—French mathematician, physicist, and philosopher.

(3) $\sum_{k=0}^{n} C(n, k) = 2^n$.

(4) $\sum_{k=0}^{n} kC(n, k) = n2^{n-1}$.

(5) $\sum_{k=0}^{n} k^2 C(n, k) = n(n + 1)2^{n-2}$.

(6) $\left[\sum_{k=0}^{n} C(n, k)\right]^2 = \sum_{k=0}^{2n} C(2n, k)$.

(7) $C(n + m, r) = \sum_{k=0}^{r} C(m, r - k) C(n, k)$,

$r \geqslant 0, m \geqslant 0, r \leqslant \min(m, n)$ (**Vandermonde's**[4] **identity**).

(8) $\sum_{k=0}^{n} C(n, k)^2 = C(2n, n)$.

4.43. Show that Pascal's identity can be proved by a method different from those presented in the solution of Exercise **4.40**, namely using the relation $(1 + x)^n = (1 + x)^{n-1} + x(1 + x)^{n-1}$.

*4.44. Prove the following connection of binomial coefficients with the Fibonacci numbers [4], determined in Exercise **1.86**:

$$\sum_{i=0}^{\lfloor n/2 \rfloor} C(n - i, i) = F_{n+1},$$

where $n \geqslant 0$, $\lfloor a \rfloor$ is the integral part of the number a; see Exercise **3.79**.

*4.45. Prove the identity, establishing a connection between binomial coefficients and harmonic numbers:

$$\sum_{i=1}^{n} \frac{(-1)^{i+1}}{i} C(n, i) = H_n, \quad n = 1, 2, \ldots$$

*4.46. Having generalized the results of Exercises **1.109** and **1.110**, write the expression for an arbitrary power of the kth Lucas number L_k^n, $n \in \mathbb{N}$.

4.47. Find the coefficient:

(1) For $a^3 b^5$ after removing the brackets in the expression $(a - b)^8$;
(2) For $x^6 yz^2$ after removing the brackets in the expression $(x - 2y + z)^9$;
(3) For xyz^2 after removing the brackets in the expression $(x - y + 2z - 2)^7$.

---

[4]Alexandre–Théophile Vandermonde (1735–1796)—French mathematician and musician.

4.48. Find the coefficient:

(1) For a^4b^5 after removing the brackets in the expression $(a-b)^9$;
(2) For x^4yz^4 after removing the brackets in the expression $(x+y-z)^9$;
(3) For xyz^2 after removing the brackets in the expression $(x-3y+z-2)^6$.

4.49. Find the coefficient:

(1) For a^6b^3 after removing the brackets in the expression $(a+b)^9$;
(2) For x^3yz^3 after removing the brackets in the expression $(x-y+z)^7$;
(3) For x^2yz^3 after removing the brackets in the expression $(2x-y+z-3)^7$.

4.50. Find the coefficient for z^l in the expansion of the expressions:

(1) $\left(\sqrt[3]{z}+\sqrt[6]{z}\right)^7$, $l=2$;
(2) $\left(\sqrt[4]{z^5}+z^{-1}\right)^{10}$, $l=8$;
(3) $\left(\sqrt{z}+\sqrt[3]{z}+\sqrt[5]{z}\right)^{10}$, $l=3$.

4.51. Find the coefficient for z^l in the expansion of the expressions:

(1) $\left(\sqrt{z}+\sqrt[4]{z}\right)^{10}$, $l=3$;
(2) $\left(\sqrt{z^3}+z^{-1}\right)^9$, $l=1$;
(3) $\left(\sqrt{z}+\sqrt[3]{z}+\sqrt[7]{z}\right)^{20}$, $l=7$.

4.52. A triangular array resented in Table 4.2 is called **Pascal's triangle** (or **arithmetical triangle**) [74]. Pascal's triangle is formed as follows: The first row contains one element, the second—two, the third—three, etc.; its vertex and both sides are formed by elements equal to one; any of the rest of the elements is equal to the sum of the two nearest elements in the above row. Denote the array elements by $C_p(n,k)$, where n is the number of a row and k is the number of an element in a row. It is customary to number the elements of Pascal's triangle starting from zero, and this is why $n=0,1,2,\ldots,k=0,1,2,\ldots,n$. Make sure that Pascal's triangle represents binomial coefficients $C(n,k)$, i.e., $C_p(n,k)=C(n,k)$:

$$C(0,0)$$
$$C(1,0)\qquad C(1,1)$$
$$C(2,0)\qquad C(2,1)\qquad C(2,2)$$
$$C(3,0)\qquad C(3,1)\qquad C(3,2)\qquad C(3,3)$$
$$\vdots\qquad\qquad\vdots\qquad\qquad\vdots\qquad\qquad\vdots$$

Table 4.2 Pascal's triangle

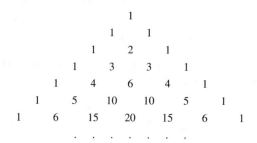

Table 4.3 To Exercise. **4.56**

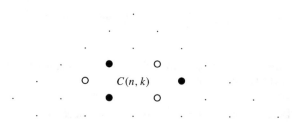

4.53. Find the sum of the elements of the nth row of Pascal's triangle.

4.54. Prove the following three properties of Pascal's triangle:

(1) The triangle is symmetric with respect to the vertical axis passing through its center.
(2) The second and the next to last number in the nth row is equal to number n of this row.
(3) The third number of each row is equal to the sum of the numbers of the preceding rows.

4.55. Let there be given Pascal's triangle, consisting of n rows. What is the sum of all its elements?

4.56. Choose an arbitrary element $C(n, k)$ of Pascal's triangle with indices $1 < k < n, n = 2, 3, \ldots$ Consider six numbers adjacent to $C(n, k)$, as shown in Table 4.3.

Prove that the product of the numbers denoted "●" is equal to the product of the numbers denoted "○" [23].

4.57. Prove that all numbers, except the first and the last, forming the row of Pascal's triangle with number 2^n, where $n = 2, 3, 4, \ldots$, are even.

4.58. Prove that all numbers forming the row of Pascal's triangle with number $2^n - 1$, where $n = 1, 2, \ldots$, are odd.

*4.59. Prove by combinatorial method the inclusion and exclusion formula for an arbitrary number of finite sets:

$$|A_1 \cup A_2 \cup \ldots \cup A_n| = \sum_{i=1}^{n} |A_i| - \sum_{\substack{i,j=1 \\ i<j}}^{n} |A_i \cap A_j| +$$

$$+ \sum_{\substack{i,j,k=1 \\ i<j<k}}^{n} |A_i \cap A_j \cap A_k| - \cdots + (-1)^{n-1}|A_1 \cap A_2 \cap \ldots \cap A_n|.$$

4.60. By "word," we will understand an arbitrary finite sequence of Latin letters. Calculate:

(1) How many different "words" can be obtained from all letters of the word "IMAGINATION".
(2) How many of them begin with letter "T".
(3) In how many of them, letters "N" are placed in a row.

4.61. Calculate:

(1) How many different "words" can be obtained from all letters of the word "INDIVISIBILITY".
(2) How many of them end with letter "Y".
(3) In how many of them, all letters "I" stand side by side.

4.62. Calculate:

(1) How many different "words" can be obtained from all letters of the word "ELECTROENCEPHALOGRAPHY".
(2) How many of them end with letter "G".
(3) In how many of them all, letters "E" stand side by side.

4.2   Answers, Hints, Solutions

4.1. *Answer:*

(1) $4 \cdot 5^3 = 500$.
(2) $2 \cdot 5^3 = 250$.

4.2. *Solution.*
As is known, there exist five odd figures: 1, 3, 5, 7, and 9. Therefore, there exist $5^6 = 15625$ possibilities for the choice of the first six figures of the 11-digit palindrome. Its remaining figures are uniquely determined by the first five. Hence, there exists the total of $15\,625$ 11-digit numbers whose decimal notation contains only odd figures and is a palindrome.

4.3. *Answer:* $2^{\lceil n/2 \rceil}$.

4.4. *Answer:* $26^2 \cdot 10^2 \cdot 36^3$.

4.5. *Solution.*
A woman may wear either earrings or clips; therefore, by the sum rule, she has $7 + 3 = 10$ pairs of decorations for ears. With them, she can assemble $10 \cdot 1 \cdot 4 = 40$ sets of decorations, consisting of a chain, a pendant, and earrings or a chain, a pendant, and clips.

4.6. *Answer:* $\widetilde{C}(6, 9) = 2002$.

4.7. *Answer:*

(1) 2^{n^2};
(2) 3^{n^2}.

4.8. *Answer:*

(1) $3^{n(n+1)/2}$;
(2) $3^{n(n-1)/2}$.

4.9. *Answer:*

(1) $2^{|A|(|A|+1)/2}$;
(2) $2^{|A|} \cdot 3^{|A|(|A|-1)/2}$.

4.10. *Answer:* 252.

4.11. *Answer:* 3168.

4.12. *Solution.*
Denote the sought value by N. Then, calculate the quantity of N^* numbers, whose decimal notation contains either figure "5" or figure "9". Since there exist seven possible options for the high order and eight options for the rest, the total quantity of such numbers according to the product rule is $N^* = 7 \cdot 8^{n-1}$.

It is easy to see that the value $N + N^*$ is equal to the quantity of all possible n-digit numbers, namely $N + N^* = 9 \cdot 10^{n-1}$. Therefore, the quantity of n-digit numbers, whose decimal notation contains at least one of the numbers "5" or "9", is $N = 9 \cdot 10^{n-1} - N^* = 9 \cdot 10^{n-1} - 7 \cdot 8^{n-1}$.

4.13. *Answer:*

(1) $P(7, 3) = 210$, $P(9, 6) = 60480$, $P(5, 2) = 20$;
(2) $C(11, 6) = 462$, $C(8, 3) = 56$, $C(10, 8) = 45$.
(3) $\widetilde{P}(3, 2) = 9$, $\widetilde{P}(2, 5) = 32$, $\widetilde{P}(5, 5) = 3125$.

(4) $\widetilde{C}(5, 1) = 5$, $\widetilde{C}(3, 5) = 21$, $\widetilde{C}(4, 14) = 680$.

4.14. *Solution.*
When choosing the prize winners, the sequence is of the essence, and repetitions are prohibited. Therefore, there exist

$$P(20, 5) = \frac{20!}{(20 - 5)!} = \frac{20!}{15!} = 1\,860\,480$$

possibilities for awarding prize-winning places.

4.15. *Answer:* $C(22, 2)$.

4.16. *Solution.*
We can choose actresses in

$$C(9, 4) = \frac{9!}{(9 - 4)!\,4!} = \frac{9!}{5!\,4!} = 126$$

ways and the actors in

$$C(17, 14) = \frac{17!}{(17 - 14)!\,14!} = \frac{17!}{3!\,14!} = 680$$

ways. Thus, by the product rule we find that there exist $126 \cdot 680 = 85\,680$ different casts of characters.

4.17. *Solution.*

(1) The sought number is equal to the combinations without repetitions of ten elements out of hundred:

$$C(100, 10) = \frac{100!}{10!\,90!}.$$

(2) After d'Artagnan is included into the detachment, there will be 99 possible candidates for 9 vacancies. This is why the number of ways in the second case will be equal to the number of combinations without repetitions of nine elements out of 99 or $C(99, 9)$.

The number of the choice options will decrease in comparison with the previous case by

$$\frac{C(100, 10)}{C(99, 9)} = \frac{100!}{10!\,90!} \cdot \frac{9!\,90!}{99!} = 10 \text{ times.}$$

4.18. *Answer:*

(1) $C(100, 25) = \dfrac{100!}{75!\,25!}.$
(2) It will decrease by four times.

Table 4.4 To Exercise **4.22**. Price of various purchase options

A	B	C	Purchase price (rubles)
5	0	0	190
4	0	1	198
2	2	0	180
1	3	0	194
1	2	1	188
1	1	2	182
0	2	2	196
0	1	3	190
0	0	4	184

4.19. *Answer:*

(1) $C(16, 5)$;
(2) $C(10, 5) + C(6, 5)$;
(3) $C(10, 2)C(6, 3) + C(10, 3)C(6, 2)$.

4.21. *Solution.*
Each bouquet is an unordered sample of fifteen tulips with repetitions of possible colors. Violet tulips can only be included in a bouquet with purple or cream ones; i.e., there exist $\widetilde{C}(3, 15) = \dfrac{(3 + 15 - 1)!}{15!\,(3 - 1)!} = \dfrac{17!}{15!\,2!} = \dfrac{16 \cdot 17}{2} = 136$ possibilities for assembling a bouquet of tulips of these three colors. It is also possible to assemble bouquets including tulips of four colors (without violet ones): $\widetilde{C}(4, 15) = \dfrac{(4 + 15 - 1)!}{15!\,(4 - 1)!} = \dfrac{18!}{15!\,3!} = \dfrac{16 \cdot 17 \cdot 18}{6} = 816$. Therefore, by the sum rule, we can assemble $136 + 816 = 952$ bouquets. But is should be noted that in such counting the bouquets consisting of only cream and purple tulips have been taken into account twice. Their number is $\widetilde{C}(2, 15) = \dfrac{(2 + 15 - 1)!}{15!\,(2 - 1)!} = \dfrac{16!}{15!\,1!} = 16$.

As a result, we obtain the final answer: There exist $952 - 16 = 936$ possibilities for assembling a bouquet.

4.22. *Solution.*
Introduce notations: A—milk, B—eggs, C—vegetable mix. In Table 4.4 are shown the possible options for selection of the products.

Therefore, you can make

$$C(7, 5) + C(7, 4) \cdot C(6, 1) + C(7, 2) \cdot C(4, 2) + C(7, 1) \cdot C(4, 3) +$$
$$+ C(7, 1) \cdot C(4, 2) \cdot C(6, 1) + C(7, 1) \cdot C(4, 1) \cdot C(6, 2) +$$
$$+ C(4, 2) \cdot C(6, 2) + C(4, 1) \cdot C(6, 3) + C(6, 4) = 1242$$

different purchases.

4.25. *Answer:*

(1) $C(19,4) = 3876$.
(2) $C(13,4) + C(14,4) - C(8,4) = 1646$.

4.26. *Answer:*

$$C(5,3)C(4,1) + C(5,2)C(7,1)C(4,1) + C(5,2)C(9,1)C(4,1) +$$
$$+ C(5,2)C(4,2) = 740.$$

4.27. *Solution.*

(1) Since all groups of employees should be represented equally, it is necessary to include exactly two people from each group. Denote the event "two engineers are elected" by A_1, event "two programmers are elected" by A_2, and event "two testers are elected" by A_3.

The number of outcomes of the event A_1 is equal to the number of $(6,2)$-combinations without repetitions, i.e., the binomial coefficient $C(6,2)$. The events A_2 and A_3 are satisfied by $(10,2)$-combinations without repetitions and $(8,2)$-combinations without repetitions, respectively.

In order to calculate the total number of outcomes of the sequence of the events A_1, A_2, A_3, the product rule should be applied:

$C(6,2)C(10,2)C(8,2) = 18900$.

(2) Introduce the following events into the consideration:

A_1—"two engineers and four other specialists are elected";
A_2—"three engineers and three other specialists are elected";
A_3—"four engineers and two other specialists are elected";
A_4—"five engineers and one other specialist are elected";
A_5—"all the six elected employees are engineers."

It is easy to see that the events A_1–A_5 form all possible ways of choosing six employees in such a manner that at least two engineers will be elected. According to the sum rule, we obtain the final answer:

$$C(6,2)C(18,4) + C(6,3)C(18,3) + C(6,4)C(18,2) +$$
$$+ C(6,5)C(18,1) + C(6,6) = 64624.$$

Note that it is possible to use another solution path and consider, for example, the following events:

B—"six employees are elected";
B_1—"six employees are elected from among programmers or testers",
B_2—"one engineer and five other employees are elected."

In this case, the answer can be presented in a more compact equivalent form:
$C(24,6) - (C(18,6) + C(6,1)C(18,5))$.

4.28. *Answer:*

$$C(8,6) + C(8,5)C(12,1) + C(8,4)C(12,2) +$$
$$+ C(8,3)C(12,3) = 17640.$$

4.29. *Answer:*

$$C(35, 8) - (C(24, 8) + C(20, 8) + C(26, 8)) +$$
$$+ (C(9, 8) + C(15, 8) + C(11, 8)) = 21\,118\,713.$$

4.30. *Answer:*

(1) $C(3, 2)C(6, 2)C(5, 2)C(4, 2) = 2700$;
(2) $C(18, 8) - C(12, 8) = 43263$.

4.33. *Solution.*

Arrange a units in a row, dividing it into n parts by symbols "$\downarrow$". A typical scheme of such a division has the form:

$$\overbrace{1111 \downarrow\, 111 \downarrow 11 \ldots 11 \downarrow 111111}^{a \text{ units, } n-1 \text{ symbol ``}\downarrow\text{''}}$$
$$\underbrace{}_{x_1}\ \underbrace{}_{x_2} \qquad\qquad \underbrace{}_{x_n}$$

As a result, we have obtained some solution of the original equation. Since the number of possible places for the symbols "$\downarrow$" is $a - 1$, then there exists a total of $C(a - 1, n - 1)$ ways of partitioning the number a into n positive summands.

4.34. *Solution.*

Add value n to both parts of the equality:

$$(x_1 + 1) + (x_2 + 1) + \cdots + (x_n + 1) = a + n.$$

Replacement of variables $\tilde{x}_i = x_i + 1$ for all $i = 1, 2, \ldots, n$ results in the equation $\tilde{x}_1 + \tilde{x}_2 + \cdots + \tilde{x}_n = a + n$, while $\forall i\ x_i \geqslant 1$, the number of whose solution, according to the result of the previous exercise, is $C(a + n - 1, n - 1)$. It is clear that the number of solutions of the original equation in nonnegative integers is $C(a + n - 1, n - 1)$, or $C(a + n - 1, a)$.

4.35. *Answer:* 10.

4.36. *Answer:* $C(24, 4) = 10626$.

4.37. *Solution.*

Consider the predicate

$$P(x_1, x_2, x_3) = \{x_1 + x_2 + x_3 = a \textbf{ and } x_1, x_2, x_3 \in \mathbb{Z}_0\}$$

and introduce notations for the following sets:

$$U = \{(x_1, x_2, x_3) \colon P(x_1, x_2, x_3)\},$$
$$A_i = \{(x_1, x_2, x_3) \colon P(x_1, x_2, x_3) \textbf{ and } x_i > c\}, \text{ where } i = 1, 2, 3.$$

In terms of the set theory, the problem reduces to finding the value $|U \setminus (A_1 \cup A_2 \cup A_3)|$. Since $A_1 \cup A_2 \cup A_3 \subset U$, then $|U \setminus (A_1 \cup A_2 \cup A_3)| = |U| - |A_1 \cup A_2 \cup A_3|$. Then, calculate cardinalities of each of the sets U and $A = A_1 \cup A_2 \cup A_3$.

The set U is formed by the solutions of the equation $x_1 + x_2 + x_3 = a$ in non-negative integers. In accordance with the result of Exercise **4.34**, $|U| = C(a + 2, 2)$ for all $a \in \mathbb{N}$.

For calculation of $|A_1 \cup A_2 \cup A_3|$, use the inclusion and exclusion formula:

$$|A| = |A_1| + |A_2| + |A_3| - |A_1 \cap A_2| - |A_1 \cap A_3| - |A_2 \cap A_3| + |A_1 \cap A_2 \cap A_3|.$$

The set A_1 is formed by the solutions of the equation $x_1 + x_2 + x_3 = a$ provided that $x_1 > c, x_1 \in \mathbb{N}, x_2, x_3 \in \mathbb{Z}_0$. Replace the variables

$$x_1 = \tilde{x}_1 + c, \quad x_2 = \tilde{x}_2 - 1, \quad x_3 = \tilde{x}_3 - 1.$$

The equation with the new variables $\tilde{x}_1 + \tilde{x}_2 + \tilde{x}_3 = a - c + 2$, $\tilde{x}_i \geqslant 1$ for $i = 1, 2, 3$, has $C(a - c + 1, 2)$ solutions if $a - c + 1 \geqslant 2$; otherwise, the set of its solutions is empty. Hence,

$$|A_1| = \begin{cases} C(a - c + 1, 2), & \text{if } c \leqslant a - 1, \\ 0, & \text{if } c > a - 1. \end{cases}$$

In view of symmetry of the problem with respect to rearrangement of variables, we have $|A_1| = |A_2| = |A_3|$.

Then, consider the set $A_1 \cap A_2$. Its elements are solutions of the equation $x_1 + x_2 + x_3 = a$ provided that $x_1, x_2 > c, x_1, x_2 \in \mathbb{N}, x_3 \in \mathbb{Z}_0$. We obtain

$$|A_1 \cap A_2| = \begin{cases} C(a - 2c, 2), & \text{if } c \leqslant \left\lfloor \dfrac{a}{2} \right\rfloor - 1, \\ 0, & \text{if } c > \left\lceil \dfrac{a}{2} \right\rceil - 1. \end{cases}$$

The symmetry properties of the problem allow writing the following equalities: $|A_1 \cap A_2| = |A_1 \cap A_3| = |A_2 \cap A_3|$.

Finally, the last summand $|A_1 \cap A_2 \cap A_3|$ is $C(a - 3c - 1, 2)$, if $c \leqslant \left\lfloor \dfrac{a}{3} \right\rfloor - 1$, and zero otherwise. We finally obtain the number of integer solutions of the original equation in the form of a piecewise-defined function:

$$|U \setminus A| = \begin{cases} \dfrac{1}{2}(a + 1)(a + 2), & \text{if } c > a - 1, \\ \dfrac{1}{2}((a + 1)(a + 2) - 3(a - c)(a - c + 1)), & \text{if } \left\lfloor \dfrac{a}{2} \right\rfloor - 1 < c \leqslant a - 1, \\ \dfrac{1}{2}(3c + 2 - a)(3c + 1 - a), & \text{if } \left\lfloor \dfrac{a}{3} \right\rfloor - 1 < c \leqslant \left\lfloor \dfrac{a}{2} \right\rfloor - 1, \\ 0, & \text{if } c \leqslant \left\lfloor \dfrac{a}{3} \right\rfloor - 1. \end{cases}$$

Fig. 4.1 Variants of spatial arrangement of the plane $x_1 + x_2 + x_3 = a$ and the cube $0 \leqslant x_1, x_2, x_3 \leqslant c$

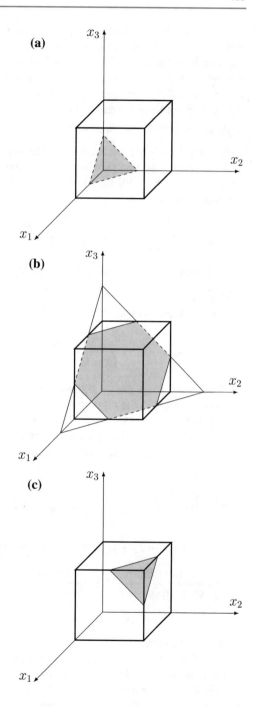

Note. The structure of the function $f(a, c) = |U \setminus A|$ becomes clear from geometrically. Actually, from the point of view of analytic geometry, the sought value is equal to the number of points lying at the intersection of the plane $x_1 + x_2 + x_3 = a$ and the cube $0 \leqslant x_1, x_2, x_3 \leqslant c$ and having integer coordinates. Intersection of these objects for different relations between the values a and c may represent: equilateral triangle—first line of the formula for $f(a, c)$, hexagon—second line, equilateral triangle again—third line. Moreover, the plane and the cube for $c \leqslant \left\lfloor \dfrac{a}{3} \right\rfloor - 1$ have no common points. Thus, partitioning of the domain $\mathbb{Z}_0^2$ of the function $f(a, c)$ into four sets reflects the four possible cases of spatial arrangement of the said geometric objects (see Fig. 4.1). We can also show that for $c = \text{const}$ the maximal value of the function $f(a, c)$ is

$$f_{\max} = f\left(\left\lfloor \frac{3c}{2} \right\rfloor, c\right) = f\left(\left\lceil \frac{3c}{2} \right\rceil, c\right) = \frac{1}{2}\left((c + 1)(c + 2) + 2\left\lfloor \frac{c}{2} \right\rfloor \left\lceil \frac{c}{2} \right\rceil\right).$$

4.38. *Answer:* $C(33, 3) - 4C(22, 3) + 6C(11, 3) = 286.$

4.39. *Proof.*
Use the definitive formula for binomial coefficient in the right side of the equality:

$$C(n, n - k) = \frac{n!}{(n - (n - k))!\,(n - k)!} = \frac{n!}{k!\,(n - k)!} = C(n, k).$$

Note. Apart from purely algebraic proofs, many identities allow combinatorial proving. For proving the suggested identity, we can reason as follows. It is known that $C(n, k)$ defines the number of methods by which we can choose k different elements from the n element set. Each such method uniquely defines another sample — $(n - k)$ elements from the n element set. Therefore, $C(n, n - k) = C(n, k)$.

4.40. *Proof.*
(1) Algebraic proof.
Transform the sum in the right side of the equality

$$C(n - 1, k - 1) + C(n - 1, k) = \frac{(n - 1)!}{(n - k)!\,(k - 1)!} + \frac{(n - 1)!}{(n - 1 - k)!\,k!} =$$

$$= \frac{(n - 1)!}{(n - k)!\,(k - 1)!} + \frac{(n - 1)!(n - k)}{(n - k)!\,k!} = \frac{(n - 1)!}{(n - k)!\,k!}(k + (n - k)) =$$

$$= C(n, k).$$

(2) Combinatorial proof.
The value $C(n, k)$ defines the number of methods for choosing k different elements of the n element set (e.g., the set A). Fix one of the elements of the set $t \in A$. The number of k element subsets of the set A, not containing t, is $C(n - 1, k)$. Then, the number of k element subsets A, containing t, is $C(n - 1, k - 1)$, since for forming such sets, to the originally chosen element t another $k - 1$ elements

should be added. The total number of k element subsets A is equal to the sum $C(n-1, k-1) + C(n-1, k)$, which proves Pascal's identity.

4.41. *Hint.* Use Pascal's identity to justify the induction transition.

4.42. *Proof.*

(1) Having presented $n!$ and $k!$ in the form $n! = n(n-1)!$, $k! = k(k-1)!$, by the definitive formula for $C(n, k)$, we obtain:

$$C(n, k) = \frac{n!}{(n-k)! \, k!} = \frac{n(n-1)!}{(n-k)! \, k \, (k-1)!} = \frac{n}{k} C(n-1, k-1).$$

(2) By the definitive formula for binomial coefficients, we have:

$$C(n, m) \, C(n-m, k-m) = \frac{n!}{(n-m)! \, m!} \cdot \frac{(n-m)!}{(n-k)! \, (k-m)!} =$$

$$= \frac{n!}{(n-k)! \, k!} \cdot \frac{k!}{(k-m)! \, m!} = C(n, k) \, C(k, m).$$

(3) Consider the expression $(1+1)^n$. Expand it by Newton binomial formula.

$$(1+1)^n = \sum_{k=0}^{n} C(n, k) 1^{n-k} 1^k = \sum_{k=0}^{n} C(n, k).$$

But, on the other hand, $(1+1)^n \equiv 2^n$. Comparing the obtained expressions, we see that the identity $\sum\limits_{k=0}^{n} C(n, k) = 2^n$ is valid.

(4) For proving the formula, consider the auxiliary function $f(x) = (1+x)^n$. Expansion of $f(x)$ into series in powers of the variable x has the following form:

$$f(x) = (1+x)^n = \sum_{k=0}^{n} C(n, k) 1^{n-k} x^k = \sum_{k=0}^{n} C(n, k) x^k.$$

Differentiate the obtained equality with respect to x.

$$f'(x) = n(1+x)^{n-1} = \sum_{k=0}^{n} C(n, k) \cdot k x^{k-1}.$$

Substituting $x = 1$ to the last equality, we obtain the identity

$$n2^{n-1} = \sum_{k=0}^{n} k C(n, k).$$

(5) Differentiation of the equality $(1 + x)^n = \sum_{k=0}^{n} C(n, k)x^k$ with respect to the variable x twice and substitution of $x = 1$ results in

$$n(n - 1)2^{n-2} = \sum_{k=0}^{n} k(k - 1)C(n, k).$$

Taking into account item (4) of this exercise, we obtain

$$n(n - 1)2^{n-2} = \sum_{k=0}^{n} k^2 C(n, k) - n2^{n-1},$$

whence it follows that $\sum_{k=0}^{n} k^2 C(n, k) = n(n + 1)2^{n-2}$.

(6) Expanding the right and the left sides of the equality $(1 + 1)^{2n} = [(1 + 1)^n]^2$, we obtain the identity from the condition.

(7) Consider the relation $(1 + x)^{m+n} = (1 + x)^m (1 + x)^n$. Expand the binomials into series, and transform the sums via introduction of a new variable $r = k_1 + k_2$ in the right side:

$$\sum_{r=0}^{m+n} C(m + n, r)x^r = \sum_{k_1=0}^{m} C(m, k_1)x^{k_1} \sum_{k_2=0}^{n} C(n, k_2)x^{k_2},$$

$$\sum_{r=0}^{m+n} C(m + n, r)x^r = \sum_{r=0}^{m+n} \sum_{k_2=0}^{r} C(m, r - k_2)C(n, k_2)x^r;$$

By equating the coefficients at factors x^r and replacing $k_2 \rightarrow k$, we obtain

$$C(m + n, r) = \sum_{k=0}^{r} C(m, r - k)C(n, k).$$

(8) The equality is a special case of Vandermonde's identity proved in the previous item.

4.44. *Proof.*
Denote the sum in the right side of the equality

$$a_{n+1} \equiv \sum_{i=0}^{\lfloor n/2 \rfloor} C(n - i, i), \quad n \geqslant 0.$$

Prove that the sequence $\{a_n\}$, $n = 1, 2, 3, \ldots$, has the definitive properties of the Fibonacci sequence

$$F_n = 1, 1, 2, 3, 5, 8, 13, 21, 34, 55, 89, 144, \ldots$$

For this, consider the first two terms $\{a_n\}$:

$$a_1 = \sum_{i=0}^{\lfloor 0/2 \rfloor} C(0-i,i) = C(0,0) = 1, \quad a_2 = \sum_{i=0}^{\lfloor 1/2 \rfloor} C(1-i,i) = C(1,0) = 1.$$

So, the first two terms $\{a_n\}$ coincide with the respective terms of the Fibonacci sequence: $a_1 = F_1, a_2 = F_2$.

Then, consider the sum of the successive terms $a_n + a_{n+1}$ and express it by a_{n+2}:

$$a_n + a_{n+1} = \sum_{i=0}^{\lfloor n/2 \rfloor} C(n-i,i) + \sum_{i=0}^{\lfloor (n+1)/2 \rfloor} C(n+1-i,i).$$

Separately, consider two cases—of even and odd values of n.
(1) Let n be even. Use the obvious equalities

$$\lfloor n/2 \rfloor = n/2, \quad \lfloor (n+1)/2 \rfloor = n/2,$$

valid for all even values of n. Then,

$$a_n + a_{n+1} = \sum_{i=0}^{n/2} C(n-i,i) + \sum_{i=0}^{n/2} C(n+1-i,i).$$

In the first sum, replace the variable $i \rightarrow i-1$ (note the change in the limits of summation!), and in the second sum explicitly isolate the term with $i = 0$:

$$a_n + a_{n+1} =$$

$$= \sum_{i=1}^{n/2+1} C(n+1-i,i-1) + \left(C(n+1,0) + \sum_{i=1}^{n/2} C(n+1-i,i) \right) =$$

$$= \left(\sum_{i=1}^{n/2} C(n+1-i,i-1) + C(n/2,n/2) \right) +$$

$$+ \left(C(n+1,0) + \sum_{i=1}^{n/2} C(n+1-i,i) \right) =$$

$$= C(n/2,n/2) + \sum_{i=1}^{n/2} \left(C(n+1-i,i-1) + C(n+1-i,i) \right) + C(n+1,0).$$

Transform the expression under the summation sign subject to Pascal's identity

$$C(n, k) = C(n - 1, k - 1) + C(n - 1, k),$$

proved in Exercise **4.40**. Taking into account the value of the binomial coefficient $C(n/2, n/2) = 1 = C(n/2 + 1, n/2 + 1)$, we obtain

$$a_n + a_{n+1} = C(n + 1, 0) + \sum_{i=1}^{n/2} C(n + 2 - i, i) + C(n/2 + 1, n/2 + 1) =$$

$$= \sum_{i=0}^{n/2+1} C(n + 2 - i, i) = \sum_{i=0}^{\lfloor (n+2)/2 \rfloor} C(n + 2 - i, i) \equiv a_{n+2}.$$

(2) Now let n be odd. Then,

$$a_n + a_{n+1} = \sum_{i=0}^{(n-1)/2} C(n - i, i) + \sum_{i=0}^{(n+1)/2} C(n + 1 - i, i).$$

In the first sum, replace $i \to i - 1$, and in the second sum explicitly isolate the term with $i = 0$:

$$a_n + a_{n+1} =$$

$$= \sum_{i=1}^{(n+1)/2} C(n + 1 - i, i - 1) + \left(C(n + 1, 0) + \sum_{i=1}^{(n+1)/2} C(n + 1 - i, i) \right) =$$

$$= C(n + 1, 0) + \sum_{i=1}^{(n+1)/2} \left(C(n + 1 - i, i - 1) + C(n + 1 - i, i) \right) =$$

$$= C(n + 2, 0) + \sum_{i=1}^{(n+1)/2} C(n + 2 - i, i) = \sum_{i=0}^{\lfloor (n+2)/2 \rfloor} C(n + 2 - i, i) \equiv a_{n+2}.$$

To sum it up, we conclude that for any $n \geqslant 3$ each term of the sequence $\{a_n\}$ is equal to the sum of the two previous, so that $a_n \equiv F_n$.

4.45. *Proof.*

Use the mathematical induction method. Denote by $P(n)$ the predicate $\sum_{i=1}^{n} \dfrac{(-1)^{i+1}}{i}$ $C(n, i) = H_n$.

  B a s i s s t e p

For $n = 1$, we have that $\dfrac{(-1)^2}{1} C(1, 1) = H_1$ is a valid equality.

Inductive step

Suppose that $P(k)$ takes the true value for some natural $k \in \mathbb{N}$. Prove the truth of $P(k+1)$:

$$\sum_{i=1}^{k+1} \frac{(-1)^{i+1}}{i} C(k+1, i) = \sum_{i=1}^{k} \frac{(-1)^{i+1}}{i} C(k+1, i) + \frac{(-1)^{k+2}}{k+1} =$$

$$= \sum_{i=1}^{k} \frac{(-1)^{i+1}}{i} (C(k, i) + C(k, i-1)) + \frac{(-1)^{k+2}}{k+1} =$$

$$= \sum_{i=1}^{k} \frac{(-1)^{i+1}}{i} C(k, i) + \sum_{i=1}^{k} \frac{(-1)^{i+1}}{i} C(k, i-1) + \frac{(-1)^{k+2}}{k+1}.$$

The total of the first summand is H_k according to the inductive supposition. Rewrite the second and the third summands in the form of a single sum

$$\sum_{i=1}^{k+1} \frac{(-1)^{i+1}}{i} C(k+1, i) = H_k + \sum_{i=1}^{k+1} \frac{(-1)^{i+1}}{i} C(k, i-1).$$

Then, use the expression of binomial coefficient by factorials of the arguments:

$$\sum_{i=1}^{k+1} \frac{(-1)^{i+1}}{i} C(k+1, i) = H_k + \sum_{i=1}^{k+1} (-1)^{i+1} \frac{k!}{i \cdot (i-1)! \, (k-i+1)!} =$$

$$= H_k + \sum_{i=1}^{k+1} \frac{(-1)^{i+1}}{k+1} \frac{(k+1)!}{i! \, (k-i+1)!} = H_k + \frac{1}{k+1} \sum_{i=1}^{k+1} (-1)^{i+1} C(k+1, i) =$$

$$= H_k + \frac{1}{k+1} \left[\sum_{i=0}^{k+1} (-1)^{i+1} C(k+1, i) - (-1) C(k+1, 0) \right] =$$

$$= H_k + \frac{1}{k+1} \left[1 - \sum_{i=0}^{k+1} (-1)^{i} C(k+1, i) \right] = H_k + \frac{1}{k+1} \left[1 - (1-1)^{k+1} \right] =$$

$$= H_k + \frac{1}{k+1} = H_{k+1}.$$

Thus, $\displaystyle\sum_{i=1}^{n} \frac{(-1)^{i+1}}{i} C(n, i) = H_n$ for all $n = 1, 2, \ldots$

4.46. *Solution.*

As is known, $L_k = \varphi^k + \widetilde{\varphi}^k$, where $\varphi = (1 + \sqrt{5})/2$ is the golden ration, $\varphi \widetilde{\varphi} = -1$. Expand L_k^n by Newton binomial formula, considering the cases of odd and even values of n.

1. Let n be odd. Then,

$$L_k^n = (\varphi^k + \widetilde{\varphi}^k)^n = \varphi^{nk} + C(n, 1)\varphi^{(n-1)k}\widetilde{\varphi}^k + C(n, 2)\varphi^{(n-2)k}\widetilde{\varphi}^{2k} + \cdots +$$

$$+ C\left(n, \frac{n-1}{2}\right)\varphi^{\frac{n+1}{2}k}\widetilde{\varphi}^{\frac{n-1}{2}k} + C\left(n, \frac{n+1}{2}\right)\varphi^{\frac{n-1}{2}k}\widetilde{\varphi}^{\frac{n+1}{2}k} + \cdots +$$

$$+ C(n, n-2)\varphi^{2k}\widetilde{\varphi}^{(n-2)k} + C(n, n-1)\varphi^k\widetilde{\varphi}^{(n-1)k} + \widetilde{\varphi}^{nk} =$$

$$= (\varphi^{nk} + \widetilde{\varphi}^{nk}) + C(n, 1)(\varphi\widetilde{\varphi})^k(\varphi^{(n-2)k} + \widetilde{\varphi}^{(n-2)k}) +$$

$$+ C(n, 2)(\varphi\widetilde{\varphi})^{2k}(\varphi^{(n-4)k} + \widetilde{\varphi}^{(n-4)k}) + \cdots + C\left(n, \frac{n-1}{2}\right)(\varphi\widetilde{\varphi})^{\frac{n-1}{2}k}(\varphi^k + \widetilde{\varphi}^k).$$

Then, use the properties $\varphi\widetilde{\varphi} = -1$, $\varphi^k + \widetilde{\varphi}^k = L_k$:

$$L_k^n = L_{nk} + C(n, 1)(-1)^k L_{(n-2)k} + C(n, 2)(-1)^{2k} L_{(n-4)k} + \cdots +$$

$$+ C\left(n, \frac{n-1}{2}\right)(-1)^{\frac{n-1}{2}k} L_1.$$

Consider that in the last summand $L_1 = 1$, and we finally obtain

$$L_k^n = L_{nk} + n(-1)^k L_{(n-2)k} + \frac{n(n-1)}{2}(-1)^{2k} L_{(n-4)k} + \cdots + C\left(n, \frac{n-1}{2}\right)(-1)^{\frac{n-1}{2}k}.$$

2. Let n be even. Then,

$$L_k^n = (\varphi^k + \widetilde{\varphi}^k)^n = \varphi^{nk} + C(n, 1)\varphi^{(n-1)k}\widetilde{\varphi}^k + C(n, 2)\varphi^{(n-2)k}\widetilde{\varphi}^{2k} + \cdots +$$

$$+ C(n, n/2 - 1)\varphi^{(n/2+1)k}\widetilde{\varphi}^{(n/2-1)k} + C(n, n/2)\varphi^{n/2k}\widetilde{\varphi}^{n/2k} +$$

$$+ C(n, n/2 + 1)\varphi^{(n/2-1)k}\widetilde{\varphi}^{(n/2+1)k} + \cdots + C(n, n-2)\varphi^{2k}\widetilde{\varphi}^{(n-2)k} +$$

$$+ C(n, n-1)\varphi^k\widetilde{\varphi}^{(n-1)k} + \widetilde{\varphi}^{nk}.$$

Taking into account the relations $\varphi\widetilde{\varphi} = -1$, $\varphi^k + \widetilde{\varphi}^k = L_k$, simplify the right side:

$$L_k^n = L_{nk} + C(n, 1)(-1)^k L_{(n-2)k} + C(n, 2)(-1)^{2k} L_{(n-4)k} + \cdots +$$

$$+ C(n, n/2 - 1)(-1)^{(n/2-1)k} L_{2k} + C(n, n/2)(-1)^{nk/2}.$$

As a result, we finally obtain

$$L_k^n = L_{nk} + n(-1)^k L_{(n-2)k} + \frac{n(n-1)}{2} L_{(n-4)k} + \cdots + C(n, n/2)(-1)^{nk/2}.$$

4.47. *Solution.*

(1) Expand $(a - b)^8$ in powers of a and b. The sought coefficient subject to sign is equal to $-C(8, 5) = -56$.

(2) The coefficient for x^6yz^2, obtained after removing the brackets in the expression $(x - 2y + z)^9$, is $-2\dfrac{9!}{6!\,1!\,2!} = -504$.

(3) The coefficient for abc^2d^3, obtained after removing the brackets in the expression $(a + b + c + d)^7$, is $\dfrac{7!}{1!\,1!\,2!\,3!} = 420$.

In other words, the seventh power of the sum $a + b + c + d$ contains the summand $420\,abc^2d^3$. We assume that $a = x$, $b = -y$, $c = 2z$, and $d = -2$.

Then, after removing the brackets in the expression $(x - y + 2z - 2)^7$, we obtain $420\,x(-y)(2z)^2(-2)^3 = 13\,440\,xyz^2$. Therefore, the coefficient for xyz^2 in the expression $(x - y + 2z - 2)^7$ is $13\,440$.

4.48. *Answer:*

(1) -126;
(2) 630;
(3) -2160.

4.49. *Answer:*

(1) 84;
(2) -140;
(3) 5040.

4.50. *Solution.*

(1) In the expansion of the expressions $\left(\sqrt[3]{z} + \sqrt[6]{z}\right)^7$, there are summands of the form $C(7, k)(\sqrt[3]{z})^{7-k}\left(\sqrt[6]{z}\right)^k$, and the index of power $l = \dfrac{7-k}{3} + \dfrac{k}{6}$ is two for $k = 2$. Therefore, the coefficient for z^2 in the expansion $\left(\sqrt[3]{z} + \sqrt[6]{z}\right)^7$ is $C(7, 2) = \dfrac{7!}{5!\,2!} = 21$.

(2) In the expansion of the expressions $(\sqrt[4]{z^5} + z^{-1})^{10}$, there are summands of the form $C(10, k)(\sqrt[4]{z^5})^{10-k}z^{-k}$, and the index of power $l = \dfrac{5}{4}(10 - k) - k$ is eight for $k = 2$. The coefficient for z^8 is $C(10, 2) = 45$.

(3) In the expansion of the expressions $\left(\sqrt{z} + \sqrt[3]{z} + \sqrt[5]{z}\right)^{10}$, there are summands of the form

$$\frac{10!}{k_1!\,k_2!\,k_3!}\left(\sqrt{z}\right)^{k_1}\left(\sqrt[3]{z}\right)^{k_2}\left(\sqrt[5]{z}\right)^{k_3}, \quad k_1 + k_2 + k_3 = 10,$$

and the index of power $l = \dfrac{k_1}{2} + \dfrac{k_2}{3} + \dfrac{k_3}{5}$ is equal to three for $k_1 = 2$, $k_2 = 3$, $k_3 = 5$. Therefore, the coefficient for z^3 is $\dfrac{10!}{2!\,3!\,5!} = 2520$.

4.51. *Answer:*

(1) $C(10, 8) = 45$;
(2) $C(9, 5) = 126$;

(3) $\dfrac{20!}{10!\,3!\,7!} = 22\,170\,720.$

4.53. *Solution.*

Write out the elements that form the nth row of Pascal's triangle: $C(n, 0)$, $C(n, 1), \ldots, C(n, n)$. The sought sum is

$$C(n, 0) + C(n, 1) + \cdots + C(n, n) = \sum_{i=0}^{n} C(n, i).$$

According to the property of binomial coefficients stated in item (3) of Exercise **4.42**, the value of this sum is 2^n.

4.54. *Proof.*

(1) Symmetry of Pascal's triangle with respect to the vertical axis passing through its center follows from the easy-to-check identity $C(n, k) = C(n, n - k)$.

(2) The second number in the nth row is $C(n, 1) = \dfrac{n!}{(n-1)!\,1!} = n$. The next to last number in the nth row is $C(n, n-1) = \dfrac{n!}{1!\,(n-1)!} = C(n, 1)$.

(3) Write the explicit expression for the third number of the nth row:

$$C(n, 2) = \dfrac{n!}{(n-2)!\,2!} = \dfrac{n(n-1)}{2}.$$

On the other hand, the sum of the numbers of the preceding rows is also equal to $1 + 2 + \cdots + (n-1) = \dfrac{n(n-1)}{2}$, which actually proves the third property.

4.55. *Solution.*

The sum of all numbers included in Pascal's triangle is determined by the expression

$$\sum_{k=0}^{n} \sum_{i=0}^{k} C(k, i) = \sum_{k=0}^{n} 2^k = 2^{n+1} - 1.$$

4.56. *Proof.*

The elements of Pascal's triangle, adjacent to $C(n, k)$, are binomial coefficients with indices that differ from n and k by one:

$$C(n-1, k-1), \ C(n, k+1), \qquad C(n+1, k), \quad \text{(they are denoted by ``\textbullet''),}$$
$$C(n-1, k), \qquad C(n+1, k+1), \ C(n, k-1), \quad \text{(they are denoted by ``o'').}$$

Therefore, the statement of the problem reduces to the algebraic identity

$$C(n-1, k-1)C(n, k+1)C(n+1, k) =$$
$$= C(n-1, k)C(n+1, k+1)C(n, k-1).$$

After rearrangement of the sequence of cofactors, we obtain the required result:

$$C(n-1, k-1)C(n, k+1)C(n+1, k) =$$

$$= \frac{(n-1)!}{(k-1)!\,(n-k)!} \cdot \frac{n!}{(k+1)!\,(n-k-1)!} \cdot \frac{(n+1)!}{k!\,(n+1-k)!} =$$

$$= \frac{(n-1)!}{k!\,(n-k-1)!} \cdot \frac{(n+1)!}{(k+1)!\,(n-k)!} \cdot \frac{n!}{(k-1)!\,(n-k+1)!} =$$

$$= C(n-1, k)C(n+1, k+1)C(n, k-1).$$

4.57. *Proof.*

The row of Pascal's triangle with number 2^n is formed by the values $C(2^n, 0) = 1, C(2^n, 1), \ldots, C(2^n, i), \ldots, C(2^n, 2^n - 1), C(2^n, 2^n) = 1$.   Prove   that   all $C(2^n, i)$, where $i = 1, \ldots, 2^n - 1$, are even numbers. Let us prove by the mathematical induction method.

Denote the predicate "$C(2^n, i)$ are even for $i = 1, \ldots, 2^n - 1$" by $P(n)$.

B a s i s   s t e p

For the minimal possible value of $n = 1$, we obtain that $C(2^n, i) = C(2, i)$ is even for $i = 1$; therefore, $P(1)$ is true.

I n d u c t i v e   s t e p

Suppose that $C(2^k, i)$ is even, and consider the values $C(2^{k+1}, i) = C(2 \cdot 2^k, i)$. Next, we need an auxiliary identity

$$C(2n, i) = \sum_{j=0}^{n} C(n, j)C(n, i-j).$$

This identity follows from $(1+x)^{2n} = (1+x)^n(1+x)^n$, where coefficients are equated for x^i in the right and the left sides of the formula:

$$C(2 \cdot 2^k, i) = \sum_{j=0}^{2^k} C(2^k, j)C(2^k, i-j).$$

According to the inductive supposition, $C(2^k, i)$ are even. The sum of even numbers is an even number; therefore, all values $C(2^{k+1}, i)$ are even for $i = 1, \ldots, 2^k - 1$.

Then, for establishing the truth of the predicate $P(k+1)$ in the case $i = 2^k + 1, \ldots, 2^{k+1} - 1$ it is easy to use the property of symmetry of the binomial coefficients, provided in Exercise **4.39**:

$$C(2 \cdot 2^k, i) = C(2 \cdot 2^k, 2 \cdot 2^k - i).$$

Therefore, $C(2^{k+1}, i)$ are also even for $i = 2^k + 1, \ldots, 2^{k+1} - 1$.

The only value of i remains that should be checked, namely $i = 2^k$:

$$C(2^{k+1}, 2^k) = \sum_{i=0}^{2^k} C(2^k, i) C(2^k, 2^k - i) = \sum_{i=0}^{2^k} C(2^k, i)^2 = 1 + \sum_{i=1}^{2^k-1} C(2^k, i)^2 + 1.$$

It is easy to see that the obtained expression is even; hence, for all values of $i = 1, \ldots, 2 \cdot 2^k - 1$ the values $C(2^{k+1}, i)$ are even.

4.58. *Proof.*
According to the result of the previous exercise, $C(2^n, i)$ are even for $i = 2, \ldots, 2^n - 1$. Consider a part of Pascal's triangle, consisting of rows with numbers $2^n - 1$ and 2^n:

$$
\begin{array}{ccccccc}
 & \cdot & & \cdot & & \cdot & \cdots \\
 & 1 & C(2^n - 1, 1) & C(2^n - 1, 2) & C(2^n - 1, 3) & \cdots \\
1 & C(2^n, 1) & C(2^n, 2) & C(2^n, 3) & C(2^n, 4) & \cdots \\
\cdot & \cdot & \cdot & \cdot & \cdot & \cdots
\end{array}
$$

According to the definition of Pascal's triangle, an arbitrary element is equal to the sum of all the above elements.

Since $C(2^n, 1) = 1 + C(2^n - 1, 1)$ and $C(2^n, 1)$ are even, then the values $C(2^n - 1, 1)$ are odd. Then, since $C(2^n, 2) = C(2^n - 1, 1) + C(2^n - 1, 2)$, and the oddness of $C(2^n - 1, 1)$ is proved, then $C(2^n - 1, 2)$ is odd. Continuing in the same manner, we obtain that all binomial coefficients in the row of Pascal's triangle with number $2^n - 1$ are odd.

4.59. *Proof.*
It is required to prove that an arbitrary element of the union of sets $A_1 \cup A_2 \cup \ldots \cup A_n$ will be calculated once and only once in the right side of the formula. Let the element x belong to the sets $A_1, A_2, \ldots, A_k$. Then, x appears $C(k, 1)$ times in the sum $|A_1| + |A_2| + \cdots + |A_n|$, $C(k, 2)$ times in the sum $\sum_{\substack{i,j=1 \\ i \neq j}}^{k} |A_i \cap A_j|$, etc. In the sum

$$\sum_{i_1, i_2, \ldots, i_m = 1}^{k} |A_{i_1} \cap A_{i_2} \cap \ldots \cap A_{i_m}|$$

the element x appears $C(k, m)$ times. Therefore, we obtain that the number of occurrences of an arbitrary x will be

$$C(k, 1) - C(k, 2) + \cdots + (-1)^{k-1} C(k, k) = 1.$$

4.60. *Solution.*

(1) The suggested word consists of eleven letters. The number of permutations of eleven different objects is $P(11, 11) = 11!$. But, if we transpose the same letters,

we do not obtain a new "word". The number of permutations of three letters "I" is
3!, of two "A" is 2!, and of two "N" is 2!. We obtain that we can compose a total of
$\frac{11!}{3!\,2!\,2!}$ "words."

(2) Fix letter "T" at the first position. After that, we obtain that it is required to
find the number of permutations of ten letters that form the word "IMAGINATION"
without the letter "T". The sought number is $\frac{10!}{3!\,2!\,2!}$.

(3) Let us consider the combination "NN" to be one letter. Then, it is easy to find
the number of different "words" having the said property: $\frac{10!}{3!\,2!}$.

4.61. *Answer:*

(1) $\frac{14!}{6!}$;   (2) $\frac{13!}{6!}$;   (3) 9!.

4.62. *Answer:*

(1) $\frac{22!}{(2!)^7\,4!}$;   (2) $\frac{21!}{(2!)^7\,4!}$;   (3) $\frac{19!}{(2!)^7}$.

Graphs

Graph is a pair $G = (V, E)$, where V is a set of vertices (nodes), and E is a set of edges (arcs), connecting some pairs of vertices (nodes) [15,28,55,84]. A graph is said to be **simple** if it contains no **loops** (edges beginning and ending at the same vertex) and **multiple edges** (several edges are called multiple if they connect the same pair of vertices), and the sets V and E are finite. The figure where the graph vertices are shown as points, and edges are shown as segments or arcs, is called a **diagram of a graph**.

Two vertices u and v of a graph are **adjacent** is they are connected by an edge $r = uv$. In this case, it is said that the vertices u and v are **endpoints** of the edge r. If the vertex v is the endpoint of the edge r, then v and r are **incident** (from Latin *incēdere*—to advance). **Degree** $d(v)$ of vertex v is the number of edges incident with this vertex.

Denote the minimal degree of vertices of the graph $G(V, E)$ by $\delta(G)$, maximal—by $\Delta(G)$:

$$\delta(G(V, E)) = \min_{v \in V} d(v), \quad \Delta(G(V, E)) = \max_{v \in V} d(v).$$

If the degrees of all vertices are equal to α, then such a graph is called a **regular graph of degree** α.

Adjacency matrix M is a logical matrix of a relation on a set of vertices of a simple graph $G(V, E)$, which is specified by its edges. A adjacency matrix has a size $|V| \times |V|$, and its elements are determined by the rule

$$M(i, j) = \begin{cases} T, & \text{if edge } ij \in E, \\ F, & \text{if edge } ij \notin E. \end{cases}$$

Incidence matrix $\widetilde{M}$ of the simple graph $G(V, E)$ is a logical matrix of size $|V| \times |E|$, that reflects incidence of vertices and edges. Elements of the matrix $\widetilde{M}$

© Springer International Publishing AG, part of Springer Nature 2018
S. Kurgalin and S. Borzunov, *The Discrete Math Workbook*,
Texts in Computer Science, https://doi.org/10.1007/978-3-319-92645-2_5

are determined by the rule

$$\widetilde{M}(i, j) = \begin{cases} \text{T,} & \text{if vertex } v_i \text{ is incident with edge } e_j, \\ \text{F,} & \text{otherwise.} \end{cases}$$

Adjacency list of a simple graph consists of a set of vertices $\{v_1, \ldots, v_n\}$ and sets that for each vertex v_i $(1 \leqslant i \leqslant n)$ contain all vertices adjacent to v_i.

Example 5.1 Consider a graph $G(V, E)$, whose set of vertices and set of edges are specified as follows:

$$V = \{a, b, c, d, e\}, \quad E = \{ab, ac, ae, bc, bd, cd\}.$$

The graph $G(V, E)$ is shown in Fig. 5.1.

The adjacency matrix M and the incidence matrix $\widetilde{M}$ of the graph G have the form:

$$M = \begin{array}{c} \\ a \\ b \\ c \\ d \\ e \end{array} \begin{array}{ccccc} a & b & c & d & e \\ \left[\begin{array}{ccccc} F & T & T & F & T \\ T & F & T & T & F \\ T & T & F & T & F \\ F & T & T & F & F \\ T & F & F & F & F \end{array}\right] \end{array}, \quad \widetilde{M} = \begin{array}{c} \\ a \\ b \\ c \\ d \\ e \end{array} \begin{array}{cccccc} ab & ac & ae & bc & bd & cd \\ \left[\begin{array}{cccccc} T & T & T & F & F & F \\ T & F & F & T & T & F \\ F & T & F & T & F & T \\ F & F & F & F & T & T \\ F & F & T & F & F & F \end{array}\right] \end{array}.$$

Adjacency list of the graph $G(V, E)$:

vertices of graph	adjacent vertices
a	b, c, e
b	a, c, d
c	a, b, d
d	b, c
e	$a.$

□

Subgraph of the graph $G = (V, E)$ is the graph $G' = (V', E')$, where $V' \subseteq V$ and $E' \subseteq E$ and for each edge $r \in E'$ both its ends belong to V'.

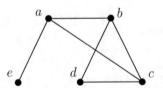

Fig. 5.1 Graph $G(V, E)$ of Example 5.1

Route of length k in the graph G is a sequence of vertices v_0, v_1, ..., v_k such that $\forall i = 1, \ldots, k$ of vertex v_{i-1} and v_i are adjacent. There are also considered **trivial** routes of the form v_i, v_i.

By the **length** of route, we will understand the number of edges in it, with regard to repetitions.

Cycle in the graph G is a closed route v_0, v_1, ..., v_0, where all vertices, except the first and the last one, are different.

A graph is **acyclic** if it contains no cycles.

A graph is referred to as **connected** if each pair of vertices is connected by some route.

Connectivity number of the graph $c(G)$ is the minimal number of connected subgraphs of the graph G. Such connected subgraphs are referred to as **connectivity components**.

Connected acyclic graph $G = (V, E)$ is called a **tree**. Acyclic graph is also called a **forest** because the connectivity components of a forest are trees.

Let $G = (V, E)$ be a graph with n vertices and m edges; then, we can state some necessary and sufficient conditions, for which G is a tree:

- any pair of vertices G is connected with the only route;
- G is connected and $m = n - 1$;
- G is connected, and removal of at least one edge breaks this property;
- G is acyclic, but if at least one edge is added, then a cycle appears.

It is often convenient to single out one of the vertices of the tree and call it a **root**.

Depth of vertex v of a tree with a root is a length of route from the root to this vertex.

Depth of tree is defined as the maximal depth of its vertices.

Vertices of the same depth form a **level** of the tree.

A vertex of the tree whose degree $d = 1$ is called a **leaf**, and the vertex of a degree greater than the first, but not a root, is called an **internal vertex**.

Spanning tree of the graph G is such its subgraph that is a tree and contains all vertices of the graph G.

In a **binary tree**, for each vertex v_i ($1 \leqslant i \leqslant |V|$), the number of vertices adjacent to it and having a depth greater by one than v_i, is no greater than two.

For a **complete binary tree**, the number of vertices on each level is maximal and equal to two.

A graph is referred to as **Eulerian**[1] graph if it contains a route beginning and ending at the same vertex and passing through each edge of the graph exactly once. Such a rout is called **Euler** route (path).

A connected graph with two or more vertices is an Eulerian graph if and only if each of its vertices has an even degree. The last statement is the first result discovered in graph theory [17].

---

[1]Leonhard Euler (1707–1783)—Swiss mathematician.

A graph is referred to as **Hamiltonian**[2] graph if it contains Hamilton cycle (circuit), i.e., a cycle passing through each vertex of the graph, and only once.

Let us introduce one more definition important for graph theory. Two graphs $G_1 = (V_1, E_1)$ and $G_2 = (V_2, E_2)$ are called **isomorphic** ($G_1 \sim G_2$) if there exists a one-to-one mapping (bijection) of vertices and edges of one of these graphs to vertices and edges of another graph respectively, for which the incidence relation is preserved.

Isomorphic graphs have the same properties and are usually not distinguished.

In scheduling problems and many other practical problems, of great importance are **chromatic** characteristics of graphs [15].

Vertex k-coloring is a function mapping a set of vertices V onto the set $\{1, 2, \ldots, k\}$ (a set of "colors"):

$$\varphi \colon V \to \{1, 2, \ldots, k\}.$$

The coloring is **regular** if $\varphi(u) \neq \varphi(v)$ for any adjacent vertices u and v, i.e., no two adjacent vertices obtain the same colors.

If it is possible to construct a regular k-coloring for a given graph, the graph is called k-**colorable**.

The minimal number k for which the graph G is k-colored is called the **chromatic number** of this graph $\chi(G)$.

A graph is **bichromatic** [55] if $\chi(G) = 2$.

Edge k-coloring is a function mapping a set of edges E onto a set of "colors" $\{1, 2, \ldots, k\}$:

$$\psi \colon E \to \{1, 2, \ldots, k\}.$$

The edge coloring is **regular** if the edges incident with one vertex obtain different colors.

The minimal number k for which the graph G has a regular edge coloring is called a **chromatic index** $\chi'(G)$.

Handshaking lemma states that *the sum of degrees of vertices of an arbitrary graph $G = (V, E)$ is double the number of its edges*:

$$\sum_{v \in V} d(v) = 2|E|.$$

5.1 Directed Graphs

Directed graph or **digraph** is a pair $D = (V, E)$, where V is a finite set of vertices, and E is a relation on V. Elements of the set E are called **directed edges** or **arcs**. An arc that connects a pair (u, v) of vertices u and v of the digraph D is denoted by

---

[2]William Rowan Hamilton (1805–1865)—Irish mathematician and physicist.

uv. A **simple** digraph contains no loops (i.e., acrs of the form uu) or multiple arcs. If $uv \in E$, then u is called an **antecedent** v (from Latin *antecēdere*) [4,8].

Path of length n in an digraph is a sequence if various vertices $v_0, v_1, \ldots, v_n$, each pair $v_{i-1}v_i$ $(i = 1, \ldots, n)$ of which forms an arc. A sequence of vertices $v_0, v_1, \ldots, v_n$, where the first vertex v_0 coincides with the last $v_0 = v_n$, and there are no other repeated vertices, forms a **circuit**. **Circuit-free** digraph contains no circuits.

Out-degree of a vertex v of the digraph D is the number of arcs $d^+(v)$ of the digraph, outgoing from v, and **in-degree** of this vertex is the number of arcs $d^-(v)$ ingoing in it.

For directed graphs, the notion of connectivity becomes more ambiguous than for undirected ones. **Connected** is an digraph from which a connected graph is obtained after abandonment of the arcs' direction. A directed graph is called **strongly connected**, if for any pair of vertices $u, v \in V$ there exists a path from u to v.

Reachability matrix M^* of the digraph $D(V, E)$ is a logical matrix of closure by transitivity of relation E. A reachability matrix stores the information about the existence of paths between the digraph vertices: at the intersection of the ith row and the jth column stands "T" if and only if there exists a path from vertex v_i to v_j. M^* can be calculated by the formula [27]

$$M^* = M \text{ or } M^2 \text{ or } \ldots \text{ or } M^n,$$

where n is the number of vertices of the directed graph, $n = |V|$. However, determination of elements of M^* by the said formula implies considerable calculation, this is why, for digraphs with a great number of vertices, **Warshall**[3] algorithm is used. The Warshall algorithm is discussed in chapter "Concept of Algorithm. Correctness of Algorithms" on p. 330.

5.2 Problems

5.1. Prove that in any nontrivial graph G (i.e., in a graph with a number of vertices $|V| > 1$) there exist two vertices of the same degree.

5.2. At the embassy reception, some guests shook hands. Prove that the number of the guests that shook an odd number of hands is even.

5.3. Write out the adjacency matrix, the incidence matrix, and the adjacency list of the graph $G(V, E)$, consisting of a set of vertices $V = \{a, b, c, d, e\}$ and a set of edges $E = \{ac, bd, be, de\}$. Draw the graph $G(V, E)$.

5.4. Write out the adjacency matrix and the adjacency list of the graph $G(V, E)$, consisting of a set of vertices $V = \{a, b, c, d, e, f, g\}$ and a set of edges $E = \{ab, bc, cd, de, ac, ce, ef, fc, ag, cg\}$. Draw the graph $G(V, E)$.

---

[3]Stephen Warshall (1935–2006)—American researcher in the field of computer sciences.

Fig. 5.2 To Exercise **5.5**

5.5. Draw the graph G, whose adjacency matrix has the form:
$$\begin{bmatrix} F & T & F & T & F & T \\ T & F & F & T & T & F \\ F & F & F & F & T & T \\ T & T & F & F & F & F \\ F & T & T & F & F & F \\ T & F & T & F & F & F \end{bmatrix}.$$
Which of the graphs shown in Fig. 5.2 can be subgraphs of the graph G?

5.6. Draw the graph G, whose adjacency matrix has the form:
$$\begin{bmatrix} F & T & T & T & F & F \\ T & F & F & F & F & T \\ T & F & F & F & F & T \\ T & F & F & F & T & F \\ F & F & F & T & F & T \\ F & T & T & F & T & F \end{bmatrix}.$$
Which of the graphs in Fig. 5.2 can be subgraphs of the graph G?

5.7. What properties does the adjacency matrix of a simple graph have?

5.8. Are adjacency matrix and the adjacency list of the simple graph uniquely defined?

$*$ **5.9.** Show that the number of various routes of length l from the graph vertex with number i to the vertex with number j is equal to the element of the matrix A_{ij}^l, where A is the matrix similar to the adjacency matrix of the graph:

$$A_{ij} = \begin{cases} 1, & \text{if the edge } ij \in E; \\ 0, & \text{if the edge } ij \notin E, \end{cases}$$

and l is the index of degree A [28,43].

5.10. Find the Hamilton cycle in the graph G in Fig. 5.3. Determine in it the cycles of length 3 and 4.

5.11. Are the graphs G_1 and G_2 in Fig. 5.4 Hamiltonian? Eulerian?

5.12. In a **complete** graph, any pair of vertices is connected with an edge. Denote a complete graph with n vertices by K_n. Is the complete graph K_n Eulerian? Hamiltonian? Consider the cases

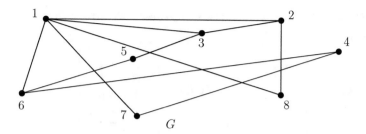

Fig. 5.3 To Exercise **5.10**

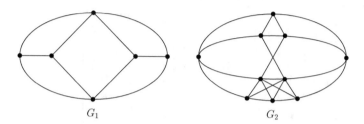

Fig. 5.4 To Exercise **5.11**

(1) of odd n;
(2) of even n.

5.13. Can a complete graph have 20, 30, or 40 edges?

5.14. Show that in the complete graph K_n for $n > 2$, there exists $\dfrac{1}{2}(n-1)!$ various Hamilton cycles.

5.15. Give an example, if possible, of a graph that simultaneously is:

(1) Eulerian and Hamiltonian;
(2) Eulerian and Non-Hamiltonian;
(3) Non-Eulerian and Hamiltonian;
(4) Non-Eulerian and Non-Hamiltonian.

∗ **5.16.** Prove that the connectivity relation

$$\mathcal{C} = \{(v_1, v_2) \in V^2 : \text{ a route exists from } v_1 \text{ to } v_2\}$$

is the equivalence relation on the set of vertices of the graph. What are different equivalence classes associated with?

∗ **5.17.** Find the isomorphism relation $\mathcal{I}$ on the set of all graphs U as follows: $G_1 \, \mathcal{I} \, G_2$ and only if the graphs $G_1 = (V_1, E_1)$ and $G_2 = (V_2, E_2)$ are isomorphic. Show that $\mathcal{I}$ is the equivalence relation [26].

5.18. Show that the graphs G_1 and G_2 will be isomorphic, specified by the following adjacency matrices:

$$M(G_1) = \begin{bmatrix} F & T & F & T & F & T \\ T & F & T & F & T & F \\ F & T & F & T & F & T \\ T & F & T & F & T & F \\ F & T & F & T & F & T \\ T & F & T & F & T & F \end{bmatrix}, \quad M(G_2) = \begin{bmatrix} F & F & F & T & T & T \\ F & F & F & T & T & T \\ F & F & F & T & T & T \\ T & T & T & F & F & F \\ T & T & T & F & F & F \\ T & T & T & F & F & F \end{bmatrix}.$$

5.19. Find whether the graphs $G_1 = (V_1, E_1)$ and $G_2 = (V_2, E_2)$ are isomorphic, being specified by the adjacency matrices:

$$M(G_1) = \begin{bmatrix} F & T & F & F & F & T & T \\ T & F & T & T & T & F & T \\ F & T & F & F & T & F & F \\ F & T & F & F & T & F & F \\ F & T & T & T & F & T & T \\ T & F & F & F & T & F & T \\ T & T & F & F & T & T & F \end{bmatrix}, \quad M(G_2) = \begin{bmatrix} F & T & T & T & T & T & F \\ T & F & T & F & F & F & F \\ T & T & F & T & T & F & T \\ T & F & T & F & F & T & T \\ T & F & T & F & F & F & F \\ T & F & F & T & F & F & T \\ F & F & T & T & F & T & F \end{bmatrix}.$$

5.20. Find the chromatic number χ:
(1) of the complete graph K_n;
(2) of the graph obtained from the complete n-vertex graph by removing one edge.

5.21. What are the chromatic number χ and the chromatic index χ' of **the Petersen**[4] **graph** P_{10} (Fig. 5.5)?

5.22. Prove the following estimate for the chromatic number:

$$\chi(G) \leqslant \Delta(G) + 1.$$

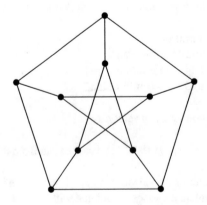

Fig. 5.5 Petersen's graph

---

[4]Julius Peter Christian Petersen (1839–1910)—Danish mathematician.

5.23. Vizing's[5] theorem states that *for the chromatic index of an arbitrary simple graph $G(V, E)$ the estimate is valid* [15]

$$\Delta(G) \leqslant \chi'(G) \leqslant \Delta(G) + 1.$$

The graphs with $\chi'(G) = \Delta(G)$ are called **graphs of class 1**, and the rest belong to **class 2**. Check Vizing's theorem for the Petersen graph and determine which class it belongs to.

5.24. The graph which is a cycle of n edges for $n \geqslant 3$ is called **n-cycle** and is denoted as C_n [26]. Check Vizing's theorem for C_n and determine which class does n-cycle belong to.

5.25. Check Vizing's theorem for the complete graph K_n and determine which class it belongs to.

5.26. In a private school, seven lessons are scheduled for Saturday: mathematics, literature, rhetoric, nature, physical education, drawing, and solfeggio. Each lesson takes exactly one hour. Some lessons can be conducted in parallel, and others, for example, those conducted by the same teacher, should be conducted in series. In Table 5.1 with "×" are marked the lessons that cannot be conducted simultaneously.

Find the minimal time required to conduct all the seven lessons on Saturday.

5.27. A database server processes the queries from the users. At some point of time, the server received nine queries $Z_1, Z_2, \ldots, Z_9$. Simultaneous processing of some of them is possible; such pairs of queries are marked in Table 5.2 with

Table 5.1 To Exercise **5.26**. Signs "×" mark the pairs of school lessons that cannot be conducted simultaneously

Subject	Mathematics	Literature	Rhetoric	Nature	Physical education	Drawing	Solfeggio
Mathematics				×	×		
Literature				×			×
Rhetoric					×	×	×
Nature	×	×					
Physical education	×		×			×	×
Drawing			×		×		×
Solfeggio		×	×		×	×	

---

[5]Vadim Georgievich Vizing) (1937–2017)—Ukrainian mathematician.

Table 5.2 To Exercise **5.27**. Sings "×" mark the pairs of queries that cannot be processes simultaneously

Query	Z_1	Z_2	Z_3	Z_4	Z_5	Z_6	Z_7	Z_8	Z_9
Z_1				×				×	×
Z_2			×	×			×		
Z_3		×					×	×	
Z_4	×	×			×		×		
Z_5				×		×	×		
Z_6					×			×	×
Z_7		×	×	×	×				
Z_8	×		×			×			×
Z_9	×					×		×	

signs "×". Supposing that processing of each query takes one second exactly, find the minimal time for processing of all nine queries.

5.28. Show the trees, specified by the following adjacency matrices:

(1) $\begin{bmatrix} F & T & F & F \\ T & F & T & F \\ F & T & F & T \\ F & F & T & F \end{bmatrix}$;

(2) $\begin{bmatrix} F & T & T & T \\ T & F & F & F \\ T & F & F & F \\ T & F & F & F \end{bmatrix}$.

5.29. Find whether the graphs specified by the following adjacency matrices are trees:

(1) $\begin{bmatrix} F & F & T & F & F & F \\ F & F & T & T & F & T \\ T & T & F & F & T & F \\ F & T & F & F & F & F \\ F & F & T & F & F & F \\ F & T & F & F & F & F \end{bmatrix}$;

(2) $\begin{bmatrix} F & F & F & T & T & F \\ F & F & T & T & F & T \\ F & T & F & T & F & F \\ T & T & T & F & F & F \\ T & F & F & F & F & F \\ F & T & F & F & F & F \end{bmatrix}$;

$$(3) \begin{bmatrix} \text{F T F T T F} \\ \text{T F F F T F} \\ \text{F F F T T T} \\ \text{T F T F F F} \\ \text{T T T F F F} \\ \text{F F T F F F} \end{bmatrix}.$$

5.30. Construct all trees with the number of vertices $n \leqslant 7$.

5.31. Relying upon the pigeonhole principle, prove that if the tree T has more than one vertex, then it has at least two vertices of the same degree.

5.32. It is known that the tree T has one vertex of degree 3, six vertices of degree 2, and seven vertices of degree 1. The rest of the vertices are of degree 4. How many vertices of degree 4 does the tree T have?

5.33. It is known that the tree T has two vertices of degree 4, four vertices of degree 3 and ten vertices of degree 2. The rest of the tree's vertices are of degree 1. How many vertices of degree 1 does the tree T have?

5.34. It is known that the tree T has two vertices of degree 4, four vertices of degree 3 and ten vertices of degree 1. The rest of the tree's vertices are of degree 2. How many vertices of degree 2 does the tree T have?

5.35. It is known that the tree T has one vertex of degree 4, seven vertices of degree 2 and six vertices of degree 1. The rest of the tree's vertices are of degree 3. How many vertices of degree 3 does the tree T have?

5.36. How many connected components has the forest containing:

(1) 12 vertices and 10 edges;
(2) n vertices and m edges $(m < n)$?

5.37. Prove that the chromatic number of the nontrivial tree is $\chi(T) = 2$.

5.38. Prove that the chromatic number of the forest F satisfies the inequality $\chi(F) \leqslant 2$.

5.39. Draw the following binary trees:

(1) complete binary tree with a root of depth 3;
(2) with a root of depth 2, where the number of leaves is odd and greater by two than the number of the internal vertices;
(3) the tree with a root of depth 4; the number of vertices of the tree is 6, and the root degree is equal to 2.

∗ **5.40.** Prove that a binary tree with N vertices has a depth no less than $\lfloor \log_2 N \rfloor$ ($\lfloor a \rfloor$ means the integer part of the number a, see Exercise **3.79**).

5.41. **Flat** graph is a graph positioned in a plane without intersection of edges. If a given graph can be positioned like this in a plane, it is called **planar** (from Latin *plānārius*—plain) [28]. Are the following graphs G_1 and G_2 planar?

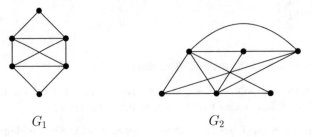

$$G_1 \qquad\qquad\qquad\qquad G_2$$

5.42. Let us call the area in a flat graph, limited by edges and containing no vertices and edges inside it, a **face**. Check, that the number of vertices n, the number of edges m and the number of faces r for the graphs G_1 and G_2 from Exercise **5.41** satisfy the relation $n - m + r = 2$ (Euler's formula).

5.43. Let in the connected planar graph $G(V, E)$, the number of vertices be $|V| = n$, the number of edges be $|E| = m$, and the number of faces be r. Prove the **Euler's formula** [28]

$$n - m + r = 2.$$

∗ **5.44.** Show that in the arbitrary simple planar graph, there exists a vertex of a degree no more than $d \leqslant 5$.

∗ **5.45.** Prove the **five color theorem**: *any planar graph can be colored with five colors* [84].

5.46. **Hypercube** Q_d of dimension d, where $d \in \mathbb{Z}_0$, contains 2^d vertices and is constructed as follows. The vertices of the graph Q_d are assigned bit strings $\mathbf{b} = (b_1, \ldots, b_t, \ldots, b_d)$, where $\forall t\ b_t \in \{0, 1\}$. Then, vertices v_i and v_j are adjacent, if and only if $\mathbf{b}(v_i)$ and $\mathbf{b}(v_j)$ differ by only one component. The main property of hypercube of dimension d is that it can be obtained through connecting with edges the respective vertices of two copies of hypercube of dimension $d - 1$ (see Fig. 5.6, where Q_d are shown for $d \leqslant 3$).

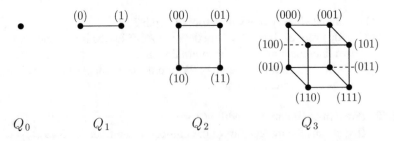

Fig. 5.6 Hypercubes of dimension $d = 0, 1, 2, 3$

Show hypercube of dimension $d = 4$.

5.47. Prove that hypercube Q_d is a d-regular graph for all $d \in \mathbb{Z}_0$.

5.48. How many edges does the hypercube Q_d contain?

5.49. Find, for what d the hypercube Q_d is:

(1) Hamiltonian;
(2) Eulerian.

∗ **5.50.** By k-**face** of the hypercube Q_d, we will understand some subgraph of the graph Q_d, having the properties of hypercube of dimension k, where $0 \leqslant k \leqslant d$. (1) Find the number of k-faces of the hypercube Q_d of dimension $d \geqslant 0$. (2) What is the total number of k-faces Q_d for all $0 \leqslant k \leqslant d$?

5.51. What are the chromatic number χ and the chromatic index χ' of hypercube Q_d for $d > 0$?

5.52. Show digraph D, specified by the adjacency matrix

$$
\begin{bmatrix}
F & T & F & F & F \\
F & F & T & F & F \\
F & F & F & T & T \\
F & F & F & F & F \\
T & T & F & T & F
\end{bmatrix}.
$$

Find out-degree and in-degree of each vertex. Find in D a directed circuit of length 4.

5.53. Write out the adjacency matrix and the adjacency list of the directed graphs D_1–D_4, which are shown in Fig. 5.7.

5.54. Which of the graphs D_1–D_4, specified in the previous exercise, are strongly connected?

5.55. Is the adjacency matrix of the directed graph uniquely defined? Is the adjancency list uniquely defined?

5.56. Digraphs $D_1 = (V_1, E_1)$ and $D_2 = (V_2, E_2)$ are isomorphic if there exists the bijection $H: V_1 \to V_2$ such that $(u, v) \in E_1$ if and only if $(H(u), H(v)) \in E_2$. Find whether the following digraphs D_1 and D_2 are isomorphic, defined by adjacency matrices:

$$
M(D_1) = \begin{bmatrix}
F & T & F & T & F & T \\
F & F & F & F & F & F \\
F & T & F & T & F & T \\
F & F & F & F & F & F \\
F & T & F & T & F & T \\
F & F & F & F & F & F
\end{bmatrix},
M(D_2) = \begin{bmatrix}
F & F & F & T & T & T \\
F & F & F & T & T & T \\
F & F & F & T & T & T \\
F & F & F & F & F & F \\
F & F & F & F & F & F \\
F & F & F & F & F & F
\end{bmatrix}.
$$

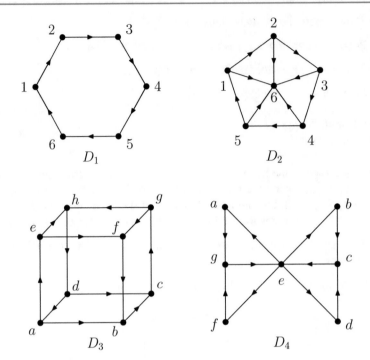

Fig. 5.7 To Exercise **5.53**

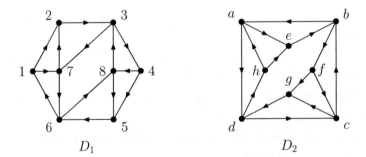

Fig. 5.8 To Exercise **5.57**

5.57. Find whether the following digraphs D_1 and D_2 are isomorphic, which are shown in Fig. 5.8.

∗ **5.58.** Consider the complete undirected graph K_n and show the direction for each edge. The so-obtained digraph we will call a **tournament** and denote by DK_n. Prove the following statements:

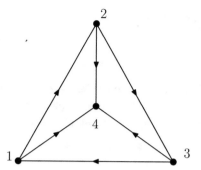

Fig. 5.9 To Exercise **5.61**

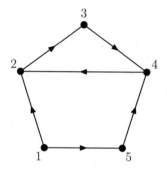

Fig. 5.10 To Exercise **5.62**

(1) The sum of the out-degrees of all vertices of the directed graph DK_n is equal to the sum of the in-degrees.

(2) The sum of squares of the out-degrees of all vertices is equal to the sum of squares of the out-degrees of all vertices in-degrees.

5.59. Show that the digraph is Eulerian, then it is strongly connected.

5.60. Prove that if the tournament DK_n is strongly connected, then it is Hamiltonian [15].

5.61. Find the reachability matrix M^* of the digraph D, which is shown in Fig. 5.9.

5.62. Find the reachability matrix M^* of the digraph D, which is shown in Fig. 5.10.

5.3 Answers, Hints, Solutions

5.1. *Proof.*

In a nontrivial graph, the number of vertices is $n \geqslant 2$. Use the mathematical induction method. Let $P(n)$ be the predicate "in any graph with n vertices, there exist two vertices of the same degree". For $n_0 = 2$, two graphs are possible:

$$\bullet \quad \bullet \qquad \text{and} \qquad \bullet\!\!-\!\!\!-\!\!\bullet \quad ,$$

for which, obviously, $P(n_0)$ is true.

The inductive supposition consists in that $P(k)$, $k \geqslant n_0$ is true. Let us show that in the graph $G'(V', E')$ with $k + 1$ vertices, there exist vertices of the same degree.

The degree of any vertex $v \in V'$ satisfies the inequalities $0 \leqslant \leqslant d(v) \leqslant k$. If in G' there exists an isolated vertex, i.e., a vertex with $d = 0$, then, according to the inductive supposition, in the subgraph G', consisting of all vertices except the isolated one, there must exist two vertices u and v, for which $d(u) = d(v)$, and $P(k + 1)$ is true. If there is no isolated vertex, then $\forall v \in V'$ $1 \leqslant d(v) \leqslant k$. But the graph G' contains a total of $k + 1$ vertices, this is why, according to the pigeonhole principle, the degrees of at least two of them coincide, $P(k + 1)$ is true in this case as well.

5.2. *Proof.*
Consider the graph $G = (V, E)$, whose vertices will be associated with the guests, and the edge is $uv \in E$, if and only if u and v shook hands. $|E|$ is the number of edges, in other words—the total number of all handshakes. Let us use the handshaking lemma

$$\sum_{\text{all vertices}} d(v_i) = 2|E|,$$

where summation is performed over all vertices of the graph G. Partition the set of vertices into two subsets, depending on the evenness of the degree of this vertex v_i:

$$\sum_{\substack{\text{all vertices} \\ \text{of odd degree}}} d(v_i) + \sum_{\substack{\text{all vertices} \\ \text{of even degree}}} d(v_i) = 2|E|,$$

$$\sum_{\substack{\text{all vertices} \\ \text{of odd degree}}} d(v_i) = 2|E| - \sum_{\substack{\text{all vertices} \\ \text{of even degree}}} d(v_i).$$

Therefore, the sum over all vertices of odd degree is even. In total, in the left side of the last equality, all values $d(v_i)$ are odd; write them in the form $d(v_i) = 2d_i + 1$, where $d_i = 0, 1, 2, \ldots$:

$$\sum_{\substack{\text{all vertices} \\ \text{of odd degree}}} d(v_i) = \sum_{\substack{\text{all vertices} \\ \text{of odd degree}}} 2d_i + \sum_{\substack{\text{all vertices} \\ \text{of odd degree}}} 1.$$

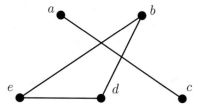

Fig. 5.11 To Exercise **5.3**

From the last equality, it follows that the number of vertices of odd degree of the graph G is even. Hence, the number of the guests that shook an odd number of hands is even.

5.3. *Answer.*

The adjacency matrix M and the incidence matrix $\tilde{M}$ of the graph G have the form:

$$M = \begin{array}{c} \\ a \\ b \\ c \\ d \\ e \end{array} \begin{array}{c} a\ b\ c\ d\ e \\ \begin{bmatrix} F & F & T & F & F \\ F & F & F & T & T \\ T & F & F & F & F \\ F & T & F & F & T \\ F & T & F & T & F \end{bmatrix} \end{array}, \quad \tilde{M} = \begin{array}{c} \\ a \\ b \\ c \\ d \\ e \end{array} \begin{array}{c} ac\ bd\ be\ de \\ \begin{bmatrix} T & F & F & F \\ F & T & T & F \\ T & F & F & F \\ F & T & F & T \\ F & F & T & T \end{bmatrix} \end{array}.$$

The adjacency list of the graph $G(V, E)$:

vertices of graph	adjacent vertices
a	c
b	d, e
c	a
d	b, e
e	$b, d.$

The graph $G(V, E)$ is shown in Fig. 5.11.

5.4. *Answer:*

The adjacency matrix of the graph $G(V, E)$:

$$\begin{array}{c} \\ a \\ b \\ c \\ d \\ e \\ f \\ g \end{array} \begin{array}{c} a\ b\ c\ d\ e\ f\ g \\ \begin{bmatrix} F & T & T & F & F & F & T \\ T & F & T & F & F & F & F \\ T & T & F & T & T & T & T \\ F & F & T & F & T & F & F \\ F & F & T & T & F & T & F \\ F & F & T & F & T & F & F \\ T & F & T & F & F & F & F \end{bmatrix} \end{array}.$$

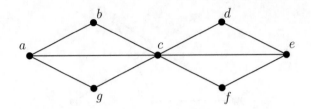

Fig. 5.12 To Exercise **5.4**

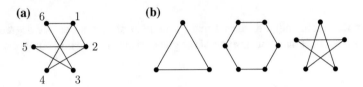

Fig. 5.13 To Exercise **5.5**

The adjacency list of the graph $G(V, E)$:
The graph $G(V, E)$ is shown in Fig. 5.12.

5.5. *Solution.*
Enumerate rows and columns of the adjacency matrix with numbers from 1 to 6, corresponding to the numbers of vertices of the graph. Draw the graph G in Fig. 5.13, (*a*). In Fig. 5.13, (*b*) are shown the graphs of Fig. 5.2, which are subgraphs of the graph G.

vertices of graph	adjacent vertices
a	b, c, g
b	a, c
c	a, b, d, e, f, g
d	c, e
e	c, d, f
f	c, e
g	$a, c.$

5.7. *Solution.*
Let G be a simple undirected graph with N vertices. Then, its adjacency matrix has N rows and N columns and is symmetric with respect to the main diagonal, since if there exists an edge v_1v_2, belonging to the set of edges E, then also the edge $v_2v_1 \in E$. On the main diagonal, we have "F", since the simple graph has no loops. Moreover, the number of logical vertices "T" in an arbitrary row or column of the matrix is equal to the degree of the respective vertex.

5.8. *Solution.*
Transformation of vertices of the graph G results in permutation of rows and columns of the matrix $M(G)$. Hence, there is ambiguity in the choice of the adjacency matrix.

The adjacency list of the graph also depends on the sequence of the vertices' notation.

5.9. *Proof.*
Use the mathematical induction method. Denote the statement of the problem by $P(l), l = 1, 2, \ldots$

B a s i s   s t e p
For $l = 1$, we obtain that $A^1_{ij} = A_{ij}$ is simply an element of the adjacency matrix. It is equal to 1, if there exists an edge ij, and is equal to 0 otherwise. This value is equal to the number of routes of length $l = 1$ from the ith vertex to the jth.

I n d u c t i v e   s t e p
Suppose that the statement $P(l)$ for some $l \in \mathbb{N}$ is true. Consider elements of the matrix A^{l+1}. Since $A^{l+1} = A^l \cdot A$, then we have

$$A^{l+1}_{ij} = \sum_{k=1}^{n} A^l_{ik} A_{kj},$$

where n is the size of the matrix A, i.e., the number of vertices of the graph.

It is known that A^l_{ik} is equal to the number of routes of length l from the graph vertex with number i to the vertex with number k, while the value A_{kj} is equal to 1, if there exists an edge kj, and is equal to 0 otherwise. Any route of length $l + 1$ from the ith vertex to the jth can be represented as a union of the route of length l of the form $i, \ldots, k$ and one edge kj. The expression for A^{l+1}_{ik} contains the sum over all auxiliary vertices k; therefore, the statement $P(l + 1)$ is true.

Therefore, in view of the mathematical induction method, the value A^l_{ij} is equal to the number of routes from the vertex with number i to the vertex with number j.

5.10. *Answer:*
The Hamilton cycle—1,8,2,3,5,6,4,7,1. As examples of cycles of length 3 and 4, we can offer 1,2,3,1 and 1,7,4,6,1, respectively.

5.11. *Answer:*
The graph G_1 is Hamiltonian, but is not Eulerian. The graph G_2 is both Hamiltonian and Eulerian.

5.12. *Solution.*
In any complete graph K_n for $n > 2$, there exists the Hamilton cycle, constructed as follows. Choose the initial vertex, for example, v_1, and pass to v_2, this is possible, since in a complete graph for any vertices $v_1, v_2 \in V$ there exists the edge $v_1 v_2 \in E$. Then, we include into the route the vertices $v_3, v_4, \ldots$ and so on up to v_n. In the route $v_1, \ldots, v_n$, each vertex of the graph K_n occurs exactly once; hence, the cycle $v_1, v_2, \ldots, v_n, v_1$ is Hamilton.

As is known, the criterion of existence of the Euler cycle is as follows: each vertex of the graph should have an even degree. In case of the complete graph K_n, all its vertices have the same degree equal to $d(v_i) = n - 1$. Hence, it follows that the graph K_n is Eulerian for odd values of n.

5.13. *Answer*: no, it cannot.

5.14. *Hint.*
Perform the procedure of constructing the Hamilton cycle similar to the one described in the solution of Exercise **5.12**.

5.15. *Solution.*
As examples, we can offer the following graphs shown in Fig. 5.14.

Note that the graph suggested in item (2) is not simple. We are leaving the construction of a simple graph satisfying the requirements of item (2) for independent solving.

5.16. *Proof.*
Prove reflexivity, symmetry, and transitivity of the connectivity relation $\mathcal{C}$.

The relation $\mathcal{C}$ is reflexive, since $\forall v \in V$ there exists a route (trivial) from v to the same vertex v.

If $(v_1, v_2) \in \mathcal{C}$, then there exists the route $v_1, \ldots, v_2$. Having reversed this route, we will obtain $v_2, \ldots, v_1$, which proves symmetry of the connectivity relation.

Now prove that $\mathcal{C}$ is transitive. Let $v_1 \, \mathcal{C} \, v_2$, $v_2 \, \mathcal{C} \, v_3$ for the vertices $v_1, v_2, v_3 \in V$. Therefore, there exist the routes $v_1, \ldots, v_2$ and $v_2, \ldots, v_3$. Then, two possibilities occur, as shown in Fig. 5.15:

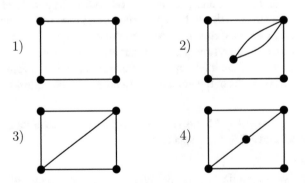

Fig. 5.14 To Exercise **5.15**

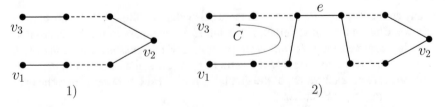

Fig. 5.15 Construction of the route C (to Exercise **5.16**)

(1) The routes contain no common edges. Then, $v_1, \ldots, v_2, \ldots, v_3$ connects the vertices v_1 and v_3, and $v_1 \, \mathcal{C} \, v_3$.

(2) There exists some edge e, common for both routes. It is always possible to construct the route C with the beginning at v_1 and the end at v_3, as shown in Fig. 5.15, (2).

Thus, it is proved that the connectivity relation is the equivalence relation. The equivalence classes are formed by subsets of the set of vertices associated with certain connected components of the graph.

5.17. *Proof.*
Prove reflexivity, symmetry, and transitivity of the isomorphism relation $\mathcal{I}$.

For any graph G, there exists $G \, \mathcal{I} \, G$, since as a bijection preserving the adjacency we can choose an identity function $f = \{(v_1, v_1), \ldots, (v_n, v_n)\}$. Hence, $\mathcal{I}$ is reflexive.

Let $G_1 \, \mathcal{I} \, G_2$ be fulfilled and the bijection $f : V_1 \rightarrow V_2$ exist, that preserves the adjacency. There exists a function inverse to f, $f^{-1} : V_2 \rightarrow V_1$. This proves: $G_2 \, \mathcal{I} \, G_1$, and the relation $\mathcal{I}$ is symmetric.

Finally, check the existence of the transitivity property. Let $G_1 \, \mathcal{I} \, G_2$ and $G_2 \, \mathcal{I} \, G_3$, corresponding to the bijection $f_1 : V_1 \rightarrow V_2$ and $f_2 : V_2 \rightarrow V_3$. Construct the composition $f_2 \circ f_1 : V_1 \rightarrow V_3$. The composition of two bijections is a bijection too, and the adjacency of the vertices is preserved. Therefore, $G_1 \, \mathcal{I} \, G_3$, and transitivity of the isomorphism relation is proved.

In summary, we conclude that isomorphism is the equivalence relation on the set of all graphs U.

5.18. *Solution.*
Permutation of rows and columns in the adjacency matrix results in an isomorphic graph. Having interchanged the second and the fifth rows in $M(G_1)$, and then the second and the fifth columns, we will obtain $M(G_2)$, which proves: $G_1 \sim G_2$. The considered graphs are shown in Fig. 5.16.

5.19. *Solution.*

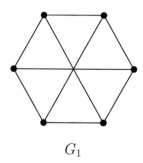
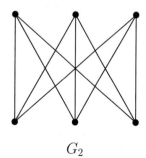

G_1 $\qquad\qquad\qquad\qquad\qquad\qquad\qquad$ G_2

Fig. 5.16 To Exercise **5.18**. Isomorphic graphs

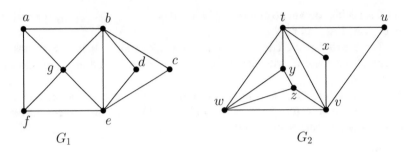

Fig. 5.17 To Exercise **5.19**

There exists the bijection $H: V_1 \to V_2$, preserving the adjacency: $H = \{(a, y),$ $(b, t), (c, x), (d, u), (e, v), (f, z), (g, w)\}$. The considered graphs are shown in Fig. 5.17.

5.20. *Solution.*

(1) Associate the first color with the vertex v_1. The vertex v_2 should be colored with the second color, since it is adjacent with v_1. For the third vertex, we need the third color, and so on. The vertex v_n will obtain color number n, and therefore for the complete graph, $\chi(K_n) = n$.

(2) Denote the graph, obtained by removing from K_n of one edge, by K'_n. Without loss of generality, we can take it that the edge $v_1 v_2$ is removed. The vertices v_1 and v_2 are not adjacent, therefore associate the first color with these vertices. The vertices $v_2, \ldots, v_n$ together with incident edges form the subgraph K_{n-1}, for regular coloring of which we need, according to the result of the previous item, $n - 1$ colors. Hence, for the graph K'_n the chromatic number is $\chi(K'_n) = n - 1$.

5.21. *Answer:*
$\chi(P_{10}) = 3$, $\chi'(P_{10}) = 4$ [31].

5.22. *Proof.*
Use the induction on the number of vertices $|V| = n$. Denote by $P(n)$ the predicate $\chi(G(V, E)) \leqslant \Delta(G(V, E)) + 1$.

B a s i s s t e p

For $n = 1$, we obtain $\chi(G) = 1$, $\Delta(G) = 0$. Therefore, the inequality $\chi(G) \leqslant \Delta(G) + 1$ is fulfilled.

I n d u c t i v e s t e p

Let for all graphs with the number of vertices no greater than k, the predicate $P(k)$ take the true value. Consider the graph $G(V, E)$ with the number of vertices $|V| = k + 1$. Remove an arbitrary vertex $v \in V$ together with the edges incident with it. Then for the obtained graph G', according to the inductive supposition, $\chi(G') \leqslant \Delta(G') + 1$.

But the degree of the removed vertex v is no greater than $\Delta(G)$, at least one color in coloring of the graph G' with the help of $\Delta(G') + 1$ colors is free for v. Having colored the vertex v in this color, we will obtain the regular coloring. Thus, we have constructed the regular coloring of the graph $G(V, E)$, that uses $\Delta(G) + 1$ colors.

5.23. *Answer.*

The Petersen graph belongs to the second class.

5.24. *Solution.*

We should consider the cases of even and odd number of the vertices n (Fig. 5.18).

It is easy to see that the edges C_{2k} can be assigned colors $\psi(v_1 v_2) = 1$, $\psi(v_2 v_3) = 2$, $\psi(v_3 v_4) = 1$, ..., $\psi(v_{2k} v_1) = 2$, hence, $\chi'(C_{2k}) = 2$.

Using the same line of reasoning, we will obtain $\chi'(C_{2k+1}) = 3$.

Figure 5.19 offers one of the possible ways of regular coloring of the considered graphs. The dashed, dotted, and dash-and-dot lines used to highlight the graphs' edges correspond to three different colors.

Since regardless of evenness of n the maximal degree of vertex of the n-cycle $\Delta(C_n)$ is equal to 2, then the graph C_{2k}, $k \in \mathbb{N}$, belongs to the first class, and the graph C_{2k+1}, $k \in \mathbb{N}$, belongs to the second class.

5.25. *Answer*:

$\chi'(K_{2k+1}) = 2k + 1$, $\chi'(K_{2k}) = 2k - 1$ for all $k \in \mathbb{N}$. Full graphs with even number of vertices K_{2k} belong to class 1, with odd number K_{2k+1}—to class 2.

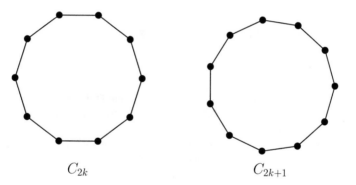

C_{2k} C_{2k+1}

Fig. 5.18 Graphs C_{2k} and C_{2k+1}, $k = 5$ (to Exercise **5.24**)

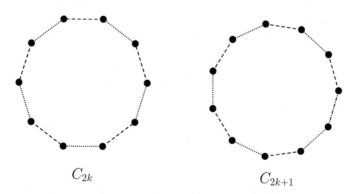

C_{2k} C_{2k+1}

Fig. 5.19 Regular coloring of the graphs C_{2k} and C_{2k+1}, $k = 5$ (to Exercise **5.24**)

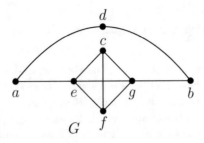

Fig. 5.20 Graph G to Exercise **5.26**

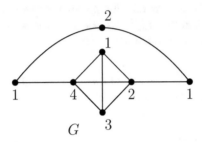

Fig. 5.21 Regular coloring of the graph G (to Exercise **5.26**)

5.26. *Solution.*
Denote the Saturday disciplines by Latin letters: mathematics—a, literature—b, rhetoric—c, nature—d, physical education—e, drawing—f, solfeggio—g. The set of disciplines $V = \{a, b, c, d, e, f, g\}$.

Construct the graph $G(V, E)$, whose vertices correspond to the lessons, where two vertices of G are adjacent if and only if these lessons cannot be conducted simultaneously (Fig. 5.20).

Then any regular coloring specifies some schedule in the sense that the lessons that have obtained the same colors can be conducted simultaneously. We obtain that the minimal number of hours required to conduct all the lessons is $\chi(G)$.

Since the graph G contains the complete subgraph K_4, then $\chi(G) \geqslant 4$.

Figure 5.21 shows one of the permissible regular colorings. So, the minimal number of hours required to conduct seven lessons on Saturday is four.

5.27. *Answer*: three seconds.

5.28. *Answer*:

5.29. *Answer*:
Only the graph with the adjacency matrix from item (1) is a tree, and the rest of the graphs contain cycles and are not trees.

5.30. *Answer*:

The tree with one vertex is unique: •

The tree with two vertices can also be shown in one way only.

•——•

The tree with three vertices is also unique.

•——•——•

The trees with four vertices:

The trees with five vertices:

The trees with six vertices:

We are leaving construction of all trees with $n = 7$ as an exercise (there must be 11 non-isomorphic trees).

5.31. *Proof.*

Let $T = (V, E)$ be the tree with $n \geqslant 2$ vertices. Denote by S the set of degrees of all its vertices. Since the number of edges in the tree is $m = n - 1$, the maximal degree of its vertices can be equal to $\Delta = n - 1$, and the minimal (due to connectivity)— $\delta = 1$. Hence, $S \subseteq \{1, 2, \ldots, n - 1\}$ and $|S| \leqslant n - 1$.

Consider the function $f : V \rightarrow S$, correlating each vertex of the graph T with its degree. Since $|V| = n$, there holds the inequality $|V| > |S|$. Therefore, according to the pigeonhole principle, there exist two vertices, at which the function f takes one and the same value, i.e., two vertices of the same degree.

5.32. *Solution*
Let n be the total number of vertices of the given tree. Then, the sum of their degrees is equal to

$$4(n - 14) + 3 \cdot 1 + 2 \cdot 6 + 1 \cdot 7 = 4n - 34.$$

On the other hand, the number of edges of the tree T must be one less than the number of vertices; i.e., the number of edges is $n - 1$. Relying upon the handshaking lemma, we obtain the relation: $4n - 34 = 2(n - 1)$. Hence, $n = 16$, whence we obtain that the number of vertices of the tree T of degree 4 is two.

5.33. *Answer*:
The tree T has ten vertices of degree $d = 1$.

5.34. *Answer*: The tree T can have any number of vertices of degree $d = 2$.

5.35. *Answer*: The tree T has two vertices of degree $d = 3$.

5.36. *Answer*:

(1) The number of connected components is $c(G) = 2$;

(2) $c(G) = n - m$.

5.37. *Proof*:
Let us produce a constructive proof performing the regular coloring of a arbitrary tree T. The root of the tree T will be of the first color, the vertices of the next level—of the second, and so on, alternating the first and the second colors. The resulting coloring is regular, since for any adjacent vertices their colors will be different, therefore, $\chi(T) = 2$.

5.38. *Hint*. See the previous exercise.

5.40. *Proof*.

Out of all possible binary trees, the complete tree has the minimal height. The number of vertices at the first k levels of the complete binary tree is

$$1 + 2 + 4 + \ldots + 2^k = 2^{k+1} - 1.$$

For an arbitrary binary tree of depth h with n vertices, the number of levels may only grow due to additional levels not completely filled; therefore, the inequality is fulfilled

$$2^{h+1} - 1 \geqslant N.$$

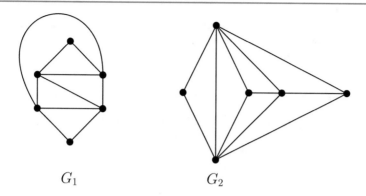

Fig. 5.22 To Exercise **5.41**

Taking the logarithm of both sides of the inequality, we will obtain

$$h + 1 \geqslant \lceil \log_2(N + 1) \rceil,$$

where $\lceil a \rceil$ is determined as the least integer, no less than a, or "rounding up" (for more details see Exercise **3.79**). Using the identity $\lceil \log_2(N + 1) \rceil = \lfloor \log_2 N \rfloor + 1$ (it is proved in the solution to Exercise **3.79**), we obtain the required estimate for the depth of the binary tree:

$$h \geqslant \lfloor \log_2 N \rfloor.$$

5.41. *Solution.*
The graphs G_1 and G_2 can be redrawn without intersection of edges, as shown in Fig. 5.22. By definition, G_1 and G_2 are planar graphs.

5.42. *Solution.*
From the diagram of the graph G_1 (see the solution to Exercise **5.41** and Fig. 5.22), we conclude that $n = 6$, $m = 10$, $r = 6$. Hence, $n - m + r = 2$. For the graph G_2, we have $n = 6$, $m = 11$, $r = 7$, and $n - m + r = 2$.

5.43. *Proof.*
The proof is performed by induction on the number of edges in the graph G. If $m = 0$, then, by virtue of connectivity, G contains only one vertex $n = 1$ and one (infinite) face $r = 1$, and the Euler's formula is fulfilled.

Now suppose that the formula is valid for any graph G with m edges. Add to G a new edge e, we will obtain the graph G' with the number of vertices, the number of edges, and the number of faces n', m', and r', respectively. Then, we should consider three cases.

(1) The edge e connects to different vertices, then one of the edges of the graph G splits into two, increasing the total number of faces by one: $r' = r + 1$. The number of vertices does not change in this case, $n' = n$:

$$n' - m' + r' = n - (m + 1) + (r + 1) = n - m + r = 2.$$

(2) The edge e is incident with only one vertex in G, i.e., e connects the existing vertex with the new one. the number of faces does not change, while the number of vertices and edges should be increased by one:

$$n' - m' + r' = (n + 1) - (m + 1) + r = n - m + r = 2.$$

(3) The edge e is a loop. A new face is formed, while the number of vertices remains unchanged:

$$n' - m' + r' = n - (m + 1) + (r + 1) = n - m + r = 2.$$

The Euler's formula remains valid in each of the considered cases.

5.44. *Proof.*
Without loss of generality, we can take it that the graph is connected and contains no less than three vertices. Let in the graph $G(V, E)$ the number of vertices be $|V| = n$, the number of edges be $|E| = m$, and the number of faces be r.

Let us use proof "by contradiction." The degree of each vertex is no less than 6. Then,

$$\sum_{v \in V} d(v) \geqslant 6n.$$

According to the handshaking lemma $\sum_{v \in V} d(v) = 2m$, hence $6n \leqslant 2m, 3n \leqslant m$.

Let us show that for the complete planar graph, the number of edges is no greater than $m \leqslant 3n - 6$. This relation follows from the fact this relation follows from the fact that in a simple graph each face is limited by at least three edges (the number of vertices $n \geqslant 3$), and each face limits no more than two edges. Then when calculating the number of edges around each face, we will obtain $3r \leqslant 2m$. Using the Euler's formula, we obtain

$$n - m + r = 2,$$
$$n - m + \frac{2}{3}m \leqslant 2,$$
$$m \leqslant 3n - 6.$$

But the supposition that $\forall v \in V \ d(v) \geqslant 6$, resulted in the inequality $m \geqslant 3n$. The obtained contradiction proves: In the planar graph, there exists a vertex of a degree no more than five.

5.45. *Proof.*

Use the induction on the number of vertices. For planar graphs with less than six vertices, the statement is obvious. Suppose that G is a planar graph with n vertices and that all planar graphs with $n - 1$ vertices can be colored five colors. According to the statement of Exercise **5.44**, it contains a vertex v, whose degree is no more than five, $d(v) \leqslant 5$. Removal of the vertex v and all edges incident with it leads us to the graph with $n - 1$ vertices, which can be colored five colors. Try to color v with one of the already used five colors.

If $d(v) < 5$, then v can be colored any color, not used for painting the adjacent vertices.

Suppose that $d(v) = 5$ and that adjacent with v vertices $v_1, \ldots, v_5$ are positioned around v as shown in Fig. 5.23. If to some two vertices v_i and v_j the same color is assigned, then for v we can find a color, not used for painting the adjacent vertices.

Finally, the case remains, when all adjacent with v vertices are assigned different colors. Let v_i be painted with color c_i, $i = 1, \ldots, 5$. Denote by C_{ij} a subgraph of

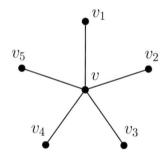

Fig. 5.23 Vertices adjacent to the vertex v

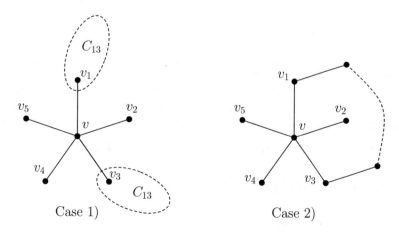

Fig. 5.24 Two possibilities of mutual positioning of vertices v_1 and v_3

the graph G, whose vertices are all vertices of color c_i or c_j, and edges—all edges, connecting the vertex of color c_i with a vertex of color c_j.

Now, we have two possibilities.

(1) The vertices v_1 and v_3 do not belong to the same component of the graph C_{13}. In this case, we will change the colors of all vertices of that component C_{13}, which contains vertex v_1 (in other words, swap the colors $c_1 \leftrightarrow c_3$). After this, v_1 receives color c_3, which allows coloring the vertex v with color c_1 thus having built the required coloring.

(2) The vertices v_1 and v_3 belong to the same component of the graph C_{13}. Then, there exists a cycle $v, v_1, \ldots, v_3, v$, a part of which, confined between v_1 and v_3, fully lies in C_{13} (see Fig. 5.24). Since v_2 is inside this cycle, and v_4 is outside of it, then there is no route from v_2 to v_4, fully lying in C_{24}; hence, we can change the colors of all vertices of component of the subgraph C_{24}, containing the vertex v_2. In this case, the vertex v_2 receives the color c_4, which allows coloring the vertex v with color c_2. The theorem of five colors is proved.

5.47. *Hint.*
Use the mathematical induction method.

5.48. *Solution.*
In accordance with the handshaking lemma, we have: $\sum_{v_i \in V} d(v_i) = 2|E|$. As is shown in the previous exercise, the graph Q_d is d-regular; hence, $|E| = \frac{1}{2}d|V|$. By virtue of the equality $|V| = 2^d$, we finally obtain the number of edges of hypercube of dimension d: $|E| = d\,2^{d-1}$.

5.49. *Answer.*

Hypercube of dimension d is Hamiltonian for all $d \geqslant 2$, Eulerian for all even $d = 2k$, where $k \in \mathbb{Z}_0$.

5.50. *Solution.*

(1) Note that we can single out some k-face of hypercube Q_d by choosing $(d-k)$ component of the vector $\mathbf{b} = (b_1, \ldots, b_d) \in \mathbb{B}^d$, and each of them takes values from the set $\{0, 1\}$. As is known, the number of $(d, d-k)$-samples without repetitions is $C(d, d-k)$. Hence, by the product rule (see chapter "Combinatorics" on p. 141) the number of k-faces of hypercube of dimension d is $C(d, d-k)2^{d-k}$, or $C(d, k)2^{d-k}$.

(2) The total number k-faces Q_d is $\sum_{k=0}^{d} C(d, k)2^k$. The obtained expression is an expansion of the binomial $(1+2)^d$; hence, $\sum_{k=0}^{d} C(d, k)2^k = 3^d$.

5.51. *Answer*:
For all $d > 0$, we have $\chi(Q_d) = 2$, $\chi'(Q_d) = d$.

5.52. *Solution.*
The digraph D is shown in Fig. 5.25.

Fig. 5.25 To Exercise **5.52**

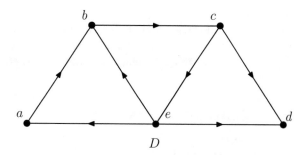

Having used the graphic presentation D, we conclude that $d^+(a) = d^+(b) = 1$, $d^+(c) = 2$, $d^+(d) = 0$, $d^+(e) = 3$, $d^-(a) = d^-(c) = d^-(e) = 1, d^-(b) = d^-(d) = 2$.

Directed circuit of length 4—a, b, c, e, a.

5.53. *Answer*:

Write out the adjacency matrix and the adjacency list of the digraphs D_1 and D_2.

$$M(D_1) = \begin{array}{c} \\ 1 \\ 2 \\ 3 \\ 4 \\ 5 \\ 6 \end{array} \begin{array}{cccccc} 1 & 2 & 3 & 4 & 5 & 6 \\ \left[\begin{array}{cccccc} F & T & F & F & F & F \\ F & F & T & F & F & F \\ F & F & F & T & F & F \\ F & F & F & F & T & F \\ F & F & F & F & F & T \\ T & F & F & F & F & F \end{array}\right] \end{array},$$

vertices of digraph D_1	adjacent vertices
1	2
2	3
3	4
4	5
5	6
6	1

;

$$M(D_2) = \begin{array}{c} \\ 1 \\ 2 \\ 3 \\ 4 \\ 5 \\ 6 \end{array} \begin{array}{cccccc} 1 & 2 & 3 & 4 & 5 & 6 \\ \left[\begin{array}{cccccc} F & T & F & F & F & T \\ F & F & T & F & F & T \\ F & F & F & T & F & T \\ F & F & F & F & T & T \\ T & F & F & F & F & T \\ F & F & F & F & F & F \end{array}\right] \end{array},$$

vertices of digraph D_2	adjacent vertices
1	2, 6
2	3, 6
3	4, 6
4	5, 6
5	1, 6
6	—

.

Representations of the digraphs D_3–D_4 are written similarly.

5.54. *Solution.*

In a strongly connected digraph for any two vertices v_1, $v_2 \in V$, there exists the path $v_1, \ldots, v_2$. It is easy to see that for D_1, D_3, D_4, this condition is fulfilled. The digraph D_2 is not strongly connected, since there is not a single arc coming out of the vertex 6.

5.55. *Hint.*

See the solution to Exercise **5.8**.

5.56. *Answer*:

digraphs are isomorphic.

5.57. *Answer*:

The presented digraphs are isomorphic. The respective bijection $H : V_1 \to V_2$ is as follows:

$$H = \{(1, h), (2, e), (3, b), (4, f), (5, g), (6, d), (7, a), (8, c)\}.$$

5.58. *Proof.*

(1) For each node v_i is fulfilled the equality

$$d^+(v_i) + d^-(v_i) = n - 1, \quad i = 1, \ldots, n.$$

Consider the matrix A, analogous to the adjacency matrix. The elements of the matrix A we will determine by the rule

$$A_{ij} = \begin{cases} 1, & \text{if the edge } ij \in E, \\ 0, & \text{if the edge } ij \notin E. \end{cases}$$

Since in the directed graph DK_n $\forall i, j \in V$ either $A_{ij} = 1$, or $A_{ji} = 1$ (but not simultaneously), then $A_{ij} + A_{ji} = 1$. Moreover, for a simple graph $A_{ii} = 0$. Then, the out-degree of the vertices v_i $d^+(v_i)$ can be expressed by elements of the matrix A

$$d^+(v_i) = \sum_{j=1}^{n} A_{ij}.$$

Then, the sum of the out-degrees of all vertices DK_n is equal to

$$\sum_{i=1}^{n} d^+(v_i) = \sum_{i=1}^{n} \sum_{j=1}^{n} A_{ij}.$$

Split the obtained sum into three parts

$$\sum_{i=1}^{n} d^+(v_i) = \sum_{i=1}^{n} A_{ii} + \sum_{\substack{i,j=1 \\ i>j}}^{n} A_{ij} + \sum_{\substack{i,j=1 \\ i<j}}^{n} A_{ij}.$$

In the right side of the equality, the first sum is equal to zero, since all its summands are equal to zero. In the third sum, pass from A_{ij} to A_{ji} and then rename the summation variables $i \leftrightarrow j$:

$$\sum_{i=1}^{n} d^+(v_i) = \sum_{\substack{i,j=1 \\ i>j}}^{n} A_{ij} + \sum_{\substack{i,j=1 \\ i<j}}^{n} (1 - A_{ji}) = \sum_{\substack{i,j=1 \\ i>j}}^{n} A_{ij} + \sum_{\substack{i,j=1 \\ j<i}}^{n} (1 - A_{ij}) =$$

$$= \sum_{\substack{i,j=1 \\ i>j}}^{n} A_{ij} + \sum_{\substack{i,j=1 \\ j<i}}^{n} 1 - \sum_{\substack{j,i=1 \\ i>j}}^{n} A_{ij} = \sum_{\substack{i,j=1 \\ j<i}}^{n} 1 = \frac{n(n-1)}{2}.$$

Having performed analogous transformations for the sum $\sum_{i=1}^{n} d^-(v_i)$, we will obtain

$$\sum_{i=1}^{n} d^-(v_i) = \frac{n(n-1)}{2}.$$

This means that the sum of the out-degrees of all vertices of the digraph DK_n is equal to the sum of the in-degrees.

(2) Let us express the sum of squares of the out-degrees by in-degree:

$$\sum_{i=1}^{n} (d^+(v_i))^2 = \sum_{i=1}^{n} ((n-1) - d^-(v_i))^2 =$$

$$= \sum_{i=1}^{n} (n-1)^2 - 2\sum_{i=1}^{n} (n-1)d^-(v_i) + \sum_{i=1}^{n} (d^-(v_i))^2.$$

Let us use the result of the previous item $\sum_{i=1}^{n} d^-(v_i) = \frac{n(n-1)}{2}$:

$$\sum_{i=1}^{n} (d^+(v_i))^2 = n(n-1)^2 - 2(n-1)\frac{n(n-1)}{2} + \sum_{i=1}^{n} (d^-(v_i))^2 =$$

$$= \sum_{i=1}^{n} (d^-(v_i))^2.$$

This proves the statement of the problem.

5.59. *Solution.*
Choose two vertices of the digraph u and v, belonging to the Euler circuit C. The latter without loss of generality can be represented as $C = A_1, u, A_2, v, A_3$, where A_i ($i = 1, 2, 3$) are the paths which are parts of the circuit C. Now we only have to note, that the path u, A_2, v is a path from the vertex u to v, and v, A_3, A_1, u from the vertex v to u. Therefore, the considered digraph is strongly connected.

Fig. 5.26 Construction of
the Hamilton circuit to
Exercise **5.60**

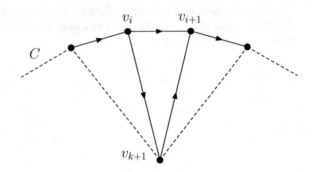

5.60. *Proof.*

Use the mathematical induction method. Let $P(n)$ be the predicate "in a strongly connected tournament with n vertices exists the Hamilton circuit." The strongly connected with the least number of vertices is DK_3; therefore, the proof of the statement $P(3)$ forms a basis step. All in all, two strongly connected tournaments with three vertices are possible, namely

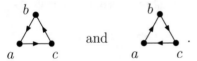

The Hamilton circuits, respectively, a, c, b, a and a, b, c, a.

Let $P(k)$ be true; i.e., in the strongly connected tournament with k vertices; there exists the Hamilton circuit, for example, $C = v_1, v_2, \ldots, v_n, v_1$. Add one more vertex with arc incident with it and consider the resulting digraph DK_{k+1}.

The strong connection of the tournament guarantees existence of at least one arc, coming out of v_{k+1}, and at least one arc coming in this vertex. Therefore, in the sequence $C = v_1, \ldots, v_n, v_1$, there will always be such vertex v_i, $1 \leqslant i \leqslant k$, that arcs $v_i v_{k+1}$ and $v_{k+1} v_{i+1}$ exist (Fig. 5.26).

The sequence $C' = v_1, \ldots, v_i, v_{k+1}, v_{i+1}, \ldots, v_n, v_1$ forms the Hamilton circuit.

So, $P(k)$ results in $P(k + 1)$, and by the mathematical induction method, we have proved that the Hamilton circuit exists in the strongly connected tournament.

5.61. *Solution.*

Write the adjacency matrix of the digraph:

$$M = \begin{bmatrix} F\,T\,F\,T \\ F\,F\,T\,T \\ T\,F\,F\,T \\ F\,F\,F\,F \end{bmatrix}.$$

Consecutively calculate the degrees M^i for $i = 2, 3, 4$:

$$M^2 = M \cdot M = \begin{bmatrix} F\,T\,F\,T \\ F\,F\,T\,T \\ T\,F\,F\,T \\ F\,F\,F\,F \end{bmatrix} \cdot \begin{bmatrix} F\,T\,F\,T \\ F\,F\,T\,T \\ T\,F\,F\,T \\ F\,F\,F\,F \end{bmatrix} = \begin{bmatrix} F\,F\,T\,T \\ T\,F\,F\,T \\ F\,T\,F\,T \\ F\,F\,F\,F \end{bmatrix},$$

$$M^3 = M^2 \cdot M = \begin{bmatrix} F & F & T & T \\ T & F & F & T \\ F & T & F & T \\ F & F & F & F \end{bmatrix} \cdot \begin{bmatrix} F & T & F & T \\ F & F & T & T \\ T & F & F & T \\ F & F & F & F \end{bmatrix} = \begin{bmatrix} T & F & F & T \\ F & T & F & T \\ F & F & T & T \\ F & F & F & F \end{bmatrix},$$

$$M^4 = M^3 \cdot M = \begin{bmatrix} T & F & F & T \\ F & T & F & T \\ F & F & T & T \\ F & F & F & F \end{bmatrix} \cdot \begin{bmatrix} F & T & F & T \\ F & F & T & T \\ T & F & F & T \\ F & F & F & F \end{bmatrix} = \begin{bmatrix} F & T & F & T \\ F & F & T & T \\ T & F & F & T \\ F & F & F & F \end{bmatrix}.$$

Note that there also exist other methods to calculate M^4: $M^4 = M \cdot M^3 = M^2 \cdot M^2$.
The reachability matrix is found by the definition formula

$$M^* = M \text{ or } M^2 \text{ or } M^3 \text{ or } M^4 = \begin{bmatrix} T & T & T & T \\ T & T & T & T \\ T & T & T & T \\ F & F & F & F \end{bmatrix}.$$

5.62. *Answer*:

$$M^* = \begin{bmatrix} F & T & T & T & T \\ F & T & T & T & F \\ F & T & T & T & F \\ F & T & T & T & F \\ F & T & T & T & F \end{bmatrix}.$$

Boolean Algebra

6

Boolean[1] **algebra** is a set $\mathbb{B} = \{0, 1\}$ with defined on it operations of disjunction ($\vee$), conjunction ($\wedge$) and negation ($^-$). **Boolean variable** p can take the values 0 or 1 (and only them), $p \in \mathbb{B}$. The action of the basic operations on the Boolean variables is presented in Table 6.1.

Boolean expression is formed by Boolean variables with the help of operations $\vee$, $\wedge$, $^-$ and brackets. The expression of the form $x_1 \wedge x_2$ is sometimes written in the form $x_1 \,\&\, x_2$ or simply $x_1 x_2$. The symbol $\vee$ derives from the first letter of Latin conjunction *vel*—"or". Denotation & is a modified spelling of the word *et*—"and".

The laws of Boolean algebra are shown in Table 6.2.

Table 6.1 Negation, disjunction, and conjunction

p	$\overline{p}$
0	1
1	0

p	q	$p \vee q$	$p \wedge q$
0	0	0	0
0	1	1	0
1	0	1	0
1	1	1	1

[1]George Boole (1815–1864)—English mathematician and logician.

© Springer International Publishing AG, part of Springer Nature 2018
S. Kurgalin and S. Borzunov, *The Discrete Math Workbook*,
Texts in Computer Science, https://doi.org/10.1007/978-3-319-92645-2_6

Table 6.2 The laws of Boolean algebra

The idempotent laws

$$p \vee p = p, \qquad\qquad\qquad p \wedge p = p$$

The commutative laws

$$p \vee q = q \vee p, \qquad\qquad\qquad p \wedge q = q \wedge p$$

The associative laws

$$p \vee (q \vee r) = (p \vee q) \vee r, \qquad p \wedge (q \wedge r) = (p \wedge q) \wedge r$$

The absorption laws

$$(p \wedge q) \vee p = p, \qquad\qquad (p \vee q) \wedge p = p$$

The distributive laws

$$p \vee (q \wedge r) = (p \vee q) \wedge (p \vee r), \qquad p \wedge (q \vee r) = (p \wedge q) \vee (p \wedge r)$$

The properties of zero and one

$$p \vee 1 = 1, \qquad\qquad\qquad p \wedge 0 = 0$$
$$p \vee 0 = p, \qquad\qquad\qquad p \wedge 1 = p$$

The negation properties

$$p \vee \overline{p} = 1, \qquad\qquad\qquad p \wedge \overline{p} = 0$$
$$\overline{\overline{p}} = p$$

De Morgan's laws

$$\overline{(p \wedge q)} = \overline{p} \vee \overline{q}, \qquad\qquad \overline{(p \vee q)} = \overline{p} \wedge \overline{q}$$

Table 6.3 Table of function values $f(\mathbf{x})$

x_1	x_2	$\cdots$	x_{n-1}	x_n	$f(x_1, x_2, \ldots, x_{n-1}, x_n)$
0	0	$\cdots$	0	0	$f(0, 0, \ldots, 0, 0)$
0	0	$\cdots$	0	1	$f(0, 0, \ldots, 0, 1)$
0	0	$\cdots$	1	0	$f(0, 0, \ldots, 1, 0)$
$\cdots$					$\cdots$
1	1	$\cdots$	1	1	$f(1, 1, \ldots, 1, 1)$

The tuple $\mathbf{x} = (x_1, x_2, \ldots, x_n)$, where $x_i \in \{0, 1\}$, $1 \leqslant i \leqslant n$, is referred to a the **Boolean tuple (vector)**. The tuple elements—**components** or **coordinates**. The function $f : \mathbb{B}^n \to \mathbb{B}$, where $f(x_1, \ldots, x_n)$—Boolean expression, is called **Boolean function**.

One of the ways to define a Boolean function $f(\mathbf{x})$ is the table of values (Table 6.3). Such a table is also referred to as the **truth table**. As is easy to see, the number of rows in the table is 2^n, where n is the number of vector components $\mathbf{x}$.

The function can also be defined by the vector of values $\alpha = (\alpha_0, \alpha_1, \ldots, \alpha_{2^n-1})$, where the coordinate α_i is equal to the value of the function on the tuple of variables located in the ith row, $0 \leqslant i \leqslant 2^n - 1$.

The set of all possible Boolean functions depending on n variables is denoted by $P_2(n)$. The number of such Boolean functions is $|P_2(n)| = 2^{2^n}$. For example, the number of various functions of one variable is 4, of two—16.

Table 6.4 lists all functions of $P_2(2)$. The symbols "$\downarrow$", "$|$", "$\oplus$", "$\leftrightarrow$", and "$\to$" are abbreviations for the combinations of disjunction, conjunction, and negation.

The functions listed in Table 6.4 are widespread, and many of them have special names as shown in Table 6.5.

Dual to the function $f(x_1, \ldots, x_n)$ is the function

$$f^*(x_1, x_2, \ldots, x_n) = \overline{f(\overline{x}_1, \overline{x}_2, \ldots, \overline{x}_n)}.$$

If the function $f(\mathbf{x})$ coincides with the function that is dual to it, i.e. $f = f^*$, then the function $f(\mathbf{x})$ is called **self-dual**.

Elementary conjunction or **minterm** is the conjunction n of different Boolean variables or their negations $x_1^{\sigma_1} \wedge x_2^{\sigma_2} \wedge \ldots \wedge x_n^{\sigma_n}$, where $n \geqslant 1$ and

$$x_i^{\sigma_i} = \begin{cases} \overline{x}_i, & \sigma_i = 0, \\ x_i, & \sigma_i = 1. \end{cases}$$

The number of variables is called a **rank** of elementary conjunction. Elementary conjunction is called **monotone**, if it contains no negations of variables. Elementary conjunction takes a value equal to 1 only on one tuple of arguments' values. Due

Table 6.4 Functions of two arguments

Notation		0	x_1	x_2	$\bar{x}_1$	$\bar{x}_2$	$x_1 \vee x_2$
x_1	x_2	f_0	f_1	f_2	f_3	f_4	f_5
0	0	0	0	0	1	1	0
0	1	0	0	1	1	0	1
1	0	0	1	0	0	1	1
1	1	0	1	1	0	0	1
Notation		$x_1 \downarrow x_2$	$x_1 \wedge x_2$	$x_1 \mid x_2$	$x_1 \oplus x_2$	$x_1 \leftrightarrow x_2$	$x_1 \rightarrow x_2$
x_1	x_2	f_6	f_7	f_8	f_9	f_{10}	f_{11}
0	0	1	0	1	0	1	1
0	1	0	0	1	1	0	1
1	0	0	0	1	1	0	0
1	1	0	1	0	0	1	1
Notation		$\overline{(x_1 \rightarrow x_2)}$	$x_2 \rightarrow x_1$	$\overline{(x_2 \rightarrow x_1)}$	1		
x_1	x_2	f_{12}	f_{13}	f_{14}	f_{15}		
0	0	0	1	0	1		
0	1	0	0	1	1		
1	0	1	1	0	1		
1	1	0	1	0	1		

to this, an arbitrary function not identically equal to zero, can be presented as a disjunction of elementary conjunctions. The expression of the form

$$\mathcal{D} = K_1 \vee K_2 \vee \ldots \vee K_l,$$

where K_i, $i = 1, 2, \ldots, l$ are pairwise different elementary conjunctions is called **disjunctive normal form** (DNF); in this case, the number l is the length of DNF.

Perfect disjunctive normal form (PDNF) consists of elementary conjunctions of rank n. Such a representation is uniquely defined accurate to permutation of elementary conjunctions.

Elementary disjunction or **maxterm** is the expression $x_1^{\sigma_1} \vee x_2^{\sigma_2} \vee \ldots \vee x_n^{\sigma_n}$, $n \geqslant 1$, where all variables within the disjunction are different. The number of variables n is called a **rank** of elementary disjunction.

Conjunctive normal form (CNF) of a given Boolean function, not identically equal to one, is its representation in the form of a conjunction of some elementary disjunctions:

Table 6.5 Names of the basic functions of Boolean algebra

Notation	Name
0	Zero constant
1	One constant
x_1	Repetition x_1
x_2	Repetition x_2
$\overline{x}_1$	Negation x_1
$\overline{x}_2$	Negation x_2
$x_1 \vee x_2$	Disjunction
$x_1 \downarrow x_2$	Peirce arrow
$x_1 \wedge x_2$	Conjunction
$x_1 \mid x_2$	Sheffer stroke
$x_1 \oplus x_2$	Addition modulo 2
$x_1 \leftrightarrow x_2$	Equivalence
$x_1 \rightarrow x_2$	Implication

$$\mathcal{K} = D_1 \wedge D_2 \wedge \ldots \wedge D_l,$$

where $D_i, i = 1, 2, \ldots, l$ are pairwise different elementary disjunctions. The number l is called the length of CNF. **Perfect conjunctive normal form** (PCNF) consists of elementary disjunctions of rank n.

Zhegalkin[2] polynomial is the sum modulo two of several elementary conjunctions of the form

$$G = K_1 \oplus K_2 \oplus \ldots \oplus K_s,$$

where $K_i, i = 1, 2, \ldots, s$, are pairwise different monotone elementary conjunctions over some set of variables $\{x_1, \ldots, x_n\}$, where $n = 1, 2, \ldots$. The greatest of ranks of elementary conjunctions included in the polynomial G is called a polynomial **degree**. It is known [37] that any Boolean function is uniquely presentable in the form of Zhegalkin polynomial accurate to the order of summands in the sum and the order of cofactors in the conjunctions.

There are several methods of constructing Zhegalkin polynomial $G(\mathbf{x})$, expressing the defined function $f(x_1, x_2, \ldots, x_n)$.

1. *Method of undetermined coefficients.*

Zhegalkin polynomial has the form

$$G(x_1, x_2, \ldots, x_n) = g_0 \wedge 1 \oplus g_1 \wedge K_1 \oplus g_2 \wedge K_2 \oplus \ldots \oplus g_{2^n-1} \wedge K_{2^n-1},$$

where K_i—monotone elementary conjunction, coefficients $g_i \in \mathbb{B}$, $i = 0, 1, \ldots, 2^n - 1$.

---

[2]Ivan Ivanovich Zhegalkin (1869–1947)—Russian and Soviet mathematician and logician.

In order to determine the unknown coefficients $g_0, g_1, \ldots, g_{2^n-1}$, we make equations $G(\mathbf{x}_j) = f(\mathbf{x}_j)$ for all sorts of tuples $\mathbf{x}_j$. We obtain a system of 2^n equations with 2^n unknowns; its solution will give us coefficients of the polynomial $G(\mathbf{x})$.

2. *Method of equivalent transformations.*

Let us write an expression for the function $f(\mathbf{x})$ and reduce it by identical transformations with the help of the equality $A \vee B = \overline{(\overline{A} \wedge \overline{B})}$ to an equivalent expression containing only operations from the set $\{^-, \wedge\}$. Then, we exclude the operations of negation, for which purpose replace everywhere the expressions of the form $\overline{A}$ for $A \oplus 1$. Then open the brackets, using the distributive law $A \wedge (B \oplus C) = (A \wedge B) \oplus (A \wedge C)$ and taking into account the equivalences

$$A \wedge A = A, \quad A \wedge 1 = A,$$
$$A \oplus A = 0, \quad A \oplus 0 = A.$$

As a result, we obtain Zhegalkin polynomial for the function $f(\mathbf{x})$.

The set K of functions of Boolean algebra is called **closed**, if an arbitrary superposition of the function of K also belongs to this set. The closed sets are called **classes**.

Let us consider the most important classes of Boolean functions.

1. *Functions that preserve constants 0 and 1.*

It is said that the function $f(x_1, \ldots, x_n)$ **0-preserves**, if $f(0, 0, \ldots, 0) = 0$. Similarly, the function $f(x_1, \ldots, x_n)$ **1-preserves**, if $f(1, 1, \ldots, 1) = 1$. The set of all functions of Boolean algebra of n variables that 0-preserve is denoted by T_0^n, respectively, the functions that preserve one, form the set T_1^n. The sets T_0^n and T_1^n are closed; their cardinalities are $|T_0^n| = |T_1^n| = 2^{2^n-1}$.

2. *Self-dual functions.*

The set of self-dual functions depending on n variables is a class and is denoted by S^n. The cardinality of the set S^n is $|S^n| = 2^{2^{n-1}}$.

3. *Linear functions.*

The function $f(x_1, x_2, \ldots, x_n)$ is called **linear**, if it can be presented in the form of Zhegalkin polynomial not higher than the first degree:

$$f(x_1, x_2, \ldots, x_n) = \gamma_0 \oplus \gamma_1 x_1 \oplus \gamma_2 x_2 \oplus \ldots \oplus \gamma_n x_n,$$

where $\gamma_i \in \{0, 1\}$, $i = 0, 1, \ldots, n$. The set of all linear functions depending on n variables is denoted by L^n and is a class. From the definition of L^n follows that the number of all linear functions is $|L^n| = 2^{n+1}$.

4. *Monotone functions.*

It is believed that the tuple α precedes the tuple β, and it is written $\alpha \preccurlyeq \beta$, if $\alpha_i \leqslant \beta_i$ for $i = 1, 2, \ldots, n$. Boolean function $f(\mathbf{x})$ is called **monotone**, if for any two tuples α and β of $\mathbb{B}^n$ such that $\alpha \preccurlyeq \beta$, the inequality is fulfilled $f(\alpha) \leqslant f(\beta)$. The set M^n of monotone functions, depending on n variables, is a class.

If from the context, the number of variables is known on which $f(\mathbf{x})$ depends, then the symbol "n" in the notation of the considered classes is omitted: T_0, T_1, and so on.

The system of functions F is called **complete**, if any Boolean function can be obtained by superposition of the functions from F. A completeness criterion is known obtained by Post[3]: *The system F of functions of $P_2(n)$ is complete if and only if it is not contained completely in any of the classes T_0, T_1, S, L, and M.* In other words, for completeness of the system F, it is necessary and sufficient that F should have: at least one function, not preserving 0; at least one function, not preserving 1; at least one non-self-dual function; at least one nonlinear function and at least one non-monotone function.

The minimal complete set of logic functions is called a **basis**. If any function is removed from the basis, the system ceases to be complete.

6.1   Karnaugh Maps

For practical applications, an important task consists in reducing the Boolean function written in disjunctive normal form. One of the methods to solve this problem was developed by Karnaugh[4] [33].

Karnaugh map is a table, each element of which represents an elementary conjunction. For the function, depending on n Boolean variables $x_1, \ldots, x_n$, there exist 2^n different elementary conjunctions of rank n; this is why the Karnaugh map of function $f(x_1, \ldots, x_n)$ contains 2^n cells, each of which corresponds to one of the combinations of the form $x_1^{\sigma_1} \wedge x_2^{\sigma_2} \wedge \ldots \wedge x_n^{\sigma_n}$. The table has $2^{\lfloor n/2 \rfloor}$ rows and $2^{\lceil n/2 \rceil}$ columns, and the adjacent cells differ in the value of one variable [4,59].

Karnaugh maps for instances $n = 2$ and $n = 3$ are shown in Fig. 6.1.

For simplification of Boolean function using the Karnaugh map, one should do the following:

(1) for each elementary conjunction, denote the respective rectangle of table, say, by the symbol "$*$";

(2) form out of the tagged cells rectangular blocks of the maximal size, fully covering in total all cells with the symbols "$*$";

(3) write the expressions, corresponding to the written blocks, and unite them by the operation of disjunction.

To make the state rules for working with the Karnaugh map more clear, consider several examples.

Example 6.1 To the Boolean function $f(p, q, r) = (p \wedge q \wedge r) \vee (p \wedge \overline{q} \wedge \overline{r}) \vee$ $\vee (\overline{p} \wedge \overline{q} \wedge \overline{r})$ corresponds the Karnaugh map

---

[3]Emil Leon Post (1897–1954)—American mathematician and logician.
[4]Maurice Karnaugh (born in 1924)—American physicist.

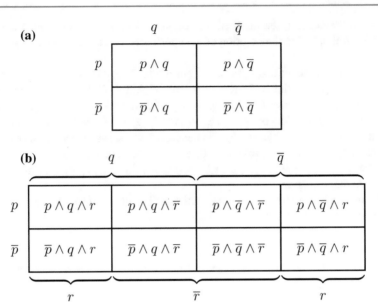

Fig. 6.1 Karnaugh maps for $n = 2$ is a panel (a) and $n = 3$ is a panel (b)

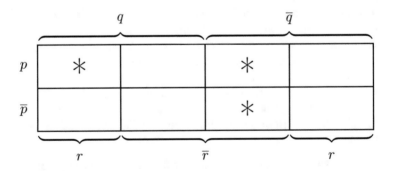

elementary conjunction, whose cells are adjacent ($p \wedge \overline{q} \wedge \overline{r}$ and $\overline{p} \wedge \overline{q} \wedge \overline{r}$), can be reduced to one conjunction, containing only two variables, namely $\overline{q} \wedge \overline{r}$:

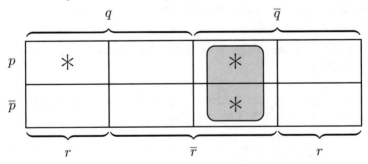

As a result, $f(p, q, r)$ can be written in the form $f(p, q, r) = (p \wedge q \wedge r) \vee (\overline{q} \wedge \overline{r})$. □

Example 6.2 Boolean function

$$g(p, q, r) = (p \wedge q \wedge \overline{r}) \vee (p \wedge \overline{q} \wedge \overline{r}) \vee (\overline{p} \wedge q \wedge \overline{r}) \vee (\overline{p} \wedge \overline{q} \wedge \overline{r})$$

contains in the Karnaugh map four order "*", that are adjacent.

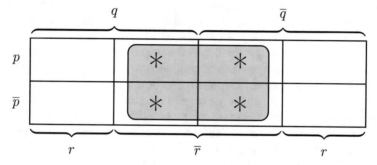

The entire expression $g(p, q, r)$ reduces to $g(p, q, r) = \overline{r}$. □

Example 6.3 Consider the Boolean function

$$h(p, q, r) = (p \wedge q \wedge r) \vee (p \wedge \overline{q} \wedge \overline{r}) \vee (p \wedge \overline{q} \wedge r) \vee (\overline{p} \wedge \overline{q} \wedge \overline{r}) \vee (\overline{p} \wedge \overline{q} \wedge r).$$

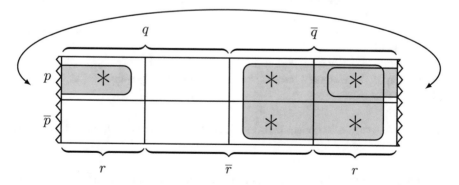

Note that the minterms $(p \wedge q \wedge r)$ and $(p \wedge \overline{q} \wedge r)$ may also be considered adjacent, if we identify ("glue") the lateral sides of the table. Therefore, $h(p, q, r)$ can be written in the form $h(p, q, r) = \overline{q} \vee (p \wedge r)$. □

6.2 Combinational Circuits

The methods of Boolean algebra are used for creation and analysis of **combinational circuits**, that allow understanding the structure and the logic of digital devices' operation, including computers. Combinational circuits consist of electronic devices (called **gates**) with a finite number of inputs and outputs, and each input and output can have only two values of the signal [27,72].

Signals 0 and 1 are set by different voltage levels. A signal of a logic zero is usually represented by lower voltage, while that of a logic one—by higher voltage; such an agreement is referred to as the **positive logic**. In the **negative logic**, it is the practice to use inverse mapping of voltages for the logic values.

The logic circuits are represented with the help of conventional graphical notations, implying the function performed by this element and not affecting the peculiarities of its physical implementation, which may be relatively complex. Presently, the literature suggests several generally accepted standards of conventional notations. The most common are the foreign standard ANSI/IEEE Std 91-1984 "IEEE Standard Graphic Symbols for Logic Functions". In other references, the conventional notations of the elements comply with the GOST 2.743-91 (State Standard) "Conventional graphic notations in schemes. Elements of digital equipment" or IEC 60617-12: 1997 "Graphical symbols for diagrams — Part 12: Binary logic elements".

Notations of the basic gates in accordance with standards IEEE and GOST are shown in Table 6.6.

It is known that with the help of the considered combinational circuits it is possible to implement any Boolean function [37].

Example 6.4 Write the Boolean expression, associated with the combinational circuit S_1 in Fig. 6.2.

For this, let us write out in Table 6.7 inputs and outputs for each binary gate that form a circuit.

We obtain the Boolean expression $(p \wedge q) \vee \overline{(q \vee r)} \vee (r \wedge \overline{p})$, which, after application of De Morgan's law, can be rewritten in the equivalent form: $(p \wedge q) \vee (\overline{q} \wedge \overline{r}) \vee (r \wedge \overline{p})$. □

The next example demonstrates how it is possible to reduce a combinational circuit using the Karnaugh map.

Example 6.5 Let combinational circuit S_2 be specified, as shown in Fig. 6.3. Using the Karnaugh map, construct an equivalent circuit consisting of no less number of gates compares to the original one.

Compile a table, containing signal values at the inputs and outputs of binary gates (Table 6.8). The function $f(p, q, r)$, generated by this circuit, has the form

$$f(p,q,r) = \overline{((p \wedge \overline{q} \wedge \overline{r}) \vee \overline{(p \vee r)})}.$$

Table 6.6 Notations of the basic gates

Name	GOST 2.743-91	ANSI/IEEE Std 91-1984	Boolean function
NOT			negation
OR			disjunction
AND			conjunction
NAND			Sheffer stroke
NOR			Peirce arrow

Table 6.7 Input and output values of the combinational circuit S_1

Gate	Inputs	Output
1	p, q	$p \wedge q$
2	q, r	$q \vee r$
3	$r, \overline{p}$	$r \wedge \overline{p}$
4	$\overline{(q \vee r)}, r \wedge \overline{p}$	$\overline{(q \vee r)} \vee (r \wedge \overline{p})$
5	$p \wedge q,$ $\overline{(q \vee r)} \vee (r \wedge \overline{p})$	$(p \wedge q) \vee \overline{(q \vee r)} \vee (r \wedge \overline{p})$

Taking into account De Morgan's laws and negation properties, we obtain:

$$f(p, q, r) = \overline{(p \wedge \overline{q} \wedge \overline{r})} \wedge (p \vee r).$$

Calculate the vector of values of the function $f(p, q, r)$: $\alpha_f = (0101\ 0111)$. Then, find the perfect disjunctive normal form f. For this, we should note that the

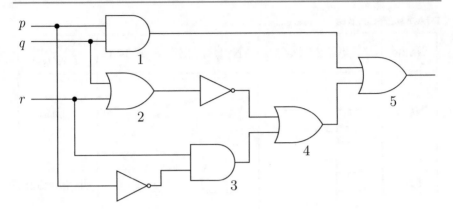

Fig. 6.2 Combinational circuit S_1 with three inputs

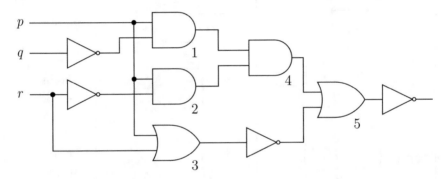

Fig. 6.3 Combinational circuit S_2

Table 6.8 Input and output values of the combinational circuit S_2

Gate	Inputs	Output
1	$p, \overline{q}$	$p \wedge \overline{q}$
2	$p, \overline{r}$	$p \wedge \overline{r}$
3	p, r	$p \vee r$
4	$p \wedge \overline{q}, p \wedge \overline{r}$	$(p \wedge \overline{q}) \wedge (p \wedge \overline{r}) =$ $p \wedge \overline{q} \wedge \overline{r}$
5	$p \wedge \overline{q} \wedge \overline{r}, \overline{(p \vee r)}$	$(p \wedge \overline{q} \wedge \overline{r}) \vee \overline{(p \vee r)}$

function f takes the value equal to 1, on the following tuples of arguments (p, q, r):

$$(0, 0, 1), \ (0, 1, 1), \ (1, 0, 1), \ (1, 1, 0), \ (1, 1, 1).$$

Uniting the respective elementary conjunctions, we obtain perfect disjunctive normal form of the function f:

$$\mathcal{D}_f = (\overline{p} \wedge \overline{q} \wedge r) \vee (\overline{p} \wedge q \wedge r) \vee (p \wedge \overline{q} \wedge r) \vee (p \wedge q \wedge \overline{r}) \vee (p \wedge q \wedge r).$$

We form the Karnaugh map based on the obtained $\mathcal{D}_f$:

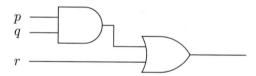

The tagged cells form two blocks, and for $f(p, q, r)$, we obtain the following expression: $f(p, q, r) = (p \wedge q) \vee r$. Thus, the combinational circuit of the form

is equivalent to S_2 and contains less number of gates compared to it. □

Based on the considered gates, it is possible to design a device that calculates the sum of two binary numbers $a +_2 b$, where $a, b \in \mathbb{B}$, and the sign "$+_2$" denotes a binary operation of addition in the binary system. Such a device is called "**half-bit adder**" because when adding binary numbers with more than one digit, only the low-order digits are taken into account by the half-bit adder [4].

For different binary numbers a and b, we obtain

$$0 +_2 0 = 00,$$
$$0 +_2 1 = 01,$$
$$1 +_2 0 = 01,$$
$$1 +_2 1 = 10.$$

Hence, the result of operation $a +_2 b$ will be binary number $(cd)_2$, whose orders are determined in accordance with Table 6.9.

From the provided Table 6.9 follows that the Boolean functions $c = c(a, b)$, and $d = d(a, b)$ have analytical representation: $c = a \wedge b$, $d = a \oplus b$. We can suggest a combinational circuit for implementation of a half-bit adder, shown in Fig. 6.4.

Similarly, a device is designed that calculates a sum of two-digit binary numbers (**2-bit adder**). Denote by a and b the digits of the fist summand of the sum (a—high order, b—low order), and by c and d—the digits of the second summand (c— high order, d—low order). The result of binary addition has three digits: $(efg)_2$.

One of possible implementations of a 2-bit adder is shown in Fig. 6.5 [27].

Table 6.9 Operation of	a	b	c	d
binary addition	0	0	0	0
$a +_2 b = (cd)_2$	0	1	0	1
	1	0	0	1
	1	1	1	0

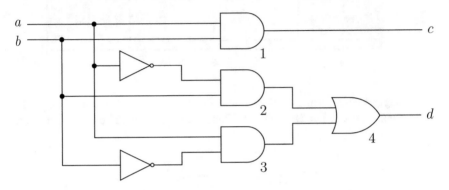

Fig. 6.4 Half-bit adder

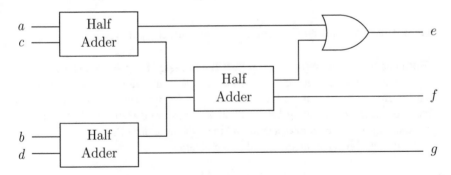

Fig. 6.5 2-bit adder

6.3 Problems

6.1. Compile the truth table of Boolean function of three variables, specified by the logic expression $f(x_1, x_2, x_3) = ((x_1 \wedge \overline{x}_2) \vee \overline{x}_3) \leftrightarrow (x_2 \oplus x_3)$. Write out the vector of values α_f of the function $f(\mathbf{x})$.

6.2. Compile the truth table of Boolean function of three variables, specified by the logic expression $f(x_1, x_2, x_3) = ((x_1 \vee \overline{x}_2) \wedge x_3) \rightarrow (x_1 \oplus x_3)$. Write out the vector of values α_f of the function $f(\mathbf{x})$.

6.3. The series of Boolean values a_i, $i = 1, 2, \ldots$, is specified by the recurrence relation (see chapter "Recurrence Relations" on p. 273)

$$\begin{cases} a_n = (a_{n-1} \wedge a_{n-2}) \oplus a_{n-3}, & n > 3, \\ a_1 = 1, \ a_2 = 0, \ a_3 = 1. \end{cases}$$

Calculate a_n for all natural values n.

6.4. The series of Boolean values a_i, $i = 1, 2, \ldots$, is specified by the recurrence relation

$$\begin{cases} a_n = (a_{n-1} \oplus a_{n-2}) \vee a_{n-3}, & n > 3, \\ a_1 = 0, \ a_2 = 1, \ a_3 = 1. \end{cases}$$

Calculate a_n for all natural values n.

6.5. The series of Boolean values a_i, $i = 1, 2, \ldots$, is specified by the recurrence relation

$$\begin{cases} a_n = (a_{n-1} \leftrightarrow a_{n-2}) \wedge a_{n-3}, & n > 3, \\ a_1 = 0, \ a_2 = 0, \ a_3 = 0. \end{cases}$$

Calculate a_n for all natural values n.

6.6. The series of Boolean values a_i, $i = 1, 2, \ldots$, is specified by the recurrence relation

$$\begin{cases} a_n = (a_{n-1} \leftrightarrow a_{n-3}) \oplus a_{n-2}, & n > 3, \\ a_1 = 0, \ a_2 = 1, \ a_3 = 1. \end{cases}$$

Calculate a_n for all natural values n.

6.7. Prove that Boolean expressions $x_1 \leftrightarrow x_2$ and $1 \oplus x_1 \oplus x_2$ determine the same Boolean function.

6.8. Prove that Boolean expressions $x_1 \mid x_2$ and $1 \oplus (x_1 \wedge x_2)$ determine the same Boolean function.

6.9. Find whether the Boolean functions f and g are equivalent:

(1) $f(x_1, x_2, x_3) = x_1 \rightarrow (x_2 \rightarrow x_3)$, $g(x_1, x_2, x_3) = (x_1 \wedge x_2) \rightarrow x_3$;
(2) $f(x_1, x_2) = (x_1 \rightarrow x_2) \wedge (\overline{x}_2 \rightarrow x_1)$, $g(x_1, x_2) = x_1$.

6.10. Using the laws of Boolean algebra, show that for the variables x_1, x_2, x_3 is fulfilled the implication distributivity over

(1) conjunction: $x_1 \rightarrow (x_2 \wedge x_3) = (x_1 \rightarrow x_2) \wedge (x_1 \rightarrow x_3)$;
(2) disjunction: $x_1 \rightarrow (x_2 \vee x_3) = (x_1 \rightarrow x_2) \vee (x_1 \rightarrow x_3)$;
(3) implication: $x_1 \rightarrow (x_2 \rightarrow x_3) = (x_1 \rightarrow x_2) \rightarrow (x_1 \rightarrow x_3)$;
(4) equivalence: $x_1 \rightarrow (x_2 \leftrightarrow x_3) = (x_1 \rightarrow x_2) \leftrightarrow (x_1 \rightarrow x_3)$.

6.11. Construct the perfect disjunctive and perfect conjunctive normal forms of the function, specified by the vector of values $\alpha_f = (1011\ 0011)$.

6.12. Construct the perfect disjunctive and perfect conjunctive normal forms of the function, specified by the vectors of values:

(1) $\alpha_{f_1} = (0011\ 0101)$;
(2) $\alpha_{f_2} = (0111\ 0111)$.

6.13. Construct PDNF and PCNF for the following functions:

(1) $f_1(x_1, x_2, x_3) = (x_1 \oplus x_3) \vee \overline{x}_2$;
(2) $f_2(x_1, x_2, x_3) = (x_1 \wedge x_2) \vee (x_2 \oplus x_3) \vee (x_1 \rightarrow x_3)$.

6.14. Construct PDNF and PCNF for the following functions:

(1) $f_1(x_1, x_2, x_3) = x_1 \rightarrow (x_2 \mid x_3)$;
(2) $f_2(x_1, x_2, x_3) = (x_1 \downarrow x_2) \leftrightarrow \overline{(x_2 \mid x_3)}$.

6.15. Using the method of undetermined coefficients, construct Zhegalkin polynomial over the vector of values of the function:

(1) $\alpha_g = (0011\ 0110)$;
(2) $\alpha_h = (1100\ 1100\ 0011\ 0000)$.

6.16. Using the method of equivalent transformations, construct Zhegalkin polynomial for each of the following functions:

(1) $g(x_1, x_2, x_3) = (x_1 \wedge x_2) \oplus (x_1 \downarrow x_3)$;
(2) $h(x_1, x_2, x_3, x_4) = \overline{(x_1 \rightarrow x_2)} \rightarrow (x_3 \leftrightarrow x_4)$.

6.17. Find out whether the function $f(x_1, x_2)$ preserves the 0 constant, if

(1) $f(x_1, x_2) = x_1 \leftrightarrow x_2$;
(2) $f(x_1, x_2) = x_1 \oplus x_2$.

6.18. Find out whether the function $f(x_1, x_2)$ preserves the 0 constant, if

(1) $f(x_1, x_2) = x_1 \mid x_2$;
(2) $f(x_1, x_2) = x_1 \downarrow x_2$.

6.19. Find out whether the function $f(x_1, x_2)$ preserves the 1 constant, if

(1) $f(x_1, x_2) = x_1 \oplus x_2 \oplus 1$;
(2) $f(x_1, x_2) = (x_1 \rightarrow x_2) \vee (x_2 \leftrightarrow \overline{x}_1)$.

6.20. Find out whether the function $f(x_1, x_2)$ preserves the 1 constant, if

(1) $f(x_1, x_2) = (x_1 \downarrow x_2) \leftrightarrow ((x_1 \oplus 1) \rightarrow (x_2 \oplus 1))$;
(2) $f(x_1, x_2) = ((x_1 \wedge x_2) \oplus ((x_1 \oplus 1) \wedge x_2)) \rightarrow (x_1 \mid x_2)$.

6.21. Find out whether the function h is dual to the function g:

(1) $g(x_1, x_2) = x_1 \vee x_2$, $\quad h(x_1, x_2) = x_1 \wedge x_2$;
(2) $g(x_1, x_2) = (x_1 \wedge \overline{x}_2) \vee (\overline{x}_1 \rightarrow x_2)$, $\quad h(x_1, x_2) = \overline{x}_1$?

6.22. Is the function f, specified by the vector of values α_f, self-dual, if

(1) $\alpha_f = (0110)$;
(2) $\alpha_f = (1011\ 0010)$?

6.23. Is the function f, specified by the vector of values α_f, self-dual, if

(1) $\alpha_f = (1000\ 0101\ 0101\ 1110)$;
(2) $\alpha_f = (0101\ 1100\ 0011\ 1010)$?

6.24. Check whether the following functions are self-dual:

(1) $f(x_1, x_2) = x_1 \oplus x_2$;
(2) $f(x_1, x_2) = x_1 \leftrightarrow x_2$.

6.25. Prove that the function $f(x_1, x_2, x_3) = x_1 \oplus x_2 \oplus x_3$ is self-dual.

6.26. Let the function $f(\mathbf{x})$ be specified by the vector of its values

$$\alpha_f = (\alpha_0, \alpha_1, \ldots, \alpha_{2^n - 2}, \alpha_{2^n - 1}).$$

Write out the vector of values of the function $f^*(\mathbf{x})$.

6.27. Weight $w(\alpha)$ of the Boolean tuple $\alpha = (\alpha_1, \alpha_2, \ldots, \alpha_n)$ is the number of its components equal to one. Find the number of vectors of length n, having weight $w(\alpha) = k$, where $k \leqslant n$.

6.28. Calculate weight of the vector of values of the arbitrary self-dual function $s(\mathbf{x})$, depending on n variables, $n \geqslant 1$.

6.29. Is the function f linear if

(1) $f(x_1, x_2) = x_1 \vee \overline{x}_2$;
(2) $f(x_1, x_2) = x_1 \mid x_2$?

6.30. Is the function f linear if

(1) $f(x_1) = \overline{(x_1 \oplus 1)}$;
(2) $f(x_1, x_2) = x_1 \rightarrow x_2$?

6.31. Check whether the function f is monotone, if

(1) $f(x_1, x_2) = x_1 \vee x_2$;
(2) $f(x_1, x_2) = x_1 \wedge x_2$.

6.32. Is the function f monotone, if

(1) $f(x_1, x_2) = x_1 \oplus x_2$;
(2) $f(x_1, x_2) = x_1 \rightarrow x_2$?

6.33. Is the function f monotone, if

(1) $f(x_1, x_2) = (x_1 \downarrow x_2) \vee x_1$;
(2) $f(x_1, x_2) = (x_1 \leftrightarrow x_2) \wedge (\overline{x}_1 \vee x_2)$?

6.34. A student conducting an experiment in a chemical laboratory by accident spilled some reagent over the page from the notes on discrete mathematics, where the vector of values of some monotone Boolean function $f(\mathbf{x})$ was written. As a result, he was unable to read some components of this vector. If we denote those components by symbol "?", the notation will look like

$$\alpha_f = (???0\ ?10?\ ?001\ 1???).$$

Restore the unknown components of α_f and write the analytical expression for $f(\mathbf{x})$.

6.35. With what minimal n the number of Boolean functions depending on n variables will exceed a billion?

6.36. Fill in the empty cells in the function class cardinalities table of Boolean algebra:

n	$|T_0^n|$	$|T_1^n|$	$|S^n|$	$|L^n|$	$|M^n|$
1	2	2	2	4	3
2					
3					
4					

∗6.37. Find the number of Boolean functions that depend on n variables and belong to the set:

(1) $X = T_0 \cup T_1$;
(2) $Y = T_0 \cup L$.

∗6.38. Find the number of Boolean functions that depend on n variables and belong to the set:

(1) $X = T_0 \cap S$;
(2) $Y = L \setminus T_1$.

∗6.39. Find the number of Boolean functions that depend on n variables and belong to the set:

(1) $X = M \setminus T_0$;
(2) $Y = L \cap M$.

∗6.40. Find the number of Boolean functions that depend on n variables and belong to the set:

(1) $T_0 \cup T_1 \cup L$;
(2) $L \setminus (T_0 \setminus T_1)$;
(3) $T_0 \cup T_1 \cup S$;
(4) $S \setminus (T_0 \setminus T_1)$.

*6.41. Prove the following estimates for the number of monotone functions of n variables, where $n \geqslant 1$:

$$2^{C(n, \lfloor n/2 \rfloor)} \leqslant |M^n| \leqslant 2 + n^{C(n, \lfloor n/2 \rfloor)}.$$

6.42. Prove that the system of functions $\{^-, \vee, \wedge\}$ is complete in P_2, where P_2 is the set of all functions of Boolean algebra.

6.43. Check whether the system of functions $\{^-, \wedge\}$ is complete in P_2.

6.44. Check whether the system of functions $\{^-, \vee\}$ is complete in P_2.

6.45. Check for completeness in P_2 each of the systems of functions

(1) $\{\downarrow\}$;
(2) $\{|\}$.

6.46. Using Post's criterion, prove that the system of functions $\{\oplus, \wedge, 1\}$ (**Zhegalkin's basis**) is complete in P_2.

6.47. Using the Post criterion, prove that each of the following systems of functions is complete in P_2:

(1) $\{0, \vee, \leftrightarrow\}$;
(2) $\{0, \rightarrow\}$.

6.48. Find which of the following systems of functions are bases in P_2:

(1) $\{\rightarrow, \oplus\}$;
(2) $\{\oplus, \leftrightarrow, \wedge\}$;
(3) $\{\oplus, \leftrightarrow, \vee\}$;
(4) $\{\leftrightarrow, \rightarrow, ^-\}$.

*6.49. Prove using "by contradiction" method, that a basis cannot contain more than four functions.

*6.50. What minimal number of functions can a basis contain?

6.51. Simplify the Boolean expression, represented by the Karnaugh map

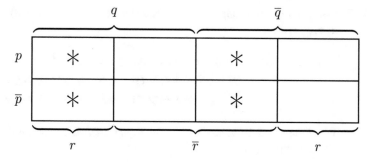

6.52. Simplify the Boolean expression, represented by the Karnaugh map

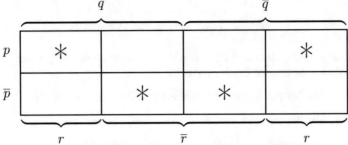

6.53. Simplify the following expression using the Karnaugh map:

(1) $S_1 = (p \wedge q \wedge r) \vee (p \wedge q \wedge \bar{r}) \vee (p \wedge \bar{q} \wedge \bar{r}) \vee (p \wedge \bar{q} \wedge r)$;

(2) $S_2 = (p \wedge q \wedge r) \vee (p \wedge q \wedge \bar{r}) \vee (\bar{p} \wedge \bar{q} \wedge \bar{r}) \vee (\bar{p} \wedge \bar{q} \wedge r)$.

6.54. Simplify the following expression using the Karnaugh map:

(1) $S_1 = (p \wedge q \wedge r) \vee (\bar{p} \wedge q \wedge r) \vee (p \wedge \bar{q} \wedge r) \vee (\bar{p} \wedge \bar{q} \wedge r)$;

(2) $S_2 = (p \wedge q \wedge r) \vee (\bar{p} \wedge q \wedge r) \vee (\bar{p} \wedge q \wedge \bar{r}) \vee (p \wedge \bar{q} \wedge r) \vee (\bar{p} \wedge \bar{q} \wedge r)$.

6.55. Construct the Karnaugh map for Boolean function $t(p, q, r, s)$ of four variables and simplify $t(p, q, r, s)$:

$$t(p, q, r, s) = (p \wedge q \wedge r \wedge s) \vee (p \wedge q \wedge r \wedge \bar{s}) \vee (p \wedge \bar{q} \wedge \bar{r} \wedge \bar{s}) \vee$$
$$\vee (p \wedge \bar{q} \wedge r \wedge \bar{s}) \vee (\bar{p} \wedge \bar{q} \wedge \bar{r} \wedge \bar{s}) \vee (\bar{p} \wedge \bar{q} \wedge r \wedge \bar{s}).$$

6.56. Construct the Karnaugh map for Boolean function $g(p, q, r, s)$ and simplify $g(p, q, r, s)$:

$$g(p, q, r, s) = (p \wedge q \wedge r \wedge \bar{s}) \vee (p \wedge \bar{q} \wedge r \wedge \bar{s}) \vee (\bar{p} \wedge q \wedge r \wedge s) \vee$$
$$\vee (\bar{p} \wedge q \wedge \bar{r} \wedge s) \vee (\bar{p} \wedge \bar{q} \wedge \bar{r} \wedge s) \vee (\bar{p} \wedge \bar{q} \wedge r \wedge s).$$

6.57. Construct the Karnaugh map for Boolean function $h(p, q, r, s)$ and simplify $h(p, q, r, s)$:

$$h(p, q, r, s) = (p \wedge \bar{q} \wedge \bar{r} \wedge s) \vee (p \wedge \bar{q} \wedge r \wedge s) \vee (p \wedge q \wedge r \wedge \bar{s}) \vee$$
$$\vee (p \wedge \bar{q} \wedge \bar{r} \wedge \bar{s}) \vee (p \wedge \bar{q} \wedge r \wedge \bar{s}) \vee (\bar{p} \wedge q \wedge r \wedge \bar{s}) \vee$$
$$\vee (\bar{p} \wedge q \wedge \bar{r} \wedge \bar{s}) \vee (\bar{p} \wedge q \wedge \bar{r} \wedge s) \vee (\bar{p} \wedge \bar{q} \wedge \bar{r} \wedge s).$$

6.58. Write the vector of values for the function $f(\mathbf{x})$, construct the Karnaugh map and simplify $f(\mathbf{x})$:

(1) $f(x_1, x_2, x_3, x_4) = (x_1 \oplus x_3) \vee (\overline{x}_2 \wedge x_4)$;
(2) $f(x_1, x_2, x_3, x_4) = (x_1 \wedge x_3) \vee (\overline{x}_2 \rightarrow x_4)$.

6.59. Write the Boolean function generated by the combinational circuit

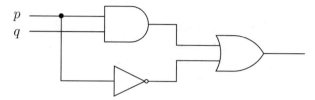

6.60. Write the Boolean function generated by the combinational circuit

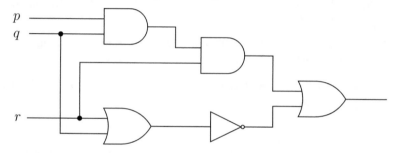

6.61. Construct combinational circuits, associated with the following Boolean expressions:

(1) $\overline{((p \vee q) \wedge r)}$;
(2) $\overline{((p \wedge q) \vee (q \wedge \overline{r}))} \wedge (q \vee r)$.

6.62. Construct a combinational circuit, implementing the implication $p \rightarrow q$, using only the following gates:

(1) **NAND**;
(2) **NOR**.

6.63. Construct a combinational circuit, implementing the equivalence function $p \leftrightarrow q$, using only the following gates:

(1) **NAND**;
(2) **NOR**.

∗**6.64.** Construct, using the gates **NAND** a combinational circuit, equivalent to the gate **NOR**.

∗**6.65.** Construct, using the gates **NOR** a combinational circuit, equivalent to the gate **NAND**.

6.66. Three signals are inputted to the combinational circuit: s_1, s_2, and s_3. The value of the output signal is equal to the value of the most of variables s_1, s_2, s_3. Draw a combinational circuit satisfying these conditions.

***6.67.** Three signals are inputted to the combinational circuit s_1, s_2, ..., s_n. The value of the output signal is equal to 1, if the number of nonzero signals among s_i, where $1 \leqslant i \leqslant n$, is even, and equal to 0 otherwise. Draw a combinational circuit satisfying these conditions.

6.68. Verify that the 2-bit adder shown in Fig. 6.5 correctly calculates the sum $01 +_2 10 = 011$.

6.69. Verify that the 2-bit adder shown in Fig. 6.5 correctly calculates the sum $10 +_2 10 = 100$.

6.70. Construct **3-bit adder**, calculating the binary sum of three-digit numbers $(abc)_2$ and $(def)_2$ [27,59].

***6.71.** Construct **full adder**, calculating the sum of two numbers, written in the binary system.

6.4   Answers, Hints, Solutions

6.1. *Solution.*
In order to compile the truth table of the function $f(x_1, x_2, x_3)$ calculate its values on all sorts of sets of values of the vector of arguments $\mathbf{x} = (x_1, x_2, x_3)$.

x_1	x_2	x_3	$x_1 \wedge \overline{x}_2$	$(x_1 \wedge \overline{x}_2) \vee \overline{x}_3$	$x_2 \oplus x_3$	$f(x_1, x_2, x_3)$
0	0	0	0	1	0	0
0	0	1	0	0	1	0
0	1	0	0	1	1	1
0	1	1	0	0	0	1
1	0	0	1	1	0	0
1	0	1	1	1	1	1
1	1	0	0	1	1	1
1	1	1	0	0	0	1

The vector of values of the function $f(\mathbf{x})$ is $\alpha_f = (0011\ 0111)$.

6.2. *Solution.*
Compiling the table of values by a regular method, we obtain

x_1	x_2	x_3	$x_1 \vee \bar{x}_2$	$(x_1 \vee \bar{x}_2) \wedge x_3$	$x_1 \oplus x_3$	$f(x_1, x_2, x_3)$
0	0	0	1	0	0	1
0	0	1	1	1	1	1
0	1	0	0	0	0	1
0	1	1	0	0	1	1
1	0	0	1	0	1	1
1	0	1	1	1	0	0
1	1	0	1	0	1	1
1	1	1	1	1	0	0

The vector of values of the function $f(\mathbf{x})$ is $\alpha_f = (1111\ 1010)$.

6.3. *Solution.*

Write out several first members of the sequence $\{a_n\}$:

a_1	a_2	a_3	a_4	a_5	a_6	a_7	a_8	a_9	a_{10}
1	0	1	1	1	0	1	1	1	0

Using the mathematical induction method, prove that for all natural values n

$$\begin{cases} a_n = 1, \ \text{if } n = 4k - 3, 4k - 1, 4k, \quad k \in \mathbb{N}, \\ a_n = 0, \ \text{if } n = 4k - 2, \quad k \in \mathbb{N}. \end{cases}$$

Denote by $P(k)$ the predicate associated with the above relation for a_n.

B a s i s   s t e p

Check the truth of $P(1)$. We have $a_0 = 1$, $a_2 = 0$, $a_3 = 1$, $a_4 = 1$ according to the statement.

I n d u c t i v e   s t e p

Let $P(k)$ be true for some natural value k. Show the truth of $P(k + 1)$, for this, calculate a_{k+1}.

Consider four possible cases.

(1) $k = 4k'$, $k' \in \mathbb{N}$,

$$a_{4k'+1} = (a_{4k'} \wedge a_{4k'-1}) \oplus a_{4k'-2} = (1 \wedge 1) \oplus 0 = 1.$$

(2) $k = 4k' + 1$, $k' \in \mathbb{N}$,

$$a_{4k'+2} = (a_{4k'+1} \wedge a_{4k'}) \oplus a_{4k'-1} = (1 \wedge 1) \oplus 1 = 0.$$

(3) $k = 4k' + 2$, $k' \in \mathbb{N}$,

$$a_{4k'+3} = (a_{4k'+2} \wedge a_{4k'+1}) \oplus a_{4k'} = (0 \wedge 1) \oplus 1 = 1.$$

(4) $k = 4k' + 3$, $k' \in \mathbb{N}$,

$$a_{4k'+4} = (a_{4k'+3} \wedge a_{4k'+2}) \oplus a_{4k'+1} = (1 \wedge 0) \oplus 1 = 1.$$

Thus, we have proved the truth of $P(k+1)$. Therefore,

$$\begin{cases} a_n = 1, & \text{if } n = 4k - 3, 4k - 1, 4k, \quad k \in \mathbb{N}, \\ a_n = 0, & \text{if } n = 4k - 2, \quad k \in \mathbb{N}. \end{cases}$$

6.4. *Answer*:

$$\begin{cases} a_n = 1, & \text{if } n = 3k - 1, 3k, \quad k \in \mathbb{N}, \\ a_n = 0, & \text{if } n = 3k - 2, \quad k \in \mathbb{N}. \end{cases}$$

6.5. *Answer*: $a_n = 0$ for $n = 1, 2, 3, \ldots$

6.6. *Answer*:

$$\begin{cases} a_n = 1, & \text{if } n = 4k - 2, 4k - 1, 4k, \quad k \in \mathbb{N}, \\ a_n = 0, & \text{if } n = 4k - 3, \quad k \in \mathbb{N}. \end{cases}$$

6.7. *Proof.*
Having compiled the truth table, we find the vectors of values of the function $f_1 = x_1 \leftrightarrow x_2$ and $f_2 = x_1 \oplus x_2 \oplus 1$: $\alpha_{f_1} = (1001)$, $\alpha_{f_2} = (1001)$. Therefore, $f_1 = f_2$.

6.9. *Solution.*

(1) Having expressed implication in terms of disjunction and negation, we obtain

$$f(x_1, x_2, x_3) = x_1 \rightarrow (x_2 \rightarrow x_3) = \overline{x}_1 \vee \overline{x}_2 \vee x_3,$$
$$g(x_1, x_2, x_3) = (x_1 \wedge x_2) \rightarrow x_3 = \overline{(x_1 \wedge x_2)} \vee x_3 = \overline{x}_1 \vee \overline{x}_2 \vee x_3.$$

Since the functions f and g are specified by equivalent Boolean expressions, then $f(x_1, x_2, x_3) = g(x_1, x_2, x_3)$.
(2) $f(x_1, x_2) = (x_1 \rightarrow x_2) \wedge (\overline{x}_2 \rightarrow x_1) = x_2 \vee (\overline{x}_1 \wedge x_1) = x_2$. Hence, the functions f and g are not equivalent.

6.11. *Solution.*

The function $f(x_1, x_2, x_3)$ takes the value equal to 1, on the following tuples of arguments (x_1, x_2, x_3):

$$(0, 0, 0), \ (0, 1, 0), \ (0, 1, 1), \ (1, 1, 0), \ (1, 1, 1).$$

Unite respective elementary conjunctions, and we obtain perfect disjunctive normal form:

$$\mathcal{D}_f = \overline{x}_1\overline{x}_2\overline{x}_3 \vee \overline{x}_1 x_2 \overline{x}_3 \vee \overline{x}_1 x_2 x_3 \vee x_1 x_2 \overline{x}_3 \vee x_1 x_2 x_3.$$

The value equal to 0, the function f takes on the tuples

$$(0, 0, 1), \ (1, 0, 0), \ (1, 0, 1).$$

Conjunction of respective elementary disjunctions results in the perfect conjunctive normal form:

$$\mathcal{K}_f = (x_1 \vee x_2 \vee \overline{x}_3)(\overline{x}_1 \vee x_2 \vee x_3)(\overline{x}_1 \vee x_2 \vee \overline{x}_3).$$

6.12. *Answer*:

(1) $\mathcal{D}_{f_1} = \overline{x}_1 x_2 \overline{x}_3 \vee \overline{x}_1 x_2 x_3 \vee x_1 \overline{x}_2 x_3 \vee x_1 x_2 x_3,$
$\mathcal{K}_{f_1} = (x_1 \vee x_2 \vee x_3)(x_1 \vee x_2 \vee \overline{x}_3)(\overline{x}_1 \vee x_2 \vee x_3)(\overline{x}_1 \vee \overline{x}_2 \vee x_3);$
(2) $\mathcal{D}_{f_2} = \overline{x}_1 \overline{x}_2 x_3 \vee \overline{x}_1 x_2 \overline{x}_3 \vee \overline{x}_1 x_2 x_3 \vee x_1 \overline{x}_2 \overline{x}_3 \vee x_1 x_2 \overline{x}_3 \vee x_1 x_2 x_3,$
$\mathcal{K}_{f_2} = (x_1 \vee x_2 \vee x_3)(\overline{x}_1 \vee x_2 \vee x_3).$

6.13. *Answer*:

(1) $\mathcal{D}_{f_1} = \overline{x}_1 \overline{x}_2 \overline{x}_3 \vee \overline{x}_1 \overline{x}_2 x_3 \vee \overline{x}_1 x_2 x_3 \vee x_1 \overline{x}_2 \overline{x}_3 \vee x_1 \overline{x}_2 x_3 \vee x_1 x_2 \overline{x}_3,$
$\mathcal{K}_{f_1} = (x_1 \vee \overline{x}_2 \vee x_3)(\overline{x}_1 \vee \overline{x}_2 \vee \overline{x}_3);$
(2) $\mathcal{D}_{f_2} = \overline{x}_1 \overline{x}_2 \overline{x}_3 \vee \overline{x}_1 \overline{x}_2 x_3 \vee \overline{x}_1 x_2 \overline{x}_3 \vee \overline{x}_1 x_2 x_3 \vee x_1 \overline{x}_2 x_3 \vee x_1 x_2 \overline{x}_3 \vee x_1 x_2 x_3,$
$\mathcal{K}_{f_2} = \overline{x}_1 \vee x_2 \vee x_3.$

6.15. *Answer*:

(1) $g(x_1, x_2, x_3) = x_2 \oplus (x_1 \wedge x_3);$
(2) $h(x_1, x_2, x_3, x_4) = 1 \oplus x_1 \oplus x_3 \oplus (x_1 \wedge x_2 \wedge x_3).$

6.16. *Answer*:

(1) $g(x_1, x_2, x_3) = 1 \oplus x_1 \oplus x_3 \oplus (x_1 \wedge x_2) \oplus (x_1 \wedge x_3);$
(2) $h(x_1, x_2, x_3, x_4) = 1 \oplus (x_1 \wedge x_3) \oplus (x_1 \wedge x_4) \oplus (x_1 \wedge x_2 \wedge x_3) \oplus (x_1 \wedge x_2 \wedge x_4).$

6.17. *Answer*:

(1) $f(x_1, x_2) \notin T_0;$
(2) $f(x_1, x_2) \in T_0.$

6.18. *Answer*:

(1) $f(x_1, x_2) \notin T_0;$
(2) $f(x_1, x_2) \notin T_0.$

6.19. *Answer*:

(1) $f(x_1, x_2) \in T_1;$
(2) $f(x_1, x_2) \in T_1.$

6.20. *Answer*:

(1) $f(x_1, x_2) \notin T_1;$
(2) $f(x_1, x_2) \notin T_1.$

6.21. *Solution.*

(1) Check the validity of the equality $h = g^*$:

$$(g(x_1, x_2))^* = \overline{\overline{x}_1 \vee \overline{x}_2} = x_1 \wedge x_2 = h(x_1, x_2).$$

Hence, $h(x_1, x_2)$ is dual to $g(x_1, x_2)$.

(2) Calculate g^*:

$$(g(x_1, x_2))^* = \overline{(x_1 \wedge \overline{x}_2) \vee \overline{(x_1 \rightarrow \overline{x}_2)}} = \overline{(\overline{x}_1 \wedge x_2)} \wedge (x_1 \rightarrow \overline{x}_2) =$$
$$= (x_1 \vee \overline{x}_2) \wedge (\overline{x}_1 \vee \overline{x}_2) = (x_1 \wedge \overline{x}_1) \vee \overline{x}_2 = \overline{x}_2.$$

From the relation $h \neq g^*$ follows that the function $h(x_1, x_2)$ is not dual to $g(x_1, x_2)$.

6.22. *Answer*:

(1) f is not self-dual;
(2) f self-dual.

6.23. *Answer*:

(1) f self-dual;
(2) f is not self-dual.

6.24. *Solution.*

(1) Calculate the function f^*:

$$f^* = (x_1 \oplus x_2)^* = \overline{(\overline{x}_1 \oplus \overline{x}_2)}.$$

From the definition of operation "modulo two sum" follows the equality $\overline{p} \oplus \overline{q} = p \oplus q$. By virtue of this

$$f^* = \overline{(\overline{x}_1 \oplus \overline{x}_2)} = \overline{x_1 \oplus x_2} = x_1 \leftrightarrow x_2.$$

Since $f^* \neq f$, then the function $f(x_1, x_2)$ is not self-dual.

(2) As is shown in the previous item, $(x_1 \oplus x_2)^* = x_1 \leftrightarrow x_2$, this is why $(x_1 \leftrightarrow x_2)^* = x_1 \oplus x_2$. The function $f(x_1, x_2) = x_1 \leftrightarrow x_2$ is not self-dual.

6.25. *Proof.*

One of the simplest methods to prove self-duality of $f(x_1, x_2, x_3)$ is to use the identity $\overline{x} = x \oplus 1$:

$$f^*(x_1, x_2, x_3) = \overline{\overline{x}_1 \oplus \overline{x}_2 \oplus \overline{x}_3} = ((x_1 \oplus 1) \oplus (x_2 \oplus 1) \oplus (x_3 \oplus 1)) \oplus 1 =$$
$$= x_1 \oplus x_2 \oplus x_3 \oplus 1 \oplus 1 \oplus 1 \oplus 1 = x_1 \oplus x_2 \oplus x_3 = f(x_1, x_2, x_3).$$

Therefore, the function $f(\mathbf{x})$ is self-dual.

6.26. *Answer:* $\alpha_{f^*} = (\overline{\alpha}_{2^n-1}, \overline{\alpha}_{2^n-2}, \ldots, \overline{\alpha}_1, \overline{\alpha}_0)$.

6.27. *Answer:*
The sought number is equal to the number of k-combinations without repetitions out of n elements, namely $C(n, k)$.

6.28. *Answer:* $\forall \mathbf{x} \in \mathbb{B}^n \ w(s(\mathbf{x})) = 2^{n-1}$.

6.29. *Solution.*

(1) Represent the function f in the form of Zhegalkin polynomial. For this purpose, use the method of equivalent logical transformations:

$$f(x_1, x_2) = x_1 \vee \overline{x}_2 = \overline{(\overline{x}_1 \wedge x_2)} = (\overline{x}_1 \wedge x_2) \oplus 1 = ((x_1 \oplus 1) \wedge x_2) \oplus 1 =$$
$$= ((x_1 \wedge x_2) \oplus (x_2 \wedge 1)) \oplus 1 = 1 \oplus x_2 \oplus (x_1 \wedge x_2).$$

The degree of the obtained polynomial is greater than one, this is why $f(x_1, x_2) = x_1 \vee \overline{x}_2$ is not a linear function.

(2) Having written the function f in the form of Zhegalkin polynomial

$$f(x_1, x_2) = x_1 \mid x_2 = \overline{(x_1 \wedge x_2)} = (x_1 \wedge x_2) \oplus 1,$$

we come to the conclusion that $f(x_1, x_2) = x_1 \mid x_2$ is not a linear function.

6.30. *Answer:*

(1) $f(x_1)$—linear function;
(2) $f(x_1, x_2)$ is not linear.

6.31. *Answer:*

(1) $f(x_1, x_2)$—monotone function;
(2) $f(x_1, x_2)$—monotone function.

6.32. *Answer:*

(1) $f(x_1, x_2)$ is not monotone;
(2) $f(x_1, x_2)$ is not monotone.

6.33. *Answer:*

(1) $f(x_1, x_2)$ is not monotone;
(2) $f(x_1, x_2)$ is not monotone.

6.34. *Answer:* $\alpha_f = (0000\ 0101\ 0001\ 1111)$, $f(\mathbf{x}) = (x_1 \wedge x_2) \vee (x_2 \wedge x_4) \vee (x_1 \wedge x_3 \wedge x_4)$.

6.35. *Solution.*
As is known, the number of all Boolean functions from n arguments is $|P_2(n)| = 2^{2^n}$. The minimal value n that satisfies the inequality $2^{2^n} > 10^9$, is $n_{\min} = \lceil \log_2 \log_2 10^9 \rceil = \lceil 4,90\ldots \rceil = 5$.

6.36. *Hint.*
Cardinalities of all the mentioned sets, except M^n, can conveniently be calculated
using the analytical formulae provided in the present chapter. In order to find the
number of the monotone functions $|M^n|$, one should use the complete enumeration
of the functions of $P_2(n)$, in particular, $|M^4| = 168$ can be obtained. Some estimates
for the value $|M^n|$ are shown in Exercise **6.41**.

6.37. *Solution.*

(1) Since $T_0 \cap T_1 \neq \varnothing$, then we should use the inclusion and exclusion formula
(see chapter "Set Theory"):

$$|X| = |T_0 \cup T_1| = |T_0| + |T_1| - |T_0 \cap T_1|.$$

As is known, $|T_0| = |T_1| = 2^{2^n-1}$. The number of functions, preserving both 0,
and 1, is $|T_0 \cap T_1| = 2^{2^n-2}$, since in the table of values, the first and the last rows
associated with the vectors $(0, 0, \dots, 0)$ and $(1, 1, \dots, 1)$ are defined by the constant
preservation conditions. Therefore,

$$|X| = 2^{2^n-1} + 2^{2^n-1} - 2^{2^n-2} = (2 + 2 - 1) \cdot 2^{2^n-2} = 3 \cdot 2^{2^n-2}.$$

(2) Calculate the cardinality of the set $T_0 \cap L$. Let f be an arbitrary linear func-
tion, then $f = \gamma_0 \oplus \gamma_1 x_1 \oplus \dots \oplus \gamma_n x_n$ for some $\gamma_0, \gamma_1, \dots, \gamma_n \in \{0, 1\}$. We obtain
$f(0, 0, \dots, 0) = \gamma_0$, this is why linear function preserves 0 if and only if free term
$\gamma_0 = 0$. Therefore, $|T_0 \cap L| = \dfrac{1}{2} 2^{n+1} = 2^n$. Finally,

$$|Y| = |T_0 \cup L| = |T_0| + |L| - |T_0 \cap L| = 2^{2^n-1} + 2^{n+1} - 2^n = 2^{2^n-1} + 2^n.$$

6.38. *Answer*:

(1) $|X| = 2^{2^{n-1}-1}$;
(2) $|Y| = 2^n$.

6.39. *Solution.*

(1) The set $X = M \setminus T_0$ is formed by the monotone functions $f(\mathbf{x})$, not preserving
zero. The condition $f \notin T_0$ will be written in the form $f(\alpha_0) = 1$, where $\alpha_0 = (0, 0, \dots, 0)$. Since the vector α_0 precedes all possible tuples $\mathbf{x} \in \mathbb{B}^n$, then, according
to the monotonicity property $\forall \mathbf{x} \ f(\alpha_0) \leqslant f(\mathbf{x})$, or $f(\mathbf{x}) \geqslant 1$. Only one function
satisfies this condition, namely $f(\mathbf{x}) \equiv 1$, this is why $|M \setminus T_0| = 1$.

(2) Partition the set L into non-intercepting subsets $\mathcal{L}^{(0)}, \mathcal{L}^{(1)}, \dots, \mathcal{L}^{(n)}$ by the following rule: The set $\mathcal{L}^{(k)}$, where $k = 0, 1, \dots, n$, consists of all functions, specified
by Boolean expressions, containing exactly k of modulo two addition operations.
Note that for all $1 \leqslant i, j \leqslant n$ and $i \neq j$, the relation is fulfilled $\mathcal{L}^{(i)} \cap \mathcal{L}^{(j)} = \varnothing$ due
to uniqueness of representation $f(\mathbf{x})$ in the form of Zhegalkin polynomial.

As is easy to verify, $\forall k = 1, 2, \ldots, n$. The set $\mathcal{L}^{(k)}$ contains none of monotone functions. Indeed,

$$f(\mathbf{x}) \in \mathcal{L}^{(k)} \Leftrightarrow f(\mathbf{x}) = \underbrace{x_i \oplus x_j \oplus \ldots \oplus x_s}_{k \text{ operations } \oplus}.$$

Without loss of generality, it is possible to say that $i = 1, j = 2, \ldots, s = k + 1$. The function $f(\mathbf{x}) = x_1 \oplus x_2 \oplus \ldots \oplus x_k \oplus x_{k+1}$ has no monotonicity property. This follows, for example, from consideration of the vectors

$$\boldsymbol{\alpha} = (1, 0, 0, \ldots, 0), \quad \boldsymbol{\beta} = (1, 1, 0, \ldots, 0),$$

for which the relations are valid $\boldsymbol{\alpha} \preccurlyeq \boldsymbol{\beta}$ and $f(\boldsymbol{\alpha}) > f(\boldsymbol{\beta})$.

Hence, $|L \cap M| = |\mathcal{L}^{(0)}|$. The set $\mathcal{L}^{(0)}$ is formed by the following functions: $f(\mathbf{x}) \equiv 0$, $f(\mathbf{x}) \equiv 1$ and $f(\mathbf{x}) = x_i$ for some $1 \leqslant i \leqslant n$. Finally, we obtain $|L \cap M| = n + 2$.

6.40. *Answer*:

(1) $3 \cdot 2^{2^n - 2} + 2^{n-1}$;
(2) $3 \cdot 2^{n-1}$;
(3) $3 \cdot 2^{2^n - 2} + 2^{2^{n-1} - 1}$;
(4) $2^{2^{n-1}}$.

6.41. *Proof.*
Consider all kinds of tuples $\boldsymbol{\xi}_i \in \mathbb{B}^n$, $i = 1, 2, \ldots$, with cardinality $w(\boldsymbol{\xi}_i) = \lfloor n/2 \rfloor$ for all $i \in \mathbb{N}$. We can show, that these tuples are not pairwise comparable, and cardinality of any subset of the pairwise incomparable tuples in $\mathbb{B}^n$ does not exceed $C(n, \lfloor n/2 \rfloor)$.

Further, let some Boolean function $f(\mathbf{x}) \in P_2(n)$ be such that

(1) on the sets with weights $w < \lfloor n/2 \rfloor$ the equality is valid $f(\mathbf{x}) = 0$;
(2) on the sets with weights $w > \lfloor n/2 \rfloor$ the equality is valid $f(\mathbf{x}) = 1$.

Then, $f(\mathbf{x})$ has the monotonicity property, and on each of $\boldsymbol{\xi}_i$, it can take the value of either 0, or 1. By the product rule, we obtain the lower estimate for $|M^n|$: $|M^n| \geqslant 2^{C(n, \lfloor n/2 \rfloor)}$.

In order the find the upper estimate, use Dilworth's theorem (see chapter "Relations and Functions"), from which follows the statement: The minimal number of chains, containing all vertices of the finite ordered set $\mathbb{B}^n$ is $C(n, \lfloor n/2 \rfloor)$. Further, consider one of such chains: $C^{(k)} = \{\boldsymbol{\alpha}_1^{(k)}, \boldsymbol{\alpha}_2^{(k)}, \ldots, \boldsymbol{\alpha}_p^{(k)}\}$, where $p \leqslant n + 1$. Note that cardinality of the chain $p = |C^{(k)}| \leqslant n + 1$ and the values of the monotone function $f(\mathbf{x})$, not identically equal to the constant, at gates $\boldsymbol{\alpha}_i^{(k)} \in C^{(k)}$, $1 \leqslant i \leqslant p$, form the vector

$$\left(f(\alpha_1^{(k)}), f(\alpha_2^{(k)}), \ldots, f(\alpha_p^{(k)}) \right) = (\underbrace{00\ldots00}_{p' \text{ zeroes}} \underbrace{11\ldots11}_{p-p' \text{ ones}})$$

for some $p' = 1, 2, \ldots, p - 1$.

By the product rule, there exists no more than $(p - 1)^{C(n, \lfloor n/2 \rfloor)}$ possibilities for assigning the values of the function $f(\mathbf{x})$ at gates of all chains $C^{(k)}$. The total number of all monotone functions M^n should also include the identical 0 and the identical 1. As a result, $|M^n| \leqslant 2 + n^{C(n, \lfloor n/2 \rfloor)}$.

6.42. *Proof.*
Any Boolean algebra function, except $f_0 \equiv 0$, can be presented in disjunctive normal form, constructed with the help of operations of negation, conjunction, and disjunction. Further, $f_0(\mathbf{x})$ can be written, for example as follows: $f_0(x_1, \ldots, x_n) = x_1 \wedge \overline{x}_1$. Therefore, operations from $\{^-, \vee, \wedge\}$ can be used to write any Boolean function.

6.43. *Hint.* Take advantage of the fact that $x_1 \vee x_2 = \overline{\overline{x}_1 \wedge \overline{x}_2}$.

6.44. *Hint.* Take advantage of the fact that $x_1 \wedge x_2 = \overline{\overline{x}_1 \vee \overline{x}_2}$.

6.45. *Solution.*

(1) Express by Sheffer stroke the operations of conjunction and negation:

$$\overline{x} = \overline{x} \vee \overline{x} = x \mid x;$$
$$x_1 \wedge x_2 = \overline{\overline{x}_1 \vee \overline{x}_2} = \overline{(x_1 \mid x_2)} = (x_1 \mid x_2) \mid (x_1 \mid x_2).$$

According to the result of Exercise **6.43**, the system $\{^-, \wedge\}$ is complete. Therefore, using Sheffer stroke, we can express any function of Boolean algebra.

(2) Express by Peirce arrow the operations of disjunction and negation:

$$\overline{x} = \overline{x} \wedge \overline{x} = x \downarrow x;$$
$$x_1 \vee x_2 = \overline{\overline{x}_1 \wedge \overline{x}_2} = \overline{(x_1 \downarrow x_2)} = (x_1 \downarrow x_2) \downarrow (x_1 \downarrow x_2).$$

According to the result of Exercise **6.44**, the system $\{^-, \vee\}$ is complete. Therefore, the system $\{\downarrow\}$ is complete as well.

6.46. *Solution.*
In order to apply Post's criterion, it is convenient to use a table, where sign "$-$" marks the properties missing in this function of the system under consideration. If at least one sign "$-$" stands in each of the columns, the system is complete.

The function	$f \in T_0$	$f \in T_1$	$f \in S$	$f \in L$	$f \in M$
$x_1 \oplus x_2$		$-$	$-$		$-$
$x_1 \wedge x_2$			$-$	$-$	
1	$-$		$-$		

As follows from the table analysis, the system of functions $\{\oplus, \wedge, 1\}$ is complete.

6.48. *Answer*:

Each of the function systems (1)–(3) forms a basis. The system of item (4) $\{\leftrightarrow, \rightarrow, ^-\}$ remains complete after removal of the function $\leftrightarrow$, and, hence, is not a basis.

6.49. *Proof.*

Suppose that some basis $\mathcal{B}$ contains more that four functions. From Post's criterion follows the statement: $\mathcal{B}$ consists of five functions $\{f_1, f_2, f_3, f_4, f_5\}$; otherwise, removal of one function will not have resulted in the system's loss of completeness.

None of the functions f_i, $i = 1, \ldots, 5$, belongs to exactly one of the classes T_0, T_1, S, M, L. Consider the function $f_k \notin T_0$, i.e., $f_k(0, 0, \ldots, 0) = 1$. Since $f_k \in T_1$, then $f_k(1, 1, \ldots, 1) = 1$. But the last equality contradicts $f_k \in S$. The obtained contradiction proves: The basis $\mathcal{B}$ contains no more than four functions.

6.50. *Answer*:

one function.

6.51. *Answer*:

$(q \wedge r) \vee (\overline{q} \wedge \overline{r})$.

6.52. *Answer*:

$(p \wedge r) \vee (\overline{p} \wedge \overline{r})$.

6.53. *Answer*:

(1) $S_1 = p$;
(2) $S_2 = (p \wedge r) \vee (\overline{p} \wedge \overline{r})$.

6.54. *Answer*:

(1) $S_1 = r$;
(2) $S_2 = r \vee (\overline{p} \wedge q)$.

6.55. *Solution.*

Karnaugh map for the function of four variables $f(x_1, x_2, x_3, x_4)$ contains 16 cells.

The block of eight cells is described by one variable x_i or its negation $\overline{x}_i$, where $1 \leqslant i \leqslant 4$;

The block of four cells is described by elementary conjunction of rank 2: $x_i^{\sigma_i} \wedge x_j^{\sigma_j}$, where $1 \leqslant i, j \leqslant 4$, and $i \neq j$; The block of two cells is described by elementary conjunction of rank 3: $x_i^{\sigma_i} \wedge x_j^{\sigma_j} \wedge x_k^{\sigma_k}$, where $1 \leqslant i, j, k \leqslant 4$, and $i \neq j \neq k$.

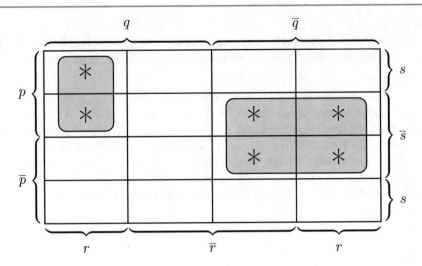

Uniting the tagged cells into two blocks, we obtain $t(p, q, r, s) = (p \wedge q \wedge r) \vee (\overline{q} \wedge \overline{s})$.

6.56. *Answer*:
Karnaugh map has the form

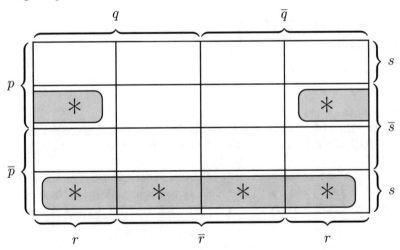

The function after simplification can be written in the form

$$g(p, q, r, s) = (\overline{p} \wedge s) \vee (p \wedge r \wedge \overline{s}).$$

6.57. *Answer*:
Karnaugh map has the form

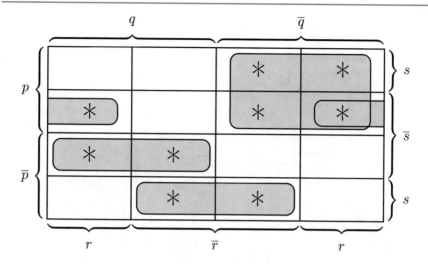

The function after simplification takes the form

$$h(p, q, r, s) = (p \wedge \overline{q}) \vee (p \wedge r \wedge \overline{s}) \vee (\overline{p} \wedge q \wedge \overline{s}) \vee (\overline{p} \wedge \overline{r} \wedge s).$$

6.58. *Answer*:

(1) $\alpha_f = (0111\ 0011\ 1101\ 1100)$, $f(\mathbf{x}) = (x_1 \wedge \overline{x}_3) \vee (\overline{x}_1 \wedge x_3) \vee (\overline{x}_2 \wedge x_4)$;
(2) $\alpha_f = (0101\ 1111\ 0111\ 1111)$, $f(\mathbf{x}) = (x_1 \wedge x_3) \vee x_2 \vee x_4$.

6.59. *Answer*:
$f(p, q) = \overline{p} \vee q$.

6.60. *Answer*:
$f(p, q, r) = (p \wedge q \wedge r) \vee (\overline{q} \wedge \overline{r})$.

6.62. *Answer*:
Combinational circuits, satisfying the conditions of items (1) and (2), are shown in
Fig. 6.6a, b respectively.

6.71. *Solution.*
Assume that we need to add up the numbers $a = (a_n a_{n-1} \ldots a_0)_2$ and $b = (b_n b_{n-1} \ldots b_0)_2$. Construct a combinational circuit, on the outputs of which we will have values
of orders of the sum $c = a +_2 b$, $c = (c_{n+1} c_n \ldots c_0)_2$.

By the rule of adding up two numbers known from algebra

$$c_i = a_i \oplus b_i \oplus t_i,$$

where $0 \leqslant i \leqslant n$, and by t_i is denoted the result of carrying from the previous orders,
and $t_0 = 0$. The higher order of the result of summation c_{n+1}, as one can easily see,
is $c_{n+1} = t_{n+1}$.

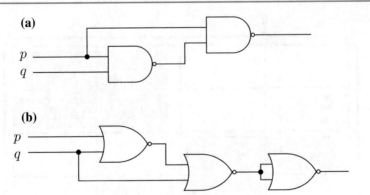

Fig. 6.6 To Exercise **6.62**

Fig. 6.7 Full adder $\mathcal{S}$

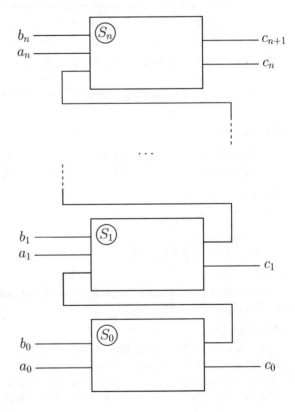

Taking into account that carrying to the $i + 1$ order occurs if and only if at least two out of three values a_i, b_i, t_i are equal to one, write the carry bit t_{i+1} in a form convenient for constructing the combinational circuit:

$$t_{i+1} = (a_i \wedge b_i) \vee (a_i \wedge t_i) \vee (b_i \wedge t_i), \quad 0 \leqslant i \leqslant n.$$

The function $c_i(a_i, b_i, t_i) = a_i \oplus b_i \oplus t_i$ is also easy to construct using the gates, namely **AND, OR, NOT**, applying the identity

$$a_i \oplus b_i \oplus t_i = \left(\overline{((a_i \wedge b_i) \vee (a_i \wedge t_i) \vee (b_i \wedge t_i))} \wedge (a_i \vee b_i \vee c_i) \right) \vee (a_i \wedge b_i \wedge c_i)$$

or, in view of the PDNF of the function $c_i(a_i, b_i, t_i)$,

$$a_i \oplus b_i \oplus t_i = (a_i \wedge b_i \wedge t_i) \vee (a_i \wedge \bar{b}_i \wedge \bar{t}_i) \vee (\bar{a}_i \wedge \bar{b}_i \wedge t_i) \vee (\bar{a}_i \wedge b_i \wedge \bar{t}_i).$$

Denote the combinational circuit

implementing the considered functions for $i = 1, \ldots, n$, by S_i. Apart from this, the half-bit adder S_0 implements the relations

$$\begin{cases} c_0 = a_0 \oplus b_0 = (a_0 \wedge \bar{b}_0) \vee (\bar{a}_0 \wedge b_0), \\ t_1 = a_0 \wedge b_0. \end{cases}$$

The full adder S is obtained combining the combinational circuits $S_i, i = 0, \ldots, n$ and is shown in Fig. 6.7.

Complex Numbers

7

Complex number z is an ordered pair of real numbers (a, b), where $a, b \in \mathbb{R}$. The first number a is called the **real part** of the complex number $z = (a, b)$ and is denoted by symbol Re z, while the second number of the pair b is called the **imaginary part** z and is denoted Im z [20].

A complex number of the form $(a, 0)$, where the imaginary part is zero, is identified with the real number a, i.e., $(a, 0) \equiv a$. This allows considering the set of all real numbers $\mathbb{R}$ as s subset of set of complex numbers $\mathbb{C}$.

Two complex numbers $z_1 = (a_1, b_1)$ and $z_2 = (a_2, b_2)$ are considered equal if and only if their real and imaginary parts are pairwise equal: $z_1 = z_2 \Leftrightarrow a_1 = a_2$, $b_1 = b_2$.

On the set $\mathbb{C}$, the operations of addition and multiplication of complex numbers are defined. **Sum** of the complex numbers $z_1 = (a_1, b_1)$ and $z_2 = (a_2, b_2)$ is the complex number z, equal to $z_1 + z_2 = (a_1 + a_2, b_1 + b_2)$. **Product** of the numbers $z_1 = (a_1, b_1)$ and $z_2 = (a_2, b_2)$ is such a complex number $z = (a, b)$, where $a = a_1 a_2 - b_1 b_2$, $b = a_1 b_2 + a_2 b_1$.

The pair $(0, 1)$ is of the greatest importance in the operations with complex numbers; it is denoted by $(0, 1) \equiv i$ and is called **imaginary unit**. The basic property of the imaginary unit consists in that $i^2 = i \cdot i = (0, 1) \cdot (0, 1) = (-1, 0)$, or $i^2 = -1$.

A complex number of the form $z = (0, b)$ is called **purely imaginary**. Since $(0, b) = (b, 0) \cdot (0, 1)$, the purely imaginary number z is presentable in the form of the product $z = bi$.

Any complex number can be presented in the form

$$z = (a, b) = (a, 0) + (0, b) = (a, 0) + (b, 0) \cdot (0, 1) = a + ib.$$

Such a notation is referred to as the **algebraic form** of a complex number. This allows considering i as a factor, whose square is equal to -1, and performing oper-

© Springer International Publishing AG, part of Springer Nature 2018
S. Kurgalin and S. Borzunov, *The Discrete Math Workbook*,
Texts in Computer Science, https://doi.org/10.1007/978-3-319-92645-2_7

Fig. 7.1 Complex plane

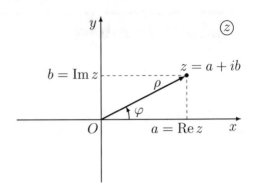

ations with complex numbers in the same manner as with algebraic polynomials, in intermediate calculations assuming $i^2 = -1$.

A complex number $z^* = (a, -b) = a - ib$ is called a **conjugate** of the complex number $z = (a, b) = a + ib$. There is one more frequent notation of a conjugate—$\bar{z}$. If the coefficients of the polynomial $p(z)$ are real, the equality is valid $(p(z))^* = p(z^*)$.

It is convenient to present the number $z = a + ib$ as the point (x, y) of a plane with Cartesian coordinates $x = a$ and $y = b$. Correlate each complex number z with a point with coordinates (x, y) (and a position vector, connecting the origin of coordinates with this point). Such a plane is denoted by $\math{z}$ and is referred to as **complex** (Fig. 7.1). Note that geometric interpretation of complex numbers is sometimes referred to as the **Argand**[1] **diagram**.

Many applications widely use a **trigonometric form** of the complex number z. Let us introduce the polar coordinate system so that the pole is at the origin of Cartesian system (x, y). The axis of the polar system will be directed along the positive direction of the axis Ox.

In this case, Cartesian and polar coordinates of an arbitrary point other than the origin of coordinates are related by the formulae

$$x = \rho \cos \varphi, \quad y = \rho \sin \varphi,$$

$$\rho = \sqrt{x^2 + y^2}, \quad \varphi = \begin{cases} \arctan \dfrac{y}{x}, & \text{if } x > 0; \\[2mm] \arctan \dfrac{y}{x} + \pi, & \text{if } x < 0, \ y \geqslant 0; \\[2mm] \arctan \dfrac{y}{x} - \pi, & \text{if } x < 0, \ y < 0; \\[2mm] \dfrac{\pi}{2} \operatorname{sign} y, & \text{if } x = 0. \end{cases}$$

---

[1] Jean-Robert Argand (1768–1822)—French mathematician.

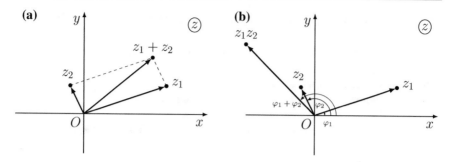

Fig. 7.2 Sum is the panel (**a**), and product is the panel (**b**) complex numbers z_1 and z_2

As a result, we obtain a trigonometric form of the number z

$$z = (x, y) = x + iy = \rho(\cos \varphi + i \sin \varphi).$$

The value ρ is called **modulus**, and φ—**argument** of the complex number z and denoted $\rho = |z|$, $\varphi = \arg z$. It should be noted that the argument φ is ambiguously determined: Instead of the value φ, we can take the value $\varphi + 2\pi k$, where $k \in \mathbb{Z}$. If $\arg z$ is chosen in such a way that $-\pi < \arg z \leqslant \pi$, then such a value is called the **principal** value of the argument.

For the numbers $z_1 = \rho_1(\cos \varphi_1 + i \sin \varphi_1)$ and $z_2 = \rho_2(\cos \varphi_2 + i \sin \varphi_2)$, specified in the trigonometric form,

$$z_1 z_2 = \rho_1 \rho_2 (\cos(\varphi_1 + \varphi_2) + i \sin(\varphi_1 + \varphi_2)),$$
$$\frac{z_1}{z_2} = \frac{\rho_1}{\rho_2} (\cos(\varphi_1 - \varphi_2) + i \sin(\varphi_1 - \varphi_2)), \quad \rho_2 \neq 0.$$

Geometric illustration of the sum and the product of complex numbers is shown in Fig. 7.2. For any $z_1, z_2 \in \mathbb{C}$, the position vector of the sum $z_1 + z_2$ is equal to the sum of the position vectors of the summands z_1 and z_2. The position vector of the product $z_1 z_2$ is obtained by rotating the position vector of the number z_1 by the angle of $\arg z_2$ counterclockwise and extending by $|z_2|$ times.

Euler's formula relates the exponential function of the imaginary argument with trigonometric functions of the imaginary part of the argument:

$$e^{i\varphi} = \cos \varphi + i \sin \varphi.$$

This is why we can introduce one more notation of the complex number, namely **exponential**: $z = \rho e^{i\varphi}$. The exponential notation is convenient for operations of multiplication, division, raising to a power and extraction of root. For example, the nth power of the number z can be presented in the form

$$z^n = (\rho e^{i\varphi})^n = \rho^n e^{in\varphi} = \rho^n(\cos n\varphi + i \sin n\varphi)$$

for all integer values n. An important consequence of the obtained formula

$$(\cos \varphi + i \sin \varphi)^n = \cos n\varphi + i \sin n\varphi$$

is associated with the name of **de Moivre**.[2]

The nth root of $z = \rho(\cos \varphi + i \sin \varphi)$ can be calculated as

$$\sqrt[n]{z} \equiv z^{1/n} = [\rho(\cos(\varphi + 2\pi k) + i \sin(\varphi + 2\pi k))]^{1/n}, \quad k \in \mathbb{Z},$$

or, after applying Euler's formula,

$$\sqrt[n]{z} = \rho^{1/n} \left(\cos \left(\frac{\varphi + 2\pi k}{n} \right) + i \sin \left(\frac{\varphi + 2\pi k}{n} \right) \right), \quad k = 0, 1, \ldots, n - 1.$$

Here, we obtain n possible values of the nth root for $k = 0, 1, \ldots, n - 1$. Other acceptable k do not result in new values of $\sqrt[n]{z}$. For example, for $k = n$ the argument is $\arg z = \varphi/n + 2\pi$ and differs from the case $k = 0$ by 2π, which corresponds to the complex number equal to it.

Fundamental theorem of algebra states that *any polynomial of a zero degree with complex coefficients has a complex root* [41]. This is why an arbitrary polynomial with real (or complex) coefficients always has some root $z \in \mathbb{C}$.

Each polynomial of degree n

$$p(z) = c_n z^n + c_{n-1} z^{n-1} + \cdots + c_0, \quad c_i \in \mathbb{C} \text{ for } i = 0, 1, \ldots, n, \ c_n \neq 0,$$

can uniquely (accurate to the order of cofactors) be expanded into the product

$$p(z) = c_n (z - z_1)^{m_1} (z - z_2)^{m_2} \ldots (z - z_k)^{m_k},$$

where z_i—root of the polynomial $p(z)$ with a multiplicity of m_i, $1 \leqslant i \leqslant k$.

For polynomials with degree lower than the fifth, we can always find the roots having expressed them by arithmetic operations or arithmetic roots of an arbitrary multiplicity, or **radicals**. The method for calculating the cubic polynomial's roots was suggested by **Cardanus**,[3] and of the fourth degree polynomial—**Ferrari**.[4] However, for finding roots of polynomials of higher degrees, there are no common methods: According to **Abel**[5]–**Ruffini**[6] **theorem** *any arbitrary equation of degree n for $n \geqslant 5$ is unsolvable in radicals.*

---

[2] Abraham de Moivre (1667–1754)—English mathematician of French origin.
[3] Hieronymus Cardanus (1501–1576)—Italian mathematician and philosopher.
[4] Lodovico Ferrari (1522–1565)—Italian mathematician.
[5] Niels Henrik Abel (1802–1829)—Norwegian mathematician.
[6] Paolo Ruffini (1765–1822)—Italian mathematician.

7.1 Problems

7.1. Perform algebraic manipulations and represent the specified complex number z in the algebraic form $z = \operatorname{Re} z + i \operatorname{Im} z$:

(1) $(2 + 3i)(5 - 4i)$;
(2) $(2 - i)(2 + i)$;
(3) $(1 + i)^4 + (1 - i)^4$;
(4) $1 + i^3$.

7.2. Perform algebraic manipulations and represent the specified complex number z in the algebraic form $z = \operatorname{Re} z + i \operatorname{Im} z$:

(1) $(3 - i)(4 + i)$;
(2) $(7 + 10i)(7 + 5i)$;
(3) $(2 + 3i)^2 + (1 - i)^2$;
(4) $i^3 - i^4$.

7.3. Given are the complex numbers $z_1 = 1 - 2i$, $z_2 = 3 + i$, $z_3 = -5 + i$. Find $z_1 z_2 + z_3^2$.

7.4. Given are the complex numbers $z_1 = 1 + i$, $z_2 = -1 - 5i$, $z_3 = 3 + 4i$. Find $z_1(z_2^3 - z_3^3)$.

7.5. Perform the actions:

(1) $(2 + 3i)^3 + (2 - 3i)^3$;
(2) $(5 - 2i)^4 - (5 + 2i)^4$.

7.6. Having performed the division, represent the complex number $z = \dfrac{a + ib}{c + id}$, where $a, b, c, d \in \mathbb{R}$, $c \neq 0$, $d \neq 0$, in the algebraic form.

7.7. Simplify the expressions:

(1) $\dfrac{1 + i}{1 - i}$;

(2) $\dfrac{2 + 3i}{2 - 3i}$;

(3) $\dfrac{i^3}{1 + i^3}$;

(4) $\dfrac{i}{5 + 4i} + \dfrac{i}{5 - 4i}$.

7.8. Given are the complex numbers $z_1 = 1 + i$, $z_2 = z_1^*$, $z_3 = z_1 + 2z_2$. Find $z_1 + (z_1 - z_3)(z_2 - z_3)/z_2$.

7.9. Given are the complex numbers $z_1 = 2 + i$, $z_2 = 1 - z_1$, $z_3 = 1 - 3z_1$. Find $(z_1 z_2 + z_2 z_3 + z_1 z_3)/(z_1 z_2 z_3)$.

7.10. Find the number conjugate with the number z, if:

(1) $z = \dfrac{5 - 3i}{5 + 3i}$;

(2) $z = 1 + i + \dfrac{1}{1 + i}$.

7.11. Find z, if $z + 2z^* = 9 + i$.

7.12. Find z, if $z^* - 5z = 4 - i$.

7.13. Prove that for arbitrary $z_1, z_2 \in \mathbb{C}$ the equalities are valid:

(1) $(z_1 + z_2)^* = z_1^* + z_2^*$;
(2) $(z_1 z_2)^* = z_1^* z_2^*$.

7.14. Prove that if $|z| = 1$, then $z^{-1} = z^*$.

***7.15.** Prove that for any $z_1, z_2 \in \mathbb{C}$ **triangle inequalities** are valid:

$$\big||z_1| - |z_2|\big| \leqslant |z_1 + z_2| \leqslant |z_1| + |z_2|.$$

***7.16.** Calculate the sums:

(1) $\displaystyle\sum_{k=1}^{100} i^k$;

(2) $\displaystyle\sum_{k=-99}^{99} i^k$.

7.17. Simplify the expression i^m for arbitrary $m \in \mathbb{Z}$.

7.18. Represent the complex number in a trigonometric form:

(1) $1 + i$;
(2) $-3 + 4i$;
(3) $-4i$;
(4) $\dfrac{5 - i}{5 + i}$.

7.19. Represent the following complex numbers in an algebraic form $z = \mathrm{Re}\, z + i\,\mathrm{Im}\, z$:

(1) $z = 3\left(\cos\dfrac{\pi}{6} + i\sin\dfrac{\pi}{6}\right)$;

(2) $z = \sqrt{2}\left(\cos(-\dfrac{\pi}{6}) + i\sin(-\dfrac{\pi}{6})\right);$

(3) $z = \dfrac{\sqrt{2}}{\cos\dfrac{\pi}{4} + i\sin\dfrac{\pi}{4}};$

(4) $z = \dfrac{2i}{\cos\left(-\dfrac{3\pi}{4}\right) + i\sin\left(-\dfrac{3\pi}{4}\right)}.$

7.20. Prove that a complex number of the type $u = \dfrac{a+ib}{a-ib}$, where $a, b \in \mathbb{R}$, can be presented in the form of an exponent with purely imaginary index, i.e., in the form

$$u = e^{i\delta}, \quad \delta \in \mathbb{R}.$$

7.21. Calculate i^i.

7.22. On the complex plane $\widehat{z}$, show a set of points satisfying the specified conditions:

(1) $|\operatorname{Re} z| \leqslant 1$;
(2) $\operatorname{Im} z < -2$;
(3) $|z| \leqslant \sqrt{2}$;
(4) $\dfrac{\pi}{8} < \arg z < \dfrac{\pi}{4}$.

7.23. On the complex plane $\widehat{z}$, show a set of points satisfying the specified conditions:

(1) $\dfrac{1}{2} \leqslant |z+i| < 1$;
(2) $z^{-1} = z^*$;
(3) $|z-1| + |z+1| \leqslant 2\sqrt{2}$;
(4) $|z+1| = |z-2+i|$.

7.24. Calculate:

(1) $\sqrt{1+i}$;
(2) $\sqrt[3]{-1}$.

7.25. Calculate:

(1) $\sqrt{8i}$;
(2) $\sqrt[6]{64}$.

∗7.26. Denote roots of the equation $z^n = 1$, where n is a natural number, by ω_k, $k = 0, \ldots, n-1$. Prove that the following statements are true [20]:

(1) On a complex plane, the points associated with the vertices ω_k are located at the vertices of a right n-gon, inscribed in a unit circle, whose center is located at the origin of coordinates;

(2) $\omega_{k+n/2} = -\omega_k$ for even n and $0 \leqslant k \leqslant n/2 - 1$;

(3) $\sum\limits_{k=0}^{n-1} \omega_k = 0$ for $n > 1$;

(4) $\prod\limits_{k=0}^{n-1} \omega_k = (-1)^{n-1}$.

∗7.27. Prove the validity of the identities for the roots of unity ω_k, where $0 \leqslant k \leqslant n - 1$, for all natural values n:

(1) $\prod\limits_{k=0}^{n-1} (z - \omega_k) = z^n - 1$;

(2) $\sum\limits_{k=0}^{n-1} (\omega_k)^d = \begin{cases} 0, & 1 \leqslant d \leqslant n - 1; \\ n, & d = n. \end{cases}$

∗7.28. Prove the validity of the identities for the nth roots of unity ω_k, where $k = 0, 1, \ldots, n - 1$, for all values $n > 2$:

(1) $\sum\limits_{k=0}^{n-2} \omega_k \omega_{k+1} = -\omega_{n-1}$;

(2) $\sum\limits_{k=1}^{n-2} \dfrac{\omega_{k-1}\omega_{k+1}}{\omega_k} = -(1 + \omega_{n-1})$;

(3) $\sum\limits_{\substack{k,k'=0 \\ k<k'}}^{n-1} \omega_k \omega_{k'} = 0$;

(4) $\sum\limits_{\substack{k,k'=0 \\ k<k'}}^{n-1} \dfrac{\omega_k \omega_{k'}}{\omega_{k'-k}} = \dfrac{n}{1 - \omega_2}$.

∗7.29. Let ω_k, where $0 \leqslant k \leqslant n - 1$, are the nth roots of unity, x is an arbitrary complex number, and $x \neq \omega_k$ for no k. Calculate the sum $\sum\limits_{k=0}^{n-1} \dfrac{\omega_k}{x - \omega_k}$.

∗7.30. Prove that for natural $n \in \mathbb{N}$ the following identities are valid:

(1) $\cos \alpha + \cos 2\alpha + \cdots + \cos n\alpha = \dfrac{\sin(n\alpha/2)}{\sin(\alpha/2)} \cos[(n + 1)\alpha/2]$;

(2) $\sin \alpha + \sin 2\alpha + \cdots + \sin n\alpha = \dfrac{\sin(n\alpha/2)}{\sin(\alpha/2)} \sin[(n + 1)\alpha/2]$.

7.31. Prove de Moivre's formula for natural values of the exponent n, using the mathematical induction method.

∗7.32. Using de Moivre's formula, express $\cos n\varphi$ and $\sin n\varphi$ by $\cos \varphi$ and $\sin \varphi$. Consider the cases:

(1) $n = 3$;
(2) $n = 4$.

∗7.33. Calculate the mean value on the segment $[0, 2\pi]$ of the hundredth power of the sine [6].

∗7.34. Calculate the mean over the period of the two hundredth power of cosine.

∗7.35. Calculate the mean over the period of the twentieth power of function

$$f(x) = \sin x + \cos x.$$

∗7.36. Calculate the mean over the period of the twentieth power of function

$$f(x) = a \sin x + b \cos x, \text{ where } a, b\text{---const}, a, b \in \mathbb{R}.$$

∗7.37. There exist relations that express polynomial's coefficients by its roots (**Viete**[7] **formulae**). If α_1, α_2, ..., α_n are roots of the polynomial $p(z) = x^n + a_1 x^{n-1} + \cdots + a_n$, and each root is taken the number of times corresponding to its multiplicity, then the following equalities are valid:

$$\alpha_1 + \alpha_2 + \cdots + \alpha_n = - a_1,$$
$$\alpha_1\alpha_2 + \alpha_2\alpha_3 + \cdots + \alpha_1\alpha_n + \alpha_2\alpha_3 + \cdots + \alpha_{n-1}\alpha_n = a_2,$$
$$\alpha_1\alpha_2\alpha_3 + \alpha_1\alpha_2\alpha_4 + \cdots + \alpha_{n-2}\alpha_{n-1}\alpha_n = - a_2,$$

$$\cdots$$

$$\alpha_1\alpha_2 \ldots \alpha_{n-1} + \alpha_1\alpha_2\alpha_{n-2}\alpha_n + \cdots + \alpha_2\alpha_3 \ldots \alpha_n = (-1)^{n-1}a_{n-1},$$
$$\alpha_1\alpha_2 \ldots \alpha_n = (-1)^n a_n.$$

Prove validity of Viete formulae.

7.38. Let z_1, z_2 be the roots of the quadratic trinomial $p(z) = z^2 + pz + q$. Express the following values by the coefficients p and q:

(1) $z_1^2 + z_2^2$;

(2) $z_1^{-2} + z_2^{-2}$.

7.39. Let z_1, z_2 be the roots of the quadratic trinomial $p(z) = z^2 + pz + q$. Express the following values by the coefficients p and q:

---

[7]François Viète, seigneur de la Bigotière (1540–1603)—French mathematician.

(1) $z_1^4 + z_2^4$;

(2) $z_1^{-4} + z_2^{-4}$.

7.40. Find the sum and the product of all roots for each equation:

(1) $z^3 + z^2 + z + 1 = 0$;

(2) $z^4 + 4z^2 + 4 = 0$.

7.41. Find the sum and the product of all roots for each equation:

(1) $z^4 - 2z = 0$;

(2) $z^5 + z^4 + z^3 = 0$.

∗**7.42**. Find the sum of square roots of the equation $(x^2 - 2x)^2 - 17(x^2 - 2x) + 35 = 0$.

∗**7.43**. Find the sum of square roots of the equation $(x^2 - 3x)^2 + (x^2 - 3x) - 2000 = 0$.

7.44. Calculate the sum of square roots of the equation $x^3 - 7x + 6 = 0$.

7.45. Calculate the sum of square roots of the equation $x^3 - 10x^2 - 20x + 6 = 0$.

∗**7.46**. In order to determine the roots of the cubic equation

$$az^3 + bz^2 + cz + d = 0, \quad a, b, c, d \in \mathbb{C},$$

proceed as follows. Using the change of the variable $z = y - \dfrac{b}{3a}$, the equation is brought to a **canonical form**

$$y^3 + py + q = 0, \quad p, q \in \mathbb{C}.$$

By **Cardano formula**, the roots of the cubic equation y_1, y_2, y_3 in the canonical form are equal [29]

$$y_1 = \alpha + \beta,$$
$$y_2 = -\frac{\alpha + \beta}{2} + i\sqrt{3} \cdot \frac{\alpha - \beta}{2},$$
$$y_3 = -\frac{\alpha + \beta}{2} - i\sqrt{3} \cdot \frac{\alpha - \beta}{2},$$

where

$$\alpha, \beta = \sqrt[3]{-\frac{q}{2} \pm \sqrt{Q}},$$

$$Q = \left(\frac{p}{3}\right)^3 + \left(\frac{q}{2}\right)^2.$$

Using these relations, one should for each of the three values of the cube root of α take that value of the root β, for which the equality $\alpha\beta = -p/3$ is valid.

Using Cardano formula, solve the equation $z^3 - 3z + 2 = 0$.

*7.47. Solve the equation $z^3 - 5z^2 + 9z - 5 = 0$, using Cardano formula.

*7.48. Solve the equation $2z^3 - 13z^2 - 17z + 70 = 0$, using Cardano formula.

7.2 Answers, Hints, Solutions

7.1. *Solution.*

(1) Operations with complex numbers should be performed similarly to the respective operations with algebraic polynomials, using the property $i^2 = -1$:

$$(2 + 3i)(5 - 4i) = 2 \cdot 5 + 2 \cdot (-4i) + 3i \cdot 5 + 3i \cdot (-4i) =$$
$$= 10 - 8i + 15i - 12i^2 = 10 + (-8 + 15)i - 12(-1) = 22 + 7i;$$

(2) $(2 - i)(2 + i) = 2^2 - i^2 = 5$;
(3) $(1 + i)^4 + (1 - i)^4 = (1 + 4i + 6i^2 + 4i^3 + i^4) + (1 - 4i + 6i^2 - 4i^3 + i^4) =$

$$= 2\left(1 + 6i^2 + (i^2)^2\right) = 2\left(1 + 6(-1) + (-1)^2\right) = -8;$$
(4) $1 + i^3 = 1 + i \cdot i^2 = 1 - i$.

7.2. *Answer:*

(1) $13 - i$;
(2) $-1 + 105i$;
(3) $-5 + 10i$;
(4) $-1 - i$.

7.3. *Answer:*
$z_1 z_2 + z_3^2 = 29 - 15i$.

7.4. *Answer:*

$z_1(z_2^3 - z_3^3) = 125 + 257i$.

7.5. *Answer*:

(1) -92;

(2) $-1680i$.

7.6. *Solution*.

Denote $z_1 = a + ib$, $z_2 = c + id$. The fraction of the form $\dfrac{z_1}{z_2}$, where $z_1, z_2 \in \mathbb{C}$, can be conveniently transformed by multiplying it by $1 \equiv \dfrac{z_2^*}{z_2^*}$:

$$\frac{z_1}{z_2} = \frac{z_1}{z_2} \cdot 1 = \frac{z_1}{z_2} \cdot \frac{z_2^*}{z_2^*} = \frac{z_1 z_2^*}{z_2 z_2^*} = \frac{z_1 z_2^*}{|z_2|^2}.$$

This is why the result of the division of two complex numbers z_1/z_2, where $z_2 \neq 0$, will be the number

$$\frac{z_1}{z_2} = \frac{ac + bd}{c^2 + d^2} + i \frac{bc - ad}{c^2 + d^2}.$$

7.7. *Solution*.

Using the hint offered in the previous exercise, we obtain:

(1) $\dfrac{1+i}{1-i} = \dfrac{1+i}{1-i} \cdot \dfrac{1+i}{1+i} = \dfrac{(1+i)^2}{1-i^2} = \dfrac{1}{2}(1 + 2i + i^2) = i$;

(2) $\dfrac{2+3i}{2-3i} = \dfrac{2+3i}{2-3i} \cdot \dfrac{2+3i}{2+3i} = \dfrac{4 + 12i + 9i^2}{4+9} = \dfrac{1}{13}(-5 + 12i)$;

(3) $\dfrac{i^3}{1+i^3} = \dfrac{i \cdot i^2}{1 + i \cdot i^2} = \dfrac{-i}{1-i} = \dfrac{-i}{1-i} \cdot \dfrac{1+i}{1+i} =$

$$= -\frac{1}{2}(i + i^2) = \frac{1}{2}(1 - i);$$

(4) $\dfrac{i}{5+4i} + \dfrac{i}{5-4i} = i \left(\dfrac{1}{5+4i} + \dfrac{1}{5-4i} \right) = i \dfrac{5 - 4i + 5 + 4i}{(5+4i)(5-4i)} =$

$$= \frac{10i}{25 + 16} = \frac{10}{41}i.$$

7.8. *Answer*:
$z_1 + (z_1 - z_3)(z_2 - z_3)/z_2 = 5 + i$.

7.9. *Answer*:
$(z_1 z_2 + z_2 z_3 + z_1 z_3)/(z_1 z_2 z_3) = \dfrac{1}{85}(-21 + 33i)$.

7.10. *Answer*:
(1) $z = \dfrac{1}{17}(8 + 15i)$;

(2) $z = \dfrac{1}{2}(3 - i)$.

7.11. *Solution*.

Let $z = a + ib$, then $z + 2z^* = (a + ib) + 2(a - ib) = 3a - ib$. Since complex numbers are equal if and only if their real and imaginary parts are equal, we obtain

$$\begin{cases} 3a = 9, \\ -b = 1; \end{cases} \quad \Leftrightarrow \quad \begin{cases} a = 3, \\ b = -1, \end{cases}$$

whence $z = 3 - i$.

7.12. *Answer*:

$z = -1 + i/6$.

7.13. *Proof*.

Let $z_1 = x_1 + iy_1$, $z_2 = x_2 + iy_2$, where $x_1, x_2, y_1, y_2 \in \mathbb{R}$.
(1) Express the left side of the equality by x_1, x_2, y_1, and y_2:

$$(z_1 + z_2)^* = [(x_1 + x_2) + i(y_1 + y_2)]^* = (x_1 + x_2) - i(y_1 + y_2).$$

Now transform its right side

$$z_1^* + z_2^* = (x_1 - iy_1) + (x_2 - iy_2) = (x_1 + x_2) - i(y_1 + y_2).$$

Therefore, $\forall z_1, z_2 \in \mathbb{C} \ (z_1 + z_2)^* = z_1^* + z_2^*$.
(2) The left side of the equality

$$(z_1 z_2)^* = [(x_1 + iy_1)(x_2 + iy_2)]^* = (x_1 x_2 - y_1 y_2) - i(x_1 y_2 + x_2 y_1).$$

The right side coincides with the left one:

$$z_1^* z_2^* = (x_1 - iy_1)(x_2 - iy_2) = (x_1 x_2 - y_1 y_2) - i(x_1 y_2 + x_2 y_1) = (z_1 z_2)^*.$$

7.14. *Proof*.
Take the number with the modulus equal to one in the exponential form: $z = e^{i\varphi}$. After algebraic transformations

$$z^{-1} = (e^{i\varphi})^{-1} = e^{-i\varphi} = (e^{i\varphi})^* = z^*$$

we obtain the equality $z^{-1} = z^*$.

7.15. *Hint*.
Use the geometric interpretation of the numbers z_1 and z_2. The length of a side of an arbitrary triangle is no greater than the sum of the lengths of the two other sides and is no less than the absolute value of their difference.

7.16. *Solution.*

Use the formula for the geometric progression sum (proved in Exercise **1.55**):

$$\sum_{k=1}^{n} q^k = q + \cdots + q^n = \frac{q^{n+1} - q}{q - 1}.$$

(1) $\displaystyle\sum_{k=1}^{100} i^k = \frac{i^{101} - i}{i - 1} = \frac{i^{100} \cdot i - i}{i - 1} = 0.$

(2) $\displaystyle\sum_{k=-99}^{99} i^k = \frac{1}{i^{99}} \sum_{k=-99}^{99} i^{(k+99)}.$

In the last sum, replace the summation index $k' = k + 99$. Then, the sought sum takes the following form:

$$\sum_{k=-99}^{99} i^k = i^{-99} \sum_{k'=0}^{198} i^{k'} = i^{-100+1} \frac{i^{199} - 1}{i - 1} = i \cdot \frac{i^{199} - 1}{i - 1} =$$

$$= i \cdot \frac{i^{196} \cdot i^3 - 1}{i - 1} = i \cdot \frac{i^3 - 1}{i - 1} = \frac{i^4 - i}{i - 1} = \frac{1 - i}{i - 1} = -1.$$

7.17. *Solution.*

The imaginary unit has the property $i^4 = 1$. Consider four cases depending on the remainder of division m by 4:

(1) $m = 4k, k \in \mathbb{Z}$,

$$i^m = i^{4k} = (i^4)^k = 1^k = 1;$$

(2) $m = 4k + 1, k \in \mathbb{Z}$,

$$i^m = i^{4k+1} = i^{4k} \cdot i = 1 \cdot i = i;$$

(3) $m = 4k + 2, k \in \mathbb{Z}$,

$$i^m = i^{4k+2} = i^{4k} \cdot i^2 = i^2 = -1;$$

(4) $m = 4k + 3, k \in \mathbb{Z}$,

$$i^m = i^{4k+3} = i^{4k} \cdot i^3 = -i.$$

Thus, we finally obtain

$$i^m = \begin{cases} 1, & \text{if } m = 4k, \ k \in \mathbb{Z}; \\ i, & \text{if } m = 4k + 1, \ k \in \mathbb{Z}; \\ -1, & \text{if } m = 4k + 2, \ k \in \mathbb{Z}; \\ -i, & \text{if } m = 4k + 3, \ k \in \mathbb{Z}. \end{cases}$$

7.18. *Solution.*

(1) For transition to a in a trigonometric form of the complex number, we must determine the modulus $\rho = |z|$ and the argument $\varphi = \arg z$. Using the formulae for ρ and φ, we obtain

$$\rho = \sqrt{x^2 + y^2} = \sqrt{1^2 + 1^2} = \sqrt{2},$$

$$\varphi = \arctan \frac{y}{x} + 2\pi k = \arctan 1 + 2\pi k = \frac{\pi}{4} + 2\pi k, \quad k \in \mathbb{Z},$$

therefore

$$1 + i = \sqrt{2}\left(\cos\left(\frac{\pi}{4} + 2\pi k\right) + i \sin\left(\frac{\pi}{4} + 2\pi k\right)\right), \quad k \in \mathbb{Z}.$$

(2) $\rho = \sqrt{(-3)^2 + 4^2} = 5,$

$$\varphi = \arctan \frac{y}{x} + \pi = \arctan\left(-\frac{4}{3}\right) + \pi + 2\pi k = \pi - \arctan \frac{4}{3} + 2\pi k, \quad k \in \mathbb{Z},$$

therefore

$$-3 + 4i = 5\left[\cos\left(\pi(2k+1) - \arctan \frac{4}{3}\right) + i \sin\left(\pi(2k+1) - \arctan \frac{4}{3}\right)\right], k \in \mathbb{Z}.$$

(3) $\rho = 4$, $\varphi = \frac{\pi}{2}\,\text{sign}\,(-4) + 2\pi k = 2\pi k - \frac{\pi}{2} = \frac{\pi}{2}(4k - 1)$, $k \in \mathbb{Z}$, therefore

$$-4i = 4\left[\cos\left(\frac{\pi}{2}(4k-1)\right) + i \sin\left(\frac{\pi}{2}(4k-1)\right)\right], \quad k \in \mathbb{Z}.$$

(4) Transform the fraction by multiplication of numerator and denominator by the number $(5 + i)^*$:

$$\frac{5-i}{5+i} = \frac{(5-i)^2}{5^2 - i^2} = \frac{1}{26}(25 - 10i + i^2) = \frac{12 - 5i}{13},$$

$$\rho = \sqrt{\left(\frac{12}{13}\right)^2 + \left(-\frac{5}{13}\right)^2} = 1,$$

$$\varphi = \arctan\left(-\frac{5}{12}\right) + 2\pi k = -\arctan\left(\frac{5}{12}\right) + 2\pi k, \quad k \in \mathbb{Z},$$

therefore

$$\frac{5-i}{5+i} = \cos\left(2\pi k - \arctan\left(\frac{5}{12}\right)\right) + i \sin\left(2\pi k - \arctan\left(\frac{5}{12}\right)\right), \quad k \in \mathbb{Z}.$$

7.19. *Answer:*

(1) $z = 3\sqrt{3}/2 + 3i/2;$

(2) $z = \sqrt{6}/2 - i/\sqrt{2}$;
(3) $z = 1 - i$;
(4) $z = -\sqrt{2}(1 + i)$.

7.20. *Proof.*
Turn to the exponential form of the number u. Let $a + ib = \rho e^{i\varphi}$, then $a - ib = \rho e^{-i\varphi}$ and

$$u = \frac{a + ib}{a - ib} = \frac{e^{i\varphi}}{e^{-i\varphi}} = e^{i\varphi} \cdot e^{i\varphi} = e^{2i\varphi}.$$

Therefore, $u = e^{i\delta}$, where $\delta = 2\varphi$.

7.21. *Solution.*
The exponential notation of the imaginary unit has the form $i = e^{i\pi/2 + 2\pi i k}$, where $k \in \mathbb{Z}$. Having written the imaginary unit in the base in the exponential form and using the identity $(e^a)^b = e^{ab}$, we obtain

$$i^i = (e^{i\pi/2 + 2\pi i k})^i = e^{i^2(\pi/2 + 2\pi k)} = e^{-\pi/2 + 2\pi k'}, \quad \text{where } k, k' \in \mathbb{Z}.$$

As is seen from the considered example, the exponential function is a multifunction on the set of complex numbers $\mathbb{C}$.

7.22. *Solution.*

(1) The complex number $z = x + iy$ on the plane $\textcircled{z}$ is associated with a point with Cartesian coordinates (x, y). The condition $|\operatorname{Re} z| \leqslant 1$ is equivalent to $-1 \leqslant x \leqslant 1$, which is satisfied by all points with abscissas not exceeding one in the absolute value.

(2) $\operatorname{Im} z < -2 \Leftrightarrow y < -2$. This condition is satisfied by the points of a complex plane with ordinates less that -2.

(3) $|z| \leqslant \sqrt{2} \Leftrightarrow \sqrt{x^2 + y^2} \leqslant \sqrt{2}$, or $x^2 + y^2 \leqslant 2$. This inequality is satisfied by the points lying at a distance not exceeding $\sqrt{2}$ from the origin of coordinates.

(4) Argument of the number z lies within the interval $(\pi/8, \pi/4)$. On the plane $\textcircled{z}$, such numbers are associated with the inner part of the angle, formed by the rays $\varphi = \pi/8$ and $\varphi = \pi/4$.

The sets of points, satisfying the conditions of items (1)–(4), are shown in Fig. 7.3, panels (a)–(d), respectively.

7.23. *Answer:*
The sets of points, satisfying the conditions of items (1)–(4), are shown in Fig. 7.4, panels (a)–(d), respectively.

7.24. *Solution.*
(1) Trigonometric form of the number $z = 1 + i$ was found in Exercise **7.18**, 1):

$$1 + i = \sqrt{2}\left(\cos\left(\frac{\pi}{4} + 2\pi k\right) + i \sin\left(\frac{\pi}{4} + 2\pi k\right)\right), \quad k \in \mathbb{Z}.$$

Square root can be considered as raising to power equal to $\dfrac{1}{2}$: $\sqrt{z} \equiv z^{1/2}$. Using de Moivre's formula, we obtain

$$\sqrt{1+i} = (1+i)^{1/2} = \left[\sqrt{2}\left(\cos\left(\frac{\pi}{4} + 2\pi k\right) + i\sin\left(\frac{\pi}{4} + 2\pi k\right)\right)\right]^{1/2} =$$
$$= 2^{1/4}\left(\cos\left(\frac{\pi}{8} + \pi k\right) + i\sin\left(\frac{\pi}{8} + \pi k\right)\right), \quad k = 0, 1.$$

Note that it is sufficient to leave only two values $k = 0, 1$. The square root value for other integer values of k coincide with the respective values for the specified k. For example, for $k = 3$ we obtain $\cos(\frac{\pi}{8} + 3 \cdot \pi) = \cos(\frac{\pi}{8} + \pi + 2\pi) = \cos(\frac{\pi}{8} + 1 \cdot \pi)$, which corresponds to the case $k = 1$.

In order to calculate $\cos(\frac{\pi}{8} + \pi k)$ and $\sin(\frac{\pi}{8} + \pi k)$, use trigonometric half-angle formulae. As is known (see Appendix "Reference data," relations (A.8)),

$$\cos\frac{\alpha}{2} = \sqrt{\frac{1 + \cos\alpha}{2}}, \quad \sin\frac{\alpha}{2} = \sqrt{\frac{1 - \cos\alpha}{2}}$$

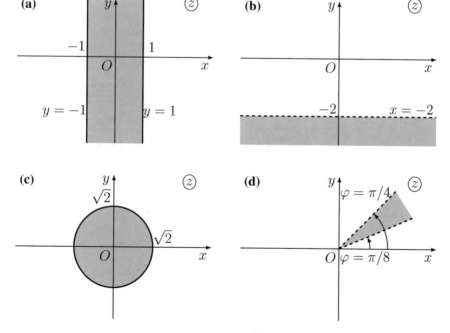

Fig. 7.3 To Exercise 7.22

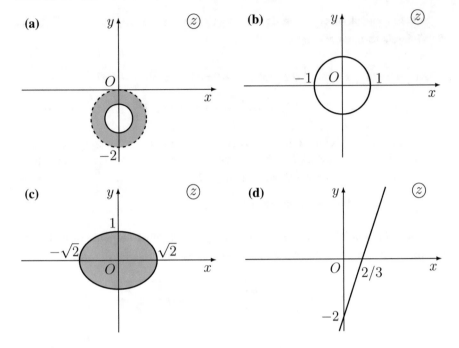

Fig. 7.4 To Exercise 7.23

Hence, for $k = 0$ we have

$$\cos\frac{\pi}{8} = \sqrt{\frac{1 + \cos\pi/4}{2}} = \sqrt{\frac{1 + \sqrt{2}/2}{2}} = \frac{1}{2}\sqrt{2 + \sqrt{2}},$$

$$\sin\frac{\pi}{8} = \sqrt{\frac{1 - \cos\pi/4}{2}} = \sqrt{\frac{1 - \sqrt{2}/2}{2}} = \frac{1}{2}\sqrt{2 - \sqrt{2}}.$$

and the first square root value $\sqrt{z}$ is

$$\sqrt{1 + i} = 2^{1/4}\left(\frac{1}{2}\sqrt{2 + \sqrt{2}} + \frac{i}{2}\sqrt{2 - \sqrt{2}}\right).$$

For $k = 1$, we obtain $\cos(\frac{\pi}{8} + \pi) = -\cos\frac{\pi}{8}$, $\sin(\frac{\pi}{8} + \pi) = -\sin\frac{\pi}{8}$, and the second square root value $\sqrt{z}$ is

$$\sqrt{1 + i} = 2^{1/4}\left(-\frac{1}{2}\sqrt{2 + \sqrt{2}} - \frac{i}{2}\sqrt{2 - \sqrt{2}}\right).$$

Represent the final answer in the form

$$\sqrt{1+i} = \pm 2^{1/4} \left(\frac{1}{2}\sqrt{2 + \sqrt{2}} + \frac{i}{2}\sqrt{2 - \sqrt{2}} \right).$$

(2) Represent the number -1 in a trigonometric form:

$$\rho = 1, \quad \varphi = \arctan\left(-\frac{0}{1}\right) + \pi = \pi;$$

$$-1 = \cos(\pi(2k+1)) + i\sin(\pi(2k+1)), \quad k \in \mathbb{Z}.$$

By de Moivre's formula, we obtain

$$\sqrt[3]{-1} = (-1)^{1/3} = \cos\left(\frac{\pi}{3}(2k+1)\right) + i\sin\left(\frac{\pi}{3}(2k+1)\right), \quad k = 0, 1, 2.$$

We are leaving only three values $k = 0, 1, 2$, since all other integer values of k will not result in new root values.

Consider the case $k = 0$. Substituting the known values $\cos(\frac{\pi}{3}) = \frac{1}{2}$, $\sin(\frac{\pi}{3}) = \frac{\sqrt{3}}{2}$, we obtain the first cube root value

$$\sqrt[3]{-1} = \cos\left(\frac{\pi}{3}\right) + i\sin\left(\frac{\pi}{3}\right) = \frac{1}{2} + i\frac{\sqrt{3}}{2}.$$

Further, for $k = 1$ we have

$$\sqrt[3]{-1} = \cos\pi + i\sin\pi = -1.$$

And, finally, for $k = 2$ we obtain

$$\sqrt[3]{-1} = \cos\left(\frac{5\pi}{3}\right) + i\sin\left(\frac{5\pi}{3}\right) = \frac{1}{2} - i\frac{\sqrt{3}}{2}.$$

So, all values of the cube root $\sqrt[3]{-1}$ belong to the set $\{-1, \frac{1}{2} \pm i\frac{\sqrt{3}}{2}\}$.

7.25. *Answer:*
(1) $\sqrt{8i} = \pm 2(1 + i)$;
(2) $\sqrt[6]{64} \in \{\pm 2, \pm 1 \pm i\sqrt{3}, \pm 1 \mp i\sqrt{3}\}$.

7.26. *Proof.*
(1) According to the introduced definition,

$$\omega_k = (e^{2\pi i})^{k/n} = e^{2\pi i k/n}, \quad k = 0, 1, \ldots, n - 1.$$

Fig. 7.5 To Exercise 7.26

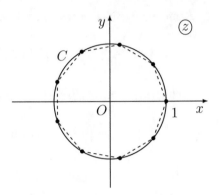

Modulus of the complex number $\omega_k = e^{2\pi i k/n}$ is equal to one for all values of the variable k, and the argument is equal to $\arg \omega_k = 2\pi k/n$, $k = 0, 1, \ldots, n-1$. Thereby, we can conclude that the nth roots of one are located on the unit circle C, and the first root ω_0, associated with $k = 0$, lies in the real axis, and ω_k divide the circle by n arcs of the same length (see the example for the instance $n = 9$ in Fig. 7.5).

(2) Transform the exponential notation of the number $\omega_{k+n/2}$:

$$\omega_{k+n/2} = e^{\frac{2\pi i (k+n/2)}{n}} = e^{\frac{2\pi i k}{n}} e^{\pi i} = \omega_k e^{\pi i}.$$

Using the equality $e^{\pi i} = -1$, we obtain: $\omega_{k+n/2} = -\omega_k$ for even n and $k = 0, 1, \ldots, n/2 - 1$.

(3) The values ω_k form geometric progression, whose denominator is $\omega_1 = e^{2\pi i/n}$. Using the formula for the geometric progression sum (see Exercise 1.55), we obtain

$$\sum_{k=0}^{n-1} \omega_k = \sum_{k=0}^{n-1} e^{2\pi i k/n} = \frac{(e^{2\pi i/n})^n - 1}{e^{2\pi i/n} - 1} = 0.$$

(4)

$$\prod_{k=0}^{n-1} \omega_k = \prod_{k=0}^{n-1} e^{2\pi i k/n} = e^{\sum\limits_{k=0}^{n-1} 2\pi i k/n} = e^{(2\pi i \sum\limits_{k=0}^{n-1} k)/n}.$$

The sum in the exponent is $\sum\limits_{k=0}^{n-1} k = \dfrac{n(n-1)}{2}$, and therefore,

$$\prod_{k=0}^{n-1} \omega_k = e^{\pi i (n-1)} = \cos \pi (n-1) + i \sin \pi (n-1) = (-1)^{n-1}.$$

7.29. *Answer*: $\dfrac{n}{x^n - 1}$.

7.30. *Proof.*

Consider the sum $\mathcal{Z} = \sum\limits_{k=1}^{n} e^{i\alpha k}$. It is easy to see that the following relations are valid:

$$\cos\alpha + \cos 2\alpha + \cdots + \cos n\alpha = \operatorname{Re}\mathcal{Z},$$
$$\sin\alpha + \sin 2\alpha + \cdots + \sin n\alpha = \operatorname{Im}\mathcal{Z}.$$

Calculate $\mathcal{Z}$, using the formula for the geometric progression sum:

$$\mathcal{Z} = \sum_{k=1}^{n} e^{i\alpha k} = \frac{e^{i\alpha(n+1)} - e^{i\alpha}}{e^{i\alpha} - 1}.$$

Simplify the obtained expression, multiplying the fraction by $1 = \dfrac{e^{-i\alpha/2}}{e^{-i\alpha/2}}$ and performing simple transformations:

$$\mathcal{Z} = \frac{e^{i\alpha(n+1)} - e^{i\alpha}}{e^{i\alpha} - 1} \cdot \frac{e^{-i\alpha/2}}{e^{-i\alpha/2}} = \frac{e^{i\alpha(n+1/2)} - e^{i\alpha/2}}{e^{i\alpha/2} - e^{-i\alpha/2}}.$$

Denominator of the obtained fraction is $e^{i\alpha/2} - e^{-i\alpha/2} = 2i\sin\alpha/2$. Rewrite the exponents in numerator, using Euler's formula:

$$\mathcal{Z} = \frac{1}{2i\sin\alpha/2}\Big[\cos[(n+1/2)\alpha] + i\sin[(n+1/2)\alpha] - (\cos\alpha/2 + i\sin\alpha/2)\Big] =$$
$$= \frac{\sin[(n+1/2)\alpha] - \sin\alpha/2}{2\sin\alpha/2} + \frac{\cos[(n+1/2)\alpha] - \cos\alpha/2}{2i\sin\alpha/2}.$$

Further, use the known trigonometric formulae (see Appendix "Reference Data," formulae (A.16) and (A.18))

$$\sin a - \sin b = 2\sin\frac{a-b}{2}\cos\frac{a+b}{2},$$
$$\cos a - \cos b = -2\sin\frac{a-b}{2}\sin\frac{a+b}{2}.$$

We obtain

$$\mathcal{Z} = \frac{\sin(n\alpha/2)}{\sin(\alpha/2)}\cos[(n+1)\alpha/2] + i\frac{\sin(n\alpha/2)}{\sin(\alpha/2)}\sin[(n+1)\alpha/2],$$

whence directly follow the expression for the sums from the condition.

7.31. *Proof.*

Denote predicate "$(\cos\varphi + i\sin\varphi)^n = \cos n\varphi + i\sin n\varphi$" by $P(n)$ and prove the statement $\forall n\, P(n)$ by the mathematical induction method.

Basis step

For $n = 1$, we obtain the valid identity $(\cos \varphi + i \sin \varphi)^1 = \cos \varphi + i \sin \varphi$, and therefore $P(1)$ is true.

Inductive step

Suppose that $P(k), k \in \mathbb{N}$ is true. Prove the truth of the proposition $P(k + 1)$. We need to prove that

$$(\cos \varphi + i \sin \varphi)^{k+1} = \cos (k + 1)\varphi + i \sin (k + 1)\varphi.$$

Consider the expression $(\cos \varphi + i \sin \varphi)^{k+1}$ and represent it in the form

$$(\cos \varphi + i \sin \varphi)^{k+1} = (\cos \varphi + i \sin \varphi)^k \cdot (\cos \varphi + i \sin \varphi).$$

According to the inductive supposition, the first factor is

$$(\cos \varphi + i \sin \varphi)^k = \cos k\varphi + i \sin k\varphi.$$

Then,

$$(\cos \varphi + i \sin \varphi)^{k+1} = (\cos k\varphi + i \sin k\varphi) \cdot (\cos \varphi + i \sin \varphi).$$

Open the brackets in the obtained expression, using the known identities for trigonometric functions, provided in Appendix Reference Data," formulae (A.11) and (A.9):

$$\cos(a + b) = \cos a \cos b - \sin a \sin b,$$
$$\sin(a + b) = \sin a \cos b + \cos a \sin b,$$

assuming $a = k\varphi, b = \varphi$. We obtain

$$(\cos \varphi + i \sin \varphi)^{k+1} = \underbrace{(\cos k\varphi \cos \varphi - \sin k\varphi \sin \varphi)}_{\cos (k+1)\varphi} +$$
$$+ i \underbrace{(\sin k\varphi \cos \varphi + \cos k\varphi \sin \varphi)}_{\sin (k+1)\varphi}.$$

Hence, according to the mathematical induction principle, de Moivre's formula

$$(\cos \varphi + i \sin \varphi)^n = \cos n\varphi + i \sin n\varphi$$

is valid for all natural values $n \in \mathbb{N}$.

7.32. *Solution.*

In the left side of de Moivre's formula stands the expression, which can be expanded by Newton binomial formula (see p. 143). Thus, represent the left side in the form

$$(\cos\varphi + i\sin\varphi)^n = \sum_{j=0}^{n} C(n,j)(\cos^{n-j}\varphi)(i\sin\varphi)^j =$$

$$= \sum_{j=0}^{n} i^j\, C(n,j)\cos^{n-j}\varphi\,\sin^j\varphi.$$

It is convenient to partition the sum into two sums—by even ($j = 2k$) and odd ($j = 2k+1$) values of j, and introduce a new summation variable $k \in \mathbb{N}$:

$$(\cos\varphi + i\sin\varphi)^n =$$

$$= \sum_{k=0}^{\lfloor n/2 \rfloor} i^{2k} C(n,2k)\cos^{n-2k}\varphi\,\sin^{2k}\varphi +$$

$$+ \sum_{k=0}^{\lfloor (n-1)/2 \rfloor} i^{2k+1} C(n,2k+1)\cos^{n-2k-1}\varphi\,\sin^{2k+1}\varphi =$$

$$= \sum_{k=0}^{\lfloor n/2 \rfloor} (-1)^k C(n,2k)\cos^{n-2k}\varphi\,\sin^{2k}\varphi +$$

$$+ i \sum_{k=0}^{\lfloor (n-1)/2 \rfloor} (-1)^k C(n,2k+1)\cos^{n-2k-1}\varphi\,\sin^{2k+1}\varphi.$$

Now we only have to take advantage of the fact that $\cos n\varphi = \mathrm{Re}\,(\cos\varphi + i\sin\varphi)^n$, $\sin n\varphi = \mathrm{Im}\,(\cos\varphi + i\sin\varphi)^n$. We obtain formulae for cosine and sine of a multiple argument:

$$\cos n\varphi = \sum_{k=0}^{\lfloor n/2 \rfloor} (-1)^k C(n,2k)\cos^{n-2k}\varphi\,\sin^{2k}\varphi,$$

$$\sin n\varphi = \sum_{k=0}^{\lfloor (n-1)/2 \rfloor} (-1)^k C(n,2k+1)\cos^{n-2k-1}\varphi\,\sin^{2k+1}\varphi.$$

Consider instances of small values of n.

(1) For $n = 3$, the obtained formulae take the form

$$\cos 3\varphi = \sum_{k=0}^{1}(-1)^k C(3,2k)\cos^{3-2k}\varphi\,\sin^{2k}\varphi = \cos^3\varphi - 3\cos\varphi\sin^2\varphi,$$

$$\sin 3\varphi = \sum_{k=0}^{1}(-1)^k C(3,2k+1)\cos^{3-2k-1}\varphi\,\sin^{2k+1}\varphi = 3\cos^2\varphi\sin\varphi - \sin^3\varphi.$$

(2) For $n = 4$, we have

$$\cos 4\varphi = \sum_{k=0}^{2}(-1)^k C(4, 2k) \cos^{4-2k}\varphi \, \sin^{2k}\varphi =$$

$$= \cos^4\varphi - 6\cos^2\varphi \sin^2\varphi + \sin^4\varphi,$$

$$\sin 4\varphi = \sum_{k=0}^{1}(-1)^k C(4, 2k+1) \cos^{4-2k-1}\varphi \, \sin^{2k+1}\varphi =$$

$$= 4\cos^3\varphi \sin\varphi - 4\cos\varphi \sin^3\varphi.$$

7.33. *Solution.*

We need to calculate $\langle \sin^{100} x \rangle = \dfrac{1}{2\pi} \int\limits_{0}^{2\pi} \sin^{100} x \, dx$. Express sine by the exponential function:

$$\sin x = \frac{e^{ix} - e^{-ix}}{2i}.$$

Then, the hundredth power of the sine can be presented in the form

$$\sin^{100} x = \left(\frac{e^{ix} - e^{-ix}}{2i}\right)^{100} = 2^{-100}(e^{ix} - e^{-ix})^{100}.$$

The mean value of the function is

$$\langle \sin^{100} x \rangle = \frac{1}{2\pi} \int\limits_{0}^{2\pi} \sin^{100} x \, dx = \frac{2^{-100}}{2\pi} \int\limits_{0}^{2\pi} (e^{ix} - e^{-ix})^{100} \, dx.$$

We will expand the expression under integral sum by Newton binomial formula:

$$\langle \sin^{100} x \rangle = \frac{2^{-100}}{2\pi} \int\limits_{0}^{2\pi} dx \sum_{k=0}^{100} C(100, k)(e^{ix})^{100-k}(-e^{-ix})^k =$$

$$= \frac{2^{-100}}{2\pi} \int\limits_{0}^{2\pi} dx \sum_{k=0}^{100} C(100, k)(-1)^k e^{i(100-2k)x}.$$

Transpose the signs of summation and integration:

$$\langle \sin^{100} x \rangle = \frac{2^{-100}}{2\pi} \sum_{k=0}^{100} C(100, k)(-1)^k \int\limits_{0}^{2\pi} e^{i(100-2k)x} \, dx.$$

Since for all integer n, except $n = 0$, the equality $\int\limits_{0}^{2\pi} e^{inx}\, dx = 0$ is valid, in the last sum, only one summand is different from zero, namely the summand with $100 - 2k = 0$, $k = 50$. We finally obtain:

$$\langle \sin^{100} x \rangle = \frac{2^{-100}}{2\pi} C(100, 50) \int\limits_{0}^{2\pi} dx = 2^{-100} C(100, 50) = \frac{100!}{2^{100}(50!)^2}.$$

Note. An alternative method of calculating similar integral with another upper limit of integration $I_n = \int\limits_{0}^{\pi/2} \sin^n x\, dx$ by the method based on the recurrence relation analysis is provided in Exercise 8.129. Concerning estimate of the value I_n for great n, see also Exercise 11.17.

7.34. *Answer:* $\langle \cos^{200} x \rangle = \dfrac{200!}{2^{200}(100!)^2}$.

7.35. *Hint.* $\sin x + \cos x = \sqrt{2} \sin(x + \pi/4)$.

7.36. *Answer:* $\langle f(x) \rangle = (a^2 + b^2)^{10} \dfrac{20!}{2^{20}(10!)^2}$.

7.37. *Hint.*

Multiply the brackets in the right side of factorization of the polynomial $p(z)$ and compare the obtained coefficients at the same powers with the coefficients $p(z)$.

7.38. *Solution.*
(1) Represent $z_1^2 + z_2^2$ in the form

$$z_1^2 + z_2^2 = (z_1 + z_2)^2 - 2z_1 z_2$$

and express the sum and the product of the roots $p(z)$ by Viete formulae, proved in the previous exercise:

$$z_1^2 + z_2^2 = (-p)^2 - 2q = p^2 - 2q.$$

(2) $\dfrac{1}{z_1^2} + \dfrac{1}{z_2^2} = \dfrac{z_1^2 + z_2^2}{z_1^2 z_2^2} = \dfrac{p^2 - 2q}{q^2}$.

7.39. *Answer:*
(1) $z_1^4 + z_2^4 = (p^2 - 2q)^2 - 2q^2$;

(2) $\dfrac{1}{z_1^4} + \dfrac{1}{z_2^4} = \dfrac{(p^2 - 2q)^2 - 2q^2}{q^4}$.

7.40. *Answer:*
(1) $\sum\limits_{k=1}^{3} z_k = \prod\limits_{k=1}^{3} z_k = -1$;

(2) $\sum\limits_{k=1}^{4} z_k = 0$, $\prod\limits_{k=1}^{4} z_k = 4$.

7.41. *Answer*:

(1) $\sum\limits_{k=1}^{4} z_k = \prod\limits_{k=1}^{4} z_k = 0$;

(2) $\sum\limits_{k=1}^{5} z_k = -1$, $\prod\limits_{k=1}^{5} z_k = 0$.

7.42. *Solution.*

Replace the variable $y = x^2 - 2x$. The obtained quadratic equation $y^2 - 17y + 35 = 0$ has two roots, which are the solutions of the system

$$\begin{cases} y_1 + y_2 = 17, \\ \quad\ y_1 y_2 = 35. \end{cases}$$

Let the equation $x^2 - 2x = y_1$ have the roots x_1 and x_2, and the equation $x^2 - 2x = y_2$—the roots x_3 and x_4. The set $\{x_1, x_2, x_3, x_4\}$ coincides with the set of solutions of the original equation.

Calculate the sum of square roots of the equation $x^2 - 2x = y_1$ and $x^2 - 2x = y_2$, using Viete formulae:

$$\sum_{i=1}^{4} x_i^2 = [(x_1 + x_2)^2 - 2x_1x_2] + [(x_3 + x_4)^2 - 2x_3x_4] =$$

$$= (2^2 + 2y_1) + (2^2 + 2y_2) = 8 + 2(y_1 + y_2).$$

As a result, we obtain $\sum\limits_{i=1}^{4} x_i^2 = 8 + 2 \cdot 17 = 42$.

7.43. *Answer*: $\sum\limits_{i=1}^{4} x_i^2 = 16$.

7.44. *Solution.*

Use the relations

$$(x_1 + x_2 + x_3)^2 = x_1^2 + x_2^2 + x_3^2 + 2x_1x_2 + 2x_1x_3 + 2x_2x_3 \Leftrightarrow$$

$$\Leftrightarrow x_1^2 + x_2^2 + x_3^2 = (x_1 + x_2 + x_3)^2 - 2(x_1x_2 + x_1x_3 + x_2x_3).$$

According to Viete formulae, $x_1 + x_2 + x_3 = 0$, $x_1x_2 + x_1x_3 + x_2x_3 = -7$. Hence,

$$\sum_{i=1}^{3} x_i^2 = 0^2 - 2(-7) = 14.$$

7.45. *Answer*: $\sum\limits_{i=1}^{3} x_i^2 = 140$.

7.46. *Answer*: $z_1 = -2$, $z_2 = z_3 = 1$.

7.47. *Solution.*

Replace the variable $z = y + \dfrac{5}{3}$. We obtain the cubic equation in the canonical form

$$y^3 + \frac{2}{3}y + \frac{20}{27} = 0,$$

here $p = \dfrac{2}{3}, q = \dfrac{20}{27}$. Further using Cardano formula:

$$Q = \left(\frac{p}{3}\right)^3 + \left(\frac{q}{2}\right)^2 = \frac{4}{27},$$

$$\alpha, \beta = \sqrt[3]{-\frac{10}{27} \pm \sqrt{\frac{4}{27}}} = \frac{1}{3}\sqrt[3]{-10 \pm 6\sqrt{3}}.$$

Let $\alpha = \dfrac{1}{3}\sqrt[3]{6\sqrt{3} - 10}$, then, in order for the condition $\alpha\beta = -p/3$ to be satisfied, we choose $\beta = -\dfrac{1}{3}\sqrt[3]{6\sqrt{3} + 10}$. The roots of the equation will have the form

$$y_1 = \frac{1}{3}\left(\sqrt[3]{6\sqrt{3} - 10} - \sqrt[3]{6\sqrt{3} + 10}\right);$$

$$y_2 = -\frac{1}{6}\left(\sqrt[3]{6\sqrt{3} - 10} - \sqrt[3]{6\sqrt{3} + 10}\right) + \frac{i\sqrt{3}}{6}\left(\sqrt[3]{6\sqrt{3} - 10} + \sqrt[3]{6\sqrt{3} + 10}\right);$$

$$y_3 = -\frac{1}{6}\left(\sqrt[3]{6\sqrt{3} - 10} - \sqrt[3]{6\sqrt{3} + 10}\right) - \frac{i\sqrt{3}}{6}\left(\sqrt[3]{6\sqrt{3} - 10} + \sqrt[3]{6\sqrt{3} + 10}\right).$$

The obtained expressions can be simplified, if we note that the equality $6\sqrt{3} \pm 10 = (\sqrt{3} \pm 1)^3$ is valid. Then, $\sqrt[3]{6\sqrt{3} \pm 10} = \sqrt{3} \pm 1$, and

$$y_1 = \frac{1}{3}\left(\sqrt{3} - 1 - (\sqrt{3} + 1)\right) = -\frac{2}{3};$$

$$y_2 = -\frac{1}{6}\left(\sqrt{3} - 1 - (\sqrt{3} + 1)\right) + \frac{i\sqrt{3}}{6}\left(\sqrt{3} - 1 + \sqrt{3} + 1\right) = \frac{1}{3} + i;$$

$$y_3 = -\frac{1}{6}\left(\sqrt{3} - 1 - (\sqrt{3} + 1)\right) - \frac{i\sqrt{3}}{6}\left(\sqrt{3} - 1 + \sqrt{3} + 1\right) = \frac{1}{3} - i.$$

Returning to the original variable $z = y + \dfrac{5}{3}$, we obtain $z_1 = 1, z_2 = 2 + i, z_3 = 2 - i$.

7.48. *Answer:* $z_1 = 7, z_2 = -\dfrac{5}{2}, z_3 = 2$.

Recurrence Relations

Recurrence relation (from Latin *recurrere*—to return) is the method of specifying the function $f(n)$, $n \in \mathbb{N}$, for which [48,60]:

(1) values of the function $f(n)$ for some subset

$$N_m = \{1, 2, \ldots, m\} \subset \mathbb{N}$$

are specified in an explicit form:

$$f(1) = a_1, \ f(2) = a_2, \ \ldots, \ f(m) = a_m;$$

(2) values of the function for the arguments $k \in \mathbb{N} \setminus N_m$ are expressed by the values $f(i)$, where $i = 1, 2, \ldots, k - 1$, in accordance with a specified rule.

The functions constructed this ways are called **recursive** or **recursively defined**, and it is the **solution** of the recurrence relation. The set of values of the function $f(n)$ is also referred to as the **recursive sequence**.

Let the recursive sequence $\{a_n\}$ be defined. The sequence $\{a_n\}$ is called the **finite history sequence**, if each subsequent term of the sequence explicitly depends on some fixed number of previous terms. If a_n depends on all a_i, $1 \leqslant i \leqslant n - 1$, then it is referred to as the **full history sequence**.

When analyzing algorithms, one quite often has to study the recursively defined sequences. Let us consider the main methods for analyzing the recurrence relations [16]:

(1) *Method of substitution;*
(2) *Method based on the solution of a characteristic equation;*
(3) *Method of generating functions.*

The method of substitution consists in establishing an explicit expression for the value a_n and subsequent proving for all n, which is performed as follows. In the right side of the recurrence relation for a_n, there exist values $a_{n-1}, a_{n-2}, \ldots$ Express

© Springer International Publishing AG, part of Springer Nature 2018

S. Kurgalin and S. Borzunov, *The Discrete Math Workbook,*

Texts in Computer Science, https://doi.org/10.1007/978-3-319-92645-2_8

them by $a_{n-2}, a_{n-3}, \ldots$, then by $a_{n-3}, a_{n-4}, \ldots$, until it is possible to establish the formula for a_n. The obtained formula should be proved, for example, using the mathematical induction.

The method based on the solution of a **characteristic equation** is used for solving linear recurrence relations.

Linear first-order homogeneous recurrence relation with constant coefficients is the relation of the form

$$a_n = ca_{n-1}, \quad c \neq 0.$$

Its solution will be $a_n = Ac^n$, where the value A is determined from the initial condition $a_1 = A$.

Linear homogeneous recurrence relation of order l with constant coefficients is the relation of the form

$$a_n = c_1 a_{n-1} + c_2 a_{n-2} + \cdots + c_l a_{n-l}, \quad c_l \neq 0.$$

In order to solve this relation, we should write out the characteristic equation

$$z^l = c_1 z^{l-1} + c_2 z^{l-2} + \cdots + c_l.$$

If the values $z_1, z_2, \ldots$ are the roots of the characteristic equation with multiplicities $m_1, m_2, \ldots$, then

$$a_n = A_1 z_1^n + A_2 n z_1^n + \cdots + A_n n^{m_1 - 1} z_1^n +$$
$$+ B_1 z_2^n + B_2 n z_2^n + \cdots + B_n n^{m_2 - 1} z_2^n + \ldots$$

Consider in more detail a practically important case of recurrence relation of the second order, $l = 2$, with the real coefficients $c_1, c_2 \in \mathbb{R}$. The characteristic equation takes the form

$$z^2 = c_1 z + c_2,$$

and we will consider three possible cases in order to write its solution.

1. The characteristic equation has two different real roots $z_1, z_2 \in \mathbb{R}$, then

$$a_n = A z_1^n + B z_2^n.$$

2. The characteristic equation has two different real roots $z_1 = z_2$, in other words, one root z_0 is of multiplicity $m = 2$. In this case,

$$a_n = A z_0^n + B n z_0^n.$$

3. The characteristic equation has two complex roots $z_1, z_2 \in \mathbb{C}$. If $c_1, c_2 \in \mathbb{R}$, then roots z_1 and z_2 will be complex conjugates: $z_2 = z_1^*$, in other words, $z_1 = a + ib$, $z_2 = a - ib$ for some real a and b. The general solution $a_n = A'(a + ib)^n + B'(a - ib)^n$ can be rewritten in the following form:

$$a_n = A\rho^n \cos n\varphi + B\rho^n \sin n\varphi,$$

where ρ and φ are polar coordinate coordinates of Cartesian plane point (a, b) (see chapter "Complex Numbers," p. 247).

Nonhomogeneous recurrence relations contain, in the right side, the additional summand $d(n)$:

$$a_n = c_1 a_{n-1} + c_2 a_{n-2} + \cdots + c_l a_{n-l} + d(n).$$

Solution of such relations are represented in the form of the sum of the general solution for the respective homogeneous equation (with $d(n) \equiv 0$) and some specific solution of nonhomogeneous

$$a_n = (a_n)_{\text{gen.}} + (a_n)_{\text{spec.}}.$$

In a frequent practical case, when $d(n)$ has the form

$$d(n) = P_m(n) \cdot t^n,$$

where P_m—polynomial of degree m, and $t = \text{const}$, the specific solution should be sought in the form

$$(a_n)_{\text{spec.}} = n^s Q_m(n) \cdot t^n,$$

where Q_m—some polynomial of degree m, s—multiplicity of the root $z = t$ [16].

Example 8.1 Find the solution of linear recurrence relation of the third order

$$\begin{cases} a_n = -3a_{n-1} + 4a_{n-3} + 18n - 33, & n > 3, \\ a_1 = 2, \ a_2 = 5, \ a_3 = 10. \end{cases}$$

Solution.
The characteristic equation has the form $z^3 = -3z^2 + 4$. Having found the roots of this equation $z_1 = 1$ and $z_2 = z_3 = -2$, write the general solution of the homogeneous relation:

$$(a_n)_{\text{gen.}} = A \cdot 1^n + B(-2)^n + C n(-2)^n,$$

where A, B, and C—real coefficients to be determined. The function $d(n) = 18n - 33$ in the right side of the recurrence relation is a first-degree polynomial; hence, the specific solution has the following form:

$$(a_n)_{\text{spec.}} = n(Dn + E) = Dn^2 + En,$$

where D and E—const.

Substituting $(a_n)_{\text{spec.}}$ into the equation $a_n = -3a_{n-1} + 4a_{n-3} + 18n - 33$, and find the values of the constant D and E:

$$Dn^2 + En = -3(D(n-1)^2 + E(n-1)) + 4(D(n-3)^2 + E(n-3)) +$$
$$+ 18n - 33, \Rightarrow D = 1, \ E = 0.$$

So, the solution of the initial recurrence relation can be represented in the form:

$$a_n = (a_n)_{\text{gen.}} + (a_n)_{\text{spec.}} = A + B(-2)^n + Cn(-2)^n + n^2,$$

and the values of the constants A, B, and C should be determined from the initial conditions $a_1 = 2$, $a_2 = 5$, $a_3 = 10$. Assuming $n = 1, 2, 3$ in the formula for a_n, write out the relations connecting A, B, and C:

$$\begin{cases} A - 2B - 2C + 1 & = 2, \\ A + 4B + 8C + 4 & = 5, \\ A - 8B - 24C + 9 & = 10. \end{cases}$$

To this linear system with respect to the values A, B, and C as is easy to see, satisfy the values $A = 1$, $B = C = 0$.

We finally obtain that $a_n = n^2 + 1$ for $n \in \mathbb{N}$.   □

Proceed to consideration of the method based on the use of the generating functions. Within the framework of this method, it is convenient to enumerate the terms of the numerical sequence under analysis, starting from zero:

$$a_0, a_1, a_2, \ldots, a_n, \ldots$$

Assume that the values a_n take complex values.

Generating function for the numerical sequence $\{a_n\}$, where $n = 0, 1, 2, \ldots$, is the series

$$G(z) = a_0 + a_1 z + a_2 z^2 + \ldots + a_n z^n + \ldots,$$

or, in abbreviated notation, $G(z) = \sum_{n=0}^{\infty} a_n z^n$. Such an expansion is considered as a formal series by integer nonnegative powers of the complex variable z. The use of this method does not envisage calculation of the values $G(z)$ at certain points, and this is why the question of convergence of series for various z is not considered in the theory of recurrence relations. In other words, $G(z)$ does not necessarily has meaning for each value of $z \in \mathbb{C}$ [42].

The numerical sequence is fully determined by its generating function. Links to the complete proof of the generating functions are shown in [35].

Let us introduce the definitions of sum and product of generating functions.

Let the numerical sequences $\{a_n\}$ and $\{b_n\}$, where $n \in \{0\} \cup \mathbb{N}$, be associated with the generating functions $G_a(z)$ and $G_b(z)$, respectively. Then **sum** and **product** of these functions are represented in the form

$$G_a(z) + G_b(z) = \sum_{n=0}^{\infty}(a_n + b_n)z^n,$$

$$G_a(z)G_b(z) = \sum_{n=0}^{\infty}\left(\sum_{i=0}^{n} a_i b_{n-i}\right)z^n.$$

Note that the introduced definitions are generalizations of the operations with polynomials known from algebra. As is easy to see, the operations of summation and multiplication of generating functions are commutative and associative.

Note that the sequence $\left\{\sum_{i=0}^{n} a_i b_{n-i}\right\}$ is called **convolution** of the sequences $\{a_n\}$ and $\{b_n\}$.

In Table 8.1 are listed the generating functions of the sequences that are often used for solutions of problems.

The method of generating functions used for solving the recurrence relations is based on finding the generating function for the recursive sequence that satisfies the condition of the problem.

In order to find the solution of $\{a_n\}$ recurrence relation of the lth degree by the method of generating functions, the following operations should be performed [60]:

(1) Multiply the left and the right sides of the recurrence relation by z^n and sum the obtained equalities over the variable n from $n = l$ to infinity;

(2) By algebraic transformations, bring the sums to the form $G(z) = \sum_{n=0}^{\infty} a_n z^n$ and solve the obtained functional equation with respect to the unknown function $G(z)$;

(3) Expand $G(z)$ in integer nonnegative powers of the variable z and thus obtain the sought coefficients a_n for $n = 0, 1, 2, \ldots$

Example 8.2 Find the solution of linear recurrence relation of the first order

$$\begin{cases} a_n = 4a_{n-1} + 6 \cdot 10^{n-1}, & n > 1, \\ a_1 = 14 \end{cases}$$

using the method of generating functions.

Table 8.1 Basic generating functions. The parameter N takes values from the natural series, c—arbitrary complex constant

Ordinal number	Generating function $G(z)$	Recursive sequence $\{a_n\}$
1.	$\dfrac{1}{1-z} = \sum\limits_{n=0}^{\infty} z^n$	$1, 1, 1, 1, \ldots$
2.	$\dfrac{1}{1-cz} = \sum\limits_{n=0}^{\infty} c^n z^n$	$1, c, c^2, c^3, \ldots$
3.	$\dfrac{1-z^{N+1}}{1-z} = \sum\limits_{n=0}^{N} z^n$	$\underbrace{1, 1, 1, \ldots, 1}_{N \text{ terms}}, 0, 0, 0, \ldots$
4.	$(1+z)^N = \sum\limits_{n=0}^{\infty} C(N,n)z^n$	$C(N,0), C(N,1), C(N,2), C(N,3), \ldots$
5.	$(1+cz)^N = \sum\limits_{n=0}^{\infty} C(N,n)c^n z^n$	$C(N,0), C(N,1)c, C(N,2)c^2,$ $C(N,3)c^3, \ldots$
6.	$\dfrac{1}{(1-z)^2} = \sum\limits_{n=0}^{\infty} (n+1)z^n$	$1, 2, 3, 4, \ldots$
7.	$\dfrac{1}{(1-z)^N} = \sum\limits_{n=0}^{\infty} C(N+n-1,n)z^n$	$C(N-1,0), C(N,1), C(N+1,2),$ $C(N+2,3), \ldots$
8.	$\dfrac{1}{(1-cz)^N} =$ $= \sum\limits_{n=0}^{\infty} C(N+n-1,n)c^n z^n$	$C(N-1,0), C(N,1)c, C(N+1,2)c^2,$ $C(N+2,3)c^3, \ldots$
9.	$\dfrac{1}{1-z^N} = \sum\limits_{n=0}^{\infty} z^{N\cdot n}$	$\underbrace{1, 0, 0, \ldots, 0}_{N \text{ terms}}, 1, 0, 0, \ldots$
10.	$e^z = \sum\limits_{n=0}^{\infty} \dfrac{z^n}{n!}$	$1, 1, \dfrac{1}{2!}, \dfrac{1}{3!}, \dfrac{1}{4!}, \ldots$
11.	$\ln(1+z) = \sum\limits_{n=1}^{\infty} \dfrac{(-1)^{n-1}z^n}{n}$	$0, 1, -\dfrac{1}{2}, \dfrac{1}{3}, -\dfrac{1}{4}, \ldots$

Solution.

Complement the sought numerical sequence $a_1, a_2, a_3, \ldots$ with the term a_0 so that the recurrence relation should also be valid for $n = 0$:

$$a_1 = 4a_0 + 6 \cdot 10^0.$$

This equality is satisfied by the value $a_0 = 2$; hence, the condition of the problem can be transformed to the following form:

$$\begin{cases} a_n = 4a_{n-1} + 6 \cdot 10^{n-1}, & n > 0, \\ a_0 = 2. \end{cases}$$

Calculate the values $a_1, a_2, a_3, \ldots$ For this, multiply the equality $a_n = 4a_{n-1} + 6 \cdot 10^{n-1}$ by z^n and sum it over the variable n:

$$\sum_{n=1}^{\infty} a_n z^n = \sum_{n=1}^{\infty} (4a_{n-1} + 6 \cdot 10^{n-1}) z^n.$$

In the obtained formal equality, perform operations of summation over n, having written them in terms of the generating function $G(z)$ for the sequence $\{a_n\}$, where $n \in \{0\} \cup \mathbb{N}$. For this, perform algebraic transformations:

$$\sum_{n=1}^{\infty} a_n z^n = 4 \sum_{n=1}^{\infty} a_{n-1} z^n + 6 \sum_{n=1}^{\infty} 10^{n-1} z^n.$$

The sums in the right side of the equality will be rewritten with the help of replacement $n \to n + 1$. Note the change in the summation limits associated with this operation.

$$\sum_{n=1}^{\infty} a_n z^n = 4 \sum_{n=0}^{\infty} a_n z^{n+1} + 6 \sum_{n=0}^{\infty} 10^n z^{n+1}, \quad \text{or}$$

$$\sum_{n=1}^{\infty} a_n z^n = 4z \sum_{n=0}^{\infty} a_n z^n + 6z \sum_{n=0}^{\infty} 10^n z^n.$$

By definition $G(z) = \sum_{n=0}^{\infty} a_n z^n = a_0 + \sum_{n=1}^{\infty} a_n z^n$ is a generating function for the sequence $\{a_n\}$, where $n = 0, 1, 2, \ldots$ Hence, we can write the functional equation, where the role of the unknown function is played by $G(z)$:

$$G(z) - a_0 = 4z G(z) + 6z \sum_{n=0}^{\infty} 10^n z^n.$$

Then take into account that $a_0 = 2$, and apply the second formula from Table 8.1:

$$G(z) - 2 = 4z G(z) + \frac{6z}{1 - 10z}, \quad \text{or}$$

$$(1 - 4z)G(z) = \frac{2 - 14z}{1 - 10z}.$$

Divide both parts of the equality by the value $(1 - 4z) \neq 0$:

$$G(z) = \frac{2 - 14z}{(1 - 4z)(1 - 10z)}.$$

Use expansion of the obtained rational function into elementary fractions:

$$G(z) = \frac{1}{1 - 4z} + \frac{1}{1 - 10z}.$$

Taking into account the second formula from Table 8.1, we obtain:

$$G(z) = \sum_{n=0}^{\infty} 4^n z^n + \sum_{n=0}^{\infty} 10^n z^n = \sum_{n=0}^{\infty} (4^n + 10^n) z^n.$$

In the expansion $G(z)$ in powers z, the coefficients for z^n correspond to the terms of the sought recursive sequence. Hence, the solution has the form $a_n = 4^n + 10^n$ for all $n \geqslant 1$. $\square$

Recurrence relation of the first order with variable coefficients

$$\begin{cases} a_n = f(n)a_{n-1} + d(n), & n > 1, \\ a_1 = \text{const} \end{cases}$$

is brought to the linear recurrence relation with constant coefficients by multiplying both sides of the equality $a_n = f(n)a_{n-1} + d(n)$ by **summation factor** [23,27]

$$F(n) = \left[\prod_{i=2}^{n} f(i) \right]^{-1}$$

and subsequent substitution of $b_n = f(n + 1)F(n + 1)a_n$, where $b_1 = a_1$. The resulting equation $b_n = b_{n-1} + F(n)d(n)$ can be solved by the method of substitution. Finally, the solution of the initial recurrence relation with variable coefficients has the form

$$a_n = [f(n + 1)F(n + 1)]^{-1} \left[a_1 + \sum_{i=2}^{n} F(i)d(i) \right], \quad n > 1.$$

Note that the numerical sequences, satisfying the linear recurrence relations, are also called **recursive sequences** [29].

Example 8.3 **Rooted strongly binary tree** will be defined as a root tree, whose root has a degree of 0 or 2, while the degree of other vertices takes values 1 or 3 [18]. For an arbitrary vertex, the vertices adjacent to it with a depth one greater than the given one will be considered to be different.

Denote by t_n the number of rooted strongly binary trees, whose height does not exceed n, where $n = 0, 1, 2, \ldots$

Prove that the values t_n satisfy the recurrence relation

$$\begin{cases} t_n = t_{n-1}^2 + 1, & n \geqslant 1, \\ t_0 = 1 \end{cases}$$

and obtain an analytical representation for t_n.

Solution.

In the proof and further analysis, we will follow the work [1]. Let us show in the figure all rooted strongly binary trees of height 0, 1, and 2:

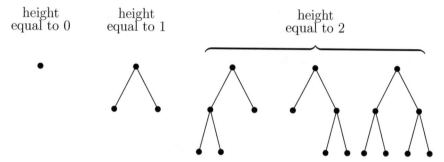

| height equal to 0 | height equal to 1 | height equal to 2 |

On the diagrams of nontrivial graphs, the root vertices are located in the upper part of the figure.

Thus, $t_0 = 1, t_1 = 2, t_2 = 5$.

Consider the set A_n of rooted strongly binary trees, whose height does not exceed n, where $n = 0, 1, 2, \ldots$ It is required to prove that the number of such trees $t_n = |A_n|$ satisfies the recurrence relation

$$\begin{cases} t_n = t_{n-1}^2 + 1, & n \geqslant 1, \\ t_0 = 1. \end{cases}$$

Fix an arbitrary tree of A_n, consisting of more than one vertex.

If we remove the root of this tree together with the incident edges, then two trees of a height not exceeding $n - 1$ will be left on the diagram. By the combinatory product rule (see chapter "Combinatorics" on p. 141), the number of pairs of respective trees is equal to $t_{n-1} \cdot t_{n-1}$. To this value, in order to obtain $|A_n|$, we should add a unity, corresponding to the trivial tree consisting of one vertex. Hence, for $n \in \mathbb{N}$ the equality is valid $t_n = t_{n-1}^2 + 1$.

Note that the resulting recurrence relation is nonlinear. Nevertheless, it appears that it is possible to find the analytical representation for t_n.

For this, take the logarithm of the equality $t_n = t_{n-1}^2 + 1$:

$$\log_2 t_n = 2\log_2 t_{n-1} + \log_2\left(1 + \frac{1}{t_{n-1}^2}\right).$$

Denote $a_{n-1} = \log_2\left(1 + \frac{1}{t_{n-1}^2}\right)$ and make the change $u_n = \log_2 t_n$ [1].

After the mentioned transformations, we obtain:

$$\begin{cases} u_n = 2u_{n-1} + a_{n-1}, & n \geqslant 1, \\ u_0 = 0. \end{cases}$$

By the method of substitution, we find u_n for $n \geqslant 1$:

$$u_n = 2^n\left(u_0 + \frac{a_0}{2} + \frac{a_1}{4} + \frac{a_2}{8} + \ldots + \frac{a_{n-1}}{2^n}\right).$$

Write the resulting expression as the difference of the series:

$$u_n = \sum_{i=0}^{\infty} 2^{n-1-i} a_i - \sum_{i=n}^{\infty} 2^{n-1-i} a_i.$$

It is easy to see that $\forall n \in \mathbb{N}$ $(a_{n-1} > a_n)$, and this, in turn, results in that the value $\varepsilon_n = \sum_{i=n}^{\infty} 2^{n-1-i} a_i$ satisfies the inequalities

$$0 < \varepsilon_n < a_n \sum_{i=n}^{\infty} 2^{n-1-i} = a_n.$$

Hence, $u_n = 2^n \sum_{i=0}^{\infty} 2^{-i-1} a_i - \varepsilon_n$. After exponentiating this equality, we obtain:

$$t_n = 2^{u_n} = \alpha^{2^n} 2^{-\varepsilon_n} = T_n 2^{-\varepsilon_n},$$

where $\alpha = 2^{\sum\limits_{i=0}^{\infty} 2^{-i-1} a_i}$, $T_n = \alpha^{2^n}$.

As was noted above, for all natural n the following inequalities are valid $0 < \varepsilon_n < a_n$. Thereby, let us provide the estimates

$$T_n = t_n 2^{\varepsilon_n} \leqslant t_n 2^{a_n} = t_n\left(1 + \frac{1}{t_n^2}\right) = t_n + \frac{1}{t_n}, \quad \text{and}$$

$$T_n = t_n 2^{\varepsilon_n} \geqslant t_n,$$

hence, $T_n \in \left[t_n, t_n + \dfrac{1}{t_n} \right]$.

For $n \geqslant 2$ the obtained inclusion allows concluding that the value t_n is equal to the integral part of the number T_n.

As a result, the answer will have the form $t_n = \lfloor T_n \rfloor = \lfloor \alpha^{2^n} \rfloor$.

Numerical calculation gives for the constant α the value $\alpha = 1{,}502\,836\,801\ldots$ For $n = 0$ and $n = 1$ the equality $t_n = \lfloor \alpha^{2^n} \rfloor$ is checked by direct calculation. So, the number of rooted strongly binary trees, whose height does not exceed n, is $t_n = \lfloor \alpha^{2^n} \rfloor$, $n = 0, 1, 2, \ldots$ □

8.1 Problems

8.1. The terms of the sequence $\{a_n\}$ are recursively defined as follows:

$$\begin{cases} a_n = 2a_{n-1} + 3a_{n-2}, & n > 2, \\ a_1 = 2, \ a_2 = 6. \end{cases}$$

(1) Write out the first eight terms of the sequence $\{a_n\}$.
(2) Prove that the explicit form of the nth term of the sequence is defined by the formula $a_n = 2 \cdot 3^{n-1}$, $n \geqslant 1$.

8.2. The terms of the sequence $\{a_n\}$ are recursively defined as follows:

$$\begin{cases} a_n = -2a_{n-1} - a_{n-2}, & n > 2, \\ a_1 = 1, \ a_2 = 2. \end{cases}$$

(1) Write out the first eight terms of the sequence $\{a_n\}$.
(2) Prove that the explicit form of the nth term of the sequence is defined by the formula $a_n = (-1)^n (3n - 4)$, $n \geqslant 1$.

8.3. The terms of the sequence $\{a_n\}$ are recursively defined as follows:

$$\begin{cases} a_n = 3a_{n-1} - 1, & n > 1, \\ a_1 = 2. \end{cases}$$

(1) Write out the first eight terms of the sequence $\{a_n\}$.
(2) Prove that the explicit form of the nth term of the sequence is defined by the formula $a_n = \dfrac{1}{2}(3^n + 1)$, $n \geqslant 1$.

8.4. The terms of the sequence $\{a_n\}$ are recursively defined as follows:

$$\begin{cases} a_n = a_{n-2} + n^2, & n > 2, \\ a_1 = 1, \; a_2 = 4. \end{cases}$$

(1) Write out the first eight terms of the sequence $\{a_n\}$.
(2) Prove that the explicit form of the nth term of the sequence is defined by the formula $a_n = \dfrac{1}{6}n(n+1)(n+2)$, $n \geqslant 1$.

8.5. Write out the recurrence relation, satisfied by the harmonic numbers $H_n = \sum_{i=1}^{n} \dfrac{1}{i}$ (see Exercise **1.82**).

8.6. The theory of **Mandelbrot**[1] **fractal** (from Latin *frāctus*—fractured) is based on the recurrence relation

$$\begin{cases} z_n = z_{n-1}^2 + c, & n \geqslant 1, \\ z_0 = 0, \end{cases}$$

where $c \in \mathbb{C}$ is a complex number [46].
Write out the first seven terms of the sequence $\{z_n\}$ of the parameter c takes the following values:

(1) $c = 1$;
(2) $c = -1$;
(3) $c = 1 + i$;
(4) $c = -i$.

8.7. By the method of substitution, solve the recurrence relation

$$\begin{cases} a_n = a_{n-1} + f_n, & n > 1, \\ a_1 = 1. \end{cases}$$

8.8. By the method of substitution, solve the recurrence relation

$$\begin{cases} a_n = f_n a_{n-1}, & n > 1, \\ a_1 = 1. \end{cases}$$

---

[1] Benoît Mandelbrot (1924–2010)—French and American mathematician of Polish origin.

8.9. Solve the nonhomogeneous recurrence relation

$$\begin{cases} a_n = a_{n-1} + n \cdot n!, & n > 1, \\ a_1 = 1. \end{cases}$$

8.10. By the method of substitution, solve the recurrence relation

$$\begin{cases} a_n = \dfrac{a_{n-1}}{a_{n-1} + c}, & n > 1, \ c = \text{const}, \\ a_1 = 1. \end{cases}$$

8.11. By the method of substitution, solve the linear recurrence relation

$$\begin{cases} a_n = 3a_{n-1} + 3, & n > 1, \\ a_1 = 2. \end{cases}$$

8.12. Solve the recurrence relation of the first order

$$\begin{cases} a_n = 2a_{n-1}, & n > 1, \\ a_1 = 3. \end{cases}$$

8.13. Solve the recurrence relation of the first order

$$\begin{cases} a_n = 7a_{n-1}, & n > 1, \\ a_1 = 1/2. \end{cases}$$

8.14. Solve the recurrence relation of the second order

$$\begin{cases} a_n = 2a_{n-1} + 3a_{n-2}, & n > 2, \\ a_1 = 4, \ a_2 = 8. \end{cases}$$

8.15. Solve the recurrence relation of the second order

$$\begin{cases} a_n = 7a_{n-1} - 12a_{n-2}, & n > 2, \\ a_1 = 1, \ a_2 = 7. \end{cases}$$

8.16. Solve the recurrence relation of the second order

$$\begin{cases} a_n = -a_{n-1} + 12a_{n-2}, & n > 2, \\ a_1 = 17, \ a_2 = -5. \end{cases}$$

8.17. Solve the recurrence relation of the second order

$$\begin{cases} a_n = 2a_{n-1} - a_{n-2}, & n > 2, \\ a_1 = 5, \ a_2 = 10. \end{cases}$$

8.18. Solve the recurrence relation of the second order

$$\begin{cases} a_n = -6a_{n-1} - 9a_{n-2}, & n > 2, \\ a_1 = -6, \ a_2 = 27. \end{cases}$$

8.19. Solve the recurrence relation of the second order

$$\begin{cases} a_n = \sqrt{2}a_{n-1} - a_{n-2}, & n > 2, \\ a_1 = \sqrt{2}, \ a_2 = 1. \end{cases}$$

8.20. Solve the recurrence relation of the second order

$$\begin{cases} a_n = a_{n-1} - a_{n-2}, & n > 2, \\ a_1 = 1, \ a_2 = 2. \end{cases}$$

8.21. Solve the recurrence relation of the second order

$$\begin{cases} a_n = -2\sqrt{2}a_{n-1} - 4a_{n-2}, & n > 2, \\ a_1 = 0, \ a_2 = 2. \end{cases}$$

8.22. Find the explicit expression for the Fibonacci numbers, defined by the recurrence relation

$$\begin{cases} F_n = F_{n-1} + F_{n-2}, & n > 2, \\ F_1 = 1, \ F_2 = 1. \end{cases}$$

8.23. Find the explicit expression of the Lucas numbers, defined by the recurrence relation

$$\begin{cases} L_n = L_{n-1} + L_{n-2}, & n > 1, \\ L_0 = 2, \ L_1 = 1. \end{cases}$$

8.24. Prove the following relation for of the Fibonacci numbers:

$$F_{2n} = F_{n+1}^2 - F_{n-1}^2 \text{ for } n = 2, 3, \ldots$$

8.25. Prove the following relation for the Fibonacci numbers and the Lucas numbers: $F_{2n} = F_n L_n$ for $n = 2, 3, \ldots$

8.26. Prove that for $n > 1$, the equality is valid $F_{n+3} = 2F_n + L_n$.

8.27. Prove that for $n \geqslant 4$, the equality is valid $F_{n-3} = -2F_n + L_n$.

8.28. Prove that $F_{n+3} - F_{n-3} = 4F_n$ for all $n \geqslant 4$.

8.29. Prove the equalities for $n > 4$:

(1) $F_{n-4} = \dfrac{1}{2}(7F_n - 3L_n);$

(2) $F_{n+4} = \dfrac{1}{2}(7F_n + 3L_n).$

8.30. Prove that $F_{n+4} + F_{n-4} = 7F_n$ for $n = 5, 6, \ldots$

8.31. Solve the recurrence relation

$$\begin{cases} a_n = a_{n-1} + a_{n-2} + 1, & n > 2, \\ a_1 = 0, \; a_2 = 0. \end{cases}$$

8.32. Solve the homogeneous recurrence relation

$$\begin{cases} a_n = a_{n-1} + 5a_{n-2} + 3a_{n-3}, & n > 3, \\ a_1 = 1, \; a_2 = 1, \; a_3 = -3. \end{cases}$$

8.33. Solve the homogeneous recurrence relation

$$\begin{cases} a_n = 3a_{n-1} + 9a_{n-2} - 27a_{n-3}, & n > 3, \\ a_1 = 7\dfrac{1}{2}, \; a_2 = 40\dfrac{1}{2}, \; a_3 = 121\dfrac{1}{2}. \end{cases}$$

8.34. Solve the homogeneous recurrence relation

$$\begin{cases} a_n = -a_{n-1} + 2a_{n-2} + 2a_{n-3}, & n > 3, \\ a_1 = 3, \; a_2 = 3, \; a_3 = 3. \end{cases}$$

8.35. Solve the homogeneous recurrence relation

$$\begin{cases} a_n = -2a_{n-1} + 3a_{n-2} + 6a_{n-3}, & n > 3, \\ a_1 = 1, \; a_2 = 2, \; a_3 = 4. \end{cases}$$

8.36. Solve the nonhomogeneous recurrence relation of the first order

$$\begin{cases} a_n = 2a_{n-1} + 3^n + 4, & n > 1, \\ a_1 = 1. \end{cases}$$

8.37. Solve the nonhomogeneous recurrence relation of the first order

$$\begin{cases} a_n = -2a_{n-1} + 10^n - 5, & n > 1, \\ a_1 = \dfrac{10}{3}. \end{cases}$$

8.38. Solve the nonhomogeneous recurrence relation of the first order

$$\begin{cases} a_n = -3a_{n-1} + (-2)^{n+1} + 1, & n > 1, \\ a_1 = 3. \end{cases}$$

8.39. Solve the nonhomogeneous recurrence relation of the first order

$$\begin{cases} a_n = 3a_{n-1} + (-1)^n - 2, & n > 1, \\ a_1 = 1. \end{cases}$$

8.40. Solve the nonhomogeneous recurrence relation of the first order

$$\begin{cases} a_n = 2a_{n-1} + 4^n, & n > 1, \\ a_1 = 2. \end{cases}$$

8.41. Solve the nonhomogeneous recurrence relation of the first order

$$\begin{cases} a_n = 16a_{n-1} + 4^n, & n > 1, \\ a_1 = 4. \end{cases}$$

8.42. Solve the nonhomogeneous recurrence relation of the first order

$$\begin{cases} a_n = 4a_{n-1} + 2^n + (-2)^n + 1, & n > 1, \\ a_1 = 1. \end{cases}$$

8.43. Solve the nonhomogeneous recurrence relation of the first order

$$\begin{cases} a_n = 4a_{n-1} + (-1)^n + 3^n + 1, & n > 1, \\ a_1 = 5. \end{cases}$$

8.44. Solve the nonhomogeneous recurrence relation of the first order

$$a_n = -2a_{n-1} + (-7)^n, \quad n > 1,$$

for different values of the initial conditions:

(1) $a_1 = 0$;
(2) $a_1 = 1$;
(3) $a_1 = 3$.

8.45. Solve the nonhomogeneous recurrence relation of the second order

$$\begin{cases} a_n = 3a_{n-1} + 4a_{n-2} + 1, & n > 2, \\ a_1 = 1, \ a_2 = 1. \end{cases}$$

8.46. Solve the nonhomogeneous recurrence relation of the second order

$$\begin{cases} a_n = -2a_{n-1} - a_{n-2} + 11, & n > 2, \\ a_1 = 1, \ a_2 = 5. \end{cases}$$

8.47. Solve the nonhomogeneous recurrence relation of the second order

$$\begin{cases} a_n = -6a_{n-1} - 9a_{n-2} + 12, & n > 2, \\ a_1 = 1, \ a_2 = 2. \end{cases}$$

8.48. Find the solution of the nonhomogeneous recurrence relation

$$\begin{cases} a_n = 5a_{n-1} - 8a_{n-2} + 4a_{n-3} + 1 + 2^n, & n > 3, \\ a_1 = 0, \ a_2 = 19, \ a_3 = 88. \end{cases}$$

8.49. Find the solution of the nonhomogeneous recurrence relation

$$a_n = 5a_{n-1} - 8a_{n-2} + 4a_{n-3} + 1 + 2^n, \quad n > 3$$

for different values of the initial conditions:

(1) $a_1 = -2, \ a_2 = 2, \ a_3 = 3$;
(2) $a_1 = 0, \ a_2 = 95, \ a_3 = 448$;
(3) $a_1 = 4, \ a_2 = 32, \ a_3 = 110$.

8.50. Find the solution of the nonhomogeneous recurrence relation

$$\begin{cases} a_n = a_{n-3} + 2, & n > 3, \\ a_1 = 1, \ a_2 = 2, \ a_3 = 3. \end{cases}$$

8.51. Find the solution of the nonhomogeneous recurrence relation

$$\begin{cases} a_n = a_{n-3} + 9, & n > 3, \\ a_1 = 2, \ a_2 = 4, \ a_3 = 9. \end{cases}$$

8.52. By the method of substitution, solve the recurrence relation

$$\begin{cases} a_n = 2a_{n/2} + 2n - 1 & \text{for } n = 2^k, \ k \in \mathbb{N}, \\ a_1 = 1 \end{cases}$$

for all n, which are the degrees of 2.

8.53. By the method of substitution, solve the recurrence relation

$$\begin{cases} a_n = 5a_{n/2} + n^3 & \text{for } n = 2^k, \ k \in \mathbb{N}, \\ a_1 = 1 \end{cases}$$

for all n, which are the degrees of 2.

8.54. By the method of substitution, solve the recurrence relation

$$\begin{cases} a_n = 4a_{n/3} + 2n + 1 & \text{for } n = 3^k, \ k \in \mathbb{N}, \\ a_1 = 1 \end{cases}$$

for all n, which are the degrees of 3.

8.55. By the method of substitution, solve the recurrence relation

$$\begin{cases} a_n = 3a_{n/3} + n^2 - n & \text{for } n = 3^k, \ k \in \mathbb{N}, \\ a_1 = 1 \end{cases}$$

for all n, which are the degrees of 3.

∗8.56. Prove that the solution of the recurrence relation

$$\begin{cases} a_n = ca_{n/d} + f(n) & \text{for } n = d^k, \ k \in \mathbb{N}, \\ a_1 = f(1), \quad c, d \ \text{—const} \end{cases}$$

is the sequence $a_n = \sum\limits_{i=0}^{k} c^i f\left(\dfrac{n}{d^i}\right)$.

8.57. Find the generating function for the sequence $\{a_n\}$, $n \geqslant 0$, defined by the relation:

(1) $a_n = 2$;
(2) $a_n = (-1)^n$;
(3) $a_n = \cos \dfrac{\pi n}{2}$;
(4) $a_n = 1 + i^n$, where i—imaginary unit.

8.58. Find the generating function for the numerical sequence:

(1) $0, 1, 4, 9, \ldots, n^2, \ldots$;
(2) $0, 1, 8, 27, \ldots, n^3, \ldots$

8.59. It is known that $G(z)$ is the generating function for the sequence $\{a_n\}$, where $n = 0, 1, 2, \ldots$. Express in terms of $G(z)$ the generating function $\tilde{G}(z)$ for the sequence:

(1) $0, a_0, a_1, a_2, a_3, \ldots$;

(2) $a_0, a_0 + a_1, a_1 + a_2, a_2 + a_3, \ldots$;
(3) $a_0, 0, a_1, 0, a_2, 0, \ldots$;
(4) $a_3, a_4, a_5, a_6, \ldots$

8.60. It is known that $G(z)$ is the generating function for the sequence $\{a_n\}$, where $n = 0, 1, 2, \ldots$. Express in terms of $G(z)$ the generating function $\tilde{G}(z)$ for the sequence:

(1) $a_0, 2a_1, 3a_2, 4a_3, \ldots$;
(2) $0, a_0, a_1/2, a_2/3, a_3/4, \ldots$;
(3) $a_0, a_2, a_4, a_6, \ldots$;
(4) $a_0, a_0 + a_1, a_0 + a_1 + a_2, a_0 + a_1 + a_2 + a_3, \ldots$

8.61. Show that the operation of multiplication of generating functions has the associativity property [42].

8.62. Calculate the convolution of the sequences $1, 2, 3, 4, \ldots$ and $1, \dfrac{1}{2}, \dfrac{1}{3}, \dfrac{1}{4}, \ldots$

8.63. Calculate the convolution $\{c_n\}$ of sequences $\{a_n\}$ and $\{b_n\}$, defined by the general terms $a_n = \dfrac{\alpha^n}{n!}$ and $b_n = \dfrac{\beta^n}{n!}$, where $n \in \{0\} \cup \mathbb{N}$, α and β are complex constants.

8.64. Calculate the convolution $\{c_n\}$ of sequences $\{a_n\}$ and $\{b_n\}$, defined by the general terms $a_n = \cos n\varphi$ and $b_n = \sin n\varphi$, where $n = 0, 1, 2, \ldots$, φ is a real constant.

8.65. By method of generating functions, solve the recurrence relation of the first order

$$\begin{cases} a_n = -a_{n-1} + 2, & n > 1, \\ a_1 = 5. \end{cases}$$

8.66. By method of generating functions, solve the recurrence relation of the first order

$$\begin{cases} a_n = -5a_{n-1} + (-5)^n, & n > 1, \\ a_1 = 7. \end{cases}$$

8.67. By method of generating functions, solve the recurrence relation of the first order

$$\begin{cases} a_n = \dfrac{1}{3}a_{n-1} + (n+1)3^{-n}, & n > 1, \\ a_1 = -1. \end{cases}$$

8.68. By method of generating functions, solve the recurrence relation of the first order

$$\begin{cases} a_n = 2a_{n-1} + 2^n + 3^n, & n > 1, \\ a_1 = 7. \end{cases}$$

8.69. By method of generating functions, solve the recurrence relation of the second order

$$\begin{cases} a_n = 6a_{n-1} - a_{n-2}, & n > 2, \\ a_1 = 2, \ a_2 = 22. \end{cases}$$

8.70. By method of generating functions, solve the recurrence relation of the second order

$$\begin{cases} a_n = 2a_{n-1} - a_{n-2} + 6n - 6, & n > 2, \\ a_1 = 1, \ a_2 = 8. \end{cases}$$

8.71. By method of generating functions, solve the recurrence relation of the second order

$$\begin{cases} a_n = 5a_{n-1} - 4a_{n-2} - 6n + 23, & n > 2, \\ a_1 = 1, \ a_2 = 0. \end{cases}$$

8.72. By method of generating functions, solve the recurrence relation of the second order

$$\begin{cases} a_n = -9a_{n-2} + 10, & n > 2, \\ a_1 = 1, \ a_2 = 10. \end{cases}$$

8.73. By method of generating functions, solve the recurrence relation of the third order

$$\begin{cases} a_n = 8a_{n-3} - 7n + 24, & n > 3, \\ a_1 = 3, \ a_2 = 6, \ a_3 = 11. \end{cases}$$

8.74. By method of generating functions, solve the recurrence relation of the third order

$$\begin{cases} a_n = 2a_{n-1} - a_{n-2} + 2a_{n-3} + 2, & n > 3, \\ a_1 = 1, \ a_2 = 2, \ a_3 = 3. \end{cases}$$

8.75. By method of generating functions, solve the recurrence relation of the third order

$$\begin{cases} a_n = 3a_{n-1} - a_{n-2} + 3a_{n-3} + 10 \cdot 3^{n-2}, & n > 3, \\ a_1 = 3, \ a_2 = 18, \ a_3 = 81. \end{cases}$$

8.76. By method of generating functions, solve the recurrence relation of the third order

$$\begin{cases} a_n = \dfrac{3}{2}a_{n-1} - \dfrac{1}{2}a_{n-3} + 3, & n > 3, \\ a_1 = \dfrac{1}{2}, \ a_2 = \dfrac{17}{4}, \ a_3 = \dfrac{71}{8}. \end{cases}$$

8.77. Show that the generating functions for the sequence of the Fibonacci numbers F_n and the sequence of the Lucas numbers L_n are determined by the expressions $G_F(z) = \dfrac{z}{1 - z - z^2}$ and $G_L(z) = \dfrac{2 - z}{1 - z - z^2}$.

8.78. Find the generating function $G_H(z)$ for the sequence of the harmonic numbers $H_n = \sum_{i=1}^{n} \dfrac{1}{i}$ taking into account the initial condition $H_0 = 0$.

8.79. Write a homogeneous recurrence relation, satisfied by the sequence of $a_n = 3^n + 4^n$, $n \geqslant 1$.

8.80. Write a homogeneous recurrence relation, satisfied by the sequence of $a_n = 2^n - (-5)^{n-1}$, $n \geqslant 1$.

8.81. Write a homogeneous recurrence relation, satisfied by the sequence of $a_n = 6^n - 1$, $n \geqslant 1$.

8.82. Write a homogeneous recurrence relation, satisfied by the sequence of $a_n = \dfrac{2^n}{3}(1 + (-1)^n)$, $n \geqslant 1$.

8.83. Suggest a nonhomogeneous recurrence relation of the first order, satisfied by the sequence of $a_n = 2^n(n^2 + n + 1)$, $n \geqslant 1$.

8.84. Suggest a nonhomogeneous recurrence relation of the first order, satisfied by the sequence of $a_n = 3^n(n + 1)^2$, $n \geqslant 1$.

8.85. Suggest a nonhomogeneous recurrence relation of the second order, satisfied by the sequence of $a_n = (n - 1)^2 2^n - 1 - (-1)^n$, $n \geqslant 1$.

8.86. Suggest a nonhomogeneous recurrence relation of the second order, satisfied by the sequence of $a_n = n4^n$, $n \geqslant 1$.

8.87. Prove that the solution of the recurrence relation

$$\begin{cases} a_n = \dfrac{2}{a_{n-1}} + 1, & n > 1, \\ a_1 = 3 \end{cases}$$

is the sequence $a_n = \dfrac{2^{n+2} - (-1)^n}{2^{n+1} + (-1)^n}$.

8.88. Prove that the solution of the recurrence relation

$$\begin{cases} a_n = \dfrac{a_{n-1} + 5}{4a_{n-1}}, & n > 1, \\ a_1 = 5 \end{cases}$$

is the sequence $a_n = \dfrac{5^{1-n}(-5(-4)^n + 8 \cdot 5^n)}{32 + (-4)^n 5^{2-n}}$.

8.89. Prove that the solution of the recurrence relation

$$\begin{cases} a_n = \dfrac{4a_{n-1} + 9}{a_{n-1} + 4}, & n > 1, \\ a_1 = 1 \end{cases}$$

is the sequence $a_n = 3 \cdot \dfrac{2 - 7^{1-n}}{2 + 7^{1-n}}$.

8.90. Solve the nonlinear recurrence relation

$$\begin{cases} a_n = \dfrac{a_{n-1} + 1}{a_{n-1}}, & n > 1, \\ a_1 = 2. \end{cases}$$

8.91. Find the solution of the nonlinear recurrence relation

$$\begin{cases} a_n^2 - a_{n-1}^2 = 2, & n > 1, \\ a_1 = 0. \end{cases}$$

8.92. Find the solution of the nonlinear recurrence relation

$$\begin{cases} a_n^3 - a_{n-1}^3 = 6, & n > 1, \\ a_1 = 1. \end{cases}$$

8.93. Find the solution of the nonlinear recurrence relation

$$\begin{cases} (2a_n + 1)^2 = (2a_{n-1} + 1)^2 + 5, & n > 1, \\ a_1 = 7. \end{cases}$$

8.94. Find the solution of the nonlinear recurrence relation

$$\begin{cases} (3a_n - 5)^3 - (3a_{n-1} - 5)^3 = 1, & n > 1, \\ a_1 = 2. \end{cases}$$

8.95. Find the solution of the nonlinear recurrence relation

$$\begin{cases} a_n^2 = 2a_{n-1}, & n > 1, \\ a_1 = 1 \end{cases}$$

by taking the logarithm of the recurrence relation.

8.96. Find the solution of the nonlinear recurrence relation

$$\begin{cases} a_n^3 = -a_{n-1}, & n > 1, \\ a_1 = 1 \end{cases}$$

by taking the logarithm of the recurrence relation.

8.97. Find the solution of the nonlinear recurrence relation

$$\begin{cases} a_n^2 = 4a_{n-1}, & n > 1, \\ a_1 = 4 \end{cases}$$

by taking the logarithm of the recurrence relation.

8.98. Find the solution of the nonlinear recurrence relation

$$\begin{cases} a_n^2 = 3a_{n-1}, & n > 1, \\ a_1 = 2 \end{cases}$$

by taking the logarithm of the recurrence relation.

8.99. The sequence $\{a_n\}$ has the following property: Each term of the sequence, except the first, is equal to the arithmetic mean of the two neighbouring terms, i.e., $a_n = (a_{n-1} + a_{n+1})/2$ for $n > 1$. Write the explicit expression a_n, if:

(1) $a_1 = 1, a_2 = 2$;
(2) $a_1 = 2, a_2 = 10$.

8.100. The sequence $\{a_n\}$ has the following property: Each term of the sequence, except the first, is equal to the geometric mean of the two neighbouring terms, i.e., $a_n = \sqrt{a_{n-1}a_{n+1}}$ for $n > 1$. Write the explicit expression a_n, if:

(1) $a_1 = 1, a_2 = \sqrt{2}$;
(2) $a_1 = 3, a_2 = 1$.

8.101. The sequence $\{a_n\}$ has the following property: Each term of the sequence, except the first, is equal to the harmonic mean of the two neighbouring terms, i.e., $a_n = \dfrac{2a_{n-1}a_{n+1}}{a_{n-1} + a_{n+1}}$ for $n > 1$. Write the explicit expression a_n, if:

(1) $a_1 = 10, a_2 = 5$;
(2) $a_1 = \dfrac{1}{2}, a_2 = \dfrac{1}{3}$.

8.102. Show that the solution of the discrete analog of the **logistic equation**

$$\begin{cases} a_n = \lambda a_{n-1}(1 - a_{n-1}), & n > 1, \\ a_1 \in [0, 1], \quad \lambda \in [0, 4], \end{cases}$$

remains limited: $0 \leqslant a_n \leqslant 1, n = 1, 2, 3, \ldots$ Write its solution $\{a_n\}$ for special instances of the parameter λ:

(1) $\lambda = 2$;
(2) $\lambda = 4$.

Note. Logistic equation is the basis of one of the simplest mathematical models used for describing the dynamics of complex systems. Nonlinearity of the mapping $x \to \lambda x(1-x)$ results in origination of bifurcations of the solutions (from Latin *bifurcum*—point of splitting) and transition to a chaotic behavior of solutions for certain values of the parameter λ. Analytical form of the solutions of a logistic equation, except the cases $\lambda = 2$ and $\lambda = 4$, is unknown, and this equation is examined by the computational mathematics methods.

*8.103. According to the known from the mathematical analysis course **Weierstrass**[2] **theorem**, *a monotone and bounded numerical sequence has the limit* [86]. Using the Weierstrass theorem, prove that the sequence $\{a_n\}$ satisfies the nonlinear recurrence relation

$$\begin{cases} a_n = \dfrac{1}{2}\left(a_{n-1} + \dfrac{x}{a_{n-1}}\right), & n > 1, \\ a_1 = \text{const}, \ a_1 > 0, \end{cases}$$

for positive values of the parameter x has the limit. Find the limit's value $\lim\limits_{n \to \infty} a_n$.

*8.104. Find $\lim\limits_{n \to \infty} a_n$, if the sequence of $\{a_n\}$ satisfies the nonlinear recurrence relation

$$\begin{cases} a_n = \dfrac{1}{3}\left(2a_{n-1} + \dfrac{x}{a_{n-1}^2}\right), & n > 1, \\ a_1 = \text{const}, \ a_1 > 0, \end{cases}$$

and the value of the parameter x is positive.

8.105. Prove that the sequence of $\{a_n\}$, defined by the nonlinear recurrence relation

$$\begin{cases} a_n = \sqrt{3 + a_{n-1}}, & n > 1, \\ a_1 = \sqrt{3}, \end{cases}$$

has the limit, equal to $\dfrac{1}{2}(1 + \sqrt{13})$.

---

[2]Karl Theodor Wilhelm Weierstrass (1815–1897)—German mathematician.

∗8.106. Prove that the sequence of $\{a_n\}$, defined by the nonlinear recurrence relation

$$\begin{cases} a_n = \sqrt{a_{n-1} + 12}, & n > 1, \\ a_1 = 3, \end{cases}$$

has the limit, equal to 4.

8.107. Solve the full history recurrence relation

$$\begin{cases} a_n = \sum_{i=1}^{n-1} a_i + 2, & n > 1, \\ a_1 = 2. \end{cases}$$

8.108. Solve the full history recurrence relation

$$\begin{cases} a_n = \sum_{i=1}^{n-1} a_i + 10, & n > 1, \\ a_1 = 2. \end{cases}$$

8.109. At the exam in discrete mathematics and mathematical logic, the student is asked to solve the full history recurrence relation

$$\begin{cases} a_n = \sum_{i=1}^{n-1} a_i + 2, & n > 1, \\ a_1 = 1. \end{cases}$$

The student decided to make use of the fact that

$$a_n - a_{n-1} = \left(\sum_{i=1}^{n-1} a_i + 2 \right) - \left(\sum_{i=1}^{n-2} a_i + 2 \right) = a_{n-1},$$

and reduced the equation to the following:

$$\begin{cases} a_n = 2a_{n-1}, & n > 1, \\ a_1 = 1. \end{cases}$$

However, he obtained a wrong answer $a_n = 2^{n-1}$, $n \geqslant 1$.
Where is the student's mistake?

8.110. Catalan[3] numbers satisfy the following full history recurrence relation:

$$\begin{cases} C_n = \sum_{i=1}^{n-1} C_k\, C_{n-k-1}, & n > 1, \\ C_0 = C_1 = 1. \end{cases}$$

Write out the first eight Catalan numbers.

8.111. Show that the generating function for the sequence of the Catalan numbers C_n is $G_C(z) = \dfrac{1 - \sqrt{1 - 4z}}{2z}$.

∗8.112. Write the explicit expression of the nth Catalan number.

8.113. Solve the recurrence relation of the first order with variable coefficients

$$\begin{cases} a_n = n a_{n-1} + n!, & n > 1, \\ a_1 = 1. \end{cases}$$

8.114. Solve the recurrence relation of the first order with variable coefficients

$$\begin{cases} a_n = n a_{n-1} + n!, & n > 1, \\ a_1 = 2. \end{cases}$$

8.115. Solve the recurrence relation of the first order with variable coefficients

$$\begin{cases} a_n = \dfrac{n+1}{n} a_{n-1} + 2, & n > 1, \\ a_1 = 0, \end{cases}$$

arising when analysing the quicksort algorithm (see chapter "Basic Algorithms" on p. 387).

8.116. Solve the recurrence relation of the first order with variable coefficients

$$\begin{cases} a_n = \left(1 + \dfrac{1}{n}\right) a_{n-1} + n, & n > 1, \\ a_1 = 1. \end{cases}$$

8.117. Solve the recurrence relation of the first order with variable coefficients

$$\begin{cases} a_n = \dfrac{n+5}{n+4} a_{n-1} + 2n + 10, & n > 1, \\ a_1 = 12. \end{cases}$$

---

[3]Eugène Charles Catalan (1814–1894)—Belgian mathematician.

8.118. Solve the recurrence relation of the first order with variable coefficients

$$\begin{cases} a_n = a_{n-1}/n + n, & n > 1, \\ a_1 = 2. \end{cases}$$

8.119. Solve the recurrence relation of the first order with variable coefficients

$$\begin{cases} a_n = \dfrac{a_{n-1}}{n-1} + n^2, & n > 1, \\ a_1 = 2. \end{cases}$$

8.120. Solve the recurrence relation of the first order with variable coefficients

$$\begin{cases} a_n = e^{2n} a_{n-1} + e^{n^2+n-2}, & n > 1, \\ a_1 = 0. \end{cases}$$

8.121. Solve the recurrence relation of the first order with variable coefficients

$$\begin{cases} a_n = e^{-2n} a_{n-1} + e^{-n^2-n+2}, & n > 1, \\ a_1 = 1. \end{cases}$$

∗8.122. Solve the homogeneous system of the recurrence relations

$$\begin{cases} a_n = 4a_{n-1} - b_{n-1}, & n > 1, \\ b_n = 2a_{n-1} + b_{n-1}, & n > 1, \\ a_1 = 7, \; b_1 = 11. \end{cases}$$

∗8.123. Solve the homogeneous system of the recurrence relations

$$\begin{cases} a_n = a_{n-1} + 2b_{n-1}, & n > 1, \\ b_n = 2a_{n-1} - b_{n-1}, & n > 1, \\ a_1 = 5, \; b_1 = \dfrac{5}{2}(\sqrt{5} - 1). \end{cases}$$

∗8.124. By method of generating functions, solve the system of recurrence relations

$$\begin{cases} a_n = 2a_{n-1} + 3b_{n-1} + 4 \cdot 3^{n-1}, & n > 1, \\ b_n = 3a_{n-1} + 6b_{n-1} - 4 \cdot 5^{n-1}, & n > 1, \\ a_1 = 8, \; b_1 = 2. \end{cases}$$

∗8.125. By method of generating functions, solve the system of recurrence relations

$$\begin{cases} a_n = 3a_{n-1} + b_{n-1} - 3n + 3, & n > 1, \\ b_n = -a_{n-1} + b_{n-1} + n + 1, & n > 1, \\ a_1 = 4, \; b_1 = -2. \end{cases}$$

∗8.126. By method of generating functions, solve the system of recurrence relations

$$\begin{cases} a_n = a_{n-1} + 3b_{n-1} + 8n - 7, & n > 1, \\ b_n = 8a_{n-1} - 4b_{n-1} - 8n^2 + 6n, & n > 1, \\ a_1 = 5, \ b_1 = 2. \end{cases}$$

∗8.127. By method of generating functions, solve the system of recurrence relations

$$\begin{cases} a_n = -3a_{n-1} + b_{n-1} + 3 \cdot 2^{n-1}, & n > 1, \\ b_n = -4a_{n-1} + 2b_{n-1} + 2^{n+1}, & n > 1, \\ a_1 = 0, \ b_1 = 2. \end{cases}$$

∗8.128. Gamma function $\Gamma(x)$ is defined for all $x, x \neq 0, -1, -2, \ldots$ For positive values of the argument, there exists an integral representation [53,87]

$$\Gamma(x) = \int_0^\infty e^{-t} t^{x-1} dt.$$

Prove the recurrence relation for the values of the gamma function $\Gamma(x + 1) = x\Gamma(x)$, and calculate $\Gamma(n)$ for natural n.

∗8.129. Obtain the recurrence relation for the integral $\int\limits_0^{\pi/2} \sin^n x \, dx, n \geqslant 1$, and find the value of the integral, having expressed it in terms of **semifactorial** of number n—product of natural values of the same parity with n up to n inclusive.

∗8.130. Obtain the recurrence relation for the integral $I_n = \int\limits_0^{\pi/2} \cos^n x \, dx, n \geqslant 1$, and find the value of the integral.

∗8.131. Obtain the recurrence relation for the integral $I_n = \int\limits_0^{\pi/4} \tan^{2n} x \, dx, n \geqslant 1$, and find the value of the integral.

∗8.132. Obtain the recurrence relation for the integral $I_n = \int\limits_0^1 (1 - x^2)^n \, dx, n \geqslant 0$, and find the value of the integral.

∗8.133. Obtain the recurrence relation for the integral $I_n = \int\limits_0^1 x^c \ln^n x \, dx, n \geqslant 0$, $c \in (-1, \infty)$, and find the value of the integral.

∗8.134. Derive the recurrence relation for the volume of a n-dimensional ball with the unit radius of $x_1^2 + x_2^2 + \ldots + x_n^2 \leqslant 1$, where $n \in \mathbb{N}$.

∗8.135. Derive the recurrence relation for the volume of the n-dimensional simplex (from Latin *simplex*) $x_1 + x_2 + \ldots + x_n \leqslant 1, x_i \geqslant 0$ for all $i = 1, 2, \ldots, n$, where $n \in \mathbb{N}$.

∗8.136. Using by the method, described in the example on p. 270, solve the nonlinear recurrence relation

$$\begin{cases} a_n = a_{n-1}^2 - 1, & n \geqslant 1, \\ a_0 = 2. \end{cases}$$

∗8.137. The problem of minimizing the number of the processor registers used when computing the arithmetic expressions is of a great importance for theoretical computer science. This problem reduces to the necessity to obtain the solution of the following nonlinear recurrence relation [19]:

$$\begin{cases} a_n = a_{n-1}^2 - 2, & n \geqslant 1, \\ a_0 = u, \end{cases}$$

where u—const. Find a_n for $n = 1, 2, \ldots$

8.2 Answers, Hints, Solutions

8.1. *Solution.*

(1) Substituting the defined initial conditions into the recurrence relation $a_n = 2a_{n-1} + 3a_{n-2}$, we sequentially obtain the first eight terms $\{a_n\}$.

a_1	a_2	a_3	a_4	a_5	a_6	a_7	a_8
2	6	18	54	162	486	1458	4374

(2) Use the mathematical induction method in the full form. Let $P(n)$ be the predicate $a_n = 2 \cdot 3^{n-1}$. Prove that $P(n)$ is true for all n, taking values out of the set of natural numbers.

B a s i s   s t e p

For $n = 1$ and $n = 2$, we have $a_1 = 2 \cdot 3^0 = 2$ and $a_2 = 2 \cdot 3^1 = 6$—true equalities.

Inductive step

Let $P(i)$ is true for all $i \leqslant k$. Prove that this results in the truth of $P(k+1)$:

$$a_{k+1} = 2a_k + 3a_{k-1}.$$

According to the inductive supposition

$$a_k = 2 \cdot 3^{k-1}, \quad a_{k-1} = 2 \cdot 3^{k-2}.$$

We obtain

$$a_{k+1} = 2(2 \cdot 3^{k-1}) + 3(2 \cdot 3^{k-2}) =$$
$$= 4 \cdot 3^{k-1} + 2 \cdot 3^{k-1} = 6 \cdot 3^{k-1} = 2 \cdot 3^{(k-1)+1}.$$

Then, the predicate $P(n)$ takes a true value for all natural n, and $a_n = 2 \cdot 3^{n-1}, n \geqslant 1$.

8.2. *Answer:*

a_1	a_2	a_3	a_4	a_5	a_6	a_7	a_8
1	2	-5	8	-11	14	-17	20

8.3. *Answer:*

a_1	a_2	a_3	a_4	a_5	a_6	a_7	a_8
2	5	14	41	122	365	1094	3281

8.4. *Answer:*

a_1	a_2	a_3	a_4	a_5	a_6	a_7	a_8
1	4	10	20	35	56	84	120

8.5. *Answer:*
$$\begin{cases} H_n = H_{n-1} + \dfrac{1}{n}, & n > 1, \\ H_1 = 1. \end{cases}$$

8.6. *Answer:*

	z_0	z_1	z_2	z_3	z_4	z_5	z_6
$c=1$	0	1	2	5	26	677	458 330
$c=-1$	0	-1	0	-1	0	-1	0
$c=1+i$	0	$1+i$	$1+3i$	$-7+7i$	$1-97i$	$-9407-193i$	$88\,454\,401 + 3\,631\,103i$
$c=-i$	0	$-i$	$-1-i$	i	$-1-i$	i	$-1-i$

8.7. *Answer:* $a_n = 1 + \sum\limits_{i=2}^{n} f_i,\ n \geqslant 1.$

8.8. *Answer:* $a_n = \prod\limits_{i=2}^{n} f_i,\ n > 1.$

8.9. *Solution.*
Write out the first terms of the sequence $\{a_n\}$:

a_1	a_2	a_3	a_4	a_5	a_6
1	5	23	119	719	5039

Analysis of the table results in the supposition $a_n = (n+1)! - 1,\ n \geqslant 1$.

For the proof, use the mathematical induction method. Let $P(n)$ be the predicate $a_n = (n+1)! - 1$ for some $n \in \mathbb{N}$. Prove that $P(n)$ is true for all $n \in \mathbb{N}$.

B a s i s   s t e p

For $n = 1$, we obtain $a_1 = 2! - 1 = 1$ is true.

I n d u c t i v e   s t e p

Suppose that $a_k = (k+1)! - 1$. Prove the truth of $P(k+1)$:

$$a_{k+1} = a_k + (k+1)(k+1)!.$$

Then write a_k, using the inductive supposition:

$$a_{k+1} = (k+1)! - 1 + (k+1)(k+1)! =$$
$$= (k+1)![1 + (k+1)] - 1 = (k+2)! - 1.$$

It is proved that $a_n = (n+1)! - 1$ for all $n \geqslant 1$.

8.10. *Solution.*
Find the first terms of the sequence a_1, a_2, a_3, a_4:

$$a_1 = 1,$$
$$a_2 = \frac{1}{1+c},$$
$$a_3 = \frac{1/(1+c)}{1/(1+c)+c} = \frac{1}{1+c+c^2},$$
$$a_4 = \frac{1/(1+c+c^2)}{1/(1+c+c^2)+c} = \frac{1}{1+c+c^2+c^3}.$$

We come to the supposition

$$a_n = \left(\sum_{i=0}^{n-1} c^i \right)^{-1}, \quad n \geqslant 1.$$

For the proof, use the mathematical induction method. Let $P(n)$ be the predicate $a_n = \left(\sum_{i=0}^{n-1} c^i \right)^{-1}$ for some $n \in \mathbb{N}$. Prove that $P(n)$ is true for all $n \in \mathbb{N}$.

Basis step

If $n = 1$, then $a_1 = c^0 = 1$—true equality.

Inductive step

Let $a_k = \left(\sum_{i=0}^{k-1} c^i \right)^{-1}$. Prove that $a_{k+1} = \left(\sum_{i=0}^{k} c^i \right)^{-1}$. According to the definition for $\{a_n\}$

$$a_{k+1} = \frac{a_k}{a_k + c}, \quad k = 1, 2, \ldots$$

Hence,

$$a_{k+1} = \frac{\left(\sum_{i=0}^{k-1} c^i \right)^{-1}}{\left(\sum_{i=0}^{k-1} c^i \right)^{-1} + c} = \frac{1}{1 + c \left(\sum_{i=0}^{k-1} c^i \right)} = \left(\sum_{i=0}^{k} c^i \right)^{-1}.$$

Thus, $P(k) \Rightarrow P(k+1)$, and $a_n = \left(\sum_{i=0}^{n-1} c^i \right)^{-1}$ for all $n \geqslant 1$.

The obtained expression can be simplified.

If $c \neq 1$, then $a_n = \dfrac{c-1}{c^n - 1}, n \geqslant 1$.

But if $c = 1$, then $a_n = \dfrac{1}{n}$, s $n \geqslant 1$.

8.11. *Solution.*

If n is sufficiently great, then

$$a_n = 3a_{n-1} + 3,$$
$$a_{n-1} = 3a_{n-2} + 3,$$
$$a_{n-2} = 3a_{n-3} + 3, \text{ and so on.}$$

Substitute into the original relation a_{n-1}, expressed in terms of a_{n-2}, then a_{n-2} express in terms of a_{n-3}, and so on.

$$
\begin{aligned}
a_n &= 3a_{n-1} + 3 = \\
&= 3(3a_{n-2} + 3) + 3 = \\
&= 9a_{n-2} + 3 \cdot 3 + 3 = \\
&= 9(3a_{n-3} + 3) + 3 \cdot 3 + 3 = \\
&= 27a_{n-3} + 3^2 \cdot 3 + 3 \cdot 3 + 3 = \\
&= 27(3a_{n-4} + 3) + 3^2 \cdot 3 + 3 \cdot 3 + 3 = \\
&= 81a_{n-4} + 3^3 \cdot 3 + 3^2 \cdot 3 + 3 \cdot 3 + 3.
\end{aligned}
$$

After the kth substitution, we will have

$$
a_n = 3^j a_{n-j} + 3 \cdot \sum_{i=0}^{j-1} 3^i, \quad j = 1, 2, \dots, n-1.
$$

The obtained relation can strictly be proved by induction. Let $P(j)$ be the predicate $a_n = 3^j a_{n-j} + 3 \cdot \sum_{i=0}^{j-1} 3^i$ for some natural j. Prove that $P(j)$ is true for all $j = 1, 2, \dots, n-1$.

Basis step

If $j = 1$, then $a_n = 3a_{n-1} + 3$—true equality.

Inductive step

Suppose that $a_n = 3^k a_{n-k} + 3 \cdot \sum_{i=0}^{k-1} 3^i$, where the variable k can take values $k = 1, 2, \dots, n-1$. Prove that $a_n = 3^{k+1} a_{n-(k+1)} + 3 \cdot \sum_{i=0}^{k} 3^i$. For this, make a substitution $a_{n-k} = 3a_{n-k-1} + 3$. We obtain

$$
a_n = 3^{k+1} a_{n-(k+1)} + 3 \cdot \sum_{i=0}^{k} 3^i.
$$

In the proved equality $a_n = 3^k a_{n-k} + 3 \cdot \sum_{i=0}^{k-1} 3^i$ we assume that $k = n-1$. Then we finally obtain

$$
a_n = 3^{n-1} a_1 + 3 \cdot \sum_{i=0}^{n-2} 3^i = \frac{1}{2}(7 \cdot 3^{n-1} - 3), \quad n \geqslant 1.
$$

8.12. *Solution.*
We have a homogeneous recurrence relation with constant coefficients of the first order of the form $a_n = ca_{n-1}, c = 2$. Its solution is $a_n = A \cdot 2^n$, where constant A is determined from the initial condition $a_1 = 2A$. As a result, we obtain $a_n = 3 \cdot 2^{n-1}$, $n \geqslant 1$.

8.13. *Answer:* $a_n = \dfrac{7^{n-1}}{2}, n \geqslant 1$.

8.14. *Solution.*
The characteristic equation has the form $z^2 = 2z + 3$, its roots are $z_1 = 3, z_2 = -1$. We obtain the general solution $a_n = A \cdot 3^n + B \cdot (-1)^n$, where A, B—const. For determining the constant A and B, use the initial conditions:

$$\begin{cases} a_1 = A \cdot 3 + B \cdot (-1) = 4, \\ a_2 = A \cdot 3^2 + B \cdot (-1)^2 = 8. \end{cases}$$

We come to the system of equations with respect to the coefficients A and B:

$$\begin{cases} 3A - B = 4, \\ 9A + B = 8. \end{cases}$$

Solving the written system, we obtain $A = 1, B = -1$, and, finally, $a_n = 3^n + (-1)^{n-1}, n \geqslant 1$.

8.15. *Answer:* $a_n = 4^n - 3^n, n \geqslant 1$.

8.16. *Answer:* $a_n = 3^{n+1} - 2(-4)^n, n \geqslant 1$.

8.17. *Solution.*
The characteristic equation $z^2 - 2z + 1 = 0$ has the root $z = 1$ of multiplicity $m = 2$. Hence, the solution is representable in the form:

$$a_n = A \cdot 1^n + B \cdot n \cdot 1^n = A + Bn.$$

For determining the constant A and B, use the initial conditions. The system of equations for the coefficients A and B has the form

$$\begin{cases} A + B = 5, \\ A + 2B = 10. \end{cases}$$

We obtain $A = 0, B = 5$, and, finally, $a_n = 5n, n \geqslant 1$.

8.18. *Answer:* $a_n = (-3)^n(n + 1), n \geqslant 1$.

8.19. *Solution.*
The characteristic equation $z^2 - \sqrt{2}z + 1 = 0$ has the roots $z = \dfrac{1}{\sqrt{2}}(1 \pm i)$. Represent z_1, z_2 in a trigonometric form

$$z_1 = \cos\left(\frac{\pi}{4}\right) + i \sin\left(\frac{\pi}{4}\right), \quad z_2 = \cos\left(-\frac{\pi}{4}\right) + i \sin\left(-\frac{\pi}{4}\right).$$

Then the general solution of the recurrence relation takes the form:

$$a_n = A \cdot 1^n \cos\left(\frac{\pi n}{4}\right) + B \cdot 1^n \sin\left(\frac{\pi n}{4}\right) = A \cos\left(\frac{\pi n}{4}\right) + B \sin\left(\frac{\pi n}{4}\right).$$

Constant A and B are determined from the initial conditions:

$$\begin{cases} (A+B)\dfrac{\sqrt{2}}{2} = \sqrt{2}, \\ A + B = 2. \end{cases}$$

We obtain $a_n = \cos(\frac{\pi n}{4}) + \sin(\frac{\pi n}{4})$, $n \geqslant 1$.

8.20. *Answer:* $a_n = \sqrt{3}\sin(\frac{\pi n}{3}) - \cos(\frac{\pi n}{3})$, $n \geqslant 1$.

8.21. *Answer:* $a_n = -2^{n-1}\left(\cos(\frac{3\pi n}{4}) - \sin(\frac{3\pi n}{4})\right)$, $n \geqslant 1$.

8.22. *Solution.*

The characteristic equation $z^2 = z + 1$, its roots $z_{1,2} = \dfrac{1 \pm \sqrt{5}}{2}$. The general solution will be written in the form:

$$a_n = A \cdot \left(\frac{1+\sqrt{5}}{2}\right)^n + B \cdot \left(\frac{1-\sqrt{5}}{2}\right)^n.$$

Find A and B from the initial conditions:

$$\begin{cases} A \cdot \left(\dfrac{1+\sqrt{5}}{2}\right) + B \cdot \left(\dfrac{1-\sqrt{5}}{2}\right) = 1, \\ A \cdot \left(\dfrac{1+\sqrt{5}}{2}\right)^2 + B \cdot \left(\dfrac{1-\sqrt{5}}{2}\right)^2 = 1. \end{cases}$$

The solution for this system will be $A = \sqrt{5}$, $B = -\sqrt{5}$.
As a result, we obtain the explicit expression of the Fibonacci numbers

$$F_n = \frac{1}{\sqrt{5}}\left(\frac{1+\sqrt{5}}{2}\right)^n - \frac{1}{\sqrt{5}}\left(\frac{1-\sqrt{5}}{2}\right)^n, \quad n \geqslant 1,$$

(see also the solution of Exercise **1.99**).

8.23. *Solution.*
Use the approach suggested in the previous exercise. Taking into account that $\cos(\pi n) = (-1)^n$ for integer values n, we obtain

$$L_n = \left(\frac{1+\sqrt{5}}{2}\right)^n + \cos(\pi n)\left(\frac{1+\sqrt{5}}{2}\right)^{-n}, \quad n \geqslant 1.$$

8.24. *Proof.*
Use the known formula for the difference of two squares $a^2 - b^2 = (a - b)(a + b)$:

$$F_{2n} = (F_{n+1} - F_{n-1})(F_{n+1} + F_{n-1}).$$

Since $F_{n+1} = F_n + F_{n-1}$ and $F_n = F_{n+1} - F_{n-1}$, then we have

$$F_{2n} = F_n(F_{n+1} + F_{n-1}) = F_n F_{n+1} + F_n F_{n-1}.$$

We obtain an instance of the formula, proved in Exercise **1.97**, for the index value $m = n$.

8.25. *Proof.*
Transform the right side of the suggested equality, having represented the Lucas number L_n in the form $L_n = F_{n-1} + F_{n+1}$ (see Exercise **1.104**):

$$F_n L_n = F_n(F_{n-1} + F_{n+1}) = F_n F_{n-1} + F_n F_{n+1}.$$

We come to the instance of the formula, proved in Exercise **1.97** , for the index value $m = n$:

$$F_{2n} = F_n F_{n-1} + F_n F_{n+1}.$$

Hence, for all $n > 1$ the following equality is valid $F_{2n} = F_n L_n$.

8.26. *Proof.*
Write the right side of the equality, using the expression of the Lucas number L_n in terms of the Fibonacci sequence $L_n = F_{n-1} + F_{n+1}$:

$$2F_n + L_n = 2F_n + (F_{n-1} + F_{n+1}) = \underbrace{(F_n + F_{n+1})}_{F_{n+2}} + \underbrace{(F_{n-1} + F_n)}_{F_{n+1}} =$$

$$= F_{n+2} + F_{n+1} = F_{n+3}.$$

8.27. *Proof.*
Transform the right side of the equality:

$$-2F_n + L_n = -2F_n + (F_{n-1} + F_{n+1}) = \underbrace{(F_{n+1} - F_n)}_{F_{n-1}} + \underbrace{(F_{n-1} - F_n)}_{-F_{n-2}} =$$

$$= F_{n-1} - F_{n-2} = F_{n-3}.$$

Since $F_{n-1} - F_{n-2} = F_{n-3}$, then we obtain true equality.

8.28. *Proof.*
Termwise summation of the identities of Exercises **8.26** and **8.27** results in the required equality.

8.29. *Hint.* Express the Lucas numbers L_n in terms of the Fibonacci sequence.

8.30. *Hint.* Use the identities proves in the previous exercise.

8.31. *Solution.*

The simplest approach to solving this exercise consists in using the substitution $b_n = a_n + 1$. Then we come to the relation

$$\begin{cases} b_n = b_{n-1} + b_{n-2}, & n > 2, \\ b_1 = 1, \ b_2 = 1, \end{cases}$$

which, as is known, is satisfied by the terms of the Fibonacci sequence, $b_n = F_n$. Finally, we obtain $a_n = F_n - 1, n \geqslant 1$.

8.32. *Solution.*

The characteristic equation has the form $z^3 = z^2 + 5z + 3$, or $z^3 - z^2 - 5z - 3 = 0$. The roots of this equation are $z_1 = 3, z_2 = z_3 = -1$. In accordance with the general rule for solution of linear recurrence relations, we represent the solution in the form

$$a_n = A \cdot 3^n + B \cdot (-1)^n + C \cdot n \cdot (-1)^n = A \cdot 3^n + (B + Cn) \cdot (-1)^n.$$

Using the original conditions, we obtain the system of linear equations

$$\begin{cases} 3A - B - C & = 1, \\ 9A + B + 2C & = 1, \\ 27A - B - 3C & = -3, \end{cases}$$

whose solutions will be the values $A = 0, B = -3, C = 2$. Thus, the solution of the initial recurrence relation is $a_n = (-1)^n(2n - 3), n \geqslant 1$.

8.33. *Answer:* $a_n = 3^n \left(n + 2 + \dfrac{(-1)^n}{2} \right), n \geqslant 1$.

8.34. *Answer:* $a_n = 3(-1)^{n+1} + 3 \cdot 2^{n/2-1} (1 + (-1)^n), n \geqslant 1$.

8.35. *Answer:* $a_n = (-2)^{n-1} + 2 \cdot 3^{n/2-1} (1 + (-1)^n), n \geqslant 1$.

8.36. *Solution.*

The characteristic equation has the form $z = 2$, and the general solution of the homogeneous equation $a_n = 2a_{n-1}$ will be written as $a_n = A \cdot 2^n$. The function $d(n) = 3^n + 4$, standing in the right side, is the sum of two summands $d_1(n) = 3^n$ and $d_2(n) = 4$.

Find specific solutions of the relations

$$a_n = 2a_{n-1} + 3^n \quad \text{and} \quad a_n = 2a_{n-1} + 4,$$

then the specific solution of the initial relation will be (due to linearity of the equation) the sought specific solution. Since $t = 3$ is not the root of the characteristic equation, then

$$(a_n)_{\text{spec.1}} = B \cdot 3^n, \quad \text{where} \ B = \text{const.}$$

Substituting into the equation $a_n = 2a_{n-1} + 3^n$, we obtain the value of the constant B:

$$B \cdot 3^n = 2B \cdot 3^{n-1} + 3^n \implies B = 3, \text{ and } (a_n)_{\text{spec.1}} = 3^{n+1}.$$

Now we find $(a_n)_{\text{spec.2}}$, $(a_n)_{\text{spec.2}} = C$, where $C = \text{const}$. Substituting into the equation $a_n = 2a_{n-1} + 4$, we obtain the value of the constant C:

$$C = 2C + 4 \Rightarrow C = -4, \text{ and } (a_n)_{\text{spec.2}} = -4.$$

As a result, the general solution of the nonhomogeneous equation takes the form

$$a_n = A \cdot 2^n + 3^{n+1} - 4.$$

Since $a_1 = 1$, $2A + 3^2 - 4 = 1 \Rightarrow A = -2$. Finally, we have

$$a_n = (-2) \cdot 2^n + 3^{n+1} - 4, \quad \text{or} \quad a_n = 3^{n+1} - 2^{n+1} - 4, \ n \geqslant 1.$$

8.37. *Answer:* $a_n = \dfrac{5}{6}\left(10^n - 2 + (-1)^n 2^{n+1}\right)$, $n \geqslant 1$.

8.38. *Answer:* $a_n = \dfrac{1}{20}(5 - 2^{n+4} - 29(-1)^n 3^n)$, $n \geqslant 1$.

8.39. *Answer:* $a_n = \dfrac{1}{4}(3^{n-1} + (-1)^n) + 1$, $n \geqslant 1$.

8.40. *Answer:* $a_n = 2^n(2^{n+1} - 3)$, $n \geqslant 1$.

8.41. *Answer:* $a_n = \dfrac{4^n}{3}(4^n - 1)$, $n \geqslant 1$.

8.42. *Answer:* $a_n = 4^n - 2^n + \dfrac{1}{3}((-2)^n - 1)$, $n \geqslant 1$.

8.43. *Answer:* $a_n = -\dfrac{1}{3} + \dfrac{(-1)^n}{5} - 3^{n+1} + \dfrac{109}{30} \cdot 4^n$, $n \geqslant 1$.

8.44. *Answer:*

(1) $a_n = \dfrac{7}{10}(-1)^{n+1}(7 \cdot 2^n - 2 \cdot 7^n)$, $n \geqslant 1$;

(2) $a_n = \dfrac{(-1)^{n+1}}{5}(27 \cdot 2^n - 7^{n+1})$, $n \geqslant 1$;

(3) $a_n = \dfrac{(-1)^{n+1}}{5}(2^{n+5} - 7^{n+1})$, $n \geqslant 1$.

8.45. *Answer:* $a_n = \dfrac{1}{60}(-10 - 42(-1)^n + 7 \cdot 4^n)$, $n \geqslant 1$.

8.46. *Answer:* $a_n = \dfrac{1}{4}(11 + (-1)^n(2n + 5))$, $n \geqslant 1$.

8.47. *Answer:* $a_n = \dfrac{1}{36}(27 + (-3)^n(8n - 11))$, $n \geqslant 1$.

8.48. *Solution.*

The roots of the characteristic equation are $z_1 = 1, z_2 = z_3 = 2$. The general solution of the respective the homogeneous equation will be written in the form

$$a_n = A \cdot 1^n + B \cdot 2^n + C \cdot n2^n.$$

The function $d(n)$ in the right side will be represented as $d(n) = d_1(n) + d_2(n)$, where $d_1(n) = 1$, $d_2(n) = 2^n$.

Find a specific solution $(a_n)_{\text{spec.1}}$ of the equation

$$a_n = 5a_{n-1} - 8a_{n-2} + 4a_{n-3} + d_1(n).$$

We have $d_1(n) = P_m(n) \cdot t^n$, where $t = 1$, $m = 0$. Multiplicity of the root $t = 1$ of the characteristic equation is $s = 1$; hence,

$$(a_n)_{\text{spec.1}} = n P_0(n) \cdot 1^n = Dn, \quad \text{where } D = \text{const.}$$

Substituting into the equation with nonhomogeneous term $d_1(n)$, we find the value of the constant D:

$$Dn = 5D(n-1) - 8D(n-2) + 4D(n-3) + 1 \;\Rightarrow\; D = 1.$$

Hence, $(a_n)_{\text{spec.1}} = n$.

Find a specific solution $(a_n)_{\text{spec.2}}$ of the equation

$$a_n = 5a_{n-1} - 8a_{n-2} + 4a_{n-3} + d_2(n).$$

We have $d_2(n) = P_m(n) \cdot t^n$, where $t = 2$, $m = 0$. Multiplicity of the root $t = 2$ of the characteristic equation is $s = 2$; hence,

$$(a_n)_{\text{spec.2}} = n^2 P_0(n) \cdot 2^n = E \cdot n^2 2^n, \quad \text{where } E = \text{const.}$$

For determining the value of the constant E substitute $(a_n)_{\text{spec.2}}$ into the equation with $d_2(n)$:

$$E \cdot n^2 2^n = 5E \cdot (n-1)^2 2^{n-1} - 8E(n-2)^2 2^{n-2} + 4E(n-3)^2 2^{n-3} + 2^n =$$
$$= E \cdot 2^{n-3}(8n^2 - 8) + 2^n \;\Rightarrow\; E = 1, \text{ and } (a_n)_{\text{spec.2}} = n^2 2^n.$$

Now we can write out the resultant solution, where the constants A, B, and C are not determined yet:

$$a_n = (a_n)_{\text{gen.}} + (a_n)_{\text{spec.1}} + (a_n)_{\text{spec.2}} = A + (B + Cn)2^n + n + n^2 2^n.$$

Substitution into the original equation results in the system of linear equations relative to A, B, and C:

$$\begin{cases} A + 2(B + C) + 1 + 2 = 0, \\ A + 4(B + 2C) + 2 + 16 = 19, \\ A + 8(B + 3C) + 3 + 72 = 88. \end{cases}$$

The solutions of this system will be the values $A = -3$, $B = -1$, $C = 1$.
As a result, we obtain $a_n = 2^n(n^2 + n - 1) + n - 3$, $n \geqslant 1$.

8.49. *Answer*:

(1) $a_n = 2^{n-1}(2n^2 - 17n + 40) + n - 28$, $n \geqslant 1$;

(2) $a_n = 2^n(n^2 + 34n - 62) + n + 53$, $n \geqslant 1$;

(3) $a_n = 2^{n-2}(4n^2 - 5n + 41) + n - 17$, $n \geqslant 1$.

8.50. *Answer*: $a_n = \dfrac{2}{3}(n+1) + \dfrac{1}{3}\cos\dfrac{2\pi n}{3} - \dfrac{1}{3\sqrt{3}}\sin\dfrac{2\pi n}{3}$, $n \geqslant 1$.

8.51. *Answer*: $a_n = 3n - 1 + \cos\dfrac{2\pi n}{3} + \dfrac{1}{\sqrt{3}}\sin\dfrac{2\pi n}{3}$, $n \geqslant 1$.

8.52. *Solution*.

$$a_n = 2a_{n/2} + 2n - 1 =$$

$$= 2(2a_{n/4} + 2\dfrac{n}{2} - 1) + 2n - 1 = 4a_{n/4} + 2n - 2 + 2n - 1 =$$

$$= 4(2a_{n/8} + 2\dfrac{n}{4} - 1) + 2n - 2 + 2n - 1 = 8a_{n/8} + 2n - 4 + 2n - 2 + 2n - 1.$$

After the kth substitution, we obtain

$$a_n = 2^k a_{n/2^k} + 2nk - \sum_{i=0}^{k-1} 2^i.$$

This formula can be proved by the mathematical induction method. Substitute into the obtained expression the value $k = \log_2 n$:

$$a_n = 2^{\log_2 n} a_1 + 2n \log_2 n - \dfrac{2^{\log_2 n} - 1}{2 - 1}.$$

As a result, we come to the explicit expression for the nth term of the sequence

$$a_n = 2n\log_2 n + 1, \ n \geqslant 1.$$

8.53. *Answer*: $a_n = \dfrac{1}{3}(8n^3 - 5n^{\log_2 5})$, $n \geqslant 1$.

8.54. *Answer*: $a_n = \dfrac{22}{3}n^{\log_3 4} - 6n - \dfrac{1}{3}$, $n \geqslant 1$.

8.55. *Answer*: $a_n = -n\log_3 n + \dfrac{n}{2}(3n - 1)$, $n \geqslant 1$.

8.56. *Hint*. Use the induction over the variable k.

8.57. *Answer*:

(1) $G(z) = \dfrac{2}{1 - z}$;

(2) $G(z) = \dfrac{1}{1 + z}$;

(3) $G(z) = \dfrac{1}{1 + z^2}$;

(4) $G(z) = \dfrac{1}{1-z} + \dfrac{1}{1-iz}$.

8.58. *Solution*:

(1) Introduce the notation $a_n = n^2$ for $n = 0, 1, 2, \ldots$ The general term of the sequence $\{a_n\}$ represents in the form $n^2 = n(n-1) + n$. By definition of generating function, we have:

$$G(z) = \sum_{n=0}^{\infty} a_n z^n = \sum_{n=0}^{\infty} n(n-1) z^n + \sum_{n=0}^{\infty} n z^n.$$

Transform the first of the obtained sums:

$$\sum_{n=0}^{\infty} n(n-1) z^n = z^2 \sum_{n=0}^{\infty} n(n-1) z^{n-2}.$$

As is known, when differentiating the power function z^n the index of power is reduced by one and the multiplication factor is taken out n: $\dfrac{d}{dz} z^n = n z^{n-1}$. Due to this, the sum $\sum_{n=0}^{\infty} n(n-1) z^{n-2}$ can be considered as the second derivative of the function $1 + z + z^2 + \ldots = \dfrac{1}{1-z}$.

The expression $\sum_{n=0}^{\infty} n z^n$ is represented similarly, using the first derivative of the mentioned function.

As a result, for $G(z)$ we obtain:

$$G(z) = z^2 \frac{d^2}{dz^2} \sum_{n=0}^{\infty} z^n + z \frac{d}{dz} \sum_{n=0}^{\infty} z^n = \frac{z(1+z)}{(1-z)^3}.$$

(2) Denote $a_n = n^3$ for $n = 0, 1, 2, \ldots$ and $G(z) = \sum_{n=0}^{\infty} a_n z^n$. Taking into account the identity $n^3 = n(n-1)(n-2) + 3n(n-1) + n$, we obtain:

$$G(z) = \sum_{n=0}^{\infty} n(n-1)(n-2) z^n + 3 \sum_{n=0}^{\infty} n(n-1) z^n + \sum_{n=0}^{\infty} n z^n =$$

$$= z^3 \frac{d^3}{dz^3} \frac{1}{1-z} + 3z^2 \frac{d^2}{dz^2} \frac{1}{1-z} + z \frac{d}{dz} \frac{1}{1-z}.$$

Perform the operations of differentiation and the subsequent algebraic transformations:

$$G(z) = \frac{z(1 + 4z + z^2)}{(1-z)^4}.$$

So, the generating function for the sequence of cubes of nonnegative integers $a_n = n^3$ is $G(z) = \dfrac{z(1 + 4z + z^2)}{(1 - z)^4}$.

8.59. *Answer*:

(1) $\widetilde{G}(z) = zG(z)$;

(2) $\widetilde{G}(z) = (1 + z)G(z)$;

(3) $\widetilde{G}(z) = G(z^2)$;

(4) $\widetilde{G}(z) = z^{-3}(G(z) - a_2 z^2 - a_1 z - a_0)$.

8.60. *Answer*:

(1) $\widetilde{G}(z) = \dfrac{d}{dz}(zG(z))$;

(2) $\widetilde{G}(z) = \int\limits_{0}^{z} G(\xi)\,d\xi$;

(3) $\widetilde{G}(z) = \dfrac{1}{2}\big(G(\sqrt{z}) + G(-\sqrt{z})\big)$;

(4) $\widetilde{G}(z) = \dfrac{G(z)}{1 - z}$.

8.61. *Proof.*

Recall the definition of the associativity property (see Exercise **2.22**):

$$\big(G_a(z) \cdot C_b(z)\big) \cdot G_c(z) = G_a(z) \cdot \big(C_b(z) \cdot G_c(z)\big),$$

where $G_a(z)$, $G_b(z)$, $G_c(z)$—the generating functions for arbitrary numerical sequences $\{a_n\}$, $\{b_n\}$, and $\{c_n\}$, $n = 0, 1, 2, \ldots$, respectively.

Prove that the associativity property, written for any numerical sequences $\{a_n\}$, $\{b_n\}$ and $\{c_n\}$, results in the algebraic identity.

For this, use the determining formulae, satisfying the products $\big(G_a(z) \cdot C_b(z)\big) \cdot G_c(z)$ and $G_a(z) \cdot \big(C_b(z) \cdot G_c(z)\big)$ taking into account the order of brackets:

$$\big(G_a(z) \cdot C_b(z)\big) \cdot G_c(z) = \sum_{n=0}^{\infty}\left(\sum_{i=0}^{n}\left(\sum_{k=0}^{i}(a_k b_{i-k}) \cdot c_{n-i}\right)\right)z^n = \sum_{n=0}^{\infty} V_n z^n,$$

$$G_a(z) \cdot \big(C_b(z) \cdot G_c(z)\big) = \sum_{n=0}^{\infty}\left(\sum_{i=0}^{n}\left(a_i \sum_{k=0}^{n-i}(b_k c_{n-i-k})\right)\right)z^n = \sum_{n=0}^{\infty} W_n z^n.$$

In the previous formulae, we have introduced the notations

$$V_n = \sum_{i=0}^{n}\sum_{k=0}^{i} a_k b_{i-k} c_{n-i}, \qquad W_n = \sum_{i=0}^{n}\sum_{k=0}^{n-i} a_i b_k c_{n-i-k}.$$

Prove that for all $n \geqslant 0$, the equality is valid $V_n = W_n$. Transform the expression for W_n, performing substitution of the variable $j = i + k$:

$$W_n = \sum_{i=0}^{n} \sum_{j=i}^{n} a_i b_{j-i} c_{n-j}.$$

Then, in the obtained expression, change the order of summation in accordance with the identity $\sum_{i=0}^{n} \sum_{j=i}^{n} a_{ij} = \sum_{j=0}^{n} \sum_{i=0}^{j} a_{ij}$, valid for any coefficients a_{ij}. The point of the applied identity consists in that the sum of the elements of an arbitrary upper triangular matrix can be calculated by two methods, both, of course, having the same result—either in rows or in columns. Accurate to renaming of the variables $i \rightarrow k$, $j \rightarrow i$ the obtained double sum

$$W_n = \sum_{j=0}^{n} \sum_{i=0}^{j} a_i b_{j-i} c_{n-j}.$$

coincides with the sum that determines the value V_n, i.e., $V_n = W_n$ for all $n \geqslant 0$.

Thus, it is proved that for the operation of multiplication of generating functions, the associativity property is valid.

8.62. *Answer*:
Convolution of sequences $1, 2, 3, 4, \ldots$ and $1, \dfrac{1}{2}, \dfrac{1}{3}, \dfrac{1}{4}, \ldots$ is the sequence $\{c_n\}$, $n \in \{0\} \cup \mathbb{N}$, with the general term

$$c_n = (n+2) H_{n+1} - (n+1),$$

where $H_{n+1} = \sum\limits_{i=1}^{n+1} \dfrac{1}{i}$ is a harmonic number.

8.63. *Answer*: $c_n = \dfrac{(\alpha + \beta)^n}{n!}$, where $n = 0, 1, 2, \ldots$

8.64. *Solution*.

The sequences $\{a_n\}$ and $\{b_n\}$ are associated with the generating functions $G_a(z) = \sum\limits_{n=0}^{\infty} \cos n\varphi \, z^n$ and $G_b(z) = \sum\limits_{n=0}^{\infty} \sin n\varphi \, z^n$.

Having written the trigonometric functions $\cos n\varphi$ and $\sin n\varphi$ in terms of the exponent with a complex index, we obtain

$$G_a(z) = \frac{1}{2}\left((1 - e^{i\varphi} z)^{-1} + (1 - e^{-i\varphi} z)^{-1}\right),$$

$$G_b(z) = \frac{1}{2i}\left((1 - e^{i\varphi} z)^{-1} - (1 - e^{-i\varphi} z)^{-1}\right).$$

As is known, the convolution of the sequences $\{a_n\}$ and $\{b_n\}$ is satisfied by the generating function $G_c(z) = G_a(z)G_b(z)$. Having applied to this product the eighth formula from Table 8.1 for the case $N = 2$

$$\frac{1}{(1 - cz)^2} = \sum_{n=0}^{\infty} C(n + 1, n)\, c^n z^n = \sum_{n=0}^{\infty} (n + 1)\, c^n z^n,$$

we obtain $G_c(z) = \dfrac{1}{2} \sum\limits_{n=0}^{\infty} (n + 1) \sin n\varphi \, z^n$.

Finally, $c_n = \dfrac{1}{2}(n + 1) \sin n\varphi$ for all $n \geqslant 0$.

8.65. *Answer:* $a_n = 1 - 4(-1)^n$, $n \geqslant 1$.

8.66. *Answer:* $a_n = (12 - 5n)(-5)^{n-1}$, $n \geqslant 1$.

8.67. *Answer:* $a_n = \dfrac{3^{-n}}{2}(n^2 + 3n - 10)$, $n \geqslant 1$.

8.68. *Answer:* $a_n = (n - 2)2^n + 3^{n+1}$, $n \geqslant 1$.

8.69. *Answer:* $a_n = (2\sqrt{8} - 5)(3 + \sqrt{8})^n - (2\sqrt{8} + 5)(3 - \sqrt{8})^n$, $n \geqslant 1$.

8.70. *Answer:* $a_n = n^3$, $n \geqslant 1$.

8.71. *Answer:* $a_n = (n - 2)^2$, $n \geqslant 1$.

8.72. *Answer:* $a_n = 1 - 3^n \cos \dfrac{\pi n}{2}$, $n \geqslant 1$.

8.73. *Answer:* $a_n = 2^n + n$, $n \geqslant 1$.

8.74. *Answer:* $a_n = \dfrac{1}{5}\left(3 \cdot 2^n - 3\cos\dfrac{\pi n}{2} + 4\sin\dfrac{\pi n}{2}\right) - 1$, $n \geqslant 1$.

8.75. *Answer:* $a_n = n \cdot 3^n$, $n \geqslant 1$.

8.76. *Answer:* $a_n = (-2)^{-n} + n^2$, $n \geqslant 1$.

8.77. *Proof.*

As is known, the sequence of the Fibonacci numbers $\{F_n\}$, where $n = 0, 1, 2, \ldots$, is determined by the recurrence relation $F_n = F_{n-1} + F_{n-2}$ for $n \geqslant 2$ with the initial conditions $F_0 = 0$ and $F_1 = 1$ (see Exercises **1.86** and **1.105** on p. 17 and 20).

Multiply the right and the left sides of the determining formula for F_n by z^n and sum it over n:

$$\sum_{n=2}^{\infty} F_n z^n = \sum_{n=2}^{\infty} F_{n-1} z^n + \sum_{n=2}^{\infty} F_{n-2} z^n.$$

Then, perform the algebraic transformations:

$$\sum_{n=0}^{\infty} F_n z^n - F_1 z - F_0 = z\left(\sum_{n=0}^{\infty} F_n z^n - F_0\right) + z^2 \sum_{n=0}^{\infty} F_n z^n.$$

Taking into account the formula for an arbitrary function of the Fibonacci numbers $G_F(z) = \sum\limits_{n=0}^{\infty} F_n z^n$, we obtain the equation

$$G_F(z) - z = z G_F(z) + z^2 G_F(z),$$

its solution is $G_F(z) = \dfrac{z}{1 - z - z^2}$. The generating function for the sequence of the

Lucas numbers $G_L(z) = \dfrac{2 - z}{1 - z - z^2}$ is derived similarly.

8.78. *Answer:* $G_H(z) = \dfrac{1}{1 - z} \ln \dfrac{1}{1 - z}$.

8.79. *Solution.*

The expression for an arbitrary term of the sequence $\{a_n\}$ has the form $a_n = Az_1^n + Bz_2^n$, where $z_1 = 3$, $z_2 = 4$. Recurrence relation of the second order with the characteristic equation $(z - 3)(z - 4) = 0$ has the solution $a_n = 3^n + 4^n$.

Hence, the solution of $a_n = 7a_{n-1} - 12a_{n-2}$ will be $a_n = A \cdot 3^n + B \cdot 4^n$. In our case $a_1 = 3 + 4 = 7$, $a_2 = 3^2 + 4^2 = 25$. As a result, we obtain that $a_n = 3^n + 4^n$ is the solution of the recurrence relation

$$\begin{cases} a_n = 7a_{n-1} - 12a_{n-2}, & n > 2, \\ a_1 = 7, \ a_2 = 25. \end{cases}$$

8.80. *Answer:* $\begin{cases} a_n = -3a_{n-1} + 10a_{n-2}, & n > 2, \\ a_1 = 1, \ a_2 = 9. \end{cases}$

8.81. *Answer:* $\begin{cases} a_n = 7a_{n-1} - 6a_{n-2}, & n > 2, \\ a_1 = 5, \ a_2 = 35. \end{cases}$

8.82. *Answer:* $\begin{cases} a_n = 4a_{n-2}, & n > 2, \\ a_1 = 0, \ a_2 = \dfrac{8}{3}. \end{cases}$

8.83. *Solution.*

The expression for an arbitrary term of the sequence $\{a_n\}$ has the form $a_n = Q_m(n)t^n$, where $t = 2$, the degree of polynomial Q_m is $m = 2$. Write the respective recurrence relation: $a_n = a_{n-1} + d(n)$, where $d(n) = (An^2 + Bn + C)2^n$, A, B, and C—const.

Calculate the values of the constants A, B, and C. For this, substitute a_n into the recurrence relation:

$$2^n(n^2 + n + 1) = 2^{n-1}((n - 1)^2 + (n - 1) + 1) + (An^2 + Bn + C)2^n.$$

Equating the coefficients of like powers n, we calculate the values A, B, and C:
$A = C = \dfrac{1}{2}$, $B = \dfrac{3}{2}$.

As a result, we obtain $\begin{cases} a_n = a_{n-1} + 2^{n-1}(n^2 + 3n + 1), & n > 1, \\ a_1 = 6. \end{cases}$

8.84. *Answer:* $\begin{cases} a_n = a_{n-1} + 3^{n-1}(2n^2 + 6n + 3), & n > 1, \\ a_1 = 12. \end{cases}$

8.85. *Answer:* $\begin{cases} a_n = a_{n-2} + 2^{n-2}(3n^2 - 2n - 5), & n > 2, \\ a_1 = 0, \ a_2 = 2. \end{cases}$

8.86. *Answer:*
$$\begin{cases} a_n = a_{n-2} + 4^{n-2}(15n+2), & n > 2, \\ a_1 = 4, \ a_2 = 32. \end{cases}$$

8.90. *Answer:* $a_n = \dfrac{F_{n+2}}{F_{n+1}}$ for all $n \geqslant 1$, where F_{n+1} and F_{n+2} are the Fibonacci numbers.

8.91. *Solution.*

Introduce a new variable $b_n = a_n^2$. The nonlinear relation is transformed to:

$$\begin{cases} b_n = b_{n-1} + 2, & n > 1, \\ b_1 = 0. \end{cases}$$

Its solution is the sequence determined by the relation $b_n = 2(n-1)$, $n = 1, 2, \ldots$, therefore, $a_n = \sqrt{2(n-1)}$, $n = 1, 2, \ldots$

8.92. *Answer:* $a_n = (6n-5)^{1/3}$, $n \geqslant 1$.

8.93. *Answer:* $a_n = \dfrac{1}{2}(\sqrt{5n+220} - 1)$, $n \geqslant 1$.

8.94. *Answer:* $a_n = \dfrac{1}{3}(\sqrt[3]{n} + 5)$, $n \geqslant 1$.

8.95. *Solution.*

Having taken the logarithm of both sides of the relation, we obtain

$$2 \ln a_n = \ln a_{n-1} + \ln 2.$$

Denote $b_n = \ln a_n$, the respective recurrence relation is linear:

$$\begin{cases} b_n = \dfrac{1}{2}(b_{n-1} + \ln 2), & n > 1, \\ b_1 = 0. \end{cases}$$

Its solution is $b_n = \ln 2(1 - 2^{-n+1})$, $n \geqslant 1$. Returning to the original value a_n, we obtain $a_n = 2^{1 - 2^{1-n}}$, $n \geqslant 1$.

Note. If a_n can take complex values, then the solution will have the form $a_n = 2^{1 - 2^{1-n}} e^{2^{2-n} \pi k i}$, where $k \in \mathbb{Z}$, $n \geqslant 1$.

8.96. *Answer:* $a_n = i^{\pm(1 - 3^{1-n})}$, $n \geqslant 1$.

8.97. *Answer:* $a_n = 4$, $n \geqslant 1$. If a_n can take complex values, then the solution will have the form $a_n = 4e^{2^{2-n} \pi k i}$, where $k \in \mathbb{Z}$, $n \geqslant 1$.

8.98. *Answer:* $a_n = 3 \left(\dfrac{2}{3} \right)^{2^{1-n}}$, $n \geqslant 1$. If a_n can take complex values, then the solution will have the form $a_n = 3 \left(\dfrac{2}{3} \right)^{2^{1-n}} e^{2^{2-n} \pi k i}$, where $k \in \mathbb{Z}$, $n \geqslant 1$.

8.99. *Answer:*

(1) $a_n = n$, $n \geqslant 1$;

(2) $a_n = 8n - 6$, $n \geqslant 1$.

8.100. *Answer*:

(1) $a_n = 2^{(n-1)/2}$, $n \geqslant 1$;

(2) $a_n = 3^{2-n}$, $n \geqslant 1$.

8.101. *Answer*:

(1) $a_n = \dfrac{10}{n}$, $n \geqslant 1$;

(2) $a_n = \dfrac{1}{n+1}$, $n \geqslant 1$.

8.102. *Solution.*

Boundedness of the sequence $\{a_n\}$, $n \in \mathbb{N}$, follows from the properties of the function $f(x) = \lambda x(1 - x)$. Calculate the derivative of $f'(x) = \lambda(1 - 2x)$, and find the point of possible extremum $f(x)$: $x = 0$, $x = \dfrac{1}{2}$, $x = 1$. The minimum and the maximum values of the function on the closed segment $[0, 1]$ are $f_{\min} = 0$, $f_{\max} = \dfrac{\lambda}{4}$.

Taking into account that $\lambda \leqslant 4$, we obtain: $a_2 = f(a_1)$ and lies in the interval $[0, 1]$, $a_3 = f(a_2)$ and also belongs to the interval $[0, 1]$, and so on; hence, $a_n \in [0, 1]$ for all $n \in \mathbb{N}$.

The equation solution method consists in applying suitable changes of the variable.

Consider the case $\lambda = 2$. The change $b_n = \dfrac{1}{2} - a_n$ results in the recurrence relation

$$\begin{cases} b_n = 2b_{n-1}^2, & n > 1, \\ b_1 = \dfrac{1}{2} - a_1. \end{cases}$$

Having taken a base 2 logarithm of both sides of the equality $b_n = 2b_{n-1}^2$ and having denoted $\widetilde{b}_n = \log_2 b_n$, we obtain a linear equation, whose solution will be $\widetilde{b}_n = A \cdot 2^n - 1$, where $A = \dfrac{1}{2} \log_2(1 - 2a_1)$. Returning to the variable a_n, we have $a_n = \dfrac{1}{2} - \dfrac{1}{2}(1 - 2a_1)^{2^{n-1}}$, $n \geqslant 1$.

Consider the case $\lambda = 4$. The change of the variable $a_n = \sin^2 b_n$, $b_n \geqslant 0$:

$$\begin{cases} \sin^2 b_n = \sin^2(2b_{n-1}), & n > 1, \\ b_1 = \arcsin \sqrt{a_1}. \end{cases}$$

Taking into account nonnegativity of the values b_n, we find the explicit expression b_n: $b_n = A \cdot 2^n - 2\pi k$, where $A = \dfrac{1}{2}(\arcsin \sqrt{a_1} + 2\pi k)$, $k \in \mathbb{Z}$.

Finally, for the case $\lambda = 4$ the solution of the equation has the form $a_n = \sin^2(2^{n-1}(\arcsin \sqrt{a_1} + 2\pi k))$, or $a_n = \sin^2(2^{n-1} \arcsin \sqrt{a_1})$, $n \geqslant 1$.

8.103. *Solution.*

Consider three cases depending on the relation of the values a_1 and $\sqrt{x}$.

(1) If $a_1 = \sqrt{x}$, then $a_2 = \dfrac{1}{2}\left(a_1 + \dfrac{x}{a_1}\right) = \dfrac{1}{2}\left(\sqrt{x} + \sqrt{x}\right) = \sqrt{x}$. Calculating the values a_3, a_4, and so on, we obtain that for all terms of the sequence, valid is the following $a_n = \sqrt{x}$, $n = 1, 2, \ldots$

(2) If $a_1 > \sqrt{x}$, then the sequence of $\{a_n\}$ monotonically decreases and is bounded from below by the value $\sqrt{x}$. Let us prove it.

Write the monotonicity condition: $a_{n+1} < a_n$. From the relation $a_n > \sqrt{x}$ follows that

$$a_{n+1} - a_n = \frac{1}{2}\left(a_n + \frac{x}{a_n}\right) - a_n = \frac{x - a_n^2}{2a_n} < 0.$$

Boundedness of the sequence $\{a_n\}$ follows from the comparison

$$a_{n+1} - \sqrt{x} = \frac{1}{2}\left(a_n + \frac{x}{a_n}\right) - \sqrt{x} = \frac{(a_n - \sqrt{x})^2}{2a_n} > 0,$$

valid for all $n = 1, 2, \ldots$

According to the Weierstrass theorem, there exists a limit $A = \lim\limits_{n \to \infty} a_n$, and from the recurrence relation follows that $A = \dfrac{1}{2}\left(A + \dfrac{x}{A}\right)$ or $A = \sqrt{x}$.

(3) In case $a_1 < \sqrt{x}$, the next term of the sequence $a_2 > \sqrt{x}$:

$$a_2 - \sqrt{x} = \frac{1}{2}\left(a_1 + \frac{x}{a_1}\right) - \sqrt{x} = \frac{(a_1 - \sqrt{x})^2}{2a_1} > 0,$$

and for the sequence $\{a_n\}$, $n = 2, 3, 4 \ldots$ the conditions of the Weierstrass theorem are checked similar to item 2).

Thus, for all possible values of the first element of a_1 the sequence of $\{a_n\}$ converges to $A = \sqrt{x}$.

8.104. *Answer:* $\lim\limits_{n \to \infty} a_n = \sqrt[3]{x}$.

8.105. *Hint.*
The sequence $\{a_n\}$ is increasing and bound from above.

8.106. *Hint.*
The sequence $\{a_n\}$ is increasing and bound from above.

8.107. *Solution.*
Calculate the value $a_n - a_{n-1}$:

$$a_n - a_{n-1} = \left(\sum_{i=1}^{n-1} a_i + 2\right) - \left(\sum_{i=1}^{n-2} a_i + 2\right) = a_{n-1},$$

whence follows that the original full history relation is equivalent to the following linear relation

$$\begin{cases} a_n = 2a_{n-1}, & n > 1, \\ a_1 = 2. \end{cases}$$

Now we only have to write the answer: $a_n = 2^n$, $n \geqslant 1$.

8.108. *Answer*: $a_n = \begin{cases} 12 \cdot 2^{n-2}, & n > 1, \\ 2, & n = 1. \end{cases}$

8.109. *Hint.*

The original recurrence relation is equivalent to the following:

$$\begin{cases} a_n = 2a_{n-1}, & n > 2, \\ a_2 = a_1 + 2, \\ a_1 = 1, \end{cases} \quad \text{and its solution is} \quad \begin{cases} a_n = 3 \cdot 2^{n-2}, & n > 1, \\ a_1 = 1. \end{cases}$$

8.110. *Answer*:

C_0	C_1	C_2	C_3	C_4	C_5	C_6	C_7
1	1	2	5	14	42	132	429

8.111. *Hint.*

From the relation for the Catalan numbers (see Exercise **8.110**) follows that the generating function $G_C(z)$ satisfies the functional equation $zG_C^2(z) = G_C(z) - C_0$.

8.112. *Answer*: $C_n = \dfrac{1}{n+1}C(2n, n)$ for all $n \geqslant 0$, $C(2n, n)$—binomial coefficient.

8.113. *Solution.*

Find the summation factor. Since $f(n) = n$, then

$$F(n) = \left[\prod_{i=2}^{n} f(i)\right]^{-1} = (n!)^{-1}.$$

Perform the change $b_n = f(n+1)F(n+1)a_n$, or

$$b_n = (n+1)\frac{1}{(n+1)!}a_n = \frac{a_n}{n!}.$$

We obtain the reciprocal equation $b_n = b_{n-1} + 1$, and $b_1 = 1$. Hence, it follows that $b_n = n$, and $a_n = n \cdot n!$ for all $n \geqslant 1$.

8.114. *Answer*: $a_n = (n+1)!$, $n \geqslant 1$.

8.115. *Solution.*

The summation factor is $F(n) = \dfrac{2}{n+1}$. After the change $b_n = \dfrac{2}{n+1}a_n$, we come to the relation

$$\begin{cases} b_n = b_{n-1} + \dfrac{4}{n+1}, & n > 1, \\ b_1 = 0, \end{cases}$$

satisfied by the sequence $b_n = 4\left(H_{n+1} - \dfrac{3}{2}\right)$. Finally, we obtain: $a_n = 2(n + 1)\left(H_{n+1} - \dfrac{3}{2}\right)$, $n \geqslant 1$.

8.116. *Answer:* $a_n = (n + 1)^2 - (n + 1)H_n, n \geqslant 1$.

8.117. *Answer:* $a_n = 2n(n + 5), n \geqslant 1$.

8.118. *Solution.*

The summation factor is

$$F(n) = \left(\prod_{i=2}^{n} \frac{1}{i}\right)^{-1} = n!,$$

and the explicit expression a_n has the form

$$a_n = \frac{1}{(n+1)^{-1}(n+1)!}\left(a_1 + \sum_{i=2}^{n} i \cdot i!\right) = \frac{1}{n!}\left(a_1 - 1 + \sum_{i=1}^{n} i \cdot i!\right).$$

In the solution of Exercise **8.9**, it is shown that $\sum_{i=1}^{n} i \cdot i! = (n + 1)! - 1$ for all $n \geqslant 1$.

Due to this, the expression for an arbitrary term of the sequence $\{a_n\}$ is simplified: $a_n = n + 1, n \geqslant 1$.

8.119. *Answer:* $a_n = n(n + 1), n \geqslant 1$.

8.120. *Answer:* $a_n = (n - 1)e^{n^2 + n - 2}, n \geqslant 1$.

8.121. *Answer:* $a_n = ne^{-n^2 - n + 2}, n \geqslant 1$.

8.122. *Solution.*

Express b_n from the first equation of the system: $b_n = 4a_n - a_{n+1}$ for $n \geqslant 1$. Substituting the obtained relations into the second equation of the system, we obtain the recurrence relation, connecting a_{n-1}, a_n and a_{n+1}:

$$a_n = 5a_{n-1} - 6a_{n-2}.$$

The solution of the obtained equation will be $a_n = A_1 2^n + B_1 3^n$, where A_1 and B_1—const.

Expressing a_{n-1} from the second equation of the system and substituting the obtained relation into the first equation, we find the recurrence relation for b_n:

$$b_n = 5b_{n-1} - 6b_{n-2}.$$

The solution of this equation is $b_n = A_2 2^n + B_2 3^n$.

In order to find the values of the constant A_1, A_2, B_1, B_2, substitute the expressions for a_n and b_n into the original system.

$$\begin{cases} -2A_1 + A_2 = 0, \\ -B_1 + B_2 = 0, \\ 2A_1 + 3B_1 = 7, \\ 2A_2 + 3B_2 = 11. \end{cases}$$

Solving the obtained nonhomogeneous system of linear equations, we find $A_1 = 2$, $A_2 = 4$, $B_1 = B_2 = 1$. Finally, we obtain $a_n = 2^{n+1} + 3^n$, $b_n = 2^{n+2} + 3^n$, $n \geqslant 1$.

8.123. *Answer:* $a_n = 5^{(n+1)/2}$, $b_n = \dfrac{1}{2}(5^{n/2+1} - 5^{(n+1)/2})$, $n \geqslant 1$.

8.124. *Solution.*
Complement the sought sequences $\{a_n\}$ and $\{b_n\}$, $n \in \mathbb{N}$, with the terms a_0 and b_0. Substituting $n = 1$ into the system, we find $a_0 = 2$, $b_0 = 0$.

Introduce into consideration the generating functions $G_a(z) = \sum\limits_{n=0}^{\infty} a_n z^n$ and $G_b(z) = \sum\limits_{n=0}^{\infty} b_n z^n$. Then, multiply the left and the right sides of the system's equations by z^n and sum it over n. As a result, we obtain a system of equations over the function $G_a(z)$ and $G_b(z)$:

$$
\begin{cases}
(1 - 2z)G_a(z) - 3z\,G_b(z) & = a_0 + \dfrac{4z}{1 - 3z}, \\[2mm]
-3z\,G_a(z) + (1 - 6z)G_b(z) & = b_0 - \dfrac{4z}{1 - 5z}.
\end{cases}
$$

Solution of this system, subject to the initial conditions $a_0 = 2$ and $b_0 = 0$, has the form: $G_a(z) = \dfrac{2 - 8z}{(1 - 3z)(1 - 5z)} = \dfrac{1}{1 - 5z} + \dfrac{1}{1 - 3z}$, $G_b(z) = \dfrac{2z}{(1 - 3z)(1 - 5z)} = $
$= \dfrac{1}{1 - 5z} - \dfrac{1}{1 - 3z}$. Using Table 8.1, we obtain the final answer: $a_n = 5^n + 3^n$, $b_n = 5^n - 3^n$ for $n = 1, 2, \ldots$

8.125. *Answer:* $a_n = n + 1 + 2^n$, $b_n = n - 1 - 2^n$ for $n = 1, 2, \ldots$
8.126. *Answer:* $a_n = 4^n + n^2$, $b_n = 4^n - 2n$ for $n \geqslant 1$.
8.127. *Answer:* $a_n = (-2)^n + 2^n$, $b_n = (-2)^n + 2^{n+1}$ for $n \geqslant 1$.
8.128. *Answer:* $\Gamma(n) = (n - 1)!$ for $n \in \mathbb{N}$.
8.129. *Solution.*
Denote $I_n = \int\limits_0^{\pi/2} \sin^n x\, dx$ and perform integrating by parts (see Appendix "Reference Data," relation (A.38)), assuming $n \geqslant 2$:

$$
\begin{aligned}
I_n &= \int_0^{\pi/2} \sin^{n-1} x \cdot \sin x\, dx = \int_0^{\pi/2} \sin^{n-1} x\, d(-\cos x) = \\
&= -\sin^{n-1} x \cdot \cos x \Big|_0^{\pi/2} + \int_0^{\pi/2} \cos x\, d\sin^{n-1} x = \\
&= (n - 1) \int_0^{\pi/2} \sin^{n-2} x \,(1 - \sin^2 x)\, dx = \\
&= (n - 1) \int_0^{\pi/2} \sin^{n-2} x\, dx - (n - 1) \int_0^{\pi/2} \sin^n x\, dx.
\end{aligned}
$$

Solving the obtained equality with respect to I_n, we obtain the recurrence relation

$$I_n = \frac{n-1}{n} I_{n-2}.$$

Calculate I_n for the values $n = 1$ and $n = 2$:

$$I_1 = \int_0^{\pi/2} \sin x \, dx = 1, \quad I_2 = \int_0^{\pi/2} \sin^2 x \, dx = \int_0^{\pi/2} \frac{1 - \cos 2x}{2} \, dx = \frac{\pi}{4}.$$

Using the initial values I_1, I_2 and the formula $I_n = \frac{n-1}{n} I_{n-2}$, we come to the relations

$$I_{2n-1} = \frac{(2n-2)!!}{(2n-1)!!}, \quad I_{2n} = \frac{\pi}{2} \frac{(2n-1)!!}{(2n)!!},$$

where $n!!$—semifactorial of the number n. Uniform expression for I_n, where $n \in \mathbb{N}$, can be written with the help of gamma function:

$$I_n = \frac{\sqrt{\pi}}{2} \frac{\Gamma((n+1)/2)}{\Gamma((n+2)/2)}.$$

8.130. *Answer*:
$$\begin{cases} I_n = \dfrac{n-1}{n} I_{n-2}, \ n > 2, \\ I_1 = 1, \ I_2 = \dfrac{\pi}{4}; \end{cases} \quad I_{2n-1} = \frac{(2n-2)!!}{(2n-1)!!}, \quad I_{2n} = \frac{\pi}{2} \frac{(2n-1)!!}{(2n)!!}, \ n \geqslant 1.$$

8.131. *Solution*.

Denote $I_{2n} = \int\limits_0^{\pi/4} \tan^{2n} x \, dx$ and perform integrating by parts, assuming $n \geqslant 1$:

$$I_{2n} = \int_0^{\pi/4} \tan^{2n-2} x \cdot \tan^2 x \, dx = \int_0^{\pi/4} \tan^{2n-2} x \, d \tan x - \int_0^{\pi/4} \tan^{2n-2} x \, dx =$$

$$= \frac{\tan^{2n-1} x}{2n-1} \Big|_0^{\pi/4} - I_{2n-2} = \frac{1}{2n-1} - I_{2n-2}.$$

Calculate I_{2n} for the values $n = 0, 1, 2, \ldots$:

$$I_0 = \frac{\pi}{4}, \quad I_2 = 1 - \frac{\pi}{4}, \quad I_4 = \frac{1}{3} - 1 + \frac{\pi}{4}, \quad \ldots,$$

$$I_{2n} = (-1)^n \left(\frac{\pi}{4} - 1 + \frac{1}{3} - \frac{1}{5} + \cdots + (-1)^n \frac{1}{2n-1} \right).$$

Finally, $I_{2n} = (-1)^n \left(\dfrac{\pi}{4} - \sum\limits_{k=1}^{n} (-1)^{k-1} \dfrac{1}{2k-1} \right), \ n \geqslant 1.$

8.132. *Answer:*
$$\begin{cases} I_n = \dfrac{2n}{2n+1}I_{n-1}, & n \geqslant 1, \\ I_0 = 1; \end{cases} \qquad I_n = \dfrac{(2n)!!}{(2n+1)!!}, \quad n \geqslant 0.$$

8.133. *Answer:*

$$\begin{cases} I_n = -\dfrac{n}{c+1}I_{n-1}, & n \geqslant 1, \\ I_0 = \dfrac{1}{c+1}; \end{cases} \qquad I_n = (-1)^n \dfrac{n!}{(c+1)^{n+1}}, \quad n \geqslant 0.$$

8.134. *Solution.*

Denote the volume of a n-dimensional ball in terms of v_n. By definition, we have [87]

$$v_n = \int \cdots \int\limits_{x_1^2 + x_2^2 + \ldots + x_n^2 \leqslant 1} dx_1 dx_2 \ldots dx_n.$$

Having chosen one of the coordinates, for example, x_n, we will write v_n in the form

$$v_n = \int_{-1}^{1} dx_n \int \cdots \int\limits_{x_1^2 + x_2^2 + \ldots + x_{n-1}^2 \leqslant 1 - x_n^2} dx_1 dx_2 \ldots dx_{n-1}.$$

The internal integral is equal to the volume of the ball with a radius of $r = (1 - x_n^2)^{1/2}$ in the space $\mathbb{R}^{n-1}$ and can be represented as

$$\int \cdots \int\limits_{x_1^2 + x_2^2 + \ldots + x_{n-1}^2 \leqslant 1 - x_n^2} dx_1 dx_2 \ldots dx_{n-1} = v_{n-1} r^{n-1} = v_{n-1}(1 - x_n^2)^{(n-1)/2}.$$

For the volume of a single n-dimensional ball, we obtain

$$v_n = v_{n-1} \int_{-1}^{1} (1 - x_n^2)^{(n-1)/2} dx_n.$$

Change of the variable $x_n = \cos \varphi$ results in

$$v_n = 2v_{n-1} \int_0^{\pi/2} \sin^n \varphi \, d\varphi.$$

The value of the integral $\int\limits_0^{\pi/2} \sin^n \varphi \, d\varphi$ for $n = 1, 2, \ldots$ is

$$\int_0^{\pi/2} \sin^n \varphi \, d\varphi = \frac{\sqrt{\pi}}{2} \frac{\Gamma((n+1)/2)}{\Gamma((n+2)/2)},$$

where $\gamma(x)$—gamma function (see Exercises **8.128** and **8.129**). Taking into account that the volume of a one-dimensional ball of a unit radius is equal to $v_1 = \int_{-1}^{1} dx_1 = 2$, write the recurrence relation for v_n:

$$\begin{cases} v_n = \sqrt{\pi} \, \dfrac{\Gamma((n+1)/2)}{\Gamma((n+2)/2)} \, v_{n-1}, & n > 1, \\ v_1 = 2. \end{cases}$$

The resulting relation is easily solved by the method of substitution:

$$v_n = \frac{\pi^{n/2}}{\Gamma(n/2 + 1)}, \quad n \in \mathbb{N}.$$

Note that as the dimensionality of the space n grows, the volume of the ball of a unit radius tends to zero. Moreover, since the following equality is valid $\lim\limits_{n \to \infty} \dfrac{r^n - (r - \varepsilon r)^n}{r^n} = 1$ for any arbitrary small positive $\varepsilon < 1$, almost the entire volume of the ball of radius r is concentrated within a thin spherical layer of width εr, adjacent to the ball's surface.

8.135. *Answer*:

Recurrence relation for the volume of the n-dimensional simplex v_n has the form

$$\begin{cases} v_n = \dfrac{1}{n} \, v_{n-1}, & n > 1, \\ v_1 = 1. \end{cases}$$

Its solution is $v_n = (n!)^{-1}, n \in \mathbb{N}$.

8.136. *Answer*: $a_n = \lceil \beta^{2^n} \rceil$, where $\beta = 1,678\,458\,965\ldots, n = 0, 1, 2, \ldots$

8.137. *Solution*.

Changing the values a_n in different ways (see below), we bring the original recurrence relation to a simpler form. Depending on the value of u, we will need different changes a_n. Let us further consider three cases: 1) $u > 2$, 2) $|u| \leqslant 2$, and 3) $u < -2$.

(1) Let the inequality be valid $u > 2$. Introduce a new numerical sequence $\{b_n\}$, defined in accordance with the relation $a_n = 2 \cosh b_n$ for $n = 0, 1, 2, \ldots$, where $\cosh z = (e^z + e^{-z})/2$ is a hyperbolic cosine.

Substitution of the values b_n into the equation $a_n = a_{n-1}^2 - 2$ subject to the valid for all real z identity $\cosh 2z = 2 \cosh^2 z - 1$ results in the equality

$$\cosh b_n = \cosh 2b_{n-1}, \text{ where } n \geqslant 0.$$

Hence, either $b_n = 2b_{n-1}$ and $b_n = 2^n b_0$ or $b_n = -2b_{n-1}$ and $b_n = (-2)^n b_0$ for all positive integer n.

The condition $a_n = 2 \cosh b_n$ in each of the considered cases results in the dependence of the form $a_n = 2 \cosh(2^n b_0)$, where b_0 is determined from the relation $2 \cosh b_0 = u$. This allows writing the explicit form for the values a_n:

$$a_n = 2 \cosh \left(2^n \text{ Arch } \frac{u}{2}\right), \quad n \geqslant 0.$$

Here $\text{Arch } z = \ln(z + \sqrt{z^2 - 1})$ is a positive branch of the inverse hyperbolic cosine.

(2) Let the inequality be valid $|u| \leqslant 2$. Make a change $a_n = 2 \cos b_n$, $n \geqslant 0$. Applying the double argument cosine formula, we come to the relation $\cos b_n = \cos(2b_{n-1})$. From this equality follows that

$$b_n = \sigma_{n-1} \cdot 2b_n + 2\pi k_{n-1},$$

where $\sigma_{n-1} = \pm 1$, $k_{n-1} \in \mathbb{Z}$ for all $n = 1, 2, 3, \ldots$ By the method of substitution, we obtain the explicit expression of the values b_n:

$$b_n = \sigma_{n-1}\sigma_{n-2} \ldots \sigma_0 \cdot 2^n b_0 +$$
$$+ 2\pi(\sigma_n\sigma_{n-1} \ldots \sigma_1 \cdot 2^{n-1} k_0 + \sigma_n\sigma_{n-1} \ldots \sigma_2 \cdot 2^{n-2} k_1 + \ldots +$$
$$+ \sigma_n \cdot 2k_{n-2} + k_{n-1}) = \pm 2^n b_0 + 2\pi k', \text{ where } k' \in \mathbb{Z}.$$

Return to the sequence $\{a_n\}$:

$$a_n = 2 \cos b_n = 2 \cos(2^n b_0) = 2 \cos \left(2^n \arccos \frac{u}{2}\right), \quad n \geqslant 0.$$

(3) Finally, consider the case $u < -2$. Since $a_1 = a_0^2 - 2 = u^2 - 2 > 2$, this case reduces to the already examined item 1), if, as the initial term, we consecutively take $u^2 - 2$:

$$a_n = 2 \cosh \left(2^{n-1} \text{ Arch } \frac{u^2 - 2}{2}\right), \quad n = 1, 2, 3 \ldots$$

The analytical expressions for the values a_n, obtained in items 1)–3), can be united into a single formula for arbitrary values of u. Using the equalities $\cosh z = (e^z + e^{-z})/2$ and $\cos z = (e^{iz} + e^{-iz})/2$, valid for all $z \in \mathbb{R}$, the final answer is represented in the form

$$a_n = \left(\frac{u + \sqrt{u^2 - 4}}{2}\right)^{2^n} + \left(\frac{u - \sqrt{u^2 - 4}}{2}\right)^{2^n}$$

where $u \in \mathbb{R}$, $n = 0, 1, 2, \ldots$

It is important to note that in the considered expression for a_n the square root is understood as a function of a complex argument (see p. 248). The formula that determines the values a_n remains correct even for complex values of the parameter u.

Concept of an Algorithm. Correctness of Algorithms

9

Algorithm is an exact prescription determining the computation process, leading from the varying source data to the sough result (data is the ordered set of characters) [47]. In other words, an algorithm describes the certain computation procedure, with the help of which the computation problem is solved. As a rule, and algorithm is used for solving some class of problems, rather than one certain problem [13,73]. The term "algorithm" derives from the name of a medieval mathematician al-Khwarizmi.[1]

Concept of an algorithm belongs to the basic, fundamental notions of mathematics. Many researchers use different wordings for the definition of an algorithm. However, all wordings explicitly or implicitly mean the following **algorithm properties** [47].

1. *Discreteness*. An algorithm shall represent a problem-solving process as a consecutive execution of separate steps. Execution of each step of an algorithm requires some time, and each operation is performed only in full and cannot be performed partially.
2. *Elementary character of steps*. The method of executing each command should be known and simple enough.
3. *Determinateness* (from Latin *dētermināre*). Each subsequent step of the algorithm operation is uniquely defined. The result should be the same for the same source data.
4. *Directedness*. It should be known - what is to be considered as the algorithm operation result.
5. *Mass character*. There should be a possibility to apply the algorithm to all collections of source data from a certain predetermined set.

---

[1] Muḥammad ibn Mūsā al-Khwārizmī (ab. 780 — ab. 850)—Persian mathematician and astronomer.

© Springer International Publishing AG, part of Springer Nature 2018
S. Kurgalin and S. Borzunov, *The Discrete Math Workbook*,
Texts in Computer Science, https://doi.org/10.1007/978-3-319-92645-2_9

Consider the algorithm $\mathcal{A}$ that solves a certain computation problem. In order to use this algorithm in a computer program, we need to justify the correct solution of the problem for all input data; i.e., we should carry out the **proof of algorithm correctness** $\mathcal{A}$. For this, we need to trace all changes in the values of variables resulting from the algorithm operation. From the mathematical point of view, we are dealing with establishing the truth values of some predicates that describe variable values.

Let P be a predicate true for the input data of the algorithm $\mathcal{A}$ and Q be a predicate taking the true value after completion of operation of $\mathcal{A}$. The introduced predicates are called **precondition** and **postcondition**, respectively.

The proposition $\{P\}\mathcal{A}\{Q\}$ means the following: "If operation of the algorithm $\mathcal{A}$ starts with the true value of the predicate P, then it will end with the true value of Q." We obtain that the proof of algorithm correctness $\mathcal{A}$ is equivalent to the proof of truth $\{P\}\mathcal{A}\{Q\}$. The pre- and postcondition together with the algorithm itself are called the **Hoare**[2] **triple**. The Hoare triple describes how the execution of a given computer program fragment changes the state of computation [59].

As an example, consider one of the important algorithms in graph theory—Warshall algorithm, which is used for calculating the reachability matrix of a specified directed graph $D(V, E)$.

The Warshall algorithm is based on forming a sequence of auxiliary logical matrices $W^{(0)}$, $W^{(1)}$, ..., $W^{(N)}$, where $N = |V|$. The first matrix is taken to be equal to the digraph adjacency matrix M. The elements $W_{ij}^{(k)}$, where $1 \leqslant i, j, k \leqslant N$, are calculated by the rule: $W_{ij}^{(k)} = T$ if there exists a path connecting the vertices v_i and v_j such that all the internal vertices belong to the set $V_k = \{v_1, v_2, \ldots, v_k\}$, and $W_{ij}^{(k)} = F$ otherwise. Note that the internal vertex of the path $P = v_i, \ldots, v_l, \ldots, v_j$ is any vertex v_l, $1 \leqslant l \leqslant N$, belonging to P, except the first v_i and the last v_j. The resulting matrix $W^{(N)}$ appears to be equal to $W^{(N)} = M^*$, since $M_{ij}^* = T$ if and only if there exists the path $v_i, \ldots, v_j$, whose all internal vertices are contained in $V = \{v_1, v_2, \ldots, v_N\}$.

The principal moment is that the matrix $W^{(k)}$ can be obtained from $W^{(k-1)}$ as follows. The path $v_i, \ldots, v_j$, containing the internal vertices from the set V_k only exists if and only if one of the conditions is met:

(1) There exists the path $v_i, \ldots, v_j$ with internal vertices from $V_{k-1} = \{v_1, v_2, \ldots, v_{k-1}\}$ only.
(2) There exist paths $v_1, \ldots, v_k$ and $v_k, \ldots, v_j$, also containing the internal vertices from V_{k-1} only.

---

[2]Charles Antony Richard Hoare (born 1934)—English scientist specializing in informatics and computing.

We obtain two cases: either $W_{ij}^{(k-1)} = T$, if v_k is included in the set of vertices allowed at this stage, or $W_{ik}^{(k-1)} = T$ and $W_{kj}^{(k-1)} = T$. Hence,

$$W_{ij}^{(k)} = W_{ij}^{(k-1)} \text{ or } \left(W_{ik}^{(k-1)} \text{ and } W_{kj}^{(k-1)}\right).$$

Let us show a respective algorithm for constructing M^* over the defined adjacency matrix M of size $N \times N$, where $N > 1$. The intermediate matrices $W^{(k)}$, where $0 \leqslant k \leqslant N - 1$, should not necessarily be stored in memory until the end of the algorithm operation; this is why, in the suggested realization, the elements $W^{(k-1)}$ are replaced by the elements of the subsequent matrix $W^{(k)}$.

```
1   const N=3;
2   type Matr = array[1..N,1..N] of boolean;
3   procedure WarshallAlgo(M:Matr; var W:Matr);
4   //calculation of the digraph reachability matrix
5   var i,j,k:integer;
6   begin
7       for i:=1 to N do
8           for j:=1 to N do
9               W[i,j]:= M[i,j];
10      for k:=1 to N do
11          for i:=1 to N do
12              for j:=1 to N do
13                  W[i,j]:= W[i,j] or W[i,k] and W[k,j];
14  end;
```

The correctness of the Warshall algorithm can be proved by the mathematical induction method [62]. Other improvements of the problem of finding M^* see in Exercise **9.15** and in [62].

Example 9.1 Let the digraph be defined D (Fig. 9.1). Construct the reachability matrix M^*, using the Warshall algorithm.
Solution.
The matrix $W^{(0)}$ coincides with the digraph adjacency matrix and has the form

$$W^{(0)} = \begin{array}{c} \\ a \\ b \\ c \\ d \end{array} \begin{array}{c} \begin{array}{cccc} a & b & c & d \end{array} \\ \left[\begin{array}{cccc} F & T & F & F \\ T & F & T & F \\ F & F & F & F \\ T & F & T & F \end{array}\right] \end{array}.$$

Calculate $W^{(1)}$. If $W_{ij}^{(0)} = T$, then the respective element $W_{ij}^{(1)}$ is also equal to T: $W_{ij}^{(1)} = T$. If $W_{ij}^{(0)} = F$, then attention should be paid to the elements of the first raw and the first column, standing at the intersection with the jth column and the ith row:

Fig. 9.1 Directed graph D

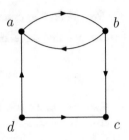

If $W_{1j}^{(0)} = W_{i1}^{(0)} = T$, then $W_{ij}^{(1)} = T$. The condition $W_{1j}^{(0)} = W_{i1}^{(0)} = T$ is fulfilled for two pairs (i, j), namely for $i = j = 2$ and $i = 4$, $j = 2$. Hence, $W_{22}^{(1)} = W_{42}^{(1)} = T$, and all the remaining elements $W^{(1)}$ coincide with the respective elements of the matrix $W^{(0)}$. For illustration purposes, in the notation of the matrix, let us highlight in bold and underline the elements $W^{(1)}$ that have changed their values at this step:

$$
W^{(1)} = \begin{array}{c} \\ a \\ b \\ c \\ d \end{array}
\begin{array}{c} \begin{array}{cccc} a & b & c & d \end{array} \\
\left[\begin{array}{cccc}
F & T & F & F \\
T & \mathbf{\underline{T}} & T & F \\
F & F & F & F \\
T & \mathbf{\underline{T}} & T & F
\end{array} \right]
\end{array}.
$$

Further, calculate $W^{(2)}$. Consider the second row and the second column of the matrix $W^{(1)}$. The elements $W^{(1)}$ that are located in the same row with the elements $W_{i2}^{(1)} = T$ from the second column and in the same row with the elements $W_{2j}^{(1)} = T$ from the second row will change their values in $W^{(1)}$ for T. Such are the elements $W_{11}^{(1)}$ and $W_{13}^{(1)}$. The rest of the elements $W^{(2)}$ coincide with the respective elements of the matrix $W^{(1)}$.

$$
W^{(2)} = \begin{array}{c} \\ a \\ b \\ c \\ d \end{array}
\begin{array}{c} \begin{array}{cccc} a & b & c & d \end{array} \\
\left[\begin{array}{cccc}
\mathbf{T} & T & \mathbf{T} & F \\
T & T & T & F \\
F & F & F & F \\
T & T & T & F
\end{array} \right]
\end{array}.
$$

At the next step, the vertex c is added into the set of possible vertices. This does not result in new elements with value T.

$$
W^{(3)} = \begin{array}{c} \\ a \\ b \\ c \\ d \end{array}
\begin{array}{c} \begin{array}{cccc} a & b & c & d \end{array} \\
\left[\begin{array}{cccc}
T & T & T & F \\
T & T & T & F \\
F & F & F & F \\
T & T & T & F
\end{array} \right]
\end{array}.
$$

At the final step, we obtain $W^{(4)} = W^{(3)}$, and the reachability matrix of the digraph D will be equal to

$$M^* = \begin{array}{c} \\ a \\ b \\ c \\ d \end{array} \begin{array}{cccc} a & b & c & d \\ \left[\begin{array}{cccc} T & T & T & F \\ T & T & T & F \\ F & F & F & F \\ T & T & T & F \end{array}\right] \end{array}.$$

□

9.1 Problems

9.1. Prove the correctness of the algorithm of exchanging two values.

```
1  procedure swap(var a, b:integer);
2  //exchanging values of the variables a and b
3  var
4      temp:integer;
5  begin
6      temp := a;
7      a := b;
8      b := temp;
9  end;
```

9.2. Prove the correctness of the algorithm of exchanging two values without using the auxiliary variable.

```
1  procedure swap2(var a, b:integer);
2  //exchanging values of the variables a and b
3  //without using the auxiliary variable
4  begin
5      a := a + b;
6      b := a - b;
7      a := a - b;
8  end;
```

9.3. Prove the correctness of the algorithm for addition of square matrices.

```
1  const N = 100;
2  type Matr = array[1..N, 1..N] of integer;
3  procedure matrix_add(a, b:Matr; var c:Matr);
4  //addition of matrices a and b
5  var i,j:integer;
6  begin
7      for i := 1 to N do
```

```
 8         for j := 1 to N do
 9             c[i,j] := a[i,j] + b[i,j];
10  end;
```

9.4. Prove the correctness of the algorithm for multiplication of square matrices.

```
 1  const N = 100;
 2  type Matr = array[1..N,1..N] of integer;
 3  procedure matrix_mult(a, b:Matr; var c:Matr);
 4  //multiplication of square matrices a and b
 5  var i,j,k,s:integer;
 6  begin
 7     for i := 1 to N do
 8        for j := 1 to N do
 9           begin
10              s := 0;
11              for k := 1 to N do
12                 s := s + a[i,k] * b[k,j];
13              c[i,j] := s;
14           end;
15  end;
```

9.5. Prove the correctness of the optimized algorithm for multiplication of matrices.

```
 1  const N=100;
 2  type Matr = array[1..N,1..N] of integer;
 3  procedure matrix_mult2(a, b:Matr; var c:Matr);
 4  //multiplication of square matrices a and b
 5  var i,j,k,s:integer;
 6          d:array[1..N] of integer;
 7  begin
 8     for i := 1 to N do
 9        for j := 1 to N do
10           begin
11              s := 0;
12              for k := 1 to N do d[k]:= b[k,j];
13              for k := 1 to N do
14                 s := s + a[i,k] * d[k];
15              c[i,j] := s;
16           end;
17  end;
```

Note. The considered version of multiplication of matrices is optimized so as to preliminarily choose the elements of the column $b(k, j)$ into the intermediate array d, which can entirely be placed into the rapid cache memory [58].

9.6. Find the number of operations of addition of two numbers, performed by the algorithm of the matrix_add(a,b,c) for matrices of size $N \times N$.

9.7. Find the number of operations of addition and multiplication, performed by the algorithm of the matrix_mult(a,b,c) for matrices of size $N \times N$.

*✱**9.8.** Suggest a method to decrease the number of addition operations performed by the algorithm of the matrix_mult(a,b,c).

9.9. Let A_1, A_2, and A_3 be numerical matrices of size 50×25, 25×30, and 30×10, respectively. Find the minimal number of multiplication operations required to calculate the product $A_1 A_2 A_3$ by a standard algorithm of the matrix_mult (whose realization for square matrices is presented in Exercise **9.4**).

9.10. Let A_1, A_2, A_3, and A_4 be numerical matrices of size $25 \times 10, 10 \times 50, 50 \times 5$, and 5×30, respectively. Find the minimal number of multiplication operations required to calculate the product $A_1 A_2 A_3 A_4$ by a standard algorithm of the matrix_mult.

9.11. Let A_1, A_2, A_3, and A_4 be numerical matrices of size 100×20, 20×15, 15×50, and 50×100, respectively. Find the minimal number of multiplication operations required to calculate the product $A_1 A_2 A_3 A_4$ by a standard algorithm of the matrix_mult.

9.12. Prove that the number of methods for calculating the product of the matrices $A_1 A_2 \ldots A_{m+1}, m \geqslant 1$, or, in other words, the number of ways to arrange the brackets in this product, where A_1, A_2, ..., A_{m+1} are numerical matrices of size $n_1 \times n_2, n_2 \times n_3, \ldots, n_{m+1} \times n_{m+2}$, respectively, is equal to the Catalan number C_m (the determining relation for the Catalan sequence and the explicit expression for C_m see in Exercises **8.110** and **8.112**).

9.13. Estimate the number of operations performed by the Warshall algorithm for obtaining the digraph reachability matrix.

9.14. Using the Warshall algorithm, calculate reachability matrix of the digraph D, presented in Fig. 5.10.

9.15. One of the ways to modify the Warshall algorithm consists in representation of the rows of the logical matrices as bit strings. In this case, for calculation of the elements of the matrices $W^{(k)}$ for $1 \leqslant i \leqslant N$ a bitwise operation **or** is used. Find the number of operations in the bit strings, performed by this realization of the Warshall algorithm.

✱ **9.16.** Write the analytical expression for the function f(N).

```
1  function  f(N:integer):integer;
2  var  i,j,k,temp:integer;
3  begin
```

```
 4 |    temp := 0;
 5 |    for  i := 1 to N do
 6 |        for  j := 1 to N do
 7 |            for  k := j to N do
 8 |                temp := temp + 1;
 9 |    result := temp;
10 | end;
```

* **9.17.** Write the analytical expression for the function g(N).

```
 1 | function  g(N:integer):integer;
 2 | var  i,j,k,temp:integer;
 3 | begin
 4 |    temp := 0;
 5 |    for  i := 1 to N do
 6 |        for  j := N downto i do
 7 |            for  k := 1 to j do
 8 |                temp := temp + 1;
 9 |    result := temp;
10 | end;
```

* **9.18.** Write the analytical expression for the function h(N).

```
 1 | function  h(N:integer):integer;
 2 | var  i,j,k,l,temp:integer;
 3 | begin
 4 |    temp := 0;
 5 |    for  i := 1 to N do
 6 |        for  j := 1 to i do
 7 |            for  k := 1 to j do
 8 |                for  l := 1 to k do
 9 |                    temp := temp + 1;
10 |    result := temp;
11 | end;
```

9.2 Answers, Hints, Solutions

9.1. *Solution.*

Let the variables a and b take the following values: $a = a_0$, $b = b_0$.

Precondition: $P = \{a = a_0, b = b_0\}$, postcondition: $Q = \{a = b_0, b = a_0\}$.

Substitute the values of the variables a and b into the body of the algorithm $\mathcal{A}$, which will result in the following values: $temp = a_0, a = b_0, b = a_0$. Therefore, the predicate $\{P\}\mathcal{A}\{Q\}$ takes the true value, and thereby we have proved the correctness of the algorithm $swap(a, b)$.

9.6. *Answer*: In order to calculate the sum of two matrices A and B, we will need N^2 additions to determine each of N^2 elements of $A + B$.

9.7. *Solution.*
Each of N^2 elements of the matrix $A \cdot B$ is calculated as a scalar product of two vectors of size N and, respectively, requires N additions and N multiplications. The total number of both additions and multiplications turns out to be equal to $N \cdot N^2 = N^3$.

9.8. *Solution.*
The body of cycle over the variable k can be rewritten in the form

$$c[i, j] := a[i, 1] * b[1, j];$$
$$\textbf{for} k := 2 \textbf{ to } N \textbf{ do}$$
$$c[i, j] := c[i, j] + a[i, k] * b[k, j];$$

The number of additions will decrease down to $N^3 - N^2$, and the number of multiplications will remain unchanged.

9.9. *Solution.*
As is known, the number of multiplication operations required to calculate the product of matrices of size $n_1 \times n_2$ and $n_2 \times n_3$ is $n_1 n_2 n_3$. (Remember that the product of two matrices is defined if the number of columns of the first one coincides with the number of rows of the second one.) Due to commutativity of the multiplication operation, the product $A_1 A_2 A_3$ can be calculated in two ways: $(A_1 A_2) A_3$ and $A_1 (A_2 A_3)$.

In the first case, we will need $50 \cdot 25 \cdot 30 + 50 \cdot 30 \cdot 10 = 52\,500$ multiplications and in the second $25 \cdot 30 \cdot 10 + 50 \cdot 25 \cdot 10 = 20\,000$. So, the minimal number of multiplication operations required to calculate the elements of the matrix $A_1 A_2 A_3$ by a standard algorithm is $20\,000$.
Note. There exists an efficient algorithm for finding the order of multiplications in the product $A_1 A_2 \ldots A_n, n > 2$, with the minimal number of operations [13].

9.10. *Answer*: $7\,500$.

9.11. *Answer*: $255\,000$.

9.13. *Solution.*
In order to calculate $W[i, k]$ in the row with number 13 (see algorithm on p. 330), two logical operations are required. Since this row is executed $N \times N \times N = N^3$ times, where N is the size of the digraph adjacency matrix, then the full number of operations for obtaining M^* is $2N^3$.

9.14. *Answer*: See answer to Exercise **5.62**.

9.16. *Answer*: $f(N) = \dfrac{N^2(N + 1)}{2}$.

9.17. *Answer*: $g(N) = \dfrac{N(N+1)(2N+1)}{6}$.

9.18. *Answer*: $h(N) = \dfrac{N(N+1)(N+2)(N+3)}{24}$.

Turing Machine

10

In order to formalize the concept of an algorithm, Turing[1] suggested a mathematical model of a computing machine [66]. **Turing machine** is an abstract-computing device consisting of a tape, a reading and printing head and a control device.

Let us see some elements of a Turing machine in more detail.

1. **Tape** is divided into similar cells and is supposed to be unlimited on both sides. In other words, the number of cells in each certain moment of time is finite, but when necessary their number can be increased.

In each cell, at the current moment of time, one symbol is written out of the finite set $\Sigma = \{a_0, a_1, \ldots, a_m\}$, $m \geqslant 1$. This set is referred to as an **alphabet**, or, to be more exact, the **external alphabet**, and its elements are referred to as **symbols**. One of the elements of the set Σ, for example a_0, is called a **blank** symbol.

2. **Reading and writing head** moves along the tape and at each moment of time observes one cell of the tape. At this moment, the head reads the content of this cell and writes into it some symbol from the alphabet Σ.

3. **Control device** at each moment of time is in some state q_i; the set of all states $Q = \{q_0, q_1, \ldots, q_n\}$, $n \geqslant 1$, is called the **internal alphabet**. Among all q_i, $0 \leqslant i \leqslant n$, initial and final states are distinguished. It is usually assumed that there exists one initial state, q_1, and one final state, q_0. At the beginning of operation, the Turing machine T is in state q_1 after passing to the final state q_0 the machine T finishes its operation.

The elements of a Turing machine are presented in Fig. 10.1.

Depending on the current state and symbol, observed by the head, the control device changes its inner state (or remains in the previous state), gives an order to the head to print some symbol in the cell, and moves the head one cell right, left or commands to stay put.

---

[1] Alan Mathison Turing (1912–1954)—English mathematician and logic.

© Springer International Publishing AG, part of Springer Nature 2018
S. Kurgalin and S. Borzunov, *The Discrete Math Workbook*,
Texts in Computer Science, https://doi.org/10.1007/978-3-319-92645-2_10

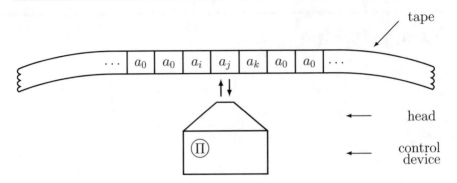

Fig. 10.1 Elements of the Turing machine

Table 10.1 Programs of the Turing machine

Π	q_1	...	q_i	...	q_n
a_0	$q_{10}a_{10}d_{10}$		$q_{i0}a_{i0}d_{i0}$		$q_{n0}a_{n0}d_{n0}$
...					
a_j	$q_{1j}a_{1j}d_{1j}$		$q_{ij}a_{ij}d_{ij}$		$q_{nj}a_{nj}d_{nj}$
...					
a_m	$q_{1m}a_{1m}d_{1m}$		$q_{im}a_{im}d_{im}$		$q_{nm}a_{nm}d_{nm}$

The operation of the control device is described by the list of fives of the form

$$q_i a_j q_{ij} a_{ij} d_{ij}, \quad q_i, q_{ij} \in Q, \ a_j, a_{ij} \in \Sigma, \ d_{ij} \in \{L, R, S\},$$

which are called **commands**.

Record $q_i a_j q_{ij} a_{ij} d_{ij}$ or $q_i a_j \rightarrow q_{ij} a_{ij} d_{ij}$ means the following. If the Turing machine T is in state q_i and observes a cell where symbol a_j is written, then the control device changes state for q_{ij}, places symbol $a_{ij} \in \Sigma$ in the current cell, and commands to the reading head to move one cell left ($d_{ij} = L$), one cell right ($d_{ij} = R$), or stay put ($d_{ij} = S$). The list of all commands is called a **program** of the Turing machine and is denoted by Π. The program is conveniently represented in the form of a table (Table 10.1).

In its upper row are enumerated all states in which the machine T can be, on the left—all symbols that can be on the tape. At the intersections of row and columns, in the table cells, respective commands are specified. Generally speaking, not for each pair (q_i, a_j) a command $q_i a_j \rightarrow q_{ij} a_{ij} d_{ij}$ is defined. If the command meant for the pair (q_i, a_j) is absent, then at the intersection of the column q_i and the row a_j appears the symbol "–". In this case, the Turing machine, being in the state q_i and observing the symbol a_j, finishes its operation.

Example 10.1 Consider the Turing machine T with $n + 1$ states, where $n \geqslant 1$, exactly one of which, q_0, is finite. Find the number of programs for the T, differing by at least one command.

Solution.

In Table 10.1, a program Π is presented for some Turing machine with the number of states equal to $n + 1$. An arbitrary command of the machine T has the form:

$$q_i a_j \rightarrow q_{ij} a_{ij} d_{ij}, \quad 1 \leqslant i \leqslant n, \ 0 \leqslant j \leqslant |\Sigma| - 1,$$

where $|\Sigma|$ is cardinality of the external alphabet. In accordance with the rule of product, the program Π contains $n|\Sigma|$ commands. In each of them, one of the regular states $q_{ij} \in \{q_0, q_1, \ldots, q_n\}$ is written, as well as some symbol of the external alphabet a_{ij} and an instruction for the next position of the head $d_{ij} \in \{L, R, S\}$, which makes $3(n + 1)|\Sigma|$ different options. Moreover, on the place of a command in the table there can be a dash, which corresponds to no command. Hence, the number of programs for the T is equal to the number of permutations with repetitions of $n|\Sigma|$ elements out of $3(n + 1)|\Sigma| + 1$ or $(3(n + 1)|\Sigma| + 1)^{n|\Sigma|}$.   □

If, at some moment of time, the leftmost nonblank cell of the tape contains the symbol a_i, and the rightmost nonblank cell contains the symbol a_s, then it is said that a **word** is written on the tape. $P = a_i \ldots a_k a_l \ldots a_s$.

Further, if at a given moment of time the control device is in the state q_t and the reading head observes the symbol a_l of the word P, then the sequence $a_i \ldots a_k q_t a_l \ldots a_s$ is called the **configuration** of the machine.

If P is the source word on the tape, then in the process of the Turing machine's operation with the set program Π there appear two possibilities.

1. The machine T stops after the finite number of steps, and the word $P' = T(P)$ appears on the tape. Then, it is said that the machine T **is applicable** to the word P.
2. The machine T never stops ("caught in a loop"). In this case, it is said that T **is not applicable** to the word P.

Example 10.2 Consider the Turing machine T, which has the following states: q_1 (initial), and q_2, q_3, and q_0 (final). Let T operate with the symbols of the external alphabet $\mathbb{B} = \{0, 1\}$ in accordance with the program Π

Π	q_1	q_2	q_3
0	$q_2 0L$	$q_1 0R$	$q_0 1L$
1	$q_2 1R$	$q_3 1R$	$q_1 0R$

Find out whether the machine T is applicable to the words $P_1 = 11$ and $P_2 = 111$. Assume that the reading and writing head at the initial moment of time observes the leftmost cell of the word P_i, $i = 1, 2$.

Solution.

Since the T is initially in the state q_1 and observes the cell with the symbol "1", then execution of the first command $q_1 1 \rightarrow q_2 1 R$ will result in a record of the symbol "1" into the current cell, passing of the T into the state q_2 and moving of the head one cell right. After that, the next command $q_2 1 \rightarrow q_3 1 R$ will be executed, then the command $q_3 0 \rightarrow q_0 1 L$. Write the sequence being formed as the configuration of the Turing machine T:

$$q_1 11, \quad 1 q_2 1, \quad 11 q_3 0, \quad 1 q_0 11.$$

Since the state q_0 is finite, then the machine T upon reaching the configuration $1 q_0 11$ will finish its operation. Hence, the Turing machine T is applicable to the word $P_1 = 11$.

Further, consider the tape with the word P_2 written on it. In this case, the sequence of configurations will have the form:

$$q_1 111, \quad 1 q_2 11, \quad 11 q_3 1, \quad 110 q_1 0, \quad 11 q_2 00, \quad 110 q_1 0, \quad 11 q_2 00, \quad \ldots$$

It is clear that two configurations, namely $110 q_1 0$ and $11 q_2 00$, pass one into the other, and in a finite number of steps the machine T will not finish its operation. Due to this, the Turing machine T is not applicable to the word $P_2 = 111$. □

The choice of information presentation on the tape is conditioned on the specific character of the problem. However, representation of numbers in binary (and especially decimal) system usually results in cumbersome programs. This is why the natural number k is often coded by a set of $k + 1$ units (**unary system**); such a sequence is denoted by 1^{k+1}. The number 0 is coded by one symbol "1". The sequence of several numbers is coded as follows.

Let $\boldsymbol{\alpha} = (\alpha_1, \alpha_2, \ldots, \alpha_n)$, and $n \geqslant 1$, —some vector of nonnegative integers. The word

$$C(\boldsymbol{\alpha}) = 1^{\alpha_1+1} 0 1^{\alpha_2+1} 0 \ldots 0 1^{\alpha_n+1}$$

is called a **code** of the vector $\boldsymbol{\alpha}$ in the alphabet $\{0, 1\}$. In other words, a code of some vector consists of the codes of its components, divided by symbol "0".

Let us consider the functions of the form $f(\mathbf{x}) = f(x_1, x_2, \ldots, x_n)$, $n \geqslant 1$, where the arguments x_i, $i = 1, 2, \ldots, n$, take values from $\mathbb{Z}_0 = \{0\} \cup \mathbb{N}$:

$$f : \underbrace{\mathbb{Z}_0 \times \mathbb{Z}_0 \times \ldots \times \mathbb{Z}_0}_{n \text{ terms}} \rightarrow \mathbb{Z}_0.$$

The Turing machine T_f **calculates the function** $f(\mathbf{x})$ if two properties are fulfilled:

(1) On condition that $f(\mathbf{x})$ is defined, the equality is valid

$$T_f(C(\mathbf{x})) = C(f(\mathbf{x}));$$

(2) On condition that $f(\mathbf{x})$ is not defined, either $T_f(C(\mathbf{x}))$ is not a code of any number out of $\mathbb{Z}_0$ or the machine T_f is not applicable to the word $C(\mathbf{x})$.

The Turing machine T_f **calculates the function** $f(\mathbf{x})$ **correctly** if the following properties are fulfilled:

(1) T_f calculates the function $f(\mathbf{x})$.
(2) If $f(\mathbf{x})$ is not defined, the machine T_f is not applicable to the word $C(\mathbf{x})$.
(3) At the initial moment of time, the reading head observes the leftmost cell of the word $C(\mathbf{x})$ after the stop observes the leftmost cell of $C(f(\mathbf{x}))$.

Church[2]–Turing thesis: *Any algorithm can be realized with the help of some Turing machine.* As experience shows, any computational operations that can be performed by a man can be represented by a sequence of actions of some Turing machine [59,60]. The Church–Turing thesis cannot be proved, since it establishes equivalence between the strictly defined class of problems solved by the Turing machine and the informal concept of an algorithm.

The Turing machine is not the only way to formalize the concept of an algorithm. Similar approaches are known, leading to equivalent results, among which we should mention the **Post machine** [75] and the **Markov[3]** [47] **normal algorithm**. The Post machine is an abstract-computing machine whose main difference from the Turing machine consists in a different scheme of commands. The normal algorithm describes the method of transformation of symbol strings of some alphabet Σ with the help of the **substitution formulae** of the form $U \rightarrow V$ or $U \rightarrow \bullet V$, where U and V are two arbitrary words in Σ, called the left and the right sides of the substitution formula, respectively, and symbol "$\bullet$" indicates the completion of the algorithm operation.

10.1 Problems

10.1. Let the Turing machine T be defined by the program Π.

Π	q_1	q_2
0	$q_2 0 R$	$q_1 0 R$
1	$q_2 1 R$	–

Find whether the machine T is applicable to the word P. If yes, then determine the final configuration. The head at the initial moment of time observes the leftmost unit of the word P. Consider the cases:

---

[2] Alonzo Church (1903–1995)—American mathematician and logician.
[3] Andrei Andreevich Markov-jr. (1903–1979)—Soviet mathematician and logician.

(1) $P = 1^3$;
(2) $P = 1001$;
(3) $P = 10001$;
(4) $P = 1^4 0 1^4$.

10.2. Let the Turing machine T be defined by the program Π.

Π	q_1	q_2	q_3
0	$q_0 1 S$	$q_1 1 R$	$q_0 1 L$
1	$q_2 1 R$	$q_3 1 R$	$q_3 1 S$

Find whether the machine T is applicable to the word P. If yes, then determine the final configuration. The head at the initial moment of time observes the leftmost unit of the word P. Consider the cases:

(1) $P = 1^2$;
(2) $P = 1^3$;
(3) $P = 1011$;
(4) $P = 1001$.

10.3. Construct in the alphabet $\mathbb{B} = \{0, 1\}$ the Turing machine, which is applicable to any word.

10.4. Construct in the alphabet $\mathbb{B} = \{0, 1\}$ the Turing machine, which is not applicable to any word out of $\mathbb{B}$.

10.5. Construct the Turing machine, which:

(1) Calculates the function $O(x)$, equal to zero for all nonnegative integer values of the argument: $O(x) = 0 \ \forall x \in \mathbb{Z}_0$, the machine head at the initial moment of time is positioned above the leftmost unit of the argument's unary code;
(2) Calculates the function correctly $O(x)$.

10.6. Construct the Turing machine, which calculates the function $f(x) = x + 1$. The machine head at the initial moment of time is positioned left of the argument's unary code.

10.7. Construct the Turing machine, which calculates the function $f(x, y) = x + y$. The machine head at the initial moment of time is positioned left of the argument vector's unary code.

10.8. Construct the Turing machine, which calculates the function correctly

$$f(x) = \begin{cases} 0, & \text{if } x - \text{even}, \\ 1, & \text{if } x - \text{odd}. \end{cases}$$

10.9. Construct the Turing machine, which calculates the function correctly

$$f(x) = \begin{cases} 0, & \text{if } x \text{ is divisible by 3,} \\ 1, & \text{if } x \text{ is not divisible by 3.} \end{cases}$$

10.10. Construct the Turing machine, which calculates the function correctly

$$f(x) = \begin{cases} x - 2, \text{ if } x \geqslant 2, \\ \text{not defined, if } x = 0 \text{ or } x = 1. \end{cases}$$

10.11. Construct the Turing machine, which calculates the function correctly

$$f(x) = \text{sign}(x) = \begin{cases} 1, & \text{if } x \neq 0, \\ 0, & \text{if } x = 0. \end{cases}$$

10.12. Construct the Turing machine, which calculates the function correctly

$$f(x) = \overline{\text{sign}}(x) = \begin{cases} 0, & \text{if } x \neq 0, \\ 1, & \text{if } x = 0. \end{cases}$$

10.13. Construct the Turing machine, which calculates the function $f(x) = \lfloor x/2 \rfloor$. The machine head is located left of the word that codes the argument x.

10.14. Construct the Turing machine, which calculates the function $f(x) = \lceil x/2 \rceil$. The machine head is located left of the word that codes the argument x.

10.15. Construct the Turing machine, which calculates the function correctly $f(x) = 2x$.

∗10.16. In the alphabet $A_{10} = \{a_0, 0, 1, 2, 3, 4, 5, 6, 7, 8, 9\}$, construct the Turing machine, correctly calculating the function $f(\alpha) = \alpha + 5$, which is defined on the set of nonnegative integers. The function $f(\alpha)$ relates to the decimal notation $\alpha_1 \alpha_2 \ldots \alpha_n$ of the number α the decimal notation of the number $\alpha + 5$.

∗10.17. In the alphabet $A_{10} = \{a_0, 0, 1, 2, 3, 4, 5, 6, 7, 8, 9\}$, construct the Turing machine, correctly calculating the function $f(\alpha) = 5\alpha$, which is defined on the set of nonnegative integers. The function $f(\alpha)$ relates to the decimal notation $\alpha_1 \alpha_2 \ldots \alpha_n$ of the number α the decimal notation of the number 5α.

10.2 Answers, Hints, Solutions

10.1. *Answer*:

(1) Applicable, the final configuration is $1q_2 11$;

(2) Applicable, the final configuration is $100q_2 1$;
(3) Not applicable;
(4) Applicable, the final configuration is $1q_2 1^3 01^4$.

10.2. *Answer*:

(1) Applicable, the final configuration is $1q_0 1^2$;
(2) Not applicable;
(3) Applicable, the final configuration is $1^3 q_0 1^2$;
(4) Applicable, the final configuration is $1^2 q_0 1^2$.

10.3. *Solution*.

It is clear that there exists an infinite number of programs, satisfying the problem statement. We can, for example, suggest a program, which stops the machine at the first symbol read:

Π	q_1
0	$q_0 1 S$
1	$q_0 1 S$

10.4. *Answer*:

The required property is satisfied, for example, by a machine with the program

Π	q_1
0	$q_1 0 R$
1	$q_1 1 R$

10.5. *Solution*.

(1) The required result can be obtained if we move the reading and writing head right, deleting all units, and in the last operation write "1" and stop. We obtain the following program Π:

Π	q_1
0	$q_0 1 S$
1	$q_1 0 R$

(2) The Turing machine, operating in accordance with the program Π from the previous item, calculates $O(x)$ correctly.

10.7. *Solution*.

On the tape, the input data will be written in the form of the word $1^{x+1} 0 1^{y+1}$. The representation of the number $x + y$, namely 1^{x+y+1}, can be obtained as follows:

Delete the first unit, change "0" between the codes of the arguments x and y for "1", and, in the resulting sequence, delete one more unit.

Π	q_1	q_2	q_3	q_4
0	$q_1 0 R$	$q_3 1 L$	$q_4 0 R$	–
1	$q_2 0 R$	$q_2 1 R$	$q_3 1 L$	$q_0 0 R$

10.8. *Solution.*

At the initial moment of time, the head is at the leftmost unit of the code of the number x. Let us move the head from left to right and put to the place "1" the symbols "0", alternating the states q_1 and q_2.

If after exhaustion of all symbols "1" the control device has passed to the state q_1, then the source number is odd, and the word "11" should be placed on the tape. If the device is in the state q_2, then out the symbol "1" on the tape, and this is where the machine finishes its operation. The program Π of the Turing machine, correctly calculating $f(x)$, is shown below.

Π	q_1	q_2
0	$q_2 1 L$	$q_0 1 S$
1	$q_2 0 R$	$q_1 0 R$

10.9. *Hint.* See previous exercise.

10.10. *Answer:*

Π	q_1	q_2	q_3
0	–	$q_2 0 S$	$q_3 0 S$
1	$q_2 0 R$	$q_3 0 R$	$q_0 1 S$

10.11. *Solution.*

Let us construct the Turing machine in compliance with the following considerations. If the word "1" is written on the tape, then it should be left unchanged, for which purpose we introduce the commands

$$q_1 1 \to q_2 1 R, \quad q_2 0 \to q_0 0 L.$$

Yet, if the word "1^n", $n \geqslant 2$, is written on the tape, then all units up to the end of the word should be deleted, except the very first; then, we should return to the first unit and place one more symbol "1", in order to obtain "1^2".

Let us introduce two more states: q_3 and q_4. Being in the state q_3, the machine moves the head up to the end of the word, encounters "0", passes to the state q_4, and returns to the beginning of the source word. The command $q_1 0 \to q_0 1 S$ writes the last unit and stops the machine. The full program of the Turing machine, correctly calculating the function $\text{sign}(x)$, is shown in the table:

Π	q_1	q_2	q_3	q_4
0	$q_0 1S$	$q_0 0L$	$q_4 0L$	$q_4 0L$
1	$q_2 1R$	$q_3 0R$	$q_3 0R$	$q_1 1L$

10.12. *Answer*:

Π	q_1	q_2	q_3	q_4
0	–	$q_0 1L$	$q_4 0L$	$q_4 0L$
1	$q_2 1R$	$q_3 0R$	$q_3 0R$	$q_0 1S$

10.13. *Solution.*
By the command $q_1 0 \to q_1 0R$, move the head to the position above the leftmost unit of the code of the number x. Delete two units on the left, move to the end of the word, and through "barrier" in the form of the symbol "0" place a unit in the code of the answer being generated. Return left and repeat the described operations until all units of the argument's code are exhausted.

Π	q_1	q_2	q_3	q_4	q_5	q_6	q_7
0	$q_1 0R$	$q_0 0R$	$q_0 1S$	$q_5 0R$	$q_6 1L$	$q_7 0L$	$q_2 0R$
1	$q_2 1S$	$q_3 0R$	$q_4 0R$	$q_4 1R$	$q_5 1R$	$q_6 1L$	$q_7 1L$

10.14. *Hint.* See previous exercise.

10.15. *Solution.*
The following problem-solving method can be suggested. If the argument x is other than zero, $x \neq 0$, organize a loop over the units that form the argument's code, from the leftmost to the last nut one on the right. Deleting the units, each time additionally write two units into the code of the answer being generated. After exiting the loop, all units of the argument's code will be exhausted, except the leftmost; delete this last unit, and additionally write one unit to the word on the tape.

Organize the respective program Π as follows. At the initial moment of time, the head is at the leftmost unit of the code of the number x. The command $q_1 1 \to q_2 0R$ deletes the leftmost unit and moves the head one position right. If in the observed cell the symbol "0" is written, then $x = 0$, and the operation ends by the command $q_2 0 \to q_0 1S$.

Yet, if in the observed cell the symbol "1" is written, then the control device passes to the state q_3 and moves the head up to the symbol "0", which divides the argument's code and the code of the answer being generated. Then, the head passes through "barrier," and in the end of the answer, by the commands

$$q_4 0 \to q_5 1R, \quad q_5 0 \to q_6 1L$$

we place two units on the tape.

Return the head back to the left, put the control device to the state q_1, and repeat the described operation until in the code located left of the "barrier," one symbol is left "1". Delete it, and place "1" on the place of the barrier; this is accomplished with the help of the states q_8 and q_9.

As a result, on the tape we obtain the code "1^{2n+1}", corresponding to the number $2n$.

Π	q_1	q_2	q_3	q_4	q_5	q_6	q_7	q_8	q_9
0	$q_0 1 S$	$q_0 1 S$	$q_4 0 R$	$q_5 1 R$	$q_6 1 L$	$q_7 0 L$	$q_1 0 R$	$q_5 0 R$	$q_1 0 R$
1	$q_2 0 R$	$q_3 1 R$	$q_3 1 R$	$q_4 1 R$	$q_1 0 R$	$q_6 1 L$	$q_8 1 L$	$q_9 1 L$	$q_9 1 L$

10.16. *Solution.*
At the initial moment of time, the Turing machine T is in the configuration $q_1 \alpha_1 \alpha_2 \ldots \alpha_n$. Move the head right; for this purpose, introduce the commands

$$q_1 k \to q_1 k R, \quad k = 0, \ldots, 9.$$

The machine should stop above the last symbol:

$$q_1 a_0 \to q_2 a_0 L.$$

Now perform addition. If α ends with one of the figures 0, 1, 2, 3, 4, the commands look like:

$$q_2 k \to q_3 (k + 5) L, \quad k = 0, 1, 2, 3, 4.$$

We only have to return the head to the position above the first figure of the number:

$$q_3 k \to q_3 k L, \quad k = 0, \ldots, 9, \quad q_3 a_0 \to q_0 a_0 R.$$

Further, consider the cases, when the last figure α_n of the number is equal to 5, 6, 7, 8, or 9. Since $5 + 5 = 10, 6 + 5 = 11, 7 + 5 = 12, 8 + 5 = 13$, and $9 + 5 = 14$, then write the commands

$$q_2 5 \to q_4 0 L,$$
$$q_2 6 \to q_4 1 L,$$
$$q_2 7 \to q_4 2 L,$$
$$q_2 8 \to q_4 3 L,$$
$$q_2 9 \to q_4 4 L.$$

But in this case it is necessary to add a unit to α_{n-1}:

$$q_4 k \to q_3 (k + 1) L, \ k = 0, \ldots, 8, \quad q_4 9 \to q_4 0 L, \quad q_4 a_0 \to q_0 1 S.$$

The last command provides exit from the formed loop.

The full program of the Turing machine, correctly calculating the function $f(\alpha)$, is shown in the table:

Π	q_1	q_2	q_3	q_4
a_0	q_2a_0L	q_30L	q_0a_0R	q_01S
0	q_10R	q_35L	q_30L	q_31L
1	q_11R	q_36L	q_31L	q_32L
2	q_12R	q_37L	q_32L	q_33L
3	q_13R	q_38L	q_33L	q_34L
4	q_14R	q_39L	q_34L	q_35L
5	q_15R	q_40L	q_35L	q_36L
6	q_16R	q_41L	q_36L	q_37L
7	q_17R	q_42L	q_37L	q_38L
8	q_18R	q_43L	q_38L	q_39L
9	q_19R	q_44L	q_39L	q_40L

10.17. *Answer*:

The following program realizes a standard multiplication algorithm. With the help of the state q_1 the reading and writing head of the Turing machine is moved to the position above the rightmost symbol of decimal code of the number α. The state q_i, where $2 \leqslant i \leqslant 6$, changes the content of the tape's cells in accordance with the table of multiplication by 5. And the states q_3, q_4, q_5, and q_6 are introduced for controlling the previous carry of the decimal notation of the argument α.

Π	q_1	q_2	q_3	q_4	q_5	q_6
a_0	q_2a_0L	q_0a_0R	q_01S	q_02S	q_03S	q_04S
0	q_10R	q_20L	q_21L	q_22L	q_23L	q_24L
1	q_11R	q_25L	q_26L	q_27L	q_28L	q_29L
2	q_12R	q_30L	q_31L	q_32L	q_33L	q_34L
3	q_13R	q_35L	q_36L	q_37L	q_38L	q_39L
4	q_14R	q_40L	q_41L	q_42L	q_43L	q_44L
5	q_15R	q_45L	q_46L	q_47L	q_48L	q_49L
6	q_16R	q_50L	q_51L	q_52L	q_53L	q_54L
7	q_17R	q_55L	q_56L	q_57L	q_58L	q_59L
8	q_18R	q_60L	q_61L	q_62L	q_63L	q_64L
9	q_19R	q_65L	q_66L	q_67L	q_68L	q_69L

Asymptotic Analysis

11

An important task of algorithm analysis is the estimation of the number of operations executed by the algorithm over a certain class of input data. The exact value of the number of elementary operations does not play any considerable role here, since it depends on the software implementation of the algorithm, the computer's architecture and other factors. This is why the algorithm's performance indicator is the growth rate of this value with the growth of the input data volume [13,49].

For analysis of the algorithm's efficiency it is necessary to estimate the operation time of the computer that solves the set problem, as well as the volume of memory used. The estimate of the computation system operation time is usually obtained by calculating the elementary operations performed during calculations (such operations are referred to as **basic**). In a supposition that one elementary operation is executed within a strictly defined period of time, the function $f(n)$, defined as the number of operations when calculating over the input data of size n, is called a **time-complexity function** [49].

Let us introduce a notation for the set of positive real numbers $\mathbb{R}^+ = (0, \infty)$ and consider the functions $f, g \colon \mathbb{N} \to \mathbb{R}^+$.

There are distinguished three classes of functions with respect to the growth rate $g(n)$.

1. *A class of functions growing no faster than* $g(n)$ is denoted by $O(g(n))$ and is defined as

$$O(g(n)) = \{f(n) \colon \exists c \in \mathbb{R}^+,\ n_0 \in \mathbb{N} \text{ such that for all } n \geqslant n_0$$
$$\text{is valid } f(n) \leqslant cg(n)\}.$$

It is said that the function $f(n)$ belongs to the class $O(g(n))$ (read as "big-O of g"), if for all values of the argument n, starting from the threshold value $n = n_0$, the inequality $f(n) \leqslant cg(n)$ is valid for some positive c.

© Springer International Publishing AG, part of Springer Nature 2018
S. Kurgalin and S. Borzunov, *The Discrete Math Workbook*,
Texts in Computer Science, https://doi.org/10.1007/978-3-319-92645-2_11

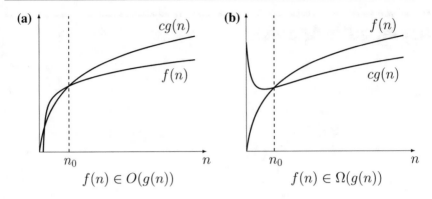

Fig. 11.1 Illustration of notations $f \in O(g(n))$ and $f \in \Omega(g(n))$

The notation $f(n) \in O(g(n))$ may be read as "the function g **majorizes** (dominates) the function f."

2. *A class of functions growing at least as fast as* $g(n)$ is denoted by $\Omega(g(n))$ and is defined as

$$\Omega(g(n)) = \{f(n)\colon \exists c \in \mathbb{R}^+,\ n_0 \in \mathbb{N} \text{ such that for all, } n \geqslant n_0$$
$$\text{is valid } f(n) \geqslant cg(n)\}.$$

It is said that $f(n)$ belongs to the class $\Omega(g(n))$ (read as "big-ω of g"), if for all values of the argument n, starting from the threshold value $n = n_0$, the inequality $f(n) \geqslant cg(n)$ is valid for some positive c.

3. *A class of functions growing at the same rate as* $g(n)$ is denoted by $\Theta(g(n))$ and is defined as the intersection of $O(g(n))$ and $\Omega(g(n))$.

It is said that $f(n)$ belongs to the class $\Theta(g(n))$ (read as "big-Theta of g"), if

$$f(n) \in O(g(n)) \cap \Omega(g(n)).$$

The introduced definitions are illustrated in Figs. 11.1 and 11.2, for illustration purposes, the argument n is assumed to be real.

Since $O(g(n))$ denotes a set of functions growing no faster than the function $g(n)$, then, in order to indicate the belonging to this set, the notation $f(n) \in O(g(n))$ is used. Another notation is rather common in literature: $f(n) = O(g(n))$, where the equals sign is understood conventionally, namely in the sense of belonging to the set. The above-listed classes of sets are referred to as the **"big O notation"**.

Let us specify some important properties of asymptotic relations [13,43].

Let $f_1, f_2, g_1, g_2\colon \mathbb{N} \to \mathbb{R}^+$. Then if $f_1(n) = O(g_1(n))$ and $f_2(n) = O(g_2(n))$, then

(1) $f_1(n) + f_2(n) = O(\max(g_1(n), g_2(n)))$;
(2) $f_1(n)f_2(n) = O(g_1(n)g_2(n))$.

The properties of the classes $O(f)$, $\Theta(f)$, $\Omega(f)$ are generalized in Table 11.1.

Fig. 11.2 Illustration of notation $f \in \Theta(g(n))$

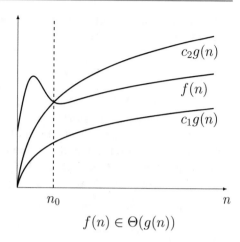

$$f(n) \in \Theta(g(n))$$

Table 11.1 The properties of the classes $O(f)$, $\Theta(f)$, $\Omega(f)$

The properties of the classes $O(f)$, $\Theta(f)$, $\Omega(f)$
Reflexivity
$f(n) = O(f(n))$ $f(n) = \Theta(f(n))$ $f(n) = \Omega(f(n))$
Symmetry
$f(n) = \Theta(g(n)) \Leftrightarrow g(n) = \Theta(f(n))$
Transitivity
$f(n) = O(g(n))$ **and** $g(n) = O(h(n)) \Rightarrow f(n) = O(h(n))$ $f(n) = \Theta(g(n))$ **and** $g(n) = \Theta(h(n)) \Rightarrow f(n) = \Theta(h(n))$ $f(n) = \Omega(g(n))$ **and** $g(n) = \Omega(h(n)) \Rightarrow f(n) = \Omega(h(n))$
Permutation of f and g
$f(n) = O(g(n)) \Leftrightarrow g(n) = \Omega(f(n))$

(Polynomial growth rate) lemma. *Polynomial of the form*

$$p(n) = a_d n^d + a_{d-1} n^{d-1} + \ldots + a_0,$$

where $a_d > 0$, belongs to the set $\Theta(n^d)$.

According to the mentioned lemma, the asymptotic behavior of a polynomial function is fully determined by the polynomial degree.

When analysing algorithms, a problem often arises to obtain an approximation for a "factorial" function, and in such cases it is convenient to use **Stirling's**[1] formula:

$$n! = \sqrt{2\pi n} \left(\frac{n}{e}\right)^n \left(1 + \Theta\left(\frac{1}{n}\right)\right),$$

where $e = 2{,}718281828\ldots$ —base of natural logarithms.

When comparing the growth rates of two set functions, it is convenient to use not the determinations of the sets O, Ω and Θ, but the consequences of the determinations, based on the calculation of the limit of the functions under consideration for infinitely great values of the argument [43].

Assume that we need to compare the asymptotic growth rates of the functions f and g, $f, g \colon \mathbb{N} \to \mathbb{R}^+$. For this purpose, calculate the limit $L = \lim\limits_{n\to\infty} \dfrac{f(n)}{g(n)}$.

Four cases are possible:

1. $L = 0$. Then $f(n) = O(g(n))$ and $f(n) \neq \Theta(g(n))$.
2. $L = c$ for some $c \in \mathbb{R}^+$. Then $f(n) = \Theta(g(n))$.
3. $L = \infty$. Then $f(n) = \Omega(g(n))$ and $f(n) \neq \Theta(g(n))$.
4. The limit $L = \lim\limits_{n\to\infty} \dfrac{f(n)}{g(n)}$ does not exist. In this case, this method cannot be used for comparing the order of growth of the functions $f(n)$ and $g(n)$.

In order to calculate the limits, we may use **L'Hopital's**[2] **rule**, according to which *the ratio limit of infinitely small at $n \to \infty$ functions (if the set $\mathbb{R}^+$ is considered as the domain) reduces to the ratio limit of their derivatives:*

$$\lim_{n\to\infty} \frac{f(n)}{g(n)} = \lim_{n\to\infty} \frac{f'(n)}{g'(n)}.$$

This ratio is valid if the functions $f(n)$ and $g(n)$ are differentiable on the interval (c, ∞) for some c, $g'(x) \neq 0$ for all points $x \in (c, \infty)$, and there exists a limit of the ratio of derivatives $\lim\limits_{n\to\infty} \dfrac{f'(n)}{g'(n)}$ (finite or infinite).

Justification of L'Hopital's rule is given in all sufficiently comprehensive mathematical analysis courses, for example [86].

For a wide and practically important class of recurrence relations of the form $T(n) = aT\left(\dfrac{n}{b}\right) + f(n)$, where a, b—const and $f(n)$—an arbitrary positive function, the asymptotic estimate of the growth rate of their solution $T(n)$ can be obtained by a universal method, based on the following theorem.

The master theorem for recurrence relations (Bentley, Haken, Saxe[3]). *Let* $T(n) = aT\left(\dfrac{n}{b}\right) + f(n)$, *where by* $\dfrac{n}{b}$ *is understood either* $\left\lfloor \dfrac{n}{b} \right\rfloor$, *or* $\left\lceil \dfrac{n}{b} \right\rceil$, $a \geqslant 1$, $b > 1$. *Then:*

---

[1] James Stirling (1692–1770)—Scottish mathematician.

[2] Guillaume François Antoine, marquis de L'Hôpital (1661–1704)—French mathematician.

[3] Jon Louis Bentley (born 1953), Dorothea Haken, James Benjamin Saxe—American mathematicians.

(1) if $f(n) = O(n^{\log_b a - \varepsilon})$ for some $\varepsilon > 0$, then $T(n) = \Theta(n^{\log_b a})$;

(2) if $f(n) = \Theta(n^{\log_b a})$, then $T(n) = \Theta(n^{\log_b a} \log_2 n)$;

(3) if $f(n) = \Omega(n^{\log_b a + \varepsilon})$ for some $\varepsilon > 0$ and $af\left(\dfrac{n}{b}\right) \leqslant cf(n)$ for some $c < 1$ and great n (**regularity condition**), *then* $T(n) = \Theta(f(n))$.

This theorem was stated in the work [9]; the comprehensive proof can be found in the fundamental guidance [13].

It should be noted that there exist recurrence relations of the form $T(n) = aT\left(\dfrac{n}{b}\right) + f(n)$, for which the considered universal method turns out to be inapplicable, and in this case one should look for a different method of estimating the growth rate of the function $T(n)$.

Note. The notation "$O(f)$" was introduced by Bachmann[4] in the textbook [7] on number theory. Note that sometimes, the "$O(f)$" symbols are called **Landau**[5] **symbols** [25]. The modern use of big O notation in the algorithm analysis is attributed to Knuth[6] [13].

The asymptotic analysis methods are used for mathematically correct estimation of the algorithm execution time and the memory used, depending on the dimension of the problem.

When analysing algorithms, the number of **basic operations** is estimated, and it is assumed that for execution of each of the below-mentioned operations the time $\Theta(1)$ [51] is required.

1. Binary arithmetic operations $(+, -, *, /)$ and operations of comparison of the real numbers $(<, \leqslant, >, \geqslant, =, \neq)$.

2. Logic operations $(\vee, \wedge, , \oplus)$.

3. Branching operations.

4. Calculation of the values of the elementary functions for relatively small values of the arguments.

For different input data, the algorithm operation time, generally speaking, may differ. Three cases are distinguished, allowing judging the algorithm's efficiency on the whole.

1. The best case

The best case for the algorithm is such a set of the input data, on which the execution time is minimal. On such a set, the number of basic operations $B(n)$ is determined by the formula

$$B(n) = \min_{\substack{\text{input} \\ \text{data}}} T(n).$$

---

[4]Paul Gustav Heinrich Bachmann (1837–1920)—German mathematician.

[5]Edmund Georg Hermann Landau (1877–1938)—German mathematician.

[6]Donald Ervin Knuth (born 1938)—American scientist, specializing in the sphere of informatics and computing.

Table 11.2 Asymptotic complexity of algorithms

Asymptotic complexity	Growth rate	Examples of algorithms
$O(1)$	Constant, not depending on the size of input data	Search in hash table
$O(\log_2 n)$	Logarithmic	Binary search
$O(n)$	Linear	Linear search
$O(n \log_2 n)$	No generally accepted name	Merge sort
$O(n^2)$	Quadratic	Insertion sort
$O(n^3)$	Cubic	Standard multiplication of matrices
$O(2^n)$	Exponential	Exhaustive search
$O(n!)$	Factorial	Generation of all permutations

2. The worst case

The worst case for the algorithm is such a set of the input data, on which the execution time is maximal:

$$W(n) = \max_{\substack{\text{input} \\ \text{data}}} T(n).$$

3. The average case

Analysis of the average case implies partitioning of the set of all possible input data into nonintersecting classes in such a manner that for each of the generated classes the number of basic operations should be the same. Then, the mathematical expectation of the number of operations is calculated based on the probability distribution for the input data:

$$A(n) = \sum_{i=1}^{k} p_i(n) t_i(n),$$

where n—input data size, k—number of partition classes, $p_i(n)$—probability that the input data belong to the class with the number i, $t_i(n)$—number of basic operations performed by the algorithm on the input data from the class with the number i.

Recall that the **probability** p of some event is the real number $0 \leqslant p \leqslant 1$. If the total number of outcomes is N and all outcomes are equally probable, then the probability of realization of each of them is $p = \dfrac{1}{N}$ [29].

Note. Notations of efficiency in the three cases described are taken from the first letters of English words *best*, *worst*, and *average*, respectively.

In Table 11.2 are shown the asymptotic complexity values most common in practice, with the examples of algorithms that have respective complexity in the average case.

11.1 Problems

11.1. Prove that $n + 1 \in O(n)$.

11.2. Prove that $5n^3 + 7n^2 - n + 7 \in O(n^3)$.

11.3. Let $f(n) = \cos n + 3$. Prove that $f(n) \in \Theta(1)$.

11.4. Let $f(n) = \sin 3n + 2$. Prove that $f(n) \in O(1)$.

11.5. Let $f(n) = \dfrac{\sin^2 3n + 2}{2}$. Prove that $f(n) \in \Omega(1)$.

11.6. Let $f(n) = \sin^4 n + \cos^4 n$. Prove that $f(n) \in \Theta(1)$.

11.7. For the defined functions f and g check whether $f \in O(g)$ is valid:

(1) $f(n) = 3n^2 + 2n + 10$, $g(n) = n^3 + 5$;
(2) $f(n) = n^3 + \sqrt{n}$, $g(n) = n^2 - \sqrt{n}$;
(3) $f(n) = \sqrt{3n}$, $g(n) = \sqrt{n^3}$;
(4) $f(n) = n^{7/5} + 3n$, $g(n) = \sqrt{n^7}$.

11.8. For the defined functions f and g check whether $f \in \Omega(g)$ is valid:

(1) $f(n) = 3n^2 - 5$, $g(n) = n(n + 2)$;
(2) $f(n) = n^3 + \sqrt{n}$, $g(n) = n(1 + n)$;
(3) $f(n) = n^{1/3}$, $g(n) = \sqrt{n^3}$;
(4) $f(n) = n^2(1 + \sqrt{n})$, $g(n) = n^2 - n$.

∗11.9. Explain what the following notations mean

(1) $f \notin O(g)$;
(2) $f \notin \Omega(g)$.

∗11.10. Prove that the relation $R_O = \{(f, g) : O(f) = O(g)\}$ on the set of real functions of the natural argument $\{f : f \in O(F)\}$, where $F : \mathbb{N} \to \mathbb{R}^+$— some function, is an equivalence relation.

∗11.11. At the examination on discrete mathematics and mathematical logic, the student states that for two arbitrary functions $f : \mathbb{N} \to \mathbb{R}^+$ and $g : \mathbb{N} \to \mathbb{R}^+$ either $f \in O(g)$, or $g \in O(f)$. Is he right?

11.12. Estimate the asymptotic behavior of the following sums:

(1) $\sum_{i=1}^{n} i$;

(2) $\sum_{i=1}^{n} i^2$.

∗**11.13.** Obtain the asymptotic estimate of $\sum_{i=1}^{n} i^d$, where $d = $ const, $d > -1$.

∗**11.14.** Estimate the asymptotic behavior of the sum $\sum_{i=1}^{n} \frac{1}{i}$.

11.15. Estimate the asymptotic behavior of the binomial coefficients $(n > 1)$:

(1) $C(n, 1)$;
(2) $C(n, 2)$.

11.16. Using Stirling's formula, prove the following estimate:

$$C(2n, n) = \Theta(4^n n^{-1/2}).$$

11.17. Using Stirling's formula, estimate the value of the expression

$$\frac{(2n - 1)!!}{(2n)!!}, n \in \mathbb{N},$$

for great n.

11.18. Obtain the asymptotic estimate for the Catalan numbers C_n $(n \geqslant 0)$.

11.19. Prove that any positive polynomial function grows faster than any power of logarithm, or, in other words, that

$$\log^k n = O(f(n)) \text{ for all } k \in \mathbb{R}^+,$$

where $f(n) = a_d n^d + a_{d-1} n^{d-1} + \ldots + a_1 n + a_0, a_d > 0, a_i = $ const, $i = 0, 1, \ldots, d, d \in \mathbb{N}$.

11.20. Prove that any exponential function with the base $c > 1$ grows faster than a polynomial function, or, in other words, that

$$c^n = \Omega(f(n)) \text{ for all } c \in \mathbb{R}^+, c > 1,$$

where $f(n) = a_d n^d + a_{d-1} n^{d-1} + \ldots + a_1 n + a_0, a_d > 0, a_i = $ const, $i = 0, 1, \ldots, d, d \in \mathbb{N}$.

11.21. Arrange the functions in order of increasing growth rate:

$$n^3 + \log_2 n; \quad 2^{n-1}; \quad n \log_2 n; \quad 6; \quad \left(\frac{3}{2}\right)^n.$$

11.22. Arrange the functions

$$\sqrt{n}; \quad 7n - 2n^2 + \frac{1}{10}n^4; \quad \sqrt{n} + \log_2(n); \quad n^2/\log_2 n; \quad e^{4\ln n}$$

in order of increasing growth rate. Separately write out the groups of functions growing at the same rate.

11.23. Arrange the functions

$$2^{2^n - 1}; \quad 2^{2n}; \quad 2^{n^2}; \quad 2^{2^{n-1}}; \quad 2^{2^n - 1/n}; \quad 2^{2\log_2 n + 1}$$

in order of increasing growth rate. Separately write out the groups of functions growing at the same rate.

11.24. Comparison of the functions' growth rates
Arrange the functions

$$(\sqrt{2})^{\log_2 n}; \quad n^2; \quad n!; \quad \lfloor \log_2 n \rfloor!; \quad \left(\frac{3}{4}\right)^n; \quad n^3;$$
$$(\log_2 n)^2; \quad \log_2(n!); \quad 2^{2^n}; \quad n^{1/\log_2 n}; \quad \ln\ln n; \quad n^3 2^n;$$
$$n^{\log_2\log_2 n}; \quad \ln n; \quad 2; \quad 2^{\log_2 n}; \quad (\log_2 n)^{\log_2 n}; \quad e^n;$$
$$4^{\lfloor \log_2 n \rfloor}; \quad \sqrt{\log_2 n}; \quad 2^{\sqrt{2\log_2 n}}; \quad n\log_2 n; \quad 2^n; \quad n$$

in order of increasing of their asymptotic growth rate. Separately write out the groups of functions growing at the same rate.

11.25. Find the asymptotic estimate of the growth rate of the solutions of the following recurrence relations:

(1) $T(n) = 2T\left(\dfrac{n}{2}\right) + \sqrt{n}$;

(2) $T(n) = 2T\left(\dfrac{n}{2}\right) + n$;

(3) $T(n) = 2T\left(\dfrac{n}{2}\right) + n^2$.

11.26. Find the asymptotic estimate of the growth rate of the solutions of the following recurrence relations:

(1) $T(n) = 3T\left(\dfrac{n}{2}\right) + n$;

(2) $T(n) = 3T\left(\dfrac{n}{2}\right) + n^{\log_2 3}$;

(3) $T(n) = 3T\left(\dfrac{n}{2}\right) + n^3$.

11.27. Find the asymptotic estimate of the growth rate of the solutions of the following recurrence relations:

(1) $T(n) = 5T\left(\dfrac{n}{8}\right) + \sqrt[3]{n}$;

(2) $T(n) = 5T\left(\dfrac{n}{8}\right) + n^{\log_8 5}$;

(3) $T(n) = 5T\left(\dfrac{n}{8}\right) + n^4 \log_2^4 n$.

11.28. The execution time of the algorithm for multiplication of matrices of size $n \times n$, developed by **Strassen,**[7] is determined by the recurrence relation [71]

$$T(n) = 7T\left(\frac{n}{2}\right) + \Theta(n^2).$$

Estimate the execution time of the Strassen algorithm.

11.29. There exists a more general statement of the master theorem, applicable to somewhat wider class of recurrence relations [13].

The master theorem (the second form). *Let* $T(n) = aT\left(\dfrac{n}{b}\right) + f(n)$, *where by* $\dfrac{n}{b}$ *is meant either* $\left\lfloor \dfrac{n}{b} \right\rfloor$, *or* $\left\lceil \dfrac{n}{b} \right\rceil$, $a \geqslant 1, b > 1$. *Then:*

(1) *if* $f(n) = O(n^{\log_b a - \varepsilon})$ *for* $\varepsilon > 0$, *then* $T(n) = \Theta\left(n^{\log_b a}\right)$;

(2) *if* $f(n) = \Theta(n^{\log_b a} \log_2^k n)$ *for some nonnegative* $k \geqslant 0$, *then* $T(n) = \Theta\left(n^{\log_b a} \log_2^{k+1} n\right)$;

(3) *if* $f(n) = \Omega(n^{\log_b a + \varepsilon})$ *for* $\varepsilon > 0$ *and for the function* $f(n)$ *the regularity condition is fulfilled, then* $T(n) = \Theta(f(n))$.

Using the mentioned theorem statement, find the asymptotic estimate of the growth rate of the solutions of the following recurrence relations.

(1) $T(n) = 4T\left(\dfrac{n}{4}\right) + n \log_2 n$;

(2) $T(n) = 5T\left(\dfrac{n}{2}\right) + n^{\log_2 5} \log_2^5 n$;

(3) $T(n) = 4T\left(\dfrac{n}{2}\right) + n^2 \log_2^2 n$.

*∗**11.30.** Arrange the functions $f_1(n)$, $f_2(n)$, $f_3(n)$, $f_4(n)$, defined by the following recurrence relations:

$$f_1(n) = 2f_1\left(\left\lfloor \frac{n}{2} \right\rfloor\right) + n \log_2 n, \quad f_1(1) = 1;$$
$$f_2(n) = f_2(n-2) + 2 \log_2 n, \quad f_2(1) = f_2(2) = 1;$$
$$f_3(n) = 3f_3\left(\lfloor \sqrt{n} \rfloor\right), \quad f_3(1) = 1;$$
$$f_4(n) = f_4\left(\left\lfloor \frac{n}{2} + \sqrt{n} - 1 \right\rfloor\right) + 1, \quad f_4(1) = 1,$$

in order of increasing of their asymptotic growth rate.

---

[7] Volker Strassen (born 1936)—German mathematician.

11.2 Answers, Hints, Solutions

11.1. *Proof*
It is necessary to prove that there exists a positive constant c such that starting from some number n_0, the following inequality is valid $n + 1 \leqslant cn$, or $(c - 1)n - 1 \geqslant 0$.

Assume that $c = 2$, then, starting from $n_0 = 1$, the last inequality is valid. Hence, $n + 1 \in O(n)$.

11.2. *Proof*
Prove that there exists a constant c such that starting from some number n_0, the following inequality is valid $5n^3 + 7n^2 - n + 7 \leqslant cn^3$, or $(c - 5)n^3 - 7n^2 + n - 7 \geqslant 0$.

Let $c = 6$, then the inequality $(c - 5)n^3 - 7n^2 + n - 7 \geqslant 0$ will be rewritten in the form

$$n^3 - 7n^2 + n - 7 \geqslant 0 \Leftrightarrow (n^2 + 1)(n - 7) \geqslant 0.$$

The obtained inequality is valid for all $n \geqslant n_0$, $n_0 = 7$. Hence, $5n^3 + 7n^2 - n + 7 \in O(n^3)$.

11.3. *Proof*
On the entire domain of the function $f(n) = \cos n + 3$, the following inequality is valid

$$2 \cdot 1 \leqslant \cos n + 3 \leqslant 4 \cdot 1 \Rightarrow f(n) \in \Theta(1).$$

11.7. *Answer*: belonging of $f \in O(g)$ is valid for the functions from items (1), (3), and (4).

11.8. *Answer*: belonging of $f \in \Omega(g)$ is valid for the functions from items (1), (2), and (4).

11.10. *Proof*
The relation R_O will be an equivalence relation, if it satisfies the conditions of reflexivity, symmetry, and transitivity. Check the fulfillment of the listed conditions.
1. Reflexivity.
  If $f = g$, then the equality is valid $O(f) = O(g)$, and the pair (f, f) belongs to R_O.
2. Symmetry.
  If $(f, g) \in R_O$, then $O(g) = O(f)$; hence, $(g, f) \in R_O$.
3. Transitivity.
  Let $(f, g) \in R_O$ and $(g, h) \in R_O$. Check whether the condition $(f, h) \in R_O$ is fulfilled:

$$\begin{cases} (f, g) \in R_O \Rightarrow O(f) = O(g), \\ (g, h) \in R_O \Rightarrow O(g) = O(h) \end{cases} \Rightarrow O(f) = O(h).$$

Hence, $(f, h) \in R_O$. The relation R_O is reflexive, symmetric, and transitive, i.e., it is an equivalence relation.

11.11. *Hint.*
Consider the functions

$$f(n) = \begin{cases} 1, & \text{if } n \text{ even,} \\ n, & \text{if } n \text{ odd} \end{cases} \quad \text{and} \quad g(n) = \begin{cases} n, & \text{if } n \text{ even,} \\ 1, & \text{if } n \text{ odd.} \end{cases}$$

11.12. *Answer*:

(1) $\sum_{i=1}^{n} i = \Theta(n^2)$;

(2) $\sum_{i=1}^{n} i^2 = \Theta(n^3)$.

11.13. *Solution.*
Consider the upper estimate. The sum contains n summands, no greater than n^d.
Hence,

$$\sum_{i=1}^{n} i^d \leqslant n \cdot n^d = n^{d+1}$$

and

$$\sum_{i=1}^{n} i^d = O(n^{d+1}).$$

Then, it is necessary to perform the lower estimate:

$$\sum_{i=1}^{n} i^d = \sum_{i=1}^{\lfloor n/2 \rfloor} i^d + \sum_{i=\lfloor n/2 \rfloor+1}^{n} i^d \geqslant \sum_{i=\lfloor n/2 \rfloor+1}^{n} i^d.$$

The last sum $\sum_{i=\lfloor n/2 \rfloor+1}^{n} i^d$ contains $n - \lfloor n/2 \rfloor = \lceil n/2 \rceil$ summands, each of which
is no smaller than the value $(\lfloor n/2 \rfloor + 1)^d$. Hence,

$$\sum_{i=1}^{n} i^d \geqslant \lceil n/2 \rceil (\lfloor n/2 \rfloor + 1)^d \geqslant \frac{n}{2} \left(\frac{n}{2}\right)^d = \frac{n^{d+1}}{2^{d+1}},$$

whence the equality follows

$$\sum_{i=1}^{n} i^d = \Omega(n^{d+1}).$$

We finally obtain

$$\sum_{i=1}^{n} i^d = O(n^{d+1}), \quad \sum_{i=1}^{n} i^d = \Omega(n^{d+1}) \quad \Rightarrow \quad \sum_{i=1}^{n} i^d = \Theta(n^{d+1}).$$

11.14. *Solution.*
For the nonincreasing function $f(n)$, the inequality is valid

$$\sum_{i=a+1}^{b} f(i) \leqslant \int_{a}^{b} f(x)\, dx \leqslant \sum_{i=a}^{b-1} f(i),$$

which directly follows from the geometric sense of the definition of an integral of piecewise smooth function as the area of the respective curvilinear trapezoid. In our case $f(n) = 1/n$.

Let $a = 1, b = n + 1$. Then

$$\int_{1}^{n+1} \frac{1}{x}\, dx \leqslant \sum_{i=1}^{n} \frac{1}{i}$$

or

$$\ln(n + 1) \leqslant \sum_{i=1}^{n} \frac{1}{i}.$$

In order to find the upper estimate, assume that $a = 1, b = n$ (we cannot directly substitute $a = 0$ because the respective integral diverges):

$$\sum_{i=2}^{n} \frac{1}{i} \leqslant \int_{1}^{n} \frac{1}{x}\, dx$$

or

$$\sum_{i=1}^{n} \frac{1}{i} \leqslant 1 + \int_{1}^{n} \frac{1}{x}\, dx.$$

From the last inequality follows

$$\sum_{i=1}^{n} \frac{1}{i} \leqslant 1 + \ln n \leqslant 2 \ln n.$$

As a result, we obtain the required estimate $\sum_{i=1}^{n} \frac{1}{i} = \Theta(\ln n)$.

Note. Let us produce a more exact estimate for the asymptotic behavior of harmonic numbers:

$$H_n = \ln n + \gamma + O\!\left(\frac{1}{n}\right),$$

where $\gamma = 0{,}5772156649\ldots$ —**Euler's constant** [23].

11.15. *Answer*:

(1) $C(n, 1) = \Theta(n)$;
(2) $C(n, 2) = \Theta(n^2)$.

11.16. *Proof*

According to the definition of the binomial coefficient $C(2n, n)$, we obtain:

$$
\begin{aligned}
C(2n, n) &= \frac{(2n)!}{n!(2n-n)!} = \frac{(2n)!}{(n!)^2} = \\
&= \frac{\sqrt{2\pi(2n)}(2n)^{2n}e^{-2n}\left(1 + \Theta(\frac{1}{n})\right)}{\left(\sqrt{2\pi n}\, n^n e^{-n}(1 + \Theta(\frac{1}{n}))\right)^2} = \frac{\sqrt{2}}{\sqrt{2\pi n}} \cdot 2^{2n} \cdot \frac{1 + \Theta(\frac{1}{n})}{\left(1 + \Theta(\frac{1}{n})\right)^2}.
\end{aligned}
$$

There exists such $c \in \mathbb{R}^+$, that $1 + \Theta(\frac{1}{n}) \leqslant 1 + c/n$, starting from some n_0. The denominator is also easy to estimate: $\left(1 + \Theta(\frac{1}{n})\right)^2 > 1$. Hence, there exists the comparison

$$
\frac{1 + \Theta(\frac{1}{n})}{\left(1 + \Theta(\frac{1}{n})\right)^2} \leqslant \frac{1 + c/n}{1} = 1 + \Theta\left(\frac{1}{n}\right).
$$

We finally obtain

$$
C(2n, n) = \frac{1}{\sqrt{\pi n}} \cdot 2^{2n} \cdot \left(1 + \Theta\left(\frac{1}{n}\right)\right) = \Theta(4^n n^{-1/2}).
$$

11.17. *Answer*:

$$
\frac{(2n-1)!!}{(2n)!!} = \frac{1}{\sqrt{\pi n}} \left(1 + \Theta\left(\frac{1}{n}\right)\right).
$$

11.18. *Answer*:

$C_n = \Theta(n^{-3/2}4^n)$.

11.19. *Hint*.

The statement of the exercise directly follows from the relation

$$
\lim_{n \to \infty} \frac{\log^k n}{n^d} = 0 \text{ for } k \in \mathbb{R}^+, \, d \in \mathbb{N},
$$

which can be proved, for example, using L'Hopital's rule $\lceil k \rceil$ times.

11.20. *Hint*.

Use the known limit $\lim\limits_{n \to \infty} \dfrac{n^d}{c^n} = 0$ for arbitrary constants $c \in (1, \infty)$ and $d \in \mathbb{N}$.

11.21. *Solution.*
Introduce the notations $f_1(n) = n^3 + \log_2 n$, $f_2(n) = 2^{n-1}$, $f_3(n) = n \log_2 n$,
$f_4(n) = 6$, $f_5(n) = \left(\dfrac{3}{2}\right)^n$. The function $f_4(n)$ does not depend on the argument,
therefore the growth rate will be the least of all suggested functions; in the final list,
$f_4(n) = 6$ will occupy the first place.

As is known, the growth rate of a polynomial function is less than that of the
exponential ones; hence, the second place will be occupied either by $f_1(n)$, or by
$f_3(n)$. Find the limit of $L = \lim\limits_{n\to\infty} \dfrac{f_3(n)}{f_1(n)}$:

$$L = \lim_{n\to\infty} \frac{f_3(n)}{f_1(n)} = \lim_{n\to\infty} \frac{n \log_2 n}{n^3 + \log_2 n} = \lim_{n\to\infty} \frac{1}{n^2/\log_2 n + 1/n} = 0.$$

Based on the obtained limit value $L = 0$, we conclude that the second place in the
list is occupied by the function $f_3(n)$, and the third by $f_1(n)$, since $f_1(n) = O(f_2(n))$
and $f_1(n) = O(f_5(n))$.

Further, compare the growth rates of $f_2(n)$ and $f_5(n)$:

$$L = \lim_{n\to\infty} \frac{f_5(n)}{f_2(n)} = \lim_{n\to\infty} \left(\frac{3}{2}\right)^n \Big/ 2^{n-1} = \lim_{n\to\infty} 2\left(\frac{3}{4}\right)^n = 0.$$

We finally obtain the list of functions in order of increasing growth rate:

$$6;\ n \log_2 n;\ n^3 + \log_2 n;\ \left(\frac{3}{2}\right)^n;\ 2^{n-1}.$$

11.22. *Answer:*
The list of functions in order of increasing growth rate:

$$\boxed{\sqrt{n};\ \sqrt{n} + \log_2(n);}\ n^2/\log_2 n;\ \boxed{e^{4\ln n};\ 7n - 2n^2 + \frac{1}{10}n^4.}$$

Ovals highlight the groups of functions with the same asymptotic growth rate.

11.23. *Answer:*
The list of functions in order of increasing growth rate:

$$\boxed{2^{2n};\ 2^{2^{\log_2 n+1}};}\ 2^{n^2};\ 2^{2^{n-1}};\ \boxed{2^{2^n-1/n};\ 2^{2^n-1}.}$$

Ovals highlight the groups of functions with the same asymptotic growth rate.

11.25. *Solution.*

(1) Let us use the method based on the master theorem for recurrence rela-
tions. In this case, $a = 2$, $b = 2$, $f(n) = n^{1/2}$. Hence, $n^{\log_b a} = n^{\log_2 2} = \Theta(n)$.
Comparing the order of growth rate of the functions $n^{\log_b a}$ and $f(n)$, we have
$f(n) = O\left(n^{\log_2 2 - \varepsilon}\right)$, where as ε can be taken $\varepsilon = \dfrac{1}{2} > 0$.

The solution of this relation is expressed by the first case of the master theorem:

$$T(n) = \Theta\left(n^{\log_2 2}\right) = \Theta(n).$$

(2) In this case, $a = 2$, $b = 2$, $f(n) = n$. Hence, $n^{\log_b a} = \Theta(n)$, $f(n) = \Theta\left(n^{\log_b a}\right)$; hence, by the second case of the master theorem

$$T(n) = \Theta\left(n^{\log_2 2} \log_2 n\right) = \Theta(n \log_2 n).$$

(3) The values a and b are equal to $a = 2$, $b = 2$, $f(n) = n^2$. We have $f(n) = \Omega\left(n^{\log_b a + \varepsilon}\right)$, whereas ε can be taken $\varepsilon = 1 > 0$.

Check the fulfillment of the regularity condition for the function $f(n)$. For sufficiently big n the condition $af\left(\dfrac{n}{b}\right) \leqslant cf(n)$ is fulfilled at $c = \dfrac{1}{2} < 1$. Hence, according to the third case, $T(n) = \Theta(n^2)$.

11.26. *Answer*:

(1) $T(n) = \Theta\left(n^{\log_2 3}\right)$;
(2) $T(n) = \Theta\left(n^{\log_2 3} \log_2 n\right)$;
(3) $T(n) = \Theta(n^3)$.

11.27. *Answer*:

(1) $T(n) = \Theta\left(n^{\log_8 5}\right)$;
(2) $T(n) = \Theta\left(n^{\log_8 5} \log_2 n\right)$;
(3) $T(n) = \Theta(n^4 \log_2^4 n)$.

11.28. *Answer*:
Time complexity $T(n) = \Theta\left(n^{\log_2 7}\right)$.
Since $\log_2 7 = 2,807\ldots < 3$, then the Strassen algorithm for sufficiently big n has smaller asymptotic complexity compared to the standard method for multiplication of matrices (see Exercise **9.4**).

11.29. *Answer*:

(1) $T(n) = \Theta(n \log_2^2 n)$;
(2) $T(n) = \Theta\left(n^{\log_2 5} \log_2^6 n\right)$;
(3) $T(n) = \Theta(n^2 \log_2^3 n)$.

11.30. *Answer*: $f_4(n)$, $f_3(n)$, $f_2(n)$, $f_1(n)$.

Basic Algorithms

12

12.1 Recursive Algorithms

Recursive is an algorithm that calls itself [61]. Such algorithms are usually easy to understand and convenient for practical realization.

The best-known example of a recursive algorithm is the algorithm for calculating the factorial of a number, based on the property $n! = n \cdot (n-1)!$, where $n = 1, 2, 3, \ldots$

```
1  function factorial(n:integer):integer;
2  //calculating the factorial of a number n
3  begin
4      if n = 1 then
5              factorial := 1
6          else
7              factorial := n*factorial(n − 1);
8  end;
```

The computational complexity of this algorithm is equal to $O(n)$.

One more example of recursion is provided by the calculation of the determinant of the matrix. The **determinant** of the square matrix A of size $n \times n$

$$A = \begin{bmatrix} a_{11} & \cdots & a_{1n} \\ \vdots & \ddots & \vdots \\ a_{n1} & \cdots & a_{nn} \end{bmatrix},$$

© Springer International Publishing AG, part of Springer Nature 2018
S. Kurgalin and S. Borzunov, *The Discrete Math Workbook*,
Texts in Computer Science, https://doi.org/10.1007/978-3-319-92645-2_12

where $n = 1, 2, \ldots$, is denoted by $\det A$ and is calculated recursively as follows ("expansion with respect to the ith row" or Laplace[1] expansion):

$$\text{if } n = 1, \text{ then } \det A = a_{11},$$

$$\text{if } n > 1, \text{ then } \det A = \sum_{j=1}^{n} (-1)^{i+j} a_{ij} \det \widetilde{A}_{ij},$$

where a_{ij}—an element standing at the intersection of the ith row and the jth column of the matrix A, and $\widetilde{A}_{ij}$—a matrix of size $(n-1) \times (n-1)$, obtained from A by deleting the ith row and the jth column. The value $(-1)^{i+j} \det \widetilde{A}_{ij}$ is called the **algebraic complement** of the element a_{ij}. The determinant $\det A$ does not depend on the choice of the row in which the expansion is performed, this is why it is usually assumed that $i = 1$.

Note that in linear algebra are known alternative methods of finding the value $\det A$, in particular, axiomatic definition [38] and definition based on the concept of permutations [41].

Write out the explicit expressions for $\det A$ in the most practically important cases $n = 2$ and $n = 3$. For the square matrix of size 2×2 from the above definition directly follows

$$\det \begin{bmatrix} a_{11} & a_{12} \\ a_{21} & a_{22} \end{bmatrix} = a_{11} \det[a_{22}] - a_{12} \det[a_{21}] = a_{11}a_{22} - a_{12}a_{21},$$

i.e., the determinant of the matrix $\begin{bmatrix} a_{11} & a_{12} \\ a_{21} & a_{22} \end{bmatrix}$ is equal to the difference of the products of the elements located on the main and the secondary diagonals.

For the matrix of size 3×3, we have

$$\det \begin{bmatrix} a_{11} & a_{12} & a_{13} \\ a_{21} & a_{22} & a_{23} \\ a_{31} & a_{32} & a_{33} \end{bmatrix} =$$

$$= a_{11} \det \begin{bmatrix} a_{22} & a_{23} \\ a_{32} & a_{33} \end{bmatrix} - a_{12} \det \begin{bmatrix} a_{21} & a_{23} \\ a_{31} & a_{33} \end{bmatrix} + a_{13} \det \begin{bmatrix} a_{21} & a_{22} \\ a_{31} & a_{32} \end{bmatrix} =$$

$$= a_{11}a_{22}a_{33} + a_{12}a_{23}a_{31} + a_{13}a_{21}a_{32} - a_{11}a_{23}a_{32} - a_{12}a_{21}a_{33} - a_{13}a_{22}a_{31}.$$

The obtained relation is sometimes called **Sarrus'**[2] **rule**.

The use of the recursive definition for calculating the determinant results in a sum consisting of $n!$ summands, which corresponds to the computational complexity $O(n!)$.

---

[1]Pierre-Simon, marquis de Laplace (1749–1827) was a French mathematician, physicist, and astronomer.
[2]Pierre Frédéric Sarrus (1798–1861)—French mathematician.

For the methodological purposes, one of the problems of the present chapter requires writing a program for calculating det A by the recursive method.

It is important to note that efficient methods for solving this task are available. The known **Gaussian**[3] **elimination** allows finding matrices of size $n \times n$ in the time $O(n^3)$. The idea of the Gaussian elimination consists in a transformation of the matrix $A = [a_{ij}]$ into the upper triangular matrix $A' = [a'_{ij}]$, having the same determinant $\det A' = \det A$, while the value $\det A'$ is equal to the product of the elements of its main diagonal $\det A' = \prod_{i=1}^{n} a'_{ii}$.

It should be emphasized that despite clarity and simplicity of recursive algorithms, the use of recursion in a program requires from a programmer to be extremely attentive to the consumption of the system resources, in particular, to controlling the system stack overflow.

12.2 Search Algorithms

A search for information in some data structure is an important programming task. We will consider a simple data structure—an array of integers. Generalization of the search algorithms for more complex structures does not usually cause any difficulty [49].

Sequential Search Algorithm

Let a problem be set to determine the number of an array element whose value is equal to a certain given value, which is called **target**. A **sequential search algorithm** is used for dealing with an unordered array and operates as follows: It sequentially scans the array elements, starting from the first, and compares them with the target. If the sought element is found, the number of this element is returned; otherwise, the result should be a number not corresponding to any element, for example, the number 0.

```
1   const N = 100;
2   type Mass = array[1..N] of integer;
3   function SequentialSearch(list :Mass;
4                                target :integer): integer;
5   // list — analysed array of N elements
6   // target — target value
7   var i :integer;
8   begin
9       result:=0;
10      for i := 1 to N do
11          if target = list[i] then
```

---

[3] Johann Carl Friedrich Gauß (1777–1855)—prominent German mathematician and astronomer.

```
12              begin
13                  result:=i;
14                      break;
15                  end;
16  end;
```

Let us analyze the sequential search algorithm.

The basic operation is the comparison of the values *target* and *list*[*i*] in the row number 11. If the sought element occupies the last place in the array or is absent, then the algorithm will perform N comparisons, where N is the number of elements in the array *list*: $W(N) = N$.

It is easy to show that $B(N) = 1$.

For analysis of the average case, suppose that the target value is known to be contained in the array and with equal probability occupies any possible position. Then, the probability to encounter the target value at the place number i, $1 \leqslant i \leqslant N$, is equal to $p = \dfrac{1}{N}$.

For the average complexity case algorithm, we obtain

$$A(N) = \frac{1}{N} \sum_{i=1}^{N} i = \frac{1}{N} \frac{N(N+1)}{2} = \frac{N+1}{2} = O(N).$$

Even if the target element is absent from the array, the result for $A(N)$ will not change: $A(N) = O(N)$.

The Binary Search Algorithm

For search in ordered arrays, the **binary search** algorithm is used [13,43].

```
1  const N = 100;
2  type Mass = array[1..N] of integer;
3  function BinarySearch(list:Mass;
4                            target:integer):integer;
5  //list — analysed array of N elements
6  //target — target value
7  var left,middle,right:integer;
8  begin
9      left:=1;
10     right:=N;
11     result:=0;
12     while left<=right do
13         begin
14             middle:=(left+right) div 2;
15             case compare(list[middle],target) of
16                 −1: left:=middle+1;
17                  0: begin
18                        result:=middle;
```

```
19 │                          break;
20 │                      end;
21 │              +1:  right:=middle−1;
22 │          end;
23 │      end;
24 │ end;
```

The auxiliary function compare (a, b) compares the integer arguments a and b and returns the value

$$\text{compare}(a, b) = \begin{cases} +1, & \text{if } a > b, \\ 0, & \text{if } a = b, \\ -1, & \text{if } a < b. \end{cases}$$

Let us consider the operation of the binary search algorithm by the example of the search for the key value $target = 5$ in the ordered array $\boxed{1\,2\,3\,4\,5\,6\,7\,8}$ (Fig. 12.1).

First, we compare with the element located in the middle of the array $list[(left + right) \text{ div } 2] = list[4]$. Since $target > list[4]$, the sought element is contained in the right side of the array $\boxed{5\,6\,7\,8}$. Then follows the comparison $(target\ ?\ list[6])$ and, finally, the equality is checked $target = list[5]$.

So, for establishing the element's number in the array $list$, equal to the key one, we have here performed three operations of comparison.

Further analysis requires introducing a new concept—"**decision tree**" [43]. Each vertex of such a tree represents an element of the array $list[i]$, where $i = 1, \ldots, N$, with which comparison is performed $(target\ ?\ list[i])$. Transition over the edge of the left subtree is performed if $target < list[i]$, and over the edge of the right subtree otherwise, if $target > list[i]$ (see Fig. 12.2, where the decision tree T_N is presented for the array of eight elements).

Fig. 12.1 Example of the binary search algorithm operation. Bold letters highlight the elements, among which, at this step of the algorithm, a search is performed; question mark marks the element being compared with the value $target$; the sought value found is underlined

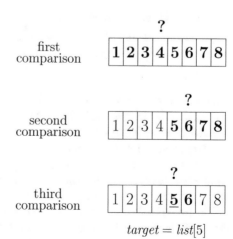

Fig. 12.2 The decision tree
T_8 for the array of eight
elements

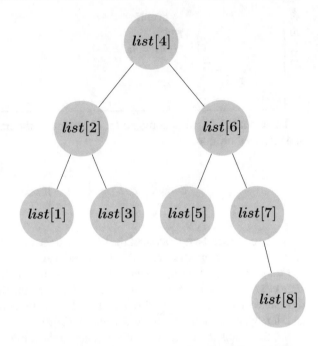

Example 12.1 Let us consider the vector $\mathbf{a} = \{a_1, a_2, \ldots, a_N\}$, where $N > 2$, whose components are formed by the elements of the non-empty set A with antisymmetric and transitive relation introduced on it, all elements of which are pairwise comparable.

Median of the vector $\mathbf{a}$ is the sth in value coordinate $\mathbf{a}$, where $s = \left\lceil \dfrac{N}{2} \right\rceil$ [13].

Based on this definition, we can formulate the basic property of the median m of an arbitrary vector: Its components contain no less than $\left\lceil \dfrac{N}{2} \right\rceil$ elements greater than m or equal to m, and no less than $\left\lceil \dfrac{N}{2} \right\rceil$ elements no greater than m.

Calculate how many comparisons will be needed for finding the median of three numbers in the average and worst cases.

Solution.

Let us represent the decision tree T for the task of ascending ordering of the numbers a_1, a_2, a_3. The median can be defined as the average element of the obtained ordered sequence. In order to construct the tree T perform the comparisons of the type $(a_i \,?\, a_j)$, where $1 \leqslant i, j \leqslant 3, i \neq j$, in the following order: $(a_1 \,?\, a_2), (a_2 \,?\, a_3)$ and, if necessary, the final comparison $(a_1 \,?\, a_3)$. In Fig. 12.3, the tree T is presented, where the circles mark the operations of comparison $(a_i \,?\, a_j)$ and the rectangles show the sought ordered sequences forms by the specified numbers.

So, from Fig. 12.3 follows that determining the order of the elements a_1, a_2, a_3 and, hence, their median, in the average case requires $(2 + 3 + 3 + 3 + 3 + 2)/6 = 8/3$ comparisons, and three comparisons are required in the worst case. $\square$

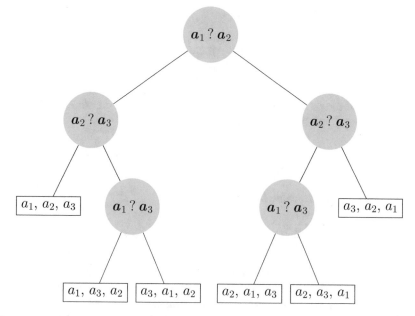

Fig. 12.3 The decision tree T for the problem of the median of three numbers

The binary search algorithm for the next comparison always chooses the middle of the array, therefore, in this case, the tree T_N either will be complete, or some leaves may be absent from its last level. For any binary tree of height h and number of leaves l the inequality is valid $2^h \geqslant l$ or $h \geqslant \lceil \log_2 l \rceil$. The number of leaves in the complete tree is $h = \lfloor \log_2 N \rfloor + 1$ (see Exercise **5.40**).

Let us estimate the asymptotic complexity of the binary search algorithm in a supposition that the target value is necessarily situated in the array *list*. The probability of finding an element equal to *target*, at an arbitrary place is equal to $p = \dfrac{1}{N}$. Construct the decision tree T_N for the binary search process. In order to search for the element on the first level, we will need one comparison, on the second level—two, on the third level—three, and so on.

The Worst Case

The maximum number of comparisons is equal to the height of the decision tree $W(N) = O(\log_2 N)$.

The Average Case

The total number of comparisons is equal to the sum over all levels of products of the number of nodes at a given level by the level's number. For the complete tree, the following equality is valid

$$A(N) = \frac{1}{N} \sum_{i=1}^{h} i 2^{i-1};$$

for the tree with unoccupied last level

$$A(N) \leqslant \frac{1}{N} \sum_{i=1}^{h} i 2^{i-1}.$$

The value of the sum in the right side of this inequality is presented in Appendix "Reference Data," relation (A.61), see also Exercise **12.21**:

$$A(N) \leqslant \frac{1}{2N}((h-1)2^{h+1}+2) = \frac{(h-1)2^h+1}{N}.$$

Substitute the known value $h = \lfloor \log_2 N \rfloor + 1$:

$$A(N) \leqslant \frac{1}{N}(\lfloor \log_2 N \rfloor \cdot 2^{\lfloor \log_2 N \rfloor + 1} + 1) = O(\log_2 N).$$

Analysis of the case when the target value can be absent from the array and the probability of search for the target value out of each of $N + 1$ intervals, formed by the array elements, is equal, also leads to the estimate $O(\log_2 N)$.

Finally, the computational complexity of the binary search algorithm is

$$A(N) = O(\log_2 N).$$

Fibonaccian Search Algorithm

Along with the binary search algorithm, the **Fibonaccian search** algorithm is used for determining the index of the target element in the ordered array. The Fibonaccian search algorithm compares the target value with the element that divides this array with respect to two sequential Fibonacci numbers. Depending on the result of this comparison, either the search stops when the sought element is found, or the search algorithm is recursively called on one of the two parts of the array where the sought element is present.

Recall that the Fibonacci sequence is defined by the recurrence relation $F_{n+2} = F_{n+1} + F_n$, where $n \geqslant 1$, with the initial conditions $F_1 = F_2 = 1$. The properties of this sequence allow organizing the algorithm operation without operations of multiplication or division.

Consider the Fibonaccian search algorithm in more detail.

Assume that the number of elements N in the ordered array *list* is F_k for some $k = 2, 3, \ldots$

The Fibonaccian search algorithm performs the following operations for finding the element *target* in this array.

First of all, the operation of comparison of the target value *target* with the element of the array *list*$[F_{k-1}]$ is performed.

Then, three cases are possible (see Fig. 12.4, which schematically shows partitioning of the array by the element *list*$[F_{k-1}]$ with respect to F_{k-1}/F_{k-2}):

Fig. 12.4 Arrangement of the array elements with index F_{k-1} in the original array $list[1..F_k]$

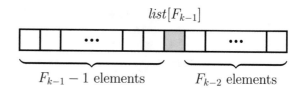

$F_{k-1} - 1$ elements F_{k-2} elements

1. If $target < list[F_{k-1}]$, then the index of the sought element is less than F_{k-1}. In this case, the search continues after changing the right bound of the considered part of the array by $(F_{k-1} - 1)$: $list[1..(F_{k-1} - 1)]$.
2. If $target = list[F_{k-1}]$, then the sought element is found and its index will be F_{k-1}.
3. If $target > list[F_{k-1}]$, then the index of the sought element is greater than F_{k-1}. In this case, the search continues after changing the left bound of the considered part of the array by $(F_{k-1} + 1)$: $list[(F_{k-1} + 1)..F_k]$.

However, if N is not the Fibonacci number F_k for any integer $k \geqslant 2$, then we shall increase the size of the array $list$ up to the nearest F_k. After that, the sequence of comparisons described above is used.

Below is the realization of the Fibonaccian search algorithm.

```
1   const N = 100;
2   type Mass = array[1..N] of integer;
3   function FibonaccianSearch(list: Mass;
4                              target: integer): integer;
5   // list — analysed array of N elements
6   // target — target value
7   var left, k, pos: integer;
8   begin
9       left:=1;
10      result:=0;
11      // find the smallest k, satisfying
12      // the condition F[k]>=N
13      k:=BinarySearch_ge();
14      while k>0 do
15          begin
16              pos:= left + F[k−1]−1;
17              if pos<= N then
18                  case compare(list[pos],target) of
19                      −1: begin
20                              left:= pos+1;
21                              k:= k−2;
22                          end;
23                      0: begin
24                              result:= pos;
```

```
25                          break;
26                      end;
27              +1: k:=k−1;
28          end
29                  else k:=k−1;
30      end;
31 end;
```

The text of the program Fibonaccian Search uses auxiliary functions compare (a, b) and BinarySearch_ge().

The first of them is described above on page 371.

The second function BinarySearch_ge() is based on the binary search algorithm and returns the least $k \in \mathbb{N} \setminus \{1\}$, satisfying the condition $F_k \geqslant N$, where N is the size of the analyzed array.

The variable *left* stores the index of the left bound of the considered part of the array; the variable *pos* contains the index of the element with which the next comparison will be performed. The global array F stores the pre-calculated Fibonacci sequence, and the array of this realization begins with zero: $F[0] = \mathbf{MinInt}$ is the least integer type value. Such an array organization method is required for correct operation of the algorithm at the values $k = 1$ and $k = 2$.

Consider the operation of the Fibonaccian search algorithm by the example of the search for the key value *target* $= 4$ in the ordered array $\boxed{1\,2\,3\,4\,5\,6\,7\,8}$.

At the beginning of the algorithm operation $left = 1$, $k = 6$.

The first comparison is performed with the element whose index is equal to $left + F_5 - 1 = 5$. Since *target* $< list[5]$, the sought element is contained in the left part of the array $\boxed{1\,2\,3\,4\,\,\,\,}$.

Then follows the comparison (*target* $?\,list[3]$) and, finally, the equality is checked *target* $= list[4]$ (see Fig. 12.5).

In order to find the position of the element equal to *target*, three operations of comparison were required.

Let us estimate the asymptotic complexity of the Fibonaccian search algorithm in a supposition that the target value is necessarily situated in the array *list*.

The Worst Case

Assume that the element equal to *target* is present at the first place of the array of size F_k. Then the worst case is being realized, and the number of the operations of comparison $W(N)$ will be subject to the recurrence relation

$$\begin{cases} W(N) = W\left(\dfrac{F_{k-1} - 1}{F_k} N\right) + 1, & N > 1, \\ W(1) = 1. \end{cases}$$

Indeed, at each subsequent step of the Fibonaccian search algorithm the size of the analyzed part of the array decreased down to the value $(F_{k-1} - 1)$.

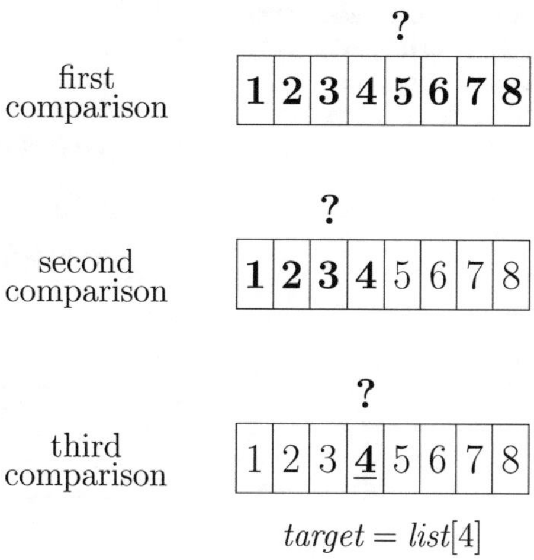

Fig. 12.5 Example of the operation of the Fibonaccian search algorithm. Bold letters highlight the elements, among which, at this step of the algorithm, a search is performed; question mark marks the element being compared with the value *target*; the sought value found is underlined

Since $\lim\limits_{k \to \infty} \dfrac{F_{k-1} - 1}{F_k} = \dfrac{1}{\varphi}$, where $\varphi = \dfrac{1 + \sqrt{5}}{2}$, then at sufficiently big N the above recurrence relation will take the form

$$\begin{cases} W(N) = W\left(\dfrac{1}{\varphi}N\right) + 1, & N > 1, \\ W(1) = 1. \end{cases}$$

By the method of substitution, it is easy to show that the asymptotic relation $W(N) = \log_\varphi N + O(1)$ takes place. However, if N is not the Fibonacci number F_k for any integer $k \geqslant 2$, then this estimate for $W(N)$ will not change.

The Average Case

Note that the recursive call of the function FibonaccianSearch will occur in ($F_{k-1} - 1$) cases on the array of size $\dfrac{F_{k-1} - 1}{F_k} N$ and in F_{k-2} cases on the array of size $\dfrac{F_{k-2}}{F_k} N$. The total number of such cases will be F_k. Hence, for the number of comparisons $A(N)$ in the average case we have:

$$\begin{cases} A(N) = \dfrac{F_{k-1} - 1}{F_k} A\left(\dfrac{F_{k-1} - 1}{F_k} N\right) + \dfrac{F_{k-2}}{F_k} A\left(\dfrac{F_{k-2}}{F_k} N\right) + 1, & N > 1, \\ A(1) = 1. \end{cases}$$

At sufficiently big N we obtain:

$$\begin{cases} A(N) = \dfrac{1}{\varphi} A\left(\dfrac{1}{\varphi} N\right) + \dfrac{1}{\varphi^2} A\left(\dfrac{1}{\varphi^2} N\right) + 1, & N > 1, \\ A(1) = 1. \end{cases}$$

Fig. 12.6 Calculation of the position of the next array element during the interpolation search

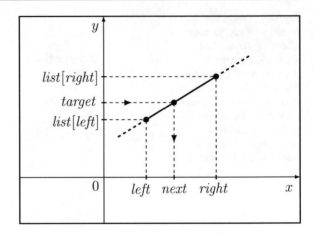

As shown in Exercise **12.24**, the computational complexity $A(N)$ of the Fibonaccian search algorithm in the average case will be

$$A(N) = \frac{\varphi}{\sqrt{5}} \log_\varphi N + O(1).$$

Interpolation Search Algorithm

Interpolation search (from Latin *interpolāre*—to refurbish) is used for ordered data. This type of search is based on the supposition that the values of the elements in an array grow linearly. The sought values is compared with the values of the element whose index is calculated as a coordinate of a point on a line connecting the points $(left, list[left])$ and $(right, list[right])$. The obtained line is defined by the equation

$$y - list[left] = \frac{list[right] - list[left]}{right - left}(x - left).$$

By this value $y = target$, we obtain the index of the element with which the next comparison will be performed (Fig. 12.6):

$$next = left + \left\lfloor \frac{right - left}{list[right] - list[left]}(target - list[left]) \right\rfloor.$$

After comparing *target* and *list*[*next*] the interpolation search algorithm continues the search among the elements $left \dots (next - 1)$ or $(next + 1) \dots right$ or stops its operation if $list[next] = target$.

Analysis of the algorithm shows [56] that the number of comparisons required for the algorithm to achieve the average-case result, is $A(N) = \Theta(\log_2 \log_2 N)$, however, in the worst case $W(N) = O(N)$.

12.3   Sort Algorithms

Quick search algorithms require ordered representation of data, this is why the problem of sorting is important from the practical point of view.

Let us consider the non-empty finite set $A = \{a_1, a_2, a_3, \ldots\}$ with antisymmetric and transitive relation introduced into it S, all elements of which are pairwise comparable. Assume that out of N elements of the set A the vector $\mathbf{a} = (a_{i_1}, a_{i_2}, \ldots, a_{i_N})$ is formed. The **sorting problem** consists in permutation (in other words, in changing the order) of the components of the vector $\mathbf{a}$, so that its components become ordered as a result:

> *input data*: $(a_{i_1}, a_{i_2}, \ldots, a_{i_N})$;
> *output data*: $(a_{j_1}, a_{j_2}, \ldots, a_{j_N})$ on condition
> $$(a_{j_1} \, S \, a_{j_2}) \textbf{ and } (a_{j_2} \, S \, a_{j_3}) \textbf{ and } \ldots \textbf{ and } (a_{j_{N-1}} \, S \, a_{j_N}).$$

The input data can be represented in the form of an array or in a more complex form, for example, as a list, each element of which has several fields. The field of the element by which sorting is performed is called a **key**. It is often required to order the data taking into account the values of other fields apart from the selected key field. In this case, we are dealing with the **secondary key** sorting [36].

Two types of sort algorithms are distinguished depending on the possibility of direct access to data: *internal* and *external* [36].

Internal sorting orders the data situated in the array whose size allows placing it into the random-access memory. In this case, there is a possibility of arbitrary access to any memory cell. Otherwise, if the array, due to its size, can only be placed in relatively slow peripheral devices, then the **external sorting** is used. The internal sorting is often used as a basis for the external ones; the latter is usually used for processing separate parts of a data array by some internal algorithm, operating in the random-access memory, and then they are united into a single array in a peripheral device.

The behavior of the sort algorithm is called **natural**, if the sorting time $T(N)$ is minimal when the elements are arranged in the required order, increases when the degree of order decreases, and becomes maximal when the elements of the array being sorted are arranged in the reverse order.

Stable is the sorting that does not change the order of the sorted elements with the same values of the key [36].

Lower bound theorem of the sorting complexity states that *any sorting based on the use of comparisons has in the worst case the complexity* $W(N) = \Omega(N \log_2 N)$.

Let us provide the proof of this statement [43]. The sort algorithm, as the output data, carries out permutation of the original N elements. The number of possible outcomes is equal to the number of permutations of the elements, i.e., $N!$ (see chapter "Combinatorics" on p. 141).

The decision tree for the sort algorithms is constructed as follows. Each vertex represents some comparison $(a \, ? \, b)$, realized by the algorithm. One of the two sub-trees, relating to a given vertex, corresponds to the outcome $a < b$, the other—to $a > b$ (for simplicity, we assume that for all elements $a \neq b$). The edges of the decision tree are the transitions between the states, and the leaves represent the possible output data of the algorithm. In the worst case, the number of comparisons is equal to the height of the tree.

For any binary tree of height h and number of leaves l the inequality is valid $2^h \geqslant l$ or $h \geqslant \lceil \log_2 l \rceil$. Hence, the height of the decision tree for the comparison-based sorting satisfies the relation $W(N) \geqslant \lceil \log_2 N! \rceil$. The expression in the right side of this inequality can be represented as:

$$\lceil \log_2 N! \rceil \geqslant \sum_{i=1}^{N} \log_2 i > \frac{N}{2} \log_2 \frac{N}{2} = \Omega(N \log_2 N).$$

Thus, the lower bound theorem is proved.

An algorithm is referred to as **asymptotically optimal**, if for its worst-case complexity the estimate is valid

$$W(N) = \Theta(N \log_2 N).$$

In the book [85], the following classification of sorting is presented, based on the comparison of the elements:

(1) sorting with the help of insertion;
(2) sorting with the help of selection;
(3) sorting with the help of exchange.

Examples of algorithms belonging to these classes are presented in Fig. 12.7.

Insertion Sort

In the process of the algorithm operation, the sorted part of the array is being generated. For this, at each step of the algorithm, an input data element is inserted into the desired position in the already sorted array until the set of the input data is exhausted.

The example of the insertion sort is presented in Table 12.1. The sorted part of the array being consecutively formed is highlighted with bold letters and underlined.

Below is the realization of the considered sort algorithm.

```
1   const N=100;
2   type Mass = array[ 1..N] of integer;
3   procedure InsertionSort(var list :Mass);
4   var i ,j ,temp:integer;
5   // list — array of N elements subject to sorting
6   begin
7       for i := 2 to N do
8           begin
9               temp:=list[i];
10              j:=i−1;
11              while (j>=1) and (list[j]>temp) do
12                  //shift all elements greater than the next
```

```
13                          begin
14                              list[j+1]:=list[j];
15                              j:=j−1;
16                          end;
17                      list[j+1]:=temp;
18                  end;
19  end;
```

Let us analyse the insertion sort algorithm. As a basic operation, choose the operation of comparison *list[j]>temp* in the row number 11.

The Worst Case

Let the added element be smaller than all other elements; then it always appears at the beginning of the array. Such a situation is realized; for example, when the elements of the original array are arranged in descending order. In this case, the number of comparisons is

$$W(N) = \sum_{i=2}^{N} (i - 1) = \frac{(N - 1)N}{2} = O(N^2).$$

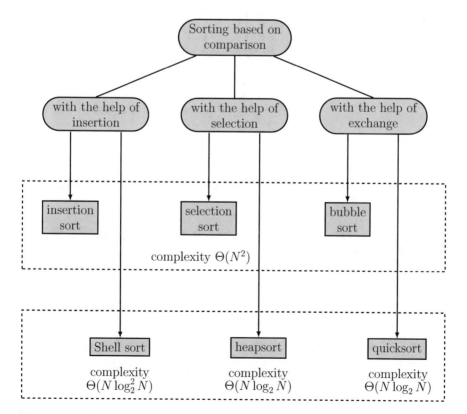

Fig. 12.7 Classification of sort algorithms based on the operations of comparison

Table 12.1 Example of the insertion sort. The sorted part of the array is highlighted with bold letters and underlined

Original array:	4	7	3	1	8	6	5	2
Before the first run:	**4**	7	3	1	8	6	5	2
Run 1:	**4**	**7**	3	1	8	6	5	2
Run 2:	**3**	**4**	**7**	1	8	6	5	2
Run 3:	**1**	**3**	**4**	**7**	8	6	5	2
Run 4:	**1**	**3**	**4**	**7**	**8**	6	5	2
Run 5:	**1**	**3**	**4**	**6**	**7**	**8**	5	2
Run 6:	**1**	**3**	**4**	**5**	**6**	**7**	**8**	2
Run 7:	**1**	**2**	**3**	**4**	**5**	**6**	**7**	**8**

Fig. 12.8 There exist i possibilities for placing the next element of the array among $(i-1)$ elements during the insertion sort

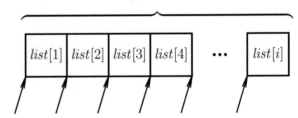

i elements of the array

the number of possibilities is equal to i

The Average Case

It is required to sum the operations of comparison for insertion of each ith element, starting from $i = 2$, i.e., $A(N) = \sum\limits_{i=2}^{N} A_i$.

The average number of comparisons for determining the position of the next element is (Fig. 12.8):

$$A_i = \frac{1}{i} \left(\sum_{j=1}^{i-1} j + (i-1) \right) = \frac{i+1}{2} - \frac{1}{i}.$$

Now it remains to sum all A_i from each of $N-1$ elements of the array.

$$A(N) = \sum_{i=2}^{N} A_i = \sum_{i=2}^{N} \left(\frac{i+1}{2} - \frac{1}{i} \right) =$$

$$= \frac{1}{2} \sum_{i=2}^{N} (i+1) - \sum_{i=2}^{N} \frac{1}{i} = \frac{1}{4}(N+4)(N-1) - \left(\sum_{i=1}^{N} \frac{1}{i} - 1 \right) =$$

$$= \frac{1}{4} \left(N^2 + 3N - 4 \right) - \left(\ln N + \gamma - 1 + O\left(\frac{1}{N} \right) \right) = O(N^2).$$

When estimating the sum $\sum_{i=2}^{N} \frac{1}{i} = H_N - 1$ the relation for asymptotics of harmonic numbers was used (A.62) (see also Exercise **11.14.**).

The simple insertion sort algorithm has a series of advantages, among which we can single out:

(1) low operation time for small datasets and on partially sorted sets as compared to the average case;
(2) use of just a limited volume of additional memory;
(3) stability;
(4) simple realization.

As a disadvantage of the algorithm shall be considered its high time complexity $\Theta(N^2)$.

Bubble Sort

The name of the algorithm appeared by analogy of displacements of the array elements $list[1..N]$ during sorting with the motion of gas bubbles in a vessel with water. The comparison operations are performed for two adjacent elements of the array, and the elements are exchanged when necessary. Here is the text of a bubble sort program.

```
1   const N=100;
2   type Mass = array[1..N] of integer;
3   procedure BubbleSort(var list:Mass);
4   var i,j: integer;
5   // list — array of N elements subject to sorting
6   begin
7       for i := 2 to N do
8           for j := N downto i do
9               if list[j−1]▷ list[j] then
10                  swap(list[j−1],list[j]);
11  end;
```

The procedure swap(a, b) exchanges the values of the variables a and b; two possible realizations of the procedure are considered in Exercises **9.1** and **9.2**.

Table 12.2 Example of the bubble sort. The elements that have occupied their places are highlighted with bold letters and underlined

Original array:	4	7	3	1	8	6	5	2
Run 1:	**1**	4	7	3	2	8	6	5
Run 2:	**1**	**2**	4	7	3	5	8	6
Run 3:	**1**	**2**	**3**	4	7	5	6	8
Run 4:	**1**	**2**	**3**	**4**	5	7	6	8
Run 5:	**1**	**2**	**3**	**4**	**5**	6	7	8
Run 6:	**1**	**2**	**3**	**4**	**5**	**6**	7	8
Run 7:	**1**	**2**	**3**	**4**	**5**	**6**	**7**	8

The example of the bubble sort algorithm operation is presented in Table 12.2.

The analysis of the bubble sort algorithm includes, as usual, the review of the best, worst, and average cases. However, for this sorting, the number of comparisons remains unchanged for any set of input data of size N, since two cycles are always performed the defined number of times regardless of the degree of order of the original array, i.e., $B(N) = W(N) = A(N)$.

During the first run, the number of comparisons is $N - 1$. Then the cycle will be repeated with $N - 2$ comparisons, etc.

$$W(N) = \sum_{i=2}^{N} \sum_{j=N}^{i} 1 = \sum_{i=2}^{N}(N - i + 1) = (N + 1)\sum_{i=2}^{N} 1 - \sum_{i=2}^{N} i =$$
$$= (N + 1)(N - 1) - \frac{N + 2}{2}(N - 1) = \frac{N^2 - N}{2} = \Theta(N^2).$$

The following improvements can be introduced into the algorithm:

(1) Remember the indices of the elements that participated in the last exchange, and in the next run consider the greatest of these indices as the lower bound of the part of the array being sorted;

(2) Not only raise the "light" elements to the beginning of the array, but also immediately lower the "heavy" ones, i.e., perform the array runs in opposite directions.

Realization of the mentioned improvements will result in **shaker** sorting.

Selection Sort

During the election sort, the element with the least value is found and exchanged for the first element of the array. Then the element with the least value out of the remaining $N - 1$ elements is selected and changed for the second element, and so on until the exchange of the two last elements.

Table 12.3 Example of the selection sort. The elements that have occupied their places are highlighted with bold letters and underlined

Original array:	4	7	3	1	8	6	5	2
Run 1:	**1**	7	3	4	8	6	5	2
Run 2:	**1**	**2**	3	4	8	6	5	7
Run 3:	**1**	**2**	**3**	4	8	6	5	7
Run 4:	**1**	**2**	**3**	**4**	8	6	5	7
Run 5:	**1**	**2**	**3**	**4**	**5**	6	8	7
Run 6:	**1**	**2**	**3**	**4**	**5**	**6**	7	8
Run 7:	**1**	**2**	**3**	**4**	**5**	**6**	**7**	8

An example of the described operation with the array is presented in Table 12.3.

```
 1  const N = 100;
 2  type Mass = array[1..N] of integer;
 3  procedure SelectSort(var list:Mass);
 4  var i,j,k,temp: integer;
 5  begin
 6     for i := 1 to N− 1 do
 7        begin
 8           k := i;
 9           temp := list[i];
10           for j := i + 1 to N do
11           //select the element with the least value
12              if list[j]<temp then
13                 begin
14                    k := j;
15                    temp := list[j];
16                 end;
17              list[k] := list[i];
18              list[i] := temp;
19        end;
20  end;
```

The number of comparisons performed by the algorithm SelectSort does not depend on the degree of order of the array and is equal to

$$W(N) = \sum_{i=1}^{N-1} \sum_{j=i+1}^{N} 1 = \Theta(N^2).$$

Despite the equal number of comparison operations performed by the bubble sort and the selection sort algorithms, the number of the exchange operations when using

Table 12.4 Example of the Shell sort operation for the increment values 1, 3, 7, ...

Original array:	4	7	3	1	8	6	5	2
After run with increment 7:	2	7	3	1	8	6	5	4
After run with increment 3:	1	4	3	2	7	6	5	8
After run with increment 1:	1	2	3	4	5	6	7	8

the selection sort does not depend on the degree of order of the array and is equal to $\Theta(N)$.

Shell Sort

The insertion sort considered above allows moving just one element to its place in one run of a cycle. Shell[4] suggested providing for a possibility of permutation of the elements positioned at a fixed distance from each other [64].

Let us introduce the definition of h-ordered array: an array *list* is h-**ordered** if the collection of the elements positioned at the distance h from each other forms a sorted array [61]. The Shell sort algorithm carries out the h-ordering of the array for several decreasing values of h, called **increments**, the last of which is equal to $h = 1$. The entire array of length N is viewed as a collection of interleaved subarrays, and the sorting reduces to multiple application of the insertion sort (Table 12.4).

Analysis of the Shell sort is rather complex and is based on the concept of **inversion** [36,49]. Inversion is a pair of elements arranged in incorrect order, i.e. a pair (a_i, a_j), for which $a_i > a_j$, where $1 \leqslant i < j \leqslant N$. For example, in an array of size N, whose element is written in the reverse order, the number of inversions is $N(N-1)/2$.

Each comparison in the insertion sort results in deleting no more than one inversion, but in the Shell sort the exchange of elements can exclude several inversions at once.

The estimate of complexity of the defined sequence of increments is a serious mathematical problem. For example, the number of comparisons in the average case for $h_s = 1, 2, 4, 8, 16, 32, \ldots$ and N, equal to the natural power of two, is expressed by the formula [21]:

$$A(N) = \frac{N}{16} \sum_{i=1}^{\log_2 N} \frac{\Gamma(2^i-1)}{2^i \Gamma(2^i)} \sum_{r=1}^{2^{i-1}} r(r+3)2^r \frac{\Gamma(2^i-r+1)}{\Gamma(2^{i-1}-r+1)} + N \log_2 N - \frac{3}{2}(N-1),$$

where $\Gamma(x)$—gamma function. Some results of the Shell algorithm analysis obtained by now are shown below [61].

Theorem of h- and k-ordering. *The result of h-sorting of the k-ordered array generates the h- and k-ordered array.*

For proving this theorem, the concept of h-**inversion** is introduced. This is the inversions, formed by the elements at the distances divisible by h. The proof of the theorem is shown in Exercise **12.45**.

---

[4]Donald Lewis Shell (1924–2015)—American researcher in the sphere of computer sciences.

There are known estimates of complexity for some sequences of increments; for example, for the sequence of the increments $1, 3, 7, 15, 31, 63, \ldots$ the worst-case asymptotic complexity is $O(N^{3/2})$. **Pratt**[5] suggested the sequence of values h, leading to the algorithm of complexity $W(N) = \Theta(N(\log_2 N)^2)$. It should be noted that the advantage of using this sequence becomes apparent only for sufficiently big size values of the array N.

Quicksort

One of the most well-studied and actively used in practice sorting algorithms is the quicksort [13,30]. The author of this algorithm, Hoare, working at the problem of machine translation, formulated an algorithm whose main property was a high information processing speed, which is reflected in its name [43].

Let us select a **pivot element** $list[v]$ (some authors call the pivot element a **partition element** [43]) and divide the array $list[left..right]$ into two parts: The first will include the elements smaller or equal to the selected one, the second will include the elements that are greater or equal, i.e., the array elements will satisfy the following properties:

(1) the $list[v]$ for some v occupies the final position;
(2) $list[k] \leqslant list[v]$ for all $k = left, left + 1, \ldots, v$;
(3) $list[k] \geqslant list[v]$ for all $k = v, v + 1, \ldots, right$,

where $left$ and $right$—lower and upper bounds of the array part being processed.

The recursive application of the partitioning procedure results in placing all elements at the final positions.

Permutation of elements is performed as follows. Let us introduce the indices i and j, whose original values are $i = left$, $j = right$. We are scanning the array moving "from left to right," i.e., increasing the index i, until we encounter the element $list[i] \geqslant v$, and then we are scanning the array "from right to left," decreasing the index j, until we encounter $list[j] \leqslant v$. When necessary, we will swap the elements $list[i]$ and $list[j]$ until all indices intersect, and $i \geqslant j$. At the final stage of the procedure, exchange $list[j]$ with the pivot element. Note that we can prevent the index i from exceeding the limits by introducing a **barrier** $list[right + 1]$, which takes a value greater than or equal to $\max_k(list[k])$.

Table 12.5 demonstrates the quicksort algorithm operation. As the pivot element of the array $list[left..right]$ we can select, for example, the element $list[left]$. From Table 12.5, we can see how the array ordering takes place in this case. The function Partition places the pivot element of the array $list$ in accordance with the above conditions.

```
1  function Partition(var list:Mass;
2                         left,right:integer):integer;
3  var i,j,v: integer;
4  begin
```

[5] Vaughan Ronald Pratt (born 1944)—American researcher specializing in the sphere of informatics and computing.

```
 5 │   v:=list[left];            // select the first element
 6 │                             // as the pivot
 7 │   i:=left;
 8 │   j:=right+1;
 9 │   repeat
10 │      repeat j:=j−1 until list[j]<=v;
11 │      repeat i:=i+1 until list[i]>=v;
12 │      swap(list[i],list[j]);  // elimination
13 │                             // of inversion
14 │   until (i>=j);
15 │   swap(list[i],list[j]);     // cancellation of
16 │                             // the last exchange
17 │   swap(list[left],list[j]);  // placement
18 │                             // of the pivot element
19 │   result:=j;
20 │ end;
```

The text of the main procedure QuickSort with recursive call of the function Partition is given below. For ordering the array $list[1..N]$, the procedure call format is as follows: QuickSort($list, 1, N$).

```
 1 │ const N=100;
 2 │ type Mass = array[1..N+1] of integer;
 3 │ procedure QuickSort(var list:Mass; left,right:integer);
 4 │ var v:integer;
 5 │ begin
 6 │   list[N+1]:=MaxInt;   // placing a barrier
 7 │   if left<right then
 8 │      begin
 9 │         v:=Partition(list,left,right);
10 │         QuickSort(list,left,v−1);
11 │         QuickSort(list,v+1,right);
12 │      end;
13 │ end;
```

Investigate the asymptotic complexity of the quicksort algorithm.

The Worst Case

Consider the case when the pivot element is either greater or smaller than all elements of the array. Then the partitioning is performed into maximally unequal parts, consisting of 1 and $N − 1$ elements:

$$\boxed{1 \quad N − 1}.$$

Each recursive call removes one element from consideration. As a result, we obtain the worst-case complexity

Table 12.5 Example of the quicksort. The elements that have occupied their places are highlighted with bold letters and underlined. The array segments for which the partitioning procedure is called are taken in square brackets

Original array:	4	7	3	1	8	6	5	2
Pivot element in cell 4:	[1	2	3]	**4**	8	6	5	7
Pivot element in cell 1:	**1**	[2	3]	**4**	8	6	5	7
Pivot element in cell 2:	**1**	**2**	**3**	**4**	[8	6	5	7]
Pivot element in cell 8:	**1**	**2**	**3**	**4**	[7	6	5]	**8**
Pivot element in cell 7:	**1**	**2**	**3**	**4**	[5	6]	**7**	**8**
Pivot element in cell 5:	**1**	**2**	**3**	**4**	**5**	**6**	**7**	**8**

$$W(N) = (N+1) + N + \cdots + 3 = O(N^2).$$

This situation is realized for the array sorted in the required order.

The Average Case

The array partitioning procedure makes $N + 1$ comparisons, partitioning the array of N elements, as described on p. 387, and $2\left\lceil\dfrac{N}{2}\right\rceil$ comparisons for the array formed by the pairwise equal elements. If the pivot element index is V, then the recursive call will be performed for the array of length $V - 1$ and $N - V$. The respective recurrence relation has the form:

$$\begin{cases} A(N) = N + 1 + \dfrac{1}{N} \sum_{i=1}^{N} (A(i-1) + A(N-i)) & \text{for } N \geqslant 2, \\ A(0) = A(1) = 0. \end{cases}$$

Following [61], rewrite this relation in the form

$$A(N) = N + 1 + \frac{1}{N} \sum_{i=0}^{N-1} (A(i) + A(N-i-1)),$$

and in the sum change the summation order $\sum_{i=0}^{N-1} A(i) = \sum_{i=0}^{N-1} A(N-i-1)$.

We obtain the equation

$$A(N) = N + 1 + \frac{2}{N} \sum_{i=0}^{N-1} A(i-1),$$

or

$$NA(N) = N(N+1) + 2 \sum_{i=0}^{N-1} A(i).$$

Change the variable $N \to N - 1$:

$$(N - 1)A(N - 1) = N(N - 1) + 2 \sum_{i=0}^{N-2} A(i).$$

From the obtained equations follows the relation

$$NA(N) = (N + 1)A(N - 1) + 2N.$$

After dividing both parts of this equality by $N(N + 1)$ we obtain

$$\frac{A(N)}{N + 1} = \frac{A(N - 1)}{N} + \frac{2}{N + 1}.$$

Change the variable $\widetilde{A}(N) = \dfrac{A(N)}{N + 1}$, following which, the recurrence relation for $\widetilde{A}(N)$ will have the form:

$$\begin{cases} \widetilde{A}(N) = \widetilde{A}(N - 1) + \dfrac{2}{N + 1} & \text{for } N \geqslant 2, \\ A(0) = A(1) = 0. \end{cases}$$

Using the method of substitution, we obtain the solution $\widetilde{A}(N) = 2 \sum_{i=3}^{N+1} \dfrac{1}{i}$.

An alternative method of solving the recurrence relation for the values $\widetilde{A}(N)$ is suggested in Exercise **8.115**.

As a result, taking into account the asymptotic relation (A.62), write the final result for the quicksort algorithm complexity:

$$A(N) = (N + 1)\widetilde{A}(N) = 2N \ln N + \Theta(N).$$

12.4 Order Statistics

Searching for the minimum, maximum, and median of several numbers represents special instances of the order statistics problem.

Definition: **the kth-order statistics** of the non-empty finite set $A = \{a_1, a_2, \ldots, a_N\}$ with antisymmetric and transitive relation introduced into it, all elements of which are pairwise comparable, is the kth component of the vector $(a_{j_1}, \ldots, a_{j_k}, \ldots, a_{j_N})$, whose components are ordered ($1 \leqslant k \leqslant N$). For example, the minimum of the numbers $a_1, a_2, \ldots, a_N$ is the first-order statistics of this set.

Finding of order statistics and, in particular, the median is required for statistical studies.

Assume that it is necessary to determine the kth biggest element of the array $list[1..N]$; denote this element by t.

One of the methods of solving the problem of the kth-order statistics consists in using the procedure partition (see p. 387).

Suppose that the pivot element v has occupied in the array $list$ the final position $list[i]$. Then the number of elements of the array $list$, smaller than or equal to v, is $i - 1$. Similarly, $N - i$ elements have the value no less than v. Then, one of the three possibilities is realized:

(1) $k < i$ and the sought element t is located in the subarray $list[1..(i - 1)]$;
(2) $k = i$ and $t = list[i]$;
(3) $k > i$ and t is located in the subarray $list[(i + 1)..N]$.

The function SelectPart determines the index of the kth-order statistics in the original array using the procedure Partition. The auxiliary function compare(k, v) performs comparison $(k\,?\,v)$ (see p. 371).

```
 1  function SelectPart(var list:Mass;k:integer):integer;
 2  var found:boolean;
 3      left,right,v:integer;
 4  begin
 5      left:=1;
 6      right:=N+1;
 7      list[N+1]:=MaxInt;
 8      found:=false; // change of logic variable
 9                    // found indicates the end of
10                    // operation
11      repeat
12          v:=Partition(list,left,right);
13          case compare(k,v) of
14              -1: right:=v-1;
15               0: begin
16                      found:=true;
17                      result:=v;
18                  end;
19              +1: left:=v+1;
20          end;
21      until found;
22  end;
```

Asymptotic analysis of the algorithm SelectPart generally copies the study of the quicksort complexity (see p. 388). The average-case complexity of the algorithm SelectPart is $A(N) = \Theta(N)$, the worst-case complexity is $W(N) = \Theta(N^2)$.

The second method of solving the problem of the kth-order statistics leads to the algorithm SelectOpt of complexity $O(N)$. Let us consider this algorithm [32].

Assume that it is necessary to determine the kth largest element of the array $list[1..N]$, where $N > 1$. For simplicity, suppose that all elements of the array $list$

are pairwise different integers. Introduce the auxiliary variables *left* and *right* with original values $left = 1$, $right = N$. Represent operation of the algorithm SelectOpt in the form of the following steps.

1. Partition the elements of the array *list*[*left*..*right*] into $\left\lfloor \dfrac{N}{w} \right\rfloor$ subarrays, w elements in each, where w is an integer greater than one.

2. Find the medians m_i of each subarray, where $1 \leqslant i \leqslant \left\lfloor \dfrac{N}{w} \right\rfloor$.

3. Determine the median μ of the set of medians $M = \left\{ m_i : 1 \leqslant i \leqslant \left\lfloor \dfrac{N}{w} \right\rfloor \right\}$.

4. The value μ is used as the pivot element for the procedure Partition (see p. 387), which is applied at this step to the array *list*[*left*..*right*].

5. If the sought element t is not found, then narrow the search range by changing the value of the variables *left* or *right* and return to the first step. If the element t is found, the algorithm stops.

Note the peculiarities of the algorithm SelectOpt.

First, partitioning of the analysed array *list*[*left*..*right*] will result in that exactly $d - w \left\lfloor \dfrac{d}{w} \right\rfloor$ elements, where $d = right - left + 1$ is the array size, will not fall into any subarray. When performing the steps 2–3, these elements are not used, but during the Partition procedure, at the fourth step, they should be taken into consideration.

Second, the medians m_i at the second step can be determined by sorting each subarray, following which the medians will be positioned at places with the index $\left\lceil \dfrac{w}{2} \right\rceil$.

Third, one more peculiarity is associated with determining the value μ. The elements of the set of medians M can be saved into the auxiliary array, but this will require allocation of additional memory. An alternative variant consists in transferring the elements m_i to the beginning of the array *list*.

Taking into account the mentioned peculiarities, let us present one of the possible realizations of the algorithm SelectOpt. Here, the value w is the algorithm parameter and is defined in the form of the global constant.

```
1    const w = 5;
2    function SelectOpt(var list :Mass;
3                                 k, left , right :integer ):integer ;
4    var i ,d,dd,v,temp:integer ;
5    begin
6        while (true) do
7            begin
8                d:=right−left+1;
9                if (d <= w) then
10                    begin
11                        InsertionSort( list , left , right );
12                        result:= left+k−1;
```

```
13                        break;
14                    end;
15                dd:= Floor(d/w);   // number of subarrays
16                for i:=1 to dd do
17                    begin
18                        InsertionSort(list ,
19                                    left+(i−1)∗w, left+i∗w−1);
20                        swap( list[left+i−1],
21                                list[left+(i−1)∗w+Ceil(w/2)−1]);
22                    end;
23                v:=SelectOpt(list ,Ceil(dd/2),
24                            left , left+dd−1);
25                swap( list[left],list[v]);
26                v := Partition(list ,left ,right );
27                temp := v−left+1;
28                case compare(k,temp) of
29                    −1: right:=v−1;
30                     0: begin
31                            result := v;
32                            break;
33                        end;
34                    +1: begin
35                            k := k−temp;
36                            left := v+1;
37                        end;
38                end;
39            end;
40  end;
```

From Fig. 12.9, we can conclude that $\left\lceil \left\lfloor \dfrac{N}{w} \right\rfloor /2 \right\rceil$ medians m_i do not exceed the value μ and $\left\lceil \left\lfloor \dfrac{N}{w} \right\rfloor /2 \right\rceil$ medians m_i are no smaller than μ.

Hence, $R = \left\lceil \dfrac{w}{2} \right\rceil \left\lceil \left\lfloor \dfrac{N}{w} \right\rfloor /2 \right\rceil$ elements of the original array *list* do not exceed μ and R elements *list* are no smaller than μ.

At sufficiently big values of N we have the asymptotic estimate: $N - R = \dfrac{3N}{4} + \Theta(1)$. Due to this, the next iteration of the cycle **while** will be applicable to the array that has no more than $\dfrac{3N}{4} + \Theta(1)$ elements.

Now we can write the recurrence relation for the number of comparisons of elements of the array *list* in the worst case:

$$\begin{cases} T(N) = T\left(\dfrac{N}{w}\right) + T\left(\dfrac{3N}{4}\right) + \Theta(N), \ N > w, \\ T(1), T(2), \ldots, T(w) \text{ — const.} \end{cases}$$

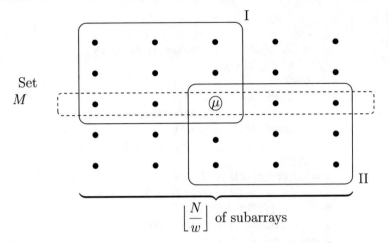

$$\left\lfloor \frac{N}{w} \right\rfloor \text{ of subarrays}$$

Fig. 12.9 The median of the set of medians μ of the array *list*. The array elements are denoted by dots and are positioned in each subarray from top to bottom in the order of non-increasing. The values *list*[i] in the range I are no greater than μ, in the range II are no smaller than μ

In Exercise **12.70**, it is shown that at $w \geqslant 5$ the obtained recurrence relation is satisfied by the function $T(N) = \Theta(N)$. So, the algorithm SelectOpt has the complexity that is linear by the number of elements of the original array.

12.5 Problems

12.1. Give the exact estimate of the asymptotic complexity of the factorial calculation algorithm given in the text of the chapter.

12.2. Estimate the complexity of the factorial calculation algorithm without recursion.

```
1  function factorial (n : integer): integer;
2  //calculating the factorial of a number n
3  //without recursion
4  var
5      i , f : integer;
6  begin
7      f := 1;
8      for i := 2 to n do
9          f := f * i;
10     factorial := f;
11 end;
```

12.3. Write the recursive algorithm for calculating the nth Fibonacci number. What is its complexity?

12.4. Verify that the following algorithm calculates the nth Fibonacci number with a memory-bounded complexity.

```
 1  function fib(n:integer):integer;
 2  //calculation of the nth Fibonacci number
 3  var a,b,i:integer;
 4  begin
 5     a := 0;
 6     b := 1;
 7     for i := 2 to n do
 8         begin
 9             b:=a+b;
10             a:=b-a;
11         end;
12     if n=1 then fib:=1
13             else fib:=b;
14  end;
```

12.5. Estimate the complexity of the recursive algorithm for calculation of the function $\lfloor \log_2 n \rfloor$, where $n \in \mathbb{N}$.

```
 1  function log2floor(n:integer):integer;
 2  //calculation of the function ⌊log₂ n⌋
 3  //of the natural number n
 4  begin
 5     if n = 1 then log2floor:=0
 6             else log2floor:=log2floor(n div 2)+1;
 7  end;
```

12.6. Using the recurrence definition, calculate the determinant of the matrix

$$A = \begin{bmatrix} 1 & 2 & 3 & 4 \\ 5 & 6 & 7 & 8 \\ 9 & 10 & 11 & 12 \\ 13 & 14 & 15 & 16 \end{bmatrix}.$$

*12.7. Calculate the determinant of a tridiagonal matrix that has n rows and n columns:

$$T(n) = \begin{bmatrix} 4 & 2 & 0 & \ldots & 0 \\ 2 & 4 & 2 & \ldots & 0 \\ 0 & 2 & 4 & \ldots & 0 \\ & & \cdots\cdots & & \\ 0 & 0 & 0 & \ldots & 4 \end{bmatrix}.$$

*12.8. Calculate the determinant of a tridiagonal matrix that has n rows and n columns:

$$T(n) = \begin{bmatrix} a & b & 0 & \dots & 0 \\ b & a & b & \dots & 0 \\ 0 & b & a & \dots & 0 \\ \multicolumn{5}{c}{\dotfill} \\ 0 & 0 & 0 & \dots & a \end{bmatrix}.$$

12.9. Calculation of the matrix determinant

In the text file `input.txt` are written, sequentially, by rows, the elements of the integer square matrix A. Using the recursion, calculate the determinant det A. Place the result into the text file `output.txt`.

*12.10. Estimate the number of multiplications performed by the recursive algorithm for calculating the determinant of the matrix of size $n \times n$ (see previous exercise).

*12.11. Prove that the exact number of multiplications performed by the recursive algorithm for calculating the determinant of the matrix of size $n \times n$ (see Exercises **12.9** and **12.10**), is equal to

$$T(n) = en\Gamma(n, 1) - n!,$$

where e is the base of natural logarithms, $\Gamma(n, x) = \int\limits_{x}^{\infty} e^{-t} t^{n-1} dt$—**incomplete gamma function** (for natural $n \in \mathbb{N}$ the following equality is valid

$$\Gamma(n, x) = (n - 1)! \, e^{-x} \sum_{k=0}^{n-1} \frac{x^k}{k!} \Big).$$

12.12. It is known that the square matrix A of size $n \times n$ contains the elements $a_{ij} \in \{0, 1\}$, where $1 \leqslant i, j \leqslant n$. Suggest a method for modification of the algorithm for calculating the determinant of the matrix A with the help of recursion (see Exercise **12.9**), decreasing the number of multiplication operations in this case.

12.13. Let the elements of the square matrix A of size $n \times n$ be equal to $a_{ij} = 1$ with the probability p and to $a_{ij} = 0$ with the probability $1 - p$, where $1 \leqslant i, j \leqslant n, 0 < p < 1$. Estimate the number of multiplications performed by the algorithm for calculation of the determinant that takes into account presence of zero elements in the matrix (see previous exercise).

*12.14. Let the square matrix A be tridiagonal:

$$A = \begin{bmatrix} a_{11} & a_{12} & 0 & 0 & \dots & 0 \\ a_{21} & a_{22} & a_{23} & 0 & \dots & 0 \\ 0 & a_{32} & a_{33} & a_{34} & \dots & 0 \\ 0 & 0 & a_{43} & a_{44} & \dots & 0 \\ \multicolumn{6}{c}{\dotfill} \\ 0 & 0 & 0 & 0 & \dots & a_{n,n} \end{bmatrix}.$$

How many multiplication operations does the algorithm for calculating the determinant of the matrix A that takes into account presence of zero elements in this matrix perform (see Exercise **12.12**)?

12.15. Give the exact estimate of the number of comparisons performed by the algorithm of sequential search for one of N elements of the array, in the case when the target element may be absent, and all of $(N + 1)$ possibilities of the target element position are equally probable.

12.16. How many comparisons will the binary search algorithm make in the array $\boxed{21\ 24\ 33\ 37\ 38\ 45\ 50}$ when searching for the value of *target* $= 22$?

12.17. How many comparisons on average will the binary search algorithm make in the array $\boxed{21\ 24\ 33\ 37\ 38\ 45\ 50}$, if the element equal to *target* is known to be present in the array?

12.18. For the array $\boxed{10\ 12\ 14\ 16\ 18\ 20\ 22\ 24}$ find:

(1) The minimal number of comparisons performed by the binary search algorithm;
(2) The maximal number of comparisons performed by the binary search algorithm.

12.19. For the array of $N = 8$ elements $\boxed{10\ 12\ 14\ 16\ 18\ 20\ 22\ 24}$ find:

(1) The average number of comparisons performed in case of a successful binary search in this array;
(2) The average number of comparisons performed in case of an unsuccessful binary search (in a supposition that the probability of the search for the target element in each of the $N + 1 = 9$ intervals, formed by the array elements, is equal).

12.20. Estimate the number of comparisons performed by the algorithm of binary search for one of N elements in the array, in the case when the target element may be absent and the probability of its search for each of the $N + 1$ intervals, formed by the array elements, is equal.

*12.21. Prove the relation (A.61), used in the analysis of the binary search algorithm.

12.22. Calculation of the variable *middle* by the binary search algorithm (see p 370) can result in overflow. This will happen if the sum of the values of the left and right bounds of the considered array part *left* and *right* exceeds the upper bound of variables type *left* and *right* in a certain programming language. Suggest an algorithm modification method that guarantees its correct operation in this case.

12.23. Construct the decision tree for the Fibonaccian search in the ordered array of size $N = 8$.

12.24. Using the method of substitution (see chapter "Recurrence Relations" on p. 273), prove that the asymptotic complexity of the Fibonaccian search algorithm in the average case has the form:

$$A(N) = \frac{\varphi}{\sqrt{5}} \log_\varphi N + O(1).$$

12.25. Search
1. In the text file `input.txt`, there is an array of N integers from 1 to N, arranged in random order without repetitions. Realize the search function in this array based on the sequential search algorithm. The main program should call the search function for each element of the array from 1 to N. Into the text file `output.txt`, enter the average number of comparisons performed by the sequential search program.
2. Perform the same for the binary search in the ordered array.
3. Perform the same for the Fibonaccian search in the ordered array.
4. Perform the same for the interpolation search in the ordered array.
Compare the calculated complexities of the algorithms considered in the problem.

12.26. Demonstrate the operation of the insertion sort algorithm by the example of the integer array $\boxed{20\,|\,12\,|\,18\,|\,16\,|\,24\,|\,10\,|\,22\,|\,14}$. Consider two cases:

(1) ascending sort of the elements;
(2) descending sort of the elements.

12.27. Is the insertion sort stable?

12.28. Insertion sort
In the text file `input.txt`, there is an array of N integers. Perform the ascending sorting of data using the insertion sort algorithm. Into the text file `output.txt` enter the sorted array (into the first row of the file) and the number of comparisons performed by the program (into the second row of the file).

12.29. In the **binary insertion sort** for finding the position of the current element *list[j]* in the ordered part of the array, the binary search is used. Will the asymptotic complexity of this sorting variant change in comparison with the standard insertion sort algorithm?

12.30. Is the binary insertion sort stable?

12.31. Demonstrate the operation of the bubble and shaker sort algorithms by the example of the array $\boxed{20\,|\,12\,|\,18\,|\,16\,|\,24\,|\,10\,|\,22\,|\,14}$. Consider the cases:

(1) ascending sort of the element values;
(2) descending sort of the element values.

12.32. Tell whether the following sorts are stable:

(1) bubble sort?
(2) shaker sort?

12.33. Shaker sort

In the text file `input.txt`, there is an array of N integers. Perform the ascending sorting of data using the shaker sort algorithm. Into the text file `output.txt` enter the sorted array (into the first row of the file) and the number of comparisons performed by the program (into the second row of the file).

12.34. Find the number of inversions in the array

$$\boxed{2}\boxed{3}\boxed{4}\boxed{5}\boxed{1}.$$

12.35. Find the number of inversions in the array $\boxed{20}\boxed{12}\boxed{18}\boxed{16}\boxed{24}\boxed{10}\boxed{22}\boxed{14}$.

12.36. Find the number of inversions in each array:

(1) $\boxed{1}\boxed{3}\boxed{5}\boxed{7}\ldots\boxed{2N-1}\boxed{2}\boxed{4}\boxed{6}\ldots\boxed{2N}$;
(2) $\boxed{2}\boxed{4}\boxed{6}\ldots\boxed{2N}\boxed{1}\boxed{3}\boxed{5}\boxed{7}\ldots\boxed{2N-1}$.

*12.37. Prove that the number of inversions in the integer array

$$\boxed{N}\boxed{1}\boxed{N-1}\boxed{2}\boxed{N-3}\ldots\boxed{\lceil N/2 \rceil},$$

where $N \geqslant 1$, is equal to $\left\lfloor \dfrac{N}{2} \right\rfloor \left\lceil \dfrac{N}{2} \right\rceil$.

12.38. Assume that a one-dimensional array is given of size N, where $N \geqslant 2$. Perform the operation of exchange of values of two elements. What maximal number of inversions can be eliminated this way?

12.39. It is known that the array *list* contains exactly $\mathcal{I}$ inversions. Find the least size that the array *list* can have.

*12.40. Suppose that the array $\boxed{a_1}\boxed{a_2}\ldots\boxed{a_{N-1}}\boxed{a_N}$ contains $\mathcal{I}$ inversions. How many inversions will be in the array $\boxed{a_N}\boxed{a_{N-1}}\ldots\boxed{a_2}\boxed{a_1}$?

12.41. Suppose that some array a of size N contains exactly $\mathcal{I}$ inversions. Prove that the number of comparisons $\mathcal{C}$, performed by the insertion sort algorithm for the ordering of this array, is equal to $\mathcal{I} + O(N)$.

12.42. Is the Shell sort stable?

12.43. Consider the sequence of increments $1, 2, 4, 8, 16, 32, \ldots$ What is the major drawback of the Shell sort algorithm based on this sequence?

12.44. In the Shell sort algorithm for the h-ordering of the subarrays of the original array, the insertion sort is used. How will the asymptotic complexity of the algorithm change, if the selection sort is used instead of the insertion sort?

*12.45. Prove the theorem of h- and k-ordering.

12.46. For the Shell sort algorithm, Pratt suggested a sequence of increments out of the set of numbers of the form $2^i 3^j$, smaller than the array size, where $i, j \in \mathbb{N} \cup \{0\}$. The increments are used in descending order. Write out the elements of the set of increments H for the case of an array of size $N = 32$.

12.47. Demonstrate the operation of the Shell sort algorithm based on the sequence suggested by Pratt (see Exercise **12.46**). Consider the array $\boxed{20\,|\,12\,|\,18\,|\,16\,|\,24\,|\,10\,|\,22\,|\,14}$.

*12.48. Prove that the worst-case asymptotic complexity of the Shell sort based on the sequence of increments suggested by Pratt is equal to $\Theta(N(\log_2 N)^2)$ [36, 45].

12.49. Shell sort

In the text file `input.txt` there is an array of N integers. Perform the ascending sorting of data using the Shell sort for the following increment values:

(1) the increment values are h_s, where $h_{s+1} = 2h_s + 1, h_0 = 1$, and $0 \leqslant s < \lfloor \log_2 N \rfloor$ (sequence 1, 3, 7, 15, 31, 63, ...);

(2) the increment values are h_s, where $h_{s+1} = 3h_s + 1$, $h_0 = 1$, and $0 \leqslant s < \lfloor \log_3(2N+1) \rfloor - 1$ (sequence 1, 4, 13, 40, 121, ...).

Into the text file `output.txt` enter the sorted array (into the first row of the file) and the number of comparisons performed by two variants of sorting (into the second row of the file).

12.50. Obtain the explicit form of the expression for the increments $h_s = h(s)$ from the previous exercise.

12.51. Find the number of comparisons performed by the quicksort algorithm when sorting the array $\boxed{21\,|\,24\,|\,33\,|\,37\,|\,38\,|\,45\,|\,50}$.

12.52. Is the quicksort stable?

12.53. As is known, if the array elements form a strictly monotone sequence, then this is the worst case for the quicksort algorithm. Give an example of a sequence that is not strictly monotone, but results in the worst case for the procedure QuickSort.

12.54. It is known that in the array $list[1..N]$ all the elements are pairwise equal. Estimate the number of comparisons performed by the quicksort algorithm when ordering the array $list$.

*12.55. In one of the quicksort options, the pivot element is chosen equal to the median of the three array elements: the leftmost, the rightmost, and the middle element. Write the recurrence relation defining the number of comparisons in this quicksort option.

12.56. Quicksort

In the text file `input.txt`, a natural number N is written, defining the array size *list*, and $N < 15$. Give all kinds of options of the array *list* and enter into the text file `output.txt` the arrays on which the quicksort algorithm performs the maximal number of comparisons.

12.57. Comparison table of the sort algorithms

Fill in the comparison table for the sort algorithms that you know (Table 12.6).

12.58. Find the median m of the array $\boxed{20\,|\,12\,|\,18\,|\,16\,|\,24\,|\,10\,|\,22\,|\,14}$.

12.59. Find the median m of each of the following arrays:

(1) $\boxed{N\,|\,1\,|\,N-1\,|\,2\,|\,N-3\,|\ldots|\,\lceil N/2\rceil}$;

(2) $\boxed{2\,|\,1\,|\,4\,|\,2\,|\,8\,|\,4\,|\ldots|\,2^N\,|\,2^{N-1}}$.

12.60. Let three real numbers a, b, c be given. Write the explicit form for calculation of their median $m(a, b, c)$.

*12.61. Let the array a contain N elements, and $N > 2$. Prove that decision tree for determining the kth largest element in this array contains no less than $2^{N-k}C(N, k-1)$ leaves.

12.62. It is known that some algorithm $\mathcal{A}$ defines the kth largest element in the array $a[1..N]$, where $N > 2$. Prove that $\mathcal{A}$ performs at least $N - k + \lceil \log_2 C(N, k-1)\rceil$ comparison operations.

12.63. Show that the algorithm defining the second largest element of the array $a[1..N]$ in less than $N + \lceil \log_2 N\rceil - 2$ comparisons does not exist.

12.64. Determine the minimal number of comparisons required to find the median of five arbitrary numbers.

12.65. Show that it is impossible to find the median of the array $a[1..N]$, where $N > 2$, in less than $\left\lceil \dfrac{3N}{2}\right\rceil - O(\log_2 N)$ comparisons.

12.66. Demonstrate the operation of the algorithm SelectPart by the example of the array median search $\boxed{20\,|\,12\,|\,18\,|\,16\,|\,24\,|\,10\,|\,22\,|\,14}$.

12.67. Find the number of comparisons $T(N)$, performed by the algorithm Select-Part, when searching for the array median

$$\boxed{N\,|\,1\,|\,N-1\,|\,2\,|\,N-3\,|\ldots|\,\lceil N/2\rceil},$$

where the number of the array elements is $N > 2$.

12.68. Demonstrate the operation of the algorithm SelectOpt by the example of the array median search from the Exercise **12.66**.

Table 12.6 Comparison table of the sort algorithms

Sort name	Author and creation date	Stability	Natural behavior	Asymptotic complexity	Advantages/ drawbacks
1. Insertion	J. von Neumann 1945[a]	Yes	Yes	$A(N)=$ $= N^2/4 + O(N),$ $W(N)=$ $= N^2/2 + O(N)$	Simplicity of the idea and realization/ inefficient for arrays of big size
2. Bubble	…	…	…	…	…
3. …	…	…	…	…	…

[a] John von Neumann (1903–1957)—American mathematician

12.69. Let $m \in \mathbb{N}$ and coefficients $\beta_i \in \mathbb{R}^+$, where $i = 1, 2, \ldots, m$, such that their sum is no greater than one: $\sum\limits_{i=1}^{m} \beta_i < 1$. Show that the recurrence relation of the form

$$\begin{cases} T(N) = \sum\limits_{i=1}^{m} T(\beta_i N) + \Theta(N), & N > N_0, \\ T(N) = \Theta(1), & 1 \leqslant N \leqslant N_0, \end{cases}$$

where N_0 is the least of the following numbers $\beta_1^{-1}, \beta_2^{-1}, \ldots, \beta_m^{-1}$, is satisfied by the function $T(N) = \Theta(N)$.

12.70. Using the previous exercise, obtain the condition for the parameter w of the algorithm SelectOpt, so that the worst-case asymptotic complexity of the algorithm should be equal to $T(N) = \Theta(N)$, where N is the number of elements of the original array.

12.71. Suggest an algorithm that determines the set of the array elements $list[1..N]$, not exceeding the kth largest element. The number of array element comparison operations should be a linear function of its size.

12.6   Answers, Hints, Solutions

12.1. *Solution.*
Assuming that the basic operation is the multiplication of integers in the row number 7, we obtain the recurrence relation for the number of operations $T(n)$:

$$\begin{cases} T(n) = T(n-1) + 1, & n > 1, \\ T(1) = 0. \end{cases}$$

It is easy to see that its solution is $T(n) = n - 1$ (it can be checked by direct substitution).

12.2. *Answer*:

The computational complexity of the algorithm is $O(n)$, as of the recursive version of the algorithm mentioned in the text of the chapter.

12.3. *Hint*.

The recurrence relation determining the number of addition operations has the form

$$\begin{cases} T(n) = T(n-1) + T(n-2) + 1, & n > 2, \\ T(1) = 0, \ T(2) = 0. \end{cases}$$

The solution of this relation is given in Exercise **8.31**: $T(n) = F_n - 1$, where F_n is the Fibonacci number. Hence, there exists estimate $T(n) = \Theta(\varphi^n)$, where $\varphi = \dfrac{1 + \sqrt{5}}{2}$.

12.5. *Solution*.

The basic operation is the addition of non-negative integers in the row number 5. Since the function log2floor at each recursive call performs exactly one addition operation, then the recurrence relation for the number of additions $T(n)$ can be written in the form:

$$\begin{cases} T(n) = T\left(\left\lfloor \dfrac{n}{2} \right\rfloor\right) + 1, & n > 1, \\ T(1) = 0. \end{cases}$$

Further, let us use the master theorem (see p. 354) taking into account the values of the parameters $a = 1$, $b = 2$ and the explicit form of the function $f(n) = 1$. We finally obtain: $T(n) = \Theta(\log_2 n)$.

Note. The exact solution is the function $T(n) = \lfloor \log_2 n \rfloor$, which can be shown, for example, by the mathematical induction method.

12.6. *Solution*.

Expand the determinant, for example, by the first row:

$$\det A = 1 \cdot \det \begin{bmatrix} 6 & 7 & 8 \\ 10 & 11 & 12 \\ 14 & 15 & 16 \end{bmatrix} - 2 \cdot \det \begin{bmatrix} 5 & 7 & 8 \\ 9 & 11 & 12 \\ 13 & 15 & 16 \end{bmatrix} +$$

$$+ 3 \cdot \det \begin{bmatrix} 5 & 6 & 8 \\ 9 & 10 & 12 \\ 13 & 14 & 16 \end{bmatrix} - 4 \cdot \det \begin{bmatrix} 5 & 6 & 7 \\ 9 & 10 & 11 \\ 13 & 14 & 15 \end{bmatrix} =$$

$$= 1 \cdot \left(6 \cdot \det \begin{bmatrix} 11 & 12 \\ 15 & 16 \end{bmatrix} - 7 \cdot \det \begin{bmatrix} 10 & 12 \\ 14 & 16 \end{bmatrix} + 8 \cdot \det \begin{bmatrix} 10 & 11 \\ 14 & 15 \end{bmatrix} \right) -$$

$$- 2 \cdot \left(5 \cdot \det \begin{bmatrix} 11 & 12 \\ 15 & 16 \end{bmatrix} - 7 \cdot \det \begin{bmatrix} 9 & 12 \\ 13 & 16 \end{bmatrix} + 8 \cdot \det \begin{bmatrix} 9 & 11 \\ 13 & 15 \end{bmatrix} \right) +$$

$$+ 3 \cdot \left(5 \cdot \det \begin{bmatrix} 10 & 12 \\ 14 & 16 \end{bmatrix} - 6 \cdot \det \begin{bmatrix} 9 & 12 \\ 13 & 16 \end{bmatrix} + 8 \cdot \det \begin{bmatrix} 9 & 10 \\ 13 & 14 \end{bmatrix} \right) -$$

$$-4 \cdot \left(5 \cdot \det \begin{bmatrix} 10 & 11 \\ 14 & 15 \end{bmatrix} - 6 \cdot \det \begin{bmatrix} 9 & 11 \\ 13 & 15 \end{bmatrix} + 7 \cdot \det \begin{bmatrix} 9 & 10 \\ 13 & 14 \end{bmatrix} \right) =$$
$$= 1 \cdot (6 \cdot (11 \cdot 16 - 12 \cdot 15) - 7 \cdot (10 \cdot 16 - 12 \cdot 14) + \ldots) - \ldots = 0.$$

Note. It is seen from the solution of the exercise that even for small size values of the matrix n the number of computational operations is rather big. In this case, it would be easier to use the known properties of the determinant for obtaining the answer. For example, the equivalence transformations, namely by subtracting the first row from the second and the third rows, we will obtain a matrix with proportional rows:

$$\begin{bmatrix} 1 & 2 & 3 & 4 \\ 5 & 6 & 7 & 8 \\ 9 & 10 & 11 & 12 \\ 13 & 14 & 15 & 16 \end{bmatrix} \sim \begin{bmatrix} 1 & 2 & 3 & 4 \\ 4 & 4 & 4 & 4 \\ 9 & 10 & 11 & 12 \\ 13 & 14 & 15 & 16 \end{bmatrix} \sim \begin{bmatrix} 1 & 2 & 3 & 4 \\ 4 & 4 & 4 & 4 \\ 8 & 8 & 8 & 8 \\ 13 & 14 & 15 & 16 \end{bmatrix}.$$

It is clear that its determinant is 0.

12.7. *Answer*: $\det T(n) = (n+1)2^n$.

12.8. *Hint.*

The determinant $d_n = \det T(n)$ satisfies the relation (see Chapter "Recurrence Relations" on p. 273)

$$\begin{cases} d_n = a d_{n-1} - b^2 d_{n-2}, & n > 2, \\ d_1 = a, \ d_2 = a^2 - b^2. \end{cases}$$

12.10. *Solution.*

For the matrix of size $n \times n$ there will be performed n recursive calls and n multiplications of type $a_{ij} \times \det \widetilde{A}_{ij}$, $j = 1, \ldots, n$. The exit from the recursion will be at $n = 1$, no multiplications are performed in this case. Due to this, the total number of multiplications satisfies the recurrence relation

$$\begin{cases} T(n) = nT(n-1) + n, & n > 1, \\ T(1) = 0. \end{cases}$$

Solve the obtained relation by the method of substitution (see p. 273):

$$\begin{aligned} T(n) = nT(n-1) + n &= \\ = n[(n-1)T(n-2) + n - 1] + n &= \\ = n(n-1)T(n-2) + n(n-1) + n &= \\ = n(n-1)[(n-2)T(n-3) + n - 2] + n(n-1) + n &= \\ = n(n-1)(n-2)T(n-3) + n(n-1)(n-2) + n(n-1) + n. \end{aligned}$$

Similarly continuing up to $T(1) = 0$, we obtain

$$T(n) = n(n-1)(n-2)\ldots 2 \cdot T(1) +$$
$$+ n(n-1)(n-2)\ldots 2 + n(n-1)(n-2)\ldots 3 + \cdots + n(n-1) + n =$$
$$= 0 + n! + \frac{n!}{2!} + \cdots + \frac{n!}{(n-2)!} + \frac{n!}{(n-1)!} =$$
$$= n! \left[1 + \frac{1}{2!} + \cdots + \frac{1}{(n-2)!} + \frac{1}{(n-1)!} \right] = n! \sum_{k=1}^{n-1} \frac{1}{k!}.$$

We can write an analytical expression for $T(n)$ in terms of non-elementary functions, but for solution of the set problem it is sufficient to estimate the asymptotic behavior of the function $T(n)$.

Note that at $n \to \infty$

$$\lim_{n \to \infty} \sum_{k=1}^{n-1} \frac{1}{k!} = \sum_{k=1}^{\infty} \frac{1}{k!} = \sum_{k=0}^{\infty} \frac{1}{k!} - 1 = e - 1,$$

where $e = 2,7182818284590\ldots$ is the base of natural logarithms. We obtain the inequality

$$n! \leqslant n! \sum_{k=1}^{n-1} \frac{1}{k!} \leqslant (e-1)n!,$$

and, finally, $T(n) = \Theta(n!)$.

12.11. *Solution.*

According to the result of the previous exercise, $T(n) = n! \sum_{k=1}^{n-1} \frac{1}{k!}$. Expand the expression $en\Gamma(n, 1) - n!$ by the definition of incomplete gamma function:

$$en\Gamma(n, 1) - n! = en \cdot (n-1)! \, e^{-1} \sum_{k=0}^{n-1} \frac{1^k}{k!} - n! = n! \sum_{k=0}^{n-1} \frac{1}{k!} - n! = n! \sum_{k=1}^{n-1} \frac{1}{k!},$$

which coincides with $T(n)$.

Note. It would be a useful exercise to prove this statement by the mathematical induction method.

12.12. *Answer:*

As the first operator of the cycle realizing the expansion in the first row of the matrix A, add a row with check for equality to zero of the element $a[1, i]$. If the following condition is satisfied $a[1, i] = 0$, then we should proceed to the next element of the first row: $i \to i + 1$. This will result in decreasing the number of recursive calls in the algorithm and, therefore, to decreasing the number of multiplication operations.

12.13. *Solution.*

Denote the total number of multiplications performed by the algorithm from Exercise **12.12**, by $T(n)$. Since during the expansion by the first row of the matrix A

there occur on average pn recursive calls, then the total number of multiplications satisfies the recurrence relation

$$\begin{cases} T(n) = p(nT(n-1)+n), & n > 1, \\ T(1) = 0. \end{cases}$$

the obtained recurrence relation is associated with the function

$$T(n) = n! \sum_{k=1}^{n-1} \frac{p^{n-k}}{k!} = p^n n! \sum_{k=1}^{n-1} \frac{p^{-k}}{k!}.$$

The estimate of the sum $\sum_{k=1}^{n-1} \dfrac{p^{-k}}{k!}$ results in the inequalities:

$$\sum_{k=1}^{n-1} \frac{p^{-k}}{k!} = p^{-1} + \frac{p^{-2}}{2!} + \ldots > p^{-1} \text{ and}$$

$$\sum_{k=1}^{n-1} \frac{p^{-k}}{k!} < \sum_{k=1}^{\infty} \frac{p^{-k}}{k!} = e^{1/p} - 1.$$

Taking into account these relations, we obtain the result: $T(n) = \Theta(p^n n!)$.

12.14. *Solution.*

Write the recurrence relation for the number of multiplication operations $T(n)$ performed by the algorithm for calculation of the determinant in this case. For this, perform the expansion in the first row of the matrix A:

$$\det A = a_{11} \det \tilde{A}_{11} - a_{12} \det \tilde{A}_{12},$$

where $\tilde{A}_{ij}$ is the matrix of size $(n-1) \times (n-1)$, obtained from A by deleting the ith row and jth column.

Note that the matrix $\tilde{A}_{11}$ is tridiagonal.

Write $\det \tilde{A}_{12}$ also in the form of the first-row expansion.

Thus, we obtain, that the solution of the problem reduces to calculating the determinants of the following three matrices of smaller size:

1. tridiagonal matrices of size $(n-1) \times (n-1)$;
2. tridiagonal matrices of size $(n-2) \times (n-2)$;
3. matrices U of size $(n-2) \times (n-2)$. This matrix is obtained from the tridiagonal one by replacing the first row with the zero one.

Calculation of the matrices' determinants of items (1)–(3) requires $T(n-1)$, $T(n-2)$ and $B(n-2)$ multiplication operations respectively, where by $B(n-2)$ is denoted an auxiliary value—the number of multiplication operations for calculating $\det U$.

The recurrence relations for the total number of multiplication operations $T(n)$ and for the value $B(n)$ can be written in the form of the system:

$$\begin{cases} T(n) = T(n-1) + T(n-2) + B(n-2) + 4, & n > 2, \\ B(n) = B(n-1) + 1, & n > 1, \\ T(1) = 0, \; T(2) = 2, \\ B(1) = 0. \end{cases}$$

The value $B(n)$ does not depend on $T(n)$, and a separate recurrence relation can be written for it:

$$\begin{cases} B(n) = B(n-1) + 1, & n > 1, \\ B(1) = 0. \end{cases}$$

It is clear that for all natural numbers n there exists the equality $B(n) = n - 1$. Returning to the system that defines the values $T(n)$ and $B(n)$, we obtain:

$$\begin{cases} T(n) = T(n-1) + T(n-2) + n + 1, & n > 1, \\ T(1) = 0, \; T(2) = 2. \end{cases}$$

In accordance with the general rule of solving linear non-homogeneous recurrence relations with constant coefficients, represent the solution in the form of the sum of general $T_{\text{gen.}}(n)$ and specific $T_{\text{spec.}}(n)$ solutions (see chapter "Recurrence Relations"):

$$T(n) = T_{\text{gen.}}(n) + T_{\text{spec.}}(n).$$

The general solution can be written in the form $T_{\text{gen.}}(n) = C_1 z_1^n + C_2 z_2^n$, where z_1, z_2 are the roots of the characteristic equation $z^2 = z + 1$:

$$z_1 = \frac{1 + \sqrt{5}}{2}, \quad z_2 = \frac{1 - \sqrt{5}}{2}.$$

We are looking for the specific solution in the form $T_{\text{spec.}}(n) = Dn + E$, where D and E are real coefficients. Direct substitution of the expression for $T_{\text{spec.}}(n)$ into the recurrence relation for $T(n)$ results in the equalities $D = -1$, $E = -4$, hence, $T_{\text{spec.}}(n) = -n - 4$.

Further, taking into account the original conditions $T(1) = 0$, $T(2) = 2$ allows calculating the values C_1 and C_2:

$$\begin{cases} C_1 \cdot \left(\dfrac{1 + \sqrt{5}}{2} \right) + C_2 \cdot \left(\dfrac{1 - \sqrt{5}}{2} \right) = 5, \\[4mm] C_1 \cdot \left(\dfrac{1 + \sqrt{5}}{2} \right)^2 + C_2 \cdot \left(\dfrac{1 - \sqrt{5}}{2} \right)^2 = 8. \end{cases}$$

The solution of the written system of linear equations has the form $C_1 = \dfrac{3\sqrt{5}+7}{2\sqrt{5}}$, $C_2 = \dfrac{3\sqrt{5}-7}{2\sqrt{5}}$. Due to this, we obtain:

$$T(n) = \frac{3\sqrt{5}+7}{2\sqrt{5}}\left(\frac{1+\sqrt{5}}{2}\right)^n + \frac{3\sqrt{5}-7}{2\sqrt{5}}\left(\frac{1-\sqrt{5}}{2}\right)^n - n - 4.$$

Simplify the obtained expression. For this, note that the following equalities are valid:

$$\frac{3\sqrt{5}+7}{2\sqrt{5}} = \left(\frac{1+\sqrt{5}}{2}\right)^4, \quad \frac{3\sqrt{5}-7}{2\sqrt{5}} = -\left(\frac{1-\sqrt{5}}{2}\right)^4.$$

Hence,

$$T(n) = \frac{1}{\sqrt{5}}\left(\frac{1+\sqrt{5}}{2}\right)^{n+4} - \frac{1}{\sqrt{5}}\left(\frac{1-\sqrt{5}}{2}\right)^{n+4} - n - 4.$$

According to Binet's formula (see Exercise **1.99**), we finally obtain the expression for the sought value $T(n)$ in terms of the Fibonacci numbers:

$$T(n) = F_{n+4} - (n+4), \quad n \geqslant 1.$$

12.15. *Hint.*

The sum should be calculated $A(N) = \dfrac{1}{N+1}\left(\displaystyle\sum_{i=1}^{N} i + N\right).$

12.16. *Answer*: 3 comparisons.

12.17. *Answer*: $\dfrac{1}{7}\displaystyle\sum_{i=1}^{3} i 2^{i-1} = \dfrac{17}{7}$ comparisons.

12.18. *Answer*:

(1) 1;
(2) 4.

12.19. *Answer*:

(1) $\dfrac{21}{8}$;

(2) $\dfrac{29}{9}$.

12.20. *Hint.*

The number of comparisons satisfying the inequality

$$A(N) \leqslant \frac{1}{2N+1}\left((N+1)h + \sum_{i=1}^{h} i 2^{i-1}\right).$$

12.21. *Proof.*

It is required to prove the equality $\sum_{i=1}^{N} i2^i = (N-1)2^{N+1} + 2$ for all $N \in \mathbb{N}$. Con-

sider the sum $\Xi(x) = \sum_{i=1}^{N} x^i = \dfrac{x^{N+1} - x}{x - 1}$ and calculate the derivative $\dfrac{d}{dx} \Xi(x)$:

$$\frac{d}{dx} \Xi(x) = \frac{d}{dx} \frac{x^{N+1} - x}{x - 1} = \frac{[(N+1)x^N - 1](x-1) - (x^{N+1} - x)}{(x-1)^2} =$$

$$= \frac{Nx^{N+1} - (N+1)x^N + 1}{(x-1)^2}.$$

On the other hand, $\dfrac{d}{dx} \Xi(x) = \dfrac{d}{dx} \sum_{i=1}^{N} x^i = \sum_{i=1}^{N} ix^{i-1} = \dfrac{1}{x} \sum_{i=1}^{N} ix^i.$

Hence,

$$\frac{1}{x} \sum_{i=1}^{N} ix^i = \frac{Nx^{N+1} - (N+1)x^N + 1}{(x-1)^2}.$$

In the obtained inequality assume that $x = 2$ and come to the relation

$$\frac{1}{2} \sum_{i=1}^{N} i2^i = \frac{N2^{N+1} - (N+1)2^N + 1}{(2-1)^2}.$$

After simplification, we obtain the equality (A.61): $\sum_{i=1}^{N} i2^i = (N-1)2^{N+1} + 2$.

12.22. *Answer*: change the row number 14 for

$$middle := left + (right - left) \textbf{ div } 2$$

12.23. *Answer*:

The decision tree for the Fibonaccian search in the ordered array, consisting of eight elements, is shown in Fig. 12.10.

12.24. *Proof.*

The recurrence relation for the number of comparison operations performed by the Fibonaccian search algorithm in the average case can be written in the form (see p 377):

$$\begin{cases} A(N) = \dfrac{1}{\varphi} A\left(\dfrac{1}{\varphi}N\right) + \dfrac{1}{\varphi^2} A\left(\dfrac{1}{\varphi^2}N\right) + 1, & N > 1, \\ A(1) = 1. \end{cases}$$

Prove that $A(N) = \dfrac{\varphi}{\sqrt{5}} \log_\varphi N + O(1)$, i.e. $A(N) \leqslant \dfrac{\varphi}{\sqrt{5}} \log_\varphi N + C$ for some $C \in \mathbb{R}$ and all natural N. Substitute the last inequality into the left and right sides of

Fig. 12.10 The decision tree for the Fibonaccian search in the array of size $N = 8$

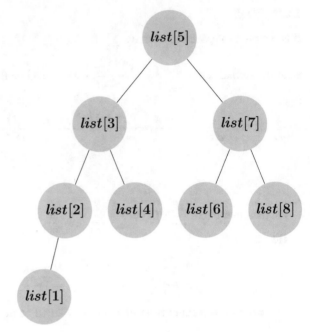

the recurrence relation:

$$\frac{\varphi}{\sqrt{5}} \log_\varphi N + C = \frac{1}{\varphi} \left(\frac{\varphi}{\sqrt{5}} \log_\varphi \left(\frac{1}{\varphi} N \right) + C \right) + \frac{1}{\varphi^2} \left(\frac{\varphi}{\sqrt{5}} \log_\varphi \left(\frac{1}{\varphi^2} N \right) + C \right) + 1.$$

After algebraic transformations, we will have:

$$\left(1 - \frac{1}{\varphi} - \frac{1}{\varphi^2} \right) \frac{\varphi}{\sqrt{5}} \log_\varphi N + \left(\frac{1}{\varphi} + \frac{2}{\varphi^2} \right) \frac{\varphi}{\sqrt{5}} + \left(1 - \frac{1}{\varphi} - \frac{1}{\varphi^2} \right) C = 1.$$

Since $\varphi^2 = \varphi + 1$ and $\varphi = \dfrac{1 + \sqrt{5}}{2}$, then, substituting these relations into the previous equality, we will obtain the identity valid for all natural N.

Hence, the asymptotic complexity of the Fibonaccian search algorithm in the average case is defined as

$$A(N) = \frac{\varphi}{\sqrt{5}} \log_\varphi N + O(1).$$

12.26. *Answer*:
Operation of the insertion sort algorithm is shown in Table 12.7.
12.27. *Answer*: yes.
12.29. *Hint*.

The number of element comparisons, performed by the binary insertion sort algorithm, belongs to the class $\Theta(N \log_2 N)$. However, when estimating the efficiency, the number of element permutations should also be taken into consideration.

12.30. *Answer*: no.

12.32. *Answer*:

(1) yes;
(2) yes.

12.34. *Answer*: 4.

12.35. *Answer*: 15.

12.36. *Answer*:

(1) $\dfrac{N(N-1)}{2}$;

(2) $\dfrac{N(N^2+1)}{2}$.

12.37. *Hint*. Consider the cases with even and odd values N.

12.38. *Answer*: $2N - 3$.

12.39. *Answer*: $\lceil (1 + \sqrt{1 + 8\mathcal{I}})/2 \rceil$.

12.40. *Solution*.

Table 12.7 Operation of the insertion sort algorithm

(1) Ascending

Original array:	20 12 18 16 24 10 22 14
Original array:	**20** 12 18 16 24 10 22 14
Run 1 :	**12 20** 18 16 24 10 22 14
Run 2 :	**12 18 20** 16 24 10 22 14
Run 3 :	**12 16 18 20** 24 10 22 14
Run 4 :	**12 16 18 20 24** 10 22 14
Run 5 :	**10 12 16 18 20 24** 22 14
Run 6 :	**10 12 16 18 20 22 24** 14
Run 7 :	**10 12 14 16 18 20 22 24**

(2) Descending

Original array:	20 12 18 16 24 10 22 14
Original array:	**20** 12 18 16 24 10 22 14
Run 1 :	**20 12** 18 16 24 10 22 14
Run 2 :	**20 18 12** 16 24 10 22 14
Run 3 :	**20 18 16 12** 24 10 22 14
Run 4 :	**24 20 18 16 12** 10 22 14
Run 5 :	**24 20 18 16 12 10** 22 14
Run 6 :	**24 22 20 18 16 12 10** 14
Run 7 :	**24 22 20 18 16 14 12 10**

Denote by $\mathcal{I}^*$ the sought number of inversions. It is convenient to calculate the value of $2(\mathcal{I}+\mathcal{I}^*)$, equal to twice the number of all inversions in the arrays $\boxed{a_1}\,\boxed{a_2}\,\ldots\,\boxed{a_{N-1}}\,\boxed{a_N}$ and $\boxed{a_N}\,\boxed{a_{N-1}}\,\ldots\,\boxed{a_2}\,\boxed{a_1}$, and then express $\mathcal{I}^*$.

When calculating the inversions, special attention should be paid to the repeated elements of the arrays. Partition the set of all elements $\{a_i : 1 \leqslant i \leqslant N\}$ into m equivalence classes E_l, $1 \leqslant l \leqslant m$. The equivalence class E_l is formed by the array elements with equal values.

Each element a_i for $1 \leqslant i \leqslant N$ makes the following contribution to $2(\mathcal{I}+\mathcal{I}^*)$: $(N-1)-(k_l-1)$, where $k_l = |E_l|$ is the cardinality of the set E_l. Summing over all elements, we obtain

$$2(\mathcal{I}+\mathcal{I}^*) = \sum_{i=1}^{N}[(N-1)-(k_l-1)] = N(N-1) - \sum_{i=1}^{N}(k_l-1) =$$

$$= N(N-1) - \sum_{l=1}^{m}\sum_{a_i \in E_l}(k_l-1) = N(N-1) - \sum_{l=1}^{m}k_l(k_l-1).$$

The final number of inversions in the array $\boxed{a_N}\,\boxed{a_{N-1}}\,\ldots\,\boxed{a_2}\,\boxed{a_1}$ is equal to

$$\mathcal{I}^* = \frac{1}{2}\left[N(N-1) - \sum_{l=1}^{m}k_l(k_l-1)\right] - \mathcal{I}.$$

If all elements of the array $\boxed{a_1}\,\boxed{a_2}\,\ldots\,\boxed{a_{N-1}}\,\boxed{a_N}$ are different, then $\forall l\ k_l = 1$ and the expression for $\mathcal{I}^*$ is simplified: $\mathcal{I}^* = \dfrac{N(N-1)}{2} - \mathcal{I}$.

12.41. *Proof.*
Consider a separate iteration of a cycle, forming rows 7–18 of the insertion sort algorithm (see p. 380), and suppose that $i = m$, where $2 \leqslant m \leqslant N$. The elements $a_1, a_2, \ldots, a_{m-1}$ for an ordered sequence. Assume that after the cycle iteration complying with the value $i = m$, the current element a_m will occupy the position with the index l_m, $1 \leqslant l_m \leqslant m$. It means that in the subarray $a[1..m]$ the element a_m participates in the formation of the inversions $(a_{l_m}, a_m), (a_{l_m+1}, a_m), \ldots, (a_{m-1}, a_m)$, only. The number of such inversions is equal to $m - l_m$. Hence, the total number of inversions in the original array of size N can be expressed by the relation

$$\mathcal{I} = \sum_{m=2}^{N}(m - l_m).$$

Now, let us look at the calculation of the number of comparisons performed by the insertion sort algorithm when ordering the original array. When, in the external cycle, the value $i = m$ is reached, the current element a_m will participate in comparisons with the elements $a_{m-1}, a_{m-2}, \ldots, a_{l_m}$, and, finally, if $l_m > 1$, then with the elements

a_{l_m-1}. Hence, the number of comparisons for this cycle iteration over the variable i is equal to $(m - l_m + 1)$, if $l_m > 1$, or $(m - l_m)$, if $l_m = 1$.

The full number of comparisons can be estimated by summation of values $(m - l_m + O(1))$ over the variable m:

$$C = \sum_{m=2}^{N}(m - l_m + O(1)) = \sum_{m=2}^{N}(m - l_m) + O(N).$$

Using the obtained above relation for $\mathcal{I}$, we come to the expression

$$C = \mathcal{I} + O(N).$$

So, the number of comparisons, performed by the insertion sort algorithm for the ordered original array, is equal to $\mathcal{I} + O(N)$.

12.42. *Answer*: no.

12.43. *Answer*:

The elements standing at even places and the elements standing at odd places are not compared between each other up to the final stage, to which corresponds the increment $h = 1$.

12.44. *Solution.*

The number of comparisons performed by the selection sort algorithm does not depend on the degree of order of the array and is equal to $\Theta(N^2)$, where N is the array size. Hence, the asymptotic complexity of the algorithm using the selection sort for the h-ordering of subarrays, will be equal to

$$W(N) = \sum_{s=0}^{s_{\max}} h_s \Theta\left(\left(\frac{N}{h_s}\right)^2\right) = \Theta\left(N^2\left(\sum_{s=0}^{s_{\max}}\frac{1}{h_s}\right)\right),$$

where h_s—applied increments ($s = 0, 1, \ldots, s_{\max}$). For the sum of values inverse to h_s, the following inequalities are valid $1 < \sum_{s=0}^{s_{\max}}\frac{1}{h_s} < H_N$, where H_N—harmonic number. Hence, $W(N) = \Omega(N^2)$.

Note that the efficiency of the Shell sort is reached due to the peculiarity of the insertion sort algorithm, namely due to a small number of comparisons in case of almost completely sorted array.

12.45. *Proof.*

Let the array *list* of size N be k-ordered. Show that h-sorting will not violate the k-ordering of this array for all $1 \leqslant h < k < N$. For simplicity, we suggest that all elements $list[i]$, where $1 \leqslant i \leqslant N$, are different.

Consider two cases [52]:

(1) h is the divisor of k;
(2) h does not divide k evenly.

It is easy to verify the validity of the theorem for the case (1). Indeed, the elements standing at a distance of $k = c \cdot h$ from each other, where $c \in \mathbb{N}$, form ordered

Table 12.8 The elements of the matrix a and the inversion $(a_{i,j}, a_{i+h,j})$

$$
\begin{array}{ccccc}
a_{i,j-1} & < & a_{i,j} & < & a_{i,j+1} \\
\vdots & & \vdots & & \vdots \\
\wedge & & \vee & & \vdots \\
\vdots & & \vdots & & \vdots \\
a_{i+h,j-1} & < & a_{i+h,j} & < & a_{i+h,j+1}
\end{array}
$$

subarrays. After the h-ordering the following inequalities remain valid $\ldots < list[j - k] < list[j] < list[j + k] < \ldots$

Proceed to case (2), when there is no such natural number c, that $k = c \cdot h$. Use the mathematical induction method over the number of h-inversions.

Arrange the elements of the k-ordered array in the form of the matrix $a_{i,j}$ with k rows and $\lceil N/k \rceil$ columns. The matrix a is filled with elements of the array *list* on columns; the last column, is necessary, is complemented by the elements with greater values, exceeding $\max_{i}(list[i])$. Note that the k-ordering means that each row of the matrix forms an ordered array. Prove that the elimination of the h-inversions will not violate this ordering.

If the array contains the h-inversion, then it is always possible to chose such one out of them, which is formed by the elements at a distance of h, for example, $a_{i,j}$ and $a_{i+h,j}$ (or $a_{i,j}$, and $a_{i+h-k,j+1}$, if $i + h > k$). Consider the case of the inversion $(a_{i,j}, a_{i+h,j})$ and try to eliminate it.

Since the matrix's rows are ordered, the following inequalities occur

$$a_{i,j-1} < a_{i,j} < a_{i,j+1}, \quad a_{i+h,j-1} < a_{i+h,j} < a_{i+h,j+1}.$$

Further, in accordance with the choice of the considered inversion

$$a_{i,j-1} < a_{i+h,j-1}, \quad a_{i,j} > a_{i+h,j}.$$

From the given inequalities, the relations follow (Table 12.8):

$$a_{i,j-1} < a_{i+h,j}, \quad a_{i+h,j-1} < a_{i,j}, \quad a_{i+h,j} < a_{i,j+1}.$$

Thus, it remains to check the relation between the elements $a_{i,j}$ and $a_{i+h,j+1}$.

If $a_{i,j+1} < a_{i+h,j+1}$, we obtain $a_{i,j} < a_{i+h,j+1}$.

If $a_{i,j+1} > a_{i+h,j+1}$, then the elements $a_{i,j}$ and $a_{i+h,j+1}$ can be arranged in the inverse order. Elimination of the inversion $(a_{i,j+1}, a_{i+h,j+1})$ will probably result in the inversion $(a_{i,j+1}, a_{i+h,j+2})$. As a result, we either finish, reaching the pair $a_{i,l} < a_{i+h,l}$ for some $1 < l \leqslant \lceil N/k \rceil$, or go beyond the matrix's limits. In any case, the rows remain ordered, and the number h-inversions decreased. The case $i + h > k$ is considered similarly.

Table 12.9 Operation of the Shell sort algorithm based on the sequence suggested by Pratt

Original array:	20	12	18	16	24	10	22	14
After the 6-ordering:	20	12	18	16	24	10	22	14
After the 4-ordering:	20	10	18	14	24	12	22	16
After the 3-ordering:	14	10	12	20	16	18	22	24
After the 2-ordering:	12	10	14	18	16	20	22	24
After the 1-ordering:	10	12	14	16	18	20	22	24

Fig. 12.11 Estimate of the number of solutions of the inequality $i \ln 2 + j \ln 3 < \ln N$ in non-negative integers

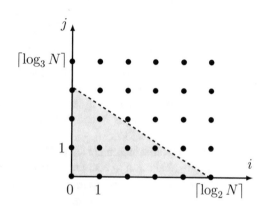

Thus, after the h-ordering, the array has not lost the property of the k-ordering.

12.46. *Answer*: $H = \{1, 2, 3, 4, 6, 8, 9, 12, 16, 18, 24, 27\}$.

12.47. *Answer*: The set of increments for an array of size $N = 8$ has the form $H = \{1, 2, 3, 4, 6\}$. Operation of the Shell sort algorithm is shown in Table 12.9.

12.48. *Proof*:

As is known, the increment variants suggested by Pratt form the set $H = \{2^i 3^j : 2^i 3^j < N$ and $i, j \in \mathbb{N} \cup \{0\}\}$, where N is the size of the array subject to ordering.

For the defined N the cardinality of the set H coincides with the number of solutions of the inequality $i \ln 2 + j \ln 3 < \ln N$ in non-negative integers. Due to this (see Fig. 12.11, where a dotted line is drawn $i \ln 2 + j \ln 3 = \ln N$), we have:

$$\frac{1}{2}(\lceil \log_2 N \rceil - 1)(\lceil \log_3 N \rceil - 1) < |H| < \frac{1}{2}(\lceil \log_2 N \rceil + 1)(\lceil \log_3 N \rceil + 1),$$

or $|H| = \Theta\big((\log_2 N)^2\big)$.

Consider the sth stage of the Shell sort, $s = 0, 1, 2, \ldots$ Let, at this stage, the increment be equal to h_s. Depending on the relation between the values h_s and $\dfrac{N}{6}$ we can single out two cases.

1. Let $h_s \geqslant \dfrac{N}{6}$. Then each of the subarrays formed at the sth stage has the size of no more than $\left\lceil \dfrac{N}{h_s} \right\rceil$ and, therefore, contains no more than $\dfrac{1}{2}\left\lceil \dfrac{N}{h_s} \right\rceil \left(\left\lceil \dfrac{N}{h_s} \right\rceil - 1\right)$ inver-

sions. Taking into account the inequality $\left\lceil \dfrac{k}{m} \right\rceil \le \dfrac{k+m-1}{m}$, valid for all $k, m \in \mathbb{N}$, we obtain the total number of comparisons at the sth stage of the Shell sort:

$$\frac{1}{2}\left\lceil \frac{N}{h_s} \right\rceil \left(\left\lceil \frac{N}{h_s} \right\rceil - 1 \right) h_s < \frac{N(N+h_s)}{2h_s} < 6N = O(N).$$

2. Let $h_s < \dfrac{N}{6}$. Then the following inequalities are valid $2h_s < N$ and $3h_s < N$. At the sth sorting stage, the considered array is $3h_s$- and $2h_s$-ordered due to the choice of the increment values. Hence, in each subarray, formed by the h_sth elements of the original array, the arbitrary element $list[k]$ forms no more than one inversion. For ordering h_s subarrays no more than $2\left\lceil \dfrac{N}{h_s} \right\rceil h_s = O(N)$ comparisons will be needed. Indeed, the elements $list[k_1]$ and $list[k_2]$ at $k_1 < k_2$ can only form an inversion if $k_2 - k_1 \ne 2c_1 + 3c_2$ for all non-negative integers c_1 and c_2. Hence, $k_2 = k_1 + 1$, and placing of some element to the sought position, will require no more than 2 comparisons. Note that the 2- and 3-ordered array $list$ of size N contains no more than $\lceil N/2 \rceil$ inversions [36].

So, to the arbitrary increment $h_s \in H$ correspond $O(N)$ array element comparison operations. Taking into account that $|H| = \Theta\left((\log_2 N)^2\right)$, we obtain the upper estimate of the number of comparisons in the worst case $O\left(N(\log_2 N)^2\right)$.

Now consider the lower estimate of the value $W(N)$. The number of comparisons performed by the auxiliary insertion sort is one less than the array size. At the sth stage, to which corresponds the increment h_s, no less than $h_s\left(\left\lceil \dfrac{N}{h_s} \right\rceil - 1\right) = \Omega(N)$ comparisons are carried out.

We finally obtain that the number of comparisons of the array elements in the worst case satisfies the relation

$$W(N) = \Theta\left(N(\log_2 N)^2\right).$$

12.50. *Answer*:

(1) $h_s = 2^{s+1} - 1, 0 \le s < \lfloor \log_2 N \rfloor$;
(2) $h_s = (3^{s+1} - 1)/2, 0 \le s < \lfloor \log_3(2N + 1) \rfloor - 1$.

12.51. *Answer*: 33.
12.52. *Answer*: no.
12.54. *Solution*.
When ordering the array, all elements of which are pairwise equal, the array partitioning procedure performs $2\left\lceil \dfrac{N}{2} \right\rceil$ comparisons of the array elements with the pivot element. Since the pivot element will be placed to a position with the index $\left\lceil \dfrac{N+1}{2} \right\rceil$, then for the number of comparisons $T(N)$ of the array elements we have

the following recurrence relation:

$$\begin{cases} T(N) = T\left(\left\lfloor \dfrac{N}{2} \right\rfloor\right) + T\left(\left\lfloor \dfrac{N-1}{2} \right\rfloor\right) + 2\left\lceil \dfrac{N}{2} \right\rceil, \ N > 2, \\ T(1), T(2) \text{ --- const,} \end{cases}$$

whence it follows that $T(N) = \Theta(N \log_2 N)$.

12.55. *Answer:*
The recurrence relation, defining the number of comparisons of the elements $A(N)$, has the form:

$$\begin{cases} A(N) = N + 1 + \displaystyle\sum_{k=1}^{N} \dfrac{(k-1)(N-k)}{C(N,3)} \, (A(k-1) + A(N-k)), \ N > 2, \\ A(0), A(1), A(2) \text{ --- const,} \end{cases}$$

where $C(N, 3)$ is a binomial coefficient equal to the number of combinations without repetitions of three elements out of N possible. Thorough analysis [63] taking into account the zero original conditions $A(0) = A(1) = A(2) = 0$ shows that $A(N) = \dfrac{12}{7} N \ln N + \Theta(N)$.

12.58. *Solution.*
Having ordered the elements of the original array in ascending order, we obtain:
| 10 | 12 | 14 | 16 | 18 | 20 | 22 | 24 |. Since the array size is $N = 8$ and $\left\lceil \dfrac{N}{2} \right\rceil = 4$, then the median m of the array is equal to the element located in the ordered sequence at the fourth place. Hence, $m = 16$.

12.59. *Answer:*

(1) $m = \left\lceil \dfrac{N}{2} \right\rceil$;

(2) $m = 2^{\lfloor N/2 \rfloor}$.

12.60. *Answer:*

$$m(a, b, c) = \dfrac{a+c}{2} - \dfrac{1}{4}\big(|a - 2b + c - |a - c|| - |a - 2b + c + |a - c||\big).$$

12.61. *Proof.*
Suppose that all elements of the array a form the finite set $A = \{a_i : 1 \leqslant i \leqslant N\}$. In order to establish the fact that some element $b \in A$ is the kth largest element, it is necessary to compare it with all other elements and determine the elements of the set $L = \{a_i : a_i < b\}$ [81].

Further, assume that with the help of some decision tree T it is possible to find the sought element b. Form the new tree T^*, where all nodes with comparisons of the form $(a_i \, ? \, a_j)$ are deleted, where $a_i \in L$, $a_j \in A \setminus L$, and, apart from this, the right subtrees of such nodes are deleted as well. Since it is known that each element of the set L is smaller than any element of the set $A \setminus L$, then this transformation of the original tree will not result in the loss of information about the value b.

Note that the tree T^* has no less than $2^{|A\backslash L|-1} = 2^{N-k}$ leaves due to the fact that with the help of T^* it is possible to find the least element in $A \backslash L$, i.e. the sought element b.

The elements of the set L from the original set A can be chosen by $C(N, k-1)$ ways. Hence, in accordance with the rule of product (see chapter "Combinatorics" on p. 141) we obtain that the tree T contains no less than $2^{N-k}C(N, k-1)$ leaves.

12.62. *Proof.*

As follows from the Exercise **12.61**, in the decision tree T for the problem of finding the kth largest element of the array, the number of leaves l satisfies the inequality $l \geqslant 2^{N-k}C(N, k-1)$. For the height h of the tree T, we obtain the estimate

$$h \geqslant \lceil \log_2 l \rceil \geqslant N - k + \lceil \log_2 C(N, k-1) \rceil.$$

12.63. *Hint.*

Use the result of the Exercise **12.62** for the case $k = 2$.

12.64. *Answer*: 6 comparisons.

12.67. *Answer*: $T(N) = 2N + 1$ for $N > 2$.

12.70. *Solution.*

As is known, the recurrence relation describing the complexity of the algorithm SelectOpt has the form

$$\begin{cases} T(N) = T\left(\dfrac{N}{w}\right) + T\left(\dfrac{3N}{4}\right) + \Theta(N), \ N > w, \\ T(1), \ T(2), \ldots, \ T(w) \ — \text{const.} \end{cases}$$

From the Exercise **12.69** follows that the sufficient condition of the relation $T(N) = \Theta(N)$ is the inequality $\dfrac{1}{w} + \dfrac{3}{4} < 1$. Hence, at $w > 4$ the algorithm for determining the kth-order statistics will have the asymptotic complexity linear in the value N.

Parallel Algorithms

13

Nowadays, the existence of labor-consuming computing tasks as well as the dynamic development of multiprocessor computer systems brings the problems of developing and analyzing parallel algorithms into focus. A great variety of existing computer system architectures leads to the necessity of their classification with respect to various parameters. Historically, one of the first ways of dividing architectures based on the criterion of multiple instruction streams and data streams was proposed by Flynn.[1]

The **stream** is defined as a sequence of instructions or data executed or processed by the processor [72]. From this point of view, the program provides the processor with a stream of instructions for execution; the data for processing also comes in the form of a stream. According to Flynn's taxonomy, instruction streams and data streams are assumed to be independent. Thus, the computing systems are divided into the following classes.

1. Single instruction stream, single data stream (SISD). These characteristics are found in standard computers with a single-core processor, which can perform only one operation at a time.
2. Single instruction stream, multiple data streams (SIMD). In such systems, the same operation is performed simultaneously over different data. This class includes, for example, vector computer systems in which a single instruction can be performed over a set of data elements.
3. Multiple instruction streams, single data stream (MISD). Despite the lack of practical significance of this approach, MISD machines can be useful for some highly specialized tasks.

---

[1]Michael J. Flynn (b. 1934) is an American researcher, expert in the field of computer system architectures.

© Springer International Publishing AG, part of Springer Nature 2018
S. Kurgalin and S. Borzunov, *The Discrete Math Workbook*,
Texts in Computer Science, https://doi.org/10.1007/978-3-319-92645-2_13

Table 13.1 Flynn's taxonomy

Computer systems

		Data stream	
		Single	Multiple
Instruction stream	Single	SISD $a_1 + b_1$	SIMD $a_1 + b_1$ $a_2 + b_2$ $a_3 + b_3$
	Multiple	MISD $a_1 + b_1$ $a_1 - b_1$ $a_1 * b_1$	MIMD $a_1 + b_1$ $a_2 - b_2$ $a_3 * b_3$

4. Multiple instruction streams, multiple data streams (MIMD). MIMD systems are
   the most extensive and diverse group in the Flynn's classification. Most modern
   multiprocessor computer systems belong to this class.

The properties of computer system classes considered above are schematically
presented in Table. 13.1. The boxes contain examples of arithmetic operations available for simultaneous execution by the corresponding systems.

There is a widely used refinement on Flynn's taxonomy, according to which the
category MIMD is divided according to the way of memory organization in the
computer system. Among MIMD systems, the **multiprocessors** (Uniform Memory
Access, UMA) and **multicomputers** (NO Remote Memory Access, NORMA) are
singled out. The interaction between processors in multicomputers is executed using
the message passing mechanism [58].

13.1 The PRAM Model

So far, in this study guide, we have used the **Random Access Machine** (RAM)
model [11,51].

Let us list the main properties of RAM.

Fig. 13.1 RAM model

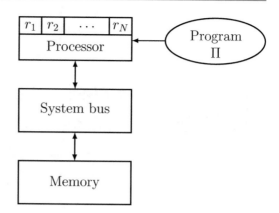

The system running within the RAM model consists of a processor, memory access device (a system bus), and the memory consisting of a finite number of cells (Fig. 13.1). The processor successively executes the instructions of program Π; it executes the main arithmetic and logical operations and reads/writes data in the memory. It is postulated that each instruction is executed for a fixed interval of time.

A random operation of the processor consists of three stages:

(1) Data reading from the memory into one of its registers r_i, where $1 \leqslant i \leqslant N$;
(2) Executing an arithmetic or logical operation with the contents of its registers;
(3) Data writing from the register r_j, where $1 \leqslant j \leqslant N$, into some memory cell.

It is assumed that the execution of the three above steps takes time $\Theta(1)$.

One of the most widely used models of parallel computer systems is Parallel Random Access Machine (PRAM) [51,58]. PRAM combines p processors, the shared memory and a control device that transmits instructions from program Π to the processors (Fig. 13.2).

An important feature of PRAM is the limited access time of any of the system's processors to a random memory cell. As in the case of RAM, the algorithm step corresponds to three processor operations:

(1) Data reading by the processor P_i from the jth memory cell;
(2) Executing an arithmetic or logical operation by the processor P_i with the contents of its registers;
(3) Data writing into the kth memory cell.

We emphasize once again that an algorithm step is executed for the time $\Theta(1)$.

Simultaneous access of two or more processors to the same memory cell leads to **access conflicts**. They are subdivided into **read** and **write conflicts**. If multiple processors attempt to read data from one cell, then two options for further operations are possible.

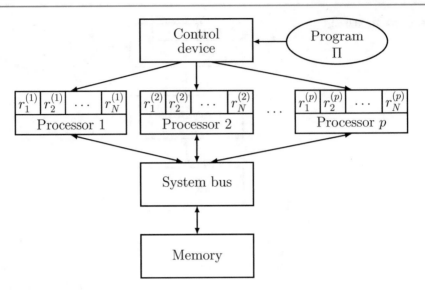

Fig. 13.2 PRAM model

1. Exclusive Read (ER). Only one processor is allowed to read at a given time, otherwise an error of the program occurs.
2. Concurrent Read (CR). The number of processors accessing the same memory cell is not limited.

If more than one processor attempts to write the data at one address, then two options exist.

1. Exclusive Write (EW). Only one processor is allowed to write into the given cell at a particular moment in time.
2. Concurrent Write (CW). Multiple processors get simultaneous access to a single memory cell.

Below are the options of what rule will a processor (or processors) follow in the latter case to write a record [39,51,58]:

— *Record of the general value.* It is assumed that all the processors ready to make a record into a single memory cell must record a general value for all of them, otherwise the recording instruction is considered to be erroneous.

— *Random choice.* The processor that executes recording is chosen randomly.

— *Prioritized recording.* Each of the competing processors is assigned a certain priority, such as its computation value, and only the value that came from the processor with the priority determined in advance (e.g., the lowest) is retained.

— *Mixed choice.* All processors provide values for the recording, from which by a certain operation the result (e.g., the sum of values, the maximum value, etc.) is created, which is then recorded.

Fig. 13.3 Conflict resolution
methods in PRAM

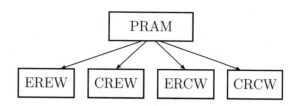

PRAM classification by conflict resolution methods is shown in Fig. 13.3.

Thus, EREW systems have significant limitations imposed on the work with the memory cells. On the other hand, CREW, ERCW, CRCW systems with a large number of processors are difficult to construct for technical reasons, since the number of cores simultaneously accessing a certain memory segment is limited. However, there is an important and somewhat unexpected result that makes it possible to simulate the work of CRCW machine on a system, built in accordance with EREW principle [67].

Emulation Theorem. *Let an algorithm for CRCW machine solve a certain problem with the parameter of size N during time $T(N)$, using p processors. Then there exists an algorithm for the same problem in EREW system with p processors which can be executed during time $O(T(N) \log_2 N)$. (The memory volume of PRAM must be increased $O(p)$ times.)*

Unlike RAM model, the main measure of the complexity of algorithms for multiprocessor computer systems is the execution time of the algorithm. We introduce the notation: $T_1(N)$ is the time required by a sequential algorithm to solve the problem, the complexity of which is estimated by the parameter N; $T_p(N)$ is the time required by a parallel algorithm on a machine with p processors, where $p > 1$. Since as it follows from the definition of RAM, each operation requires a certain time, the value $T_1(N)$ is proportional to the number of computational operations in the algorithm used.

Note that the minimum execution time of the algorithm is observed in the case of $p \to \infty$. A hypothetical computer system with an infinitely large number of available processors is called a **paracomputer**. An asymptotic complexity of the algorithm for a paracomputer is denoted by $T_\infty(N)$.

For the analysis of parallel algorithms, the concepts of **speedup, efficiency,** and **cost** are widely used. First of all, we should pay attention to how quickly the problem will be solved by comparison with the solution on a single processor machine.

Speedup $S_p(N)$ obtained by using a parallel algorithm on a machine with p processors is equal to $S_p(N) = \dfrac{T_1(N)}{T_p(N)}$. This is a measure of productivity gains compared with *the best* sequential algorithm. The greater the speedup, the greater is the difference in the problem-solving time between the system with a single processor and the running time of the algorithm on a multiprocessor system.

The **efficiency** $E_p(N)$ of the use of the processors by a particular parallel algorithm is $E_p(N) = \dfrac{T_1(N)}{pT_p(N)} = \dfrac{S_p(N)}{p}$.

The **cost** $C_p(N)$ is determined as the product of the time of the parallel solution to the problem and the number of the processors used: $C_p(N) = pT_p(N)$. The **cost-optimal algorithm** is characterized by the cost proportional to the complexity of the best sequential algorithm, and in this case $\dfrac{C_p(N)}{T_1(N)} = \Theta(1)$.

The limiting values of speedup and efficiency, as follows directly from their definitions, are $S_p = p$, $E_p = 1$. The maximum possible value of S_p is achieved when it is possible to uniformly distribute computation across all processors and no additional operations are required to provide communication between the processors at the stage of running the program and to combine the results. An attempt to increase speedup by changing the number of processors will reduce the value of E_p, and vice versa. Maximum efficiency is achieved by using only a single processor ($p = 1$).

We do not discuss here the superlinear speedup effect [83] that may arise in the models other than PRAM for several reasons:

—The sequential algorithm used for comparison is not optimal.

—The multiprocessor system architecture has specific features.

—The algorithm is nondeterministic.

—There are significant differences in the volume of available main memory when the sequential algorithm is required to access a relatively "slow" peripheral memory, and the parallel algorithm uses only "fast" memory.

Many of the algorithms described below require the presence of a sufficiently large number of processors. Fortunately, this does not limit their practical application, since for any algorithm in PRAM model there is a possibility of its modification for the system with a smaller number of processors. The last statement is called Brent's[2] Lemma [10,13,50].

Brent's Lemma. *Let a parallel algorithm* $\mathcal{A}$ *in order to solve some problem be executed on RAM during the time* T_1 *and on paracomputer during the time* T_∞. *Then there exists an algorithm* $\mathcal{A}'$ *for the solution of the given problem, such that on PRAM with* p *processors it is executed during the time* $T(p)$, *wherein* $T(p) \leqslant$
$$T_\infty + \frac{T_1 - T_\infty}{p}.$$

The proof of Brent's lemma is given in the solution to Exercise **13.7**.

Any parallel program has a sequential part which is formed by input/output operations, synchronization, etc. Assume that in comparison with a sequential way of solution:

(1) When the problem is divided into independent sub-problems, the time required for interprocessor communication and union of the results is negligibly small.

(2) The running time of the parallel part of the program decreases in proportion to the number of computing nodes.

Under these assumptions, an estimate of the value S_p is known.

---

[2]Richard Peirce Brent (b. 1946) is an Australian mathematician and computer scientist.

Amdahl's[3] Law. *Let f be a portion of sequential calculations in the algorithm $\mathcal{A}$. Then the speedup with the use of $\mathcal{A}$ on the system of p processors satisfies the inequality*

$$S_p \leqslant \frac{1}{f + (1 - f)/p}.$$

To prove this, we calculate the time required for executing the algorithm when working on multiprocessor. It consists of sequential operations $f T_1$ and the operations that can be parallelized, and is equal to $T_p = f T_1 + \dfrac{1 - f}{p} T_1$. Therefore, the upper limit for speedup can be represented as:

$$\left(S_p\right)_{\max} = \frac{T_1}{f T_1 + (1 - f) T_1/p} = \frac{1}{f + (1 - f)/p},$$

which is proved by Amdahl's Law.

Inequality $S_p \leqslant \left(S_p\right)_{\max}$ shows that the existence of sequential computations that cannot be parallelized, imposes a restriction on S_p. Even when using paracomputer, speedup does not exceed the value $S_\infty = \dfrac{1}{f}$.

Figure 13.4 shows the dependence of S_p on the number of processors p for typical values of the parameter f in computing tasks.

It is empirically found that for a wide class of computing tasks the portion of sequential computations decreases with the increasing size of the input data of the task. Therefore, in practice, speedup can be increased by increasing the computational complexity of the task being performed.

Despite the fact that PRAM model is widely spread, we should not forget about its limitations. In particular, the question of different data transfer rates for different processors in specific computer system architectures is completely ignored.

The structure of the algorithm for solving the problem may be graphically represented as a directed graph "operations–operands." It is an acyclic digraph. We denote it by $D = (V, E)$, where V is a set of vertices representing the running operations, and E is a set of edges. The edge $v_i v_j \in E$, if and only if the operation with the number j uses the result of the operation with the number i. Vertices $v_k^{(in)}$, where $k \in \mathbb{N}$, with in-degree $\forall k \; d^+(v_k^{(in)}) = 0$ are used for data input operations, vertices $v_l^{(out)}$, $l \in \mathbb{N}$, with out-degree $\forall l \; d^-(v_l^{(out)}) = 0$, correspond to the output operations (Fig. 13.5).

In the statement that computations at each vertex of the digraph D using RAM take constant time τ, the algorithm execution time on paracomputer is equal to

$$T_\infty = \tau \times \max_{(i,j)} \left(|C^{(i,j)}|\right),$$

---

[3]Gene Myron Amdahl (1922–2015) was an American computer scientist and computer technology expert.

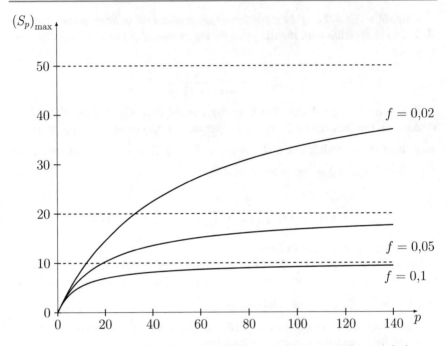

Fig. 13.4 Illustration of Amdahl's law. The dependencies of maximum speedup values $(S_p)_{max}$ on the number of processors p at different portions f of sequential computations

Fig. 13.5 Example of the "operations–operands" digraph

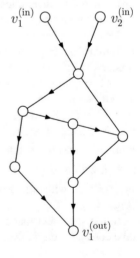

where $C^{(i,j)}$ is the path $v_i^{(in)}, \ldots, v_j^{(out)}$, and the maximum is taken with respect to all possible pairs (i, j). In other words, the time T_∞ is proportional to the number of vertices in the maximal path connecting vertices $v_i^{(in)}$ for some i and j.

13.2   The Sum and Partial Sums Problems

Consider the problem of calculating the sum of the array elements $list[1..N]$, where $N = 2^m$, for some positive integer m. The parallel version of the algorithm for solving this problem can be obtained from the following reasoning. We construct a complete binary tree, whose leaves correspond to the elements $list[1..N]$, $1 \leqslant i \leqslant N$. Into the internal vertices, we will place the sum of the values that are in the two above-standing vertices. Then at the lowest level, we obtain the required value $\sum_{i=1}^{N} list[i]$. A peculiarity of this summing sequence is the independence of the addition operation at the ith level of the tree of the remaining operations at this level [51,83].

The example of the sum calculation for $N = 8$ is shown in Fig. 13.6. Notation $[i..j]$ of the internal vertex in the figure indicates that this cell contains the sum $\sum_{k=i}^{j} list[k]$.

Note that in the program texts in this chapter, the reserved word **parallel** will be used. The record

$$\textbf{parallel for } i := 1 \textbf{ to } N \textbf{ do } \ldots$$

means that loop iterations with respect to variable i should be distributed across the available computer nodes as follows. The first processor P_1 executes instructions of the loop body using the value $i = 1$, the second processor P_2 uses $i = 2$, etc. Unless otherwise stated, it is assumed that there is a sufficient number of processors $p \geqslant N$ for a complete loop parallelization. However, in the case of $p < N$ the algorithms remain correct in view of the fact that the next value $i = p + 1$ is again processed by

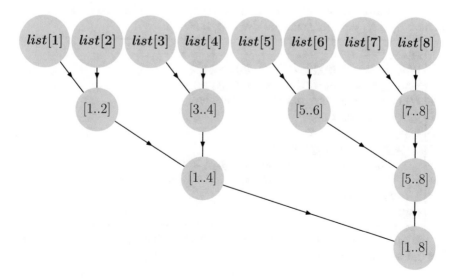

Fig. 13.6 Parallel summation of the elements at $N = 8$

the first processor, $i = p + 2$ by the second one, etc. Such procedure for distribution of the computational load is called **block-cyclic distribution**. In other words, the processor P_k, where $1 \leqslant k \leqslant p$, executes the following iterations:

$$
i = \begin{cases} k, \ p+k, \ \ldots, \ p\left(\left\lfloor \dfrac{N}{p} \right\rfloor - 1\right) + k, \ \text{if } k > N - p\left\lfloor \dfrac{N}{p} \right\rfloor, \\[4mm] k, \ p+k, \ \ldots, \ p\left\lfloor \dfrac{N}{p} \right\rfloor + k, \ \text{if } k \leqslant N - p\left\lfloor \dfrac{N}{p} \right\rfloor. \end{cases}
$$

The function of the reserved word **parallel** is similar to the directive

```
#pragma omp parallel for schedule (static, 1)
```

in OpenMP environment [12,54,58]. In particular, the code in C, corresponding to the line **parallel for** $i := 1$ **to** N **do** ..., has the form

```
#pragma omp parallel for schedule (static, 1)
    for (i=1; i<=N; i++)
       {
         ...    // loop body
       }
```

The directive #pragma omp parallel for schedule (static, 1) refers to a subsequent operator **for**, whose iterations will be executed in parallel. The parameter schedule (static, 1) indicates the block-cyclic distribution of iterations across the computer nodes.

Below is an algorithm of the parallel summation of the elements of the array *list* of the length $N = 2^m$ for some positive integer m.

```
 1  const N=8;
 2  type Mass = array[1.. N] of integer;
 3  function parSum(list: Mass):integer;
 4  //algorithm of the parallel summation
 5  var i,j:integer;
 6  begin
 7     for i:=1 to Log2(N) do
 8        parallel for j:=N downto 2 step 2^i do
 9                 list[j]:=list[j−2^(i−1)]+list[j];
10     result:=list[N];
11  end;
```

We calculate the main characteristics of this algorithm, speedup and efficiency. For the sequential version

$$
\frac{N}{2} + \frac{N}{4} + \ldots + 1 = N - 1
$$

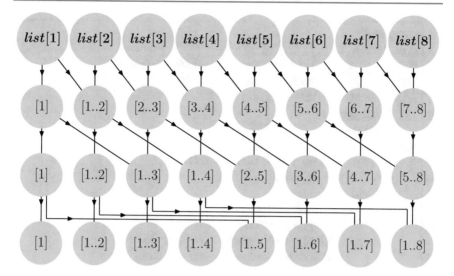

Fig. 13.7 Parallel calculation of partial sums at $N = 8$

addition operations will be required. For the multiprocessor computer system with $p = \dfrac{N}{2}$, the complexity of the algorithm for finding the sum will be proportional to the height of the digraph "operations–operands": $T_p(N) = \Theta(\log_2 N)$. Hence, $S(N) = \Theta\left(\dfrac{N}{\log_2 N}\right)$, $E(N) = \Theta(\log_2^{-1} N)$.

Partial sums calculation problem consists in determining, by the specified values $list[i], 1 \leqslant i \leqslant N$, of the sequence $S_j = \sum\limits_{i=1}^{j} list[i]$ for all $1 \leqslant j \leqslant N$. Many practical problems reduce to this statement, for example, determining of the lengths of the sequences of units in a bit string [11], as well as the continuous knapsack (rucksack) problem [3]. Assuming that $j = N$, we obtain the analyzed above problem of array elements sum.

A standard sequential algorithm requires the execution of $N - 1$ addition operations. Using PRAM with the number of processors equal to $p = N$, the result can be obtained in time $\Theta(\log_2 N)$, summing the elements as shown in Fig. 13.7. At each ith iteration of the algorithm, $1 \leqslant i \leqslant \log_2 N$ are performed operations of summation of the array and its copy shifted right by 2^{i-1} positions. The sought values S_j as a result of operation of the algorithm parPrefix will be generated in the array *prefixList*: $S_j = prefixList[j]$ for all $1 \leqslant j \leqslant N$. Note that with the purpose of differentiation of the order of access of the computation nodes to the elements of the array *prefixList* and additional array, *temp*[1..N] is introduced into the realization of the algorithm.

```
1   const N=8;
2   type Mass = array[1..N] of integer;
3   procedure parPrefix(list:Mass; var prefixList:Mass);
4   //calculation of partial sums
5   //of elements of the array list
6   var i,j:integer;
7        temp:Mass;
8   begin
9        for i:=1 to N do prefixList[i]:=list[i];
10       for i:=1 to Log2(N) do
11          begin
12             parallel for j:=1 to N do
13                temp[j]:=prefixList[j];
14             parallel for j:=N downto 2^(i−1)+1 do
15                prefixList[j]:=temp[j−2^(i−1)]+temp[j];
16          end;
17  end;
```

Since at each cycle iteration over the variable i in the ninth row of the algorithm, all $\Theta(N)$ addition operations are executed in parallel, the speedup, in comparison with the standard sequential algorithm, will have the value $S(N) = \Theta\left(\dfrac{N}{\log_2 N}\right)$. For the efficiency, we will obtain $E(N) = \Theta(\log_2^{-1} N)$.

Note. Replacement of row number 10 by the following

$$\textbf{for}\, i := 1\, \textbf{to}\, \text{Ceil}\,(\text{Log2}(N))\, \textbf{do}$$

will allow algorithm parPrefix operating with arrays of arbitrary size $N = 1, 2, 3, \ldots$

The cost of algorithm parPrefix is $O(N \log_2 N)$, which exceeds the cost of the elementary sequential algorithm. It is easy to save the logarithmic execution time and obtain the optimal solution of the parallel prefix problem, if we reduce the number of processors down to $O\left(\dfrac{N}{\log_2 N}\right)$ [51].

For this, we distribute the elements of the array $list[1..N]$ among $\left\lceil \dfrac{N}{\log_2 N} \right\rceil$ processors as follows: processor number i, where $1 \leqslant i \leqslant p$, will execute operations of the described algorithm versions for calculation of partial sums for the subarray $list[left..right]$, where $left = (i - 1) \log_2 N + 1$, $right = i \log_2 N$, and $right \leqslant N$. This will be the first step of algorithm parPrefix2.

Further, the sum of elements in subarrays is entered into the variables $temp[i]$, $1 \leqslant i \leqslant \left\lfloor \dfrac{N}{\log_2 N} \right\rfloor$, which will be the second step of the algorithm.

At the final third step, the partial sums of the values $temp[i]$, denoted by *prefix-Temp*[i], are used for calculating the values $S_j = \sum_{i=1}^{j} list[i]$ for all $1 \leqslant j \leqslant N$.

```
 1  const N=8;
 2  type Mass = array[1..N] of integer;
 3      MassTemp = array[1..Floor(N/Log2(N))] of integer;
 4  procedure parPrefix2(list:Mass; var prefixList:Mass);
 5  //cost-optimal algorithm for calculation
 6  //of partial sums of the array list
 7  var i,j,k:integer;
 8      temp,prefixTemp:MassTemp;
 9  begin
10      for i:=1 to N do prefixList[i]:=list[i];
11  //step 1
12      parallel for i:=1 to Ceil(N/Log2(N)) do
13          for j:=2 to Log2(N) do
14              begin
15                  k:=(i−1)*Log2(N)+j;
16                  if (k< =N) then
17                      prefixList[k]:=
18                      prefixList[k−1]+list[k];
19              end;
20  //step 2
21      parallel for i:=1 to Floor(N/Log2(N)) do
22          temp[i]:=prefixList[i*Log2(N)];
23      parPrefix(temp,prefixTemp);
24  //step 3
25      parallel for i:=2 to Ceil(N/Log2(N)) do
26          for j:=(i−1)*Log2(N)+1 to i*Log2(N) do
27              if (j< =N) then
28                  prefixList[j]:=
29                  prefixTemp[i−1]+prefixList[j];
30  end;
```

Let us estimate the execution time of each step of algorithm parPrefix2. It is easy to see that the first and the last steps require the time $O(\log_2 N)$, and the second step calculates the partial sums of the array *temp* and is executed in time $O\left(\log_2\left(\dfrac{N}{\log_2 N}\right)\right) = O(\log_2 N)$. Recalling that $p = O\left(\dfrac{N}{\log_2 N}\right)$, we come to the estimate of the cost of the second version of the parallel prefix calculation algorithm: $C_p = O(N)$.

In the mentioned parallel prefix calculation algorithms, the summation operation, if necessary, may be substituted by any other associative operation. Indeed, by the mathematical induction method, we can show that alteration of the row number 15 of algorithm parPrefix for

$$prefixList[j] := temp - 2^{\wedge}(i-1)] \otimes temp[j];$$

where $\otimes$ is an arbitrary binary associative operation will result in obtaining the value

$$list[1] \otimes list[2] \otimes \ldots \otimes list[j] = \overset{j}{\underset{i=1}{\bigotimes}} list[i]$$ for all $j = 1, 2, \ldots, N$. The proof of this statement is given in Exercise **13.9**.

13.3 Search Algorithms

The problem of search in a unordered sequence is easily parallelized. For simplicity, let us assume that all elements of the array *list* are different, and the element equal to the target one is known to be present in the array.

Divide the entire search range $[1..N]$ into p parts; in each of the obtained subarrays, perform sequential search in accordance with the algorithm shown on p. 369. The variables *left* and *right* are the left and right bounds of the part of the array *list*, being considered by the ith processor $(1 \leqslant left \leqslant right \leqslant N)$, and r is the order number of the sought element. Since the repeated elements are absent, one and only one processor will write into the variable r the index of the element equal to the sought one.

```
 1  const N=100;
 2  type Mass = array[1..N] of integer;
 3  function parSequentialSearch(list:Mass;
 4                                          target:integer):integer;
 5  //list — analyzed array of N elements
 6  //target — target value
 7  var i,j,left,right,r:integer;
 8  begin
 9      r:=0;
10      parallel for i:=1 to p do
11          begin
12              left:=(i−1)*Ceil(N/p)+1;
13              right:=i*Ceil(N/p);
14              if right>N then right:=N;
15              //sequential search in list[left..right]
16              for j:=left to right do
17                  if list[j]=target then r:=j;
18          end;
```

```
19 |     result:=r;
20 | end;
```

Let us define the basic characteristics of algorithm parSequentialSearch. The cycle of the variable j requires execution of $\left\lceil \dfrac{N}{p} \right\rceil$ comparisons; this is why the algorithm execution time is $O\left(\dfrac{N}{p}\right)$. The speedup is equal to $S_p = p$, and the cost is equal to $C_p = p \cdot O\left(\dfrac{N}{p}\right) = O(N)$. It means that the considered parallel algorithm for search in an unordered sequence is cost-optimal.

Now consider how we can solve the problem of search for the element equal to *target*, in a sorted array. As in the case with sequential search, divide the entire array into p parts, and perform binary search in each of the obtained subarrays (see p. 370). Let us call such an algorithm parBinarySearch. The execution time of the binary search parallel algorithm is equal to $T_p = \Theta\left(\log_2 \dfrac{N}{p}\right)$; the speedup is $S_p = O\left(\dfrac{\log_2 N}{\log_2(N/p)}\right)$; the cost is $C_p = O\left(p \log_2 \dfrac{N}{p}\right)$. Note that the cost of the described binary search parallel algorithm version exceeds that of algorithm BinarySearch. Nevertheless, an algorithm is known [3,68] that allows reducing the search cost in an ordered array down to $O(p \log_{p+1}(N+1))$.

The binary search solves the problem in time $O(\log_2 N)$. Given p processors, let us use the $(p+1)$-**ary search**, which functions as follows. At each algorithm execution step, the results of p comparisons allow dividing the array into $p+1$ subarrays, and the next step is applied recursively to one of the segments containing the key value.

Parallel search theorem. *There exists an algorithm that solves the problem of search for the target element in an ordered array on CREW machine with p processors in time* $O\left(\dfrac{\log_2(N+1)}{\log_2(p+1)}\right).$

The proof is based on determining the number of sequential steps performed by the above $(p+1)$-ary search algorithm when operating on the array of length N. Let m be the least natural number, such that $(p+1)^m \geq N+1$, or $m = \lceil \log_{p+1}(N+1) \rceil$. Show that in m steps of the $(p+1)$-ary search, it is possible to determine the position of the element equal to *target*.

Let us use the mathematical induction method. Introduce a predicate into consideration

$$P(m) = \{\text{algorithm of parallel search in an ordered array}$$
$$\text{of size } N \text{ finishes its operation in } m \text{ sequential steps}\}.$$

Basis step
In this case, $m = 1$ and $p \geq N$. It is clear that the algorithm will finish its operation in one sequential step.

Inductive step

Assume that k steps ($k > 1$) are sufficient for working with an array of size $(p + 1)^k - 1$. We will conduct the search in the array of size $(p + 1)^{k+1} - 1$ as follows. The processor number i, where $1 \leqslant i \leqslant p$, compares $list[i(p + 1)^k]$ with *target*. Based on the results of p such comparisons, performed in parallel, a conclusion is made: The sought element is contained in the subarray of size $(p + 1)^k - 1$. According to the inductive supposition, k steps will be required for finishing the algorithm's operation; i.e., the number of steps equal to $k + 1$ is sufficient for search in the array of size $(p + 1)^{k+1} - 1$.

It proves that the predicate $P(m)$ takes the true value for all natural m.

Further, consider the peculiarities of realization of the $(p + 1)$-ary search parallel algorithm. As shown above, it is enough to perform $m = \lceil \log_{p+1}(N + 1) \rceil$ steps for determining the position of the element equal to *target*. In the practical realization of the algorithm, it is convenient to substitute a **barrier** $list[N + 1]$, whose value exceeds $\max\limits_{1 \leqslant i \leqslant N} (list[i])$. In this case, no extra operations will be needed for checking the current index's exceeding the array limits.

In the array $s[i]$, where $0 \leqslant i \leqslant N + 1$, we will store 1 or 0 depending on whether *target* is positioned left or right of the current element, with which the comparison is performed by the jth processor, $1 \leqslant j \leqslant p$. The boundary values of the array are defined as $s[0] = 1$, $s[N + 1] = 0$. If $s[i - 1] \neq s[i]$ for some i ($1 \leqslant i \leqslant N + 1$), then the algorithm narrows the search range down to the segment $[(i - 1)(p + 1)^{m-1} - 1, \ldots, i(p + 1)^{m-1} - 1]$. The variables *temp*$[j]$ store the indices of the elements, which the jth processor compares with *target*. The described operations continue until the element $list[i] = target$ is found for some $1 \leqslant i \leqslant N$. If no such i is found, algorithm parPSearch returns the value 0.

```
1    const N=100;
2    type Mass = array[1..N+1] of integer;
3         MassTemp = array[0..p+1] of integer;
4    function parPSearch( list :Mass;
5                          target :integer): integer;
6    // list — analysed array of N elements
7    // target — target value
8    var temp , s :MassTemp;
9         left , right , i ,m, k :integer;
10   begin
11      left:=1;
12      right:=N;
13      k:=0;
14      m:=Ceil(Log2(N+1)/Log2(p+1));
15      s[0]:=0;    // the value of the boundary
16      s[p+1]:=1;  // elements of the array s
17      while ( left<=right ) and (k=0) do
18         begin
19            temp[0]:=left −1;
20            parallel for i:=1 to p do
```

```
21 |                        begin
22 |                        temp[i]:=(left−1)+i*(p+1)^(m−1);
23 |                        //ith processor compares
24 |                        //the value target with list[temp[i]]
25 |                        if temp[i]<=right then
26 |                            case compare(list[temp[i]],target) of
27 |                                −1: s[i]:=0;
28 |                                 0: k:=temp[i];
29 |                                +1: s[i]:=1;
30 |                            end
31 |                                            else
32 |                                                begin
33 |                                                    temp[i]:=right+1;
34 |                                                    s[i]:=1;
35 |                                                end;
36 |                            if s[i]<>s[i−1] then
37 |                                begin
38 |                                    left:=temp[i−1]+1;
39 |                                    right:=temp[i]−1;
40 |                                end;
41 |                            if (i=p) and (s[i]<>s[i+1]) then
42 |                                left:=temp[i]+1;
43 |                        end;
44 |                    m:=m−1;
45 |                end;
46 |            result:=k;
47 | end;
```

13.4 Sort Algorithms

Consider the problem of sorting the sequence *list[i]*, where $1 \leqslant i \leqslant N$, on PRAM with N processors. The simplest possible approach is generalization of the bubble sort. In the standard bubble sort method, the pairs of array values are compared strictly sequentially. Change the algorithm so that the comparisons at each iteration be independent and could be performed by different processors. This can be done if we divide the elements into two classes depending on the index parity and compare the values *list[i]* with their right neighboring elements. Such a modification of the bubble sort is called **odd–even sort**.

The odd–even sort algorithm consists of two steps, repeated $\left\lceil \dfrac{N}{2} \right\rceil$ times. At the first step, all processors with odd numbers p_i, where $i = 2k - 1$ for some positive integer k, compare *list[i]* and *list[i + 1]* and, if necessary, eliminate the inversion formed by these elements.

At the second step, similar operations are performed by processors p_i with even numbers $i = 2k'$ for some $k' = 1, 2, \ldots$

After repeating the described steps for $\left\lceil \dfrac{N}{2} \right\rceil$ times, all inversions will be eliminated and the array will become ordered [3].

```
1   const N=100;
2   type Mass = array[1..N] of integer;
3   procedure OddEvenSort(var list:Mass);
4   var i,j: integer;
5   begin
6      for i := 1 to Ceil(N/2) do
7         begin
8            parallel for j := 1 to Floor(N/2) do
9               if list[2*j−1]▷list[2*j] then
10                  swap(list[2*j−1],list[2*j]);
11           parallel for j := 1 to Floor((N−1)/2) do
12               if list[2*j]>list[2*j+1] then
13                  swap(list[2*j],list[2*j+1]);
14        end;
15  end;
```

The execution time of algorithm OddEvenSort is $O(N)$, and its cost is $C_p = O(N^2)$. The algorithm is not cost-optimal. Moreover, small speedup $S_p = O(\log_2 N)$ and considerable number of used processors $p = O(N)$ complement the list of the algorithm's drawbacks.

The Shell sort is also used for ordering data using multiprocessor computation systems. As is known, this sort is a refinement of the insertion sort. Let us form, out of the element of the array *list*, several subarrays, whose elements are distant from each other by some distance h in the array *list*. Each subarray is sorted by the simple insertion method, and then the value h decreases. The final value $h = 1$ corresponds to the insertion sort call over the entire array *list*. Thus, the sorting problem is solved.

Below is the text of the Shell sort program for the sequence of increments $h_s = 1, 4, 13, 40, 121, 364, \ldots$, or $h_{s+1} = 3h_s + 1$, $h_0 = 1$, and $0 \leqslant s < \lfloor \log_3(2N + 1) \rfloor - 1$.

```
1   const N=100;
2   type Mass = array[1..N] of integer;
3   procedure parShellSort(var list:Mass);
4   var i,j,k,h,l,s,temp: integer;
5   begin
6      h:=1;
7      s:=Floor(Log3(2*N+1))−1;
8      for i:=1 to s−1 do
9         h:=3*h+1;
10     for i:=s−1 downto 1 do
11        begin
```

```
12                    parallel for j:=1 to h do
13                        begin
14                            for k:=j+h to N do
15                                begin
16                                    temp:=list[k];
17                                    l:=k;
18                                    while (list[l− h]>temp) do
19                                        begin
20                                            list[l]:=list[l− h];
21                                            l:=l−h;
22                                            if (l<=h) then break;
23                                        end;
24                                    list[l]:=temp;
25                                end;
26                        end;
27                    h:=h div 3;
28                end;
29        InsertionSort(list);
30   end;
```

The cycle iterations over the variable j are distributed over different computation nodes. Upon completion of the parallel cycle, each processor must update the local copy of the variable h [11].

The computational complexity of the algorithm depends on the choice of the sequence h_s. In the considered case, $h_s = 1, 4, 13, 40, \ldots$, and the sequential algorithm execution time will be $O(N^{3/2})$ [81]. The asymptotic complexity of the parallel version of the Shell sort is studied in Exercise **13.22**.

13.5   Order Statistics

The parallel algorithm for finding the kth order statistics is based on algorithm SelectOpt (see chapter "Basic Algorithms").

Let us call the parallel algorithm that solves the problem of searching for order statistics parSelectOpt and represent its operation in the form of five steps. Let p processors be available, and it is required to find the kth largest element of the array $list[left..right]$, where $1 \leqslant left, right \leqslant N$. Introduce the auxiliary variable $d = right - left + 1$—array size.

1. If $d < 5$, then the processor P_1 determines the kth largest element directly. Otherwise, partition the array $list$ into p subarrays $list_j$, where $j = 1, \ldots, p$, of size $\left\lfloor \dfrac{d}{p} \right\rfloor$ each. The operations of comparison and displacement of the elements of the subarray $list_j$ will be performed by the processor P_j.

2. Each computation node P_j calls the function
   SelectOpt$\left(list_j, \left\lceil \left\lfloor \dfrac{d}{p} \right\rfloor / 2 \right\rceil \right)$ and thus determines the index of the median m_j of
   the subarray $list_j$. Thus, the set of medians $M = \{m_j : 1 \leqslant j \leqslant p\}$ is formed.
3. We recursively calculate the index of the median of the set of medians μ by calling
   the function parSelectOpt$(M, \lfloor p/2 \rfloor)$.
4. Partition the array $list$ into three parts: A, μ, and B. A will contain elements
   smaller than μ, B—elements greater than μ.
5. If the median of medians μ in the sequence $list$ has the index equal to k, then
   the answer is found—the algorithm returns the value k. Otherwise, depending on
   the size of the parts of the original array A and B obtained at the previous step,
   the function parSelectOpt is recursively called either on the first or on the second
   part of the array.

We can suppose the following method of formation of the arrays A and B at
the fourth step of the algorithm. Each processor P_j executes partitioning of the
subarray $list_j$ by procedure Partition (see p. 387), which forms the parts A_j and
B_j, $1 \leqslant j \leqslant p$. Denote by a_j the size of the array A_j. With the help of procedure
parPrefix2, we calculate the partial sums $t_i = \sum_{j=1}^{i} a_j$, and $t_0 = 0$. Further, P_j place
the copy of the array A_j into the memory area, starting from the position t_{i-1}. The
array B is constructed similarly.

Let us emphasize that algorithm parSelectOpt correctly functions in PRAM model
with the strictest conflict resolution method out of all considered—in EREW system.

Let us analyze algorithm parSelectOpt. Denote the algorithm execution time on
the array $list$ of size N by $T(N)$ and single out the values that make up $T(N)$ at each
of the steps 1–5: $T_1, T_2, \ldots, T_5$.

1. At the first step at $N \leqslant 4$, the execution time is a constant value. In the case
   $N > 4$, the subarray $list_j$ is assigned to the processor P_j, where $1 \leqslant j \leqslant p$. It
   also takes a constant time, i.e., $T_1(N) = \Theta(1)$.
2. Determining the index of the median m_j of the sequence of $\left\lfloor \dfrac{N}{p} \right\rfloor$ elements results
   in the relation $T_2(N) = \Theta\left(\dfrac{N}{p}\right)$.
3. At this step, the recursive call is performed, hence, $T_3(N) = T(p)$.
4. Partitioning of each array $list_j$ requires $\Theta\left(\dfrac{N}{p}\right)$ units of time. The partial sums
   calculation algorithm will be executed in $\Theta(\log_2 p)$ units of time. Finally, formation of the arrays A and B will require $\Theta\left(\dfrac{N}{p}\right)$ element copy operations. Thus,
   $$T_4(N) = \Theta\left(\dfrac{N}{p}\right) + \Theta(\log_2 p).$$
5. As in the sequential case, the recursive call at the fifth step requires time $T_5(N) = T\left(\dfrac{3N}{4}\right)$.

Since $T(N) = T_1(N) + T_2(N) + \ldots + T_5(N)$, as a result we obtain

$$T(N) = T(p) + T\left(\frac{3N}{4}\right) + \Theta\left(\frac{N}{p}\right) + \Theta(\log_2 p).$$

Let the number of computation nodes be $p = N^\varepsilon$, where $0 < \varepsilon < 1$. Then we can obtain the asymptotic estimate of $T(N) = \Theta(N^{1-\varepsilon})$. The cost of algorithm parSelectOpt is $C(N) = pT(N) = \Theta(N)$, and, therefore, parSelectOpt is a cost-optimal algorithm.

13.6   Fourier Transform

Fourier[4] Transform is one of the central methods of modern applied mathematics [22, 80].

Consider some function $s \colon \mathbb{R} \to \mathbb{C}$. Let $s(t)$ be absolutely integrable in the domain, i.e., $\int_{-\infty}^{\infty} |s(t)|\, dt < \infty$.

Continuous Fourier Transform of the function $s(t)$ is defined by the relation

$$S(f) = \int_{-\infty}^{\infty} s(t)e^{2\pi i f t}\, dt.$$

The function $S(f)$ is called an **image** of the original function $s(t)$. The basic properties of the arbitrary function image are established by the following lemma [22, 87].

Riemann[5]–Lebesgue[6] lemma. *For any absolutely integrable on the set of real numbers complex-valued function $s(t)$ its Fourier transform $S(f)$ is a function bounded and continuous on $\mathbb{R}$, which tends to zero at $|f| \to \infty$.*

Note. Traditionally, $s(t)$ in physical applications is considered as a **signal** and $S(f)$ as a **signal spectrum**. And the variable t has meaning of time and f of a linear frequency. Note that the notation $S(f) = \mathcal{F}[s(t)]$ is widely used.

If both functions $s(t)$ and $S(f)$ are absolutely integrable on the set of real numbers, the **inverse Fourier transform** allows, by the image, restoring the original function:

$$s(t) = \int_{-\infty}^{\infty} S(f)e^{-2\pi i f t}\, df.$$

By its definition, the Fourier transform is a linear operation. In other words, for any complex-valued functions $s_1(t)$, $s_2(t)$, defined and absolutely integrable on the

---

[4]Jean Baptiste Joseph Fourier (1786–1830)—French mathematician and physicist.
[5]Georg Friedrich Bernhard Riemann (1826–1866)—German mathematician and mechanician.
[6]Henri Léon Lebesgue (1875–1941)—French mathematician.

set of all real numbers, and for any constant $c_1, c_2 \in \mathbb{C}$ the equality is valid

$$\mathcal{F}[c_1 s_1(t) + c_2 s_2(t)] = c_1 \mathcal{F}[s_1(t)] + c_2 \mathcal{F}[s_2(t)].$$

13.7 Discrete Fourier Transform

In computational problems, the considered function is usually specified not analytically, but in the form of discrete values on some grid, usually uniform.

Let δ be the time interval between two sequential readings of the function $s(t)$:

$$x_n = s(n\delta), \quad \text{where } n \in \mathbb{Z}.$$

For any value of δ there exists the so-called **critical frequency** f_c, or the Nyquist[7] frequency, which is defined by the relation $f_c = \dfrac{1}{2\delta}$. There exists an important for information theory result: If outside the frequency range $[-f_c, f_c]$ the spectrum of some signal is equal to zero, then this signal can be completely restored by the series of reports x_n. More formally, the following theorem is valid.

Kotelnikov[8] **theorem** (in English-language literature known as the Nyquist–Shannon[9] theorem). *If the continuous and absolutely integrable on $\mathbb{R}$ function $s(t)$ has a nonzero Fourier transform for the frequencies $-f_c \leqslant f \leqslant f_c$, then $s(t)$ is completely defined by its sampling $x_n = s(n\delta)$, where $n \in \mathbb{Z}$, with the time interval δ:*

$$s(t) = \delta \sum_{n=-\infty}^{\infty} x_n \frac{\sin(2\pi f_c(t - n\delta))}{\pi(t - n\delta)}.$$

The wider the frequency band used by the communication channel, the more often the readings should be taken for correct discrete coding of the signal and its restoration. In its turn, it is compensated by the great amount of information transferred by this signal.

In practical problems, the sample length

$$\mathbf{x} = (\dots, x_{-2}, x_{-1}, x_0, x_1, x_2, \dots)$$

---

[7]Harry Theodor Nyquist (1889–1976)—American mathematician, specialist in information theory.
[8]Vladimir Aleksandrovich Kotelnikov (1908–2005)—Soviet and Russian mathematician and radio-physicist, specialist in information theory.
[9]Claude Elwood Shannon (1916–2001)—American mathematician and electrical engineer, the founder of information theory.

is usually limited, and in this connection, let us consider the transition from the continuous integral transform $\mathcal{F}[s(t)]$ to the approximate integral sum:

$$(\mathcal{F}[s(t)])_n = \int_{-\infty}^{\infty} s(t)e^{2\pi i f_n t}\,dt \rightarrow \sum_{k=-\infty}^{\infty} x_k e^{2\pi i f_n t_k}\delta = \delta \sum_{k=0}^{N-1} x_k e^{2\pi i kn/N}.$$

Thus, we come to the definition of the **discrete Fourier transform** (DFT) N of the values x_k, where $k = 0, 1, \ldots, N-1$:

$$y_n = \sum_{k=0}^{N-1} x_k e^{2\pi i kn/N}, \quad 0 \leqslant n \leqslant N-1,$$

or, using the notation $\omega = e^{2\pi i/N}$,

$$y_n = \sum_{k=0}^{N-1} \omega^{kn} x_k, \quad 0 \leqslant n \leqslant N-1.$$

Inverse discrete Fourier transform is defined by the formula

$$x_n = \frac{1}{N} \sum_{k=0}^{N-1} \omega^{-kn} y_k, \quad 0 \leqslant n \leqslant N-1.$$

In a vector notation, the considered relations have the form $\mathbf{y} = \mathcal{F}[\mathbf{x}]$ and $\mathbf{x} = \mathcal{F}^{-1}[\mathbf{y}]$.

Sequential application of the forward and inverse DFT to an arbitrary vector $\mathbf{x}$ does not change its components: $\forall \mathbf{x} \in \mathbb{R}^n \ \mathcal{F}^{-1}[\mathcal{F}[\mathbf{x}]] = \mathbf{x}$ and $\mathcal{F}[\mathcal{F}^{-1}[\mathbf{x}]] = \mathbf{x}$. Prove, for example, the first of these equalities.

$$(\mathcal{F}^{-1}[\mathcal{F}[\mathbf{x}]])_n = \frac{1}{N} \sum_{k=0}^{N-1} \omega^{-kn} (\mathcal{F}[\mathbf{x}])_k = \frac{1}{N} \sum_{k=0}^{N-1} \omega^{-kn} \sum_{i=0}^{N-1} \omega^{ik} x_i.$$

Change the summation order:

$$(\mathcal{F}^{-1}[\mathcal{F}[\mathbf{x}]])_n = \sum_{i=0}^{N-1} x_i \left(\frac{1}{N} \sum_{k=0}^{N-1} \omega^{k(i-n)} \right).$$

Further, let us use the properties of the variables $\omega = e^{2\pi i/N}$, considered in Exercise **7.27**: $\dfrac{1}{N} \displaystyle\sum_{k=0}^{N-1} \omega^{k(i-n)} = \begin{cases} 1, & \text{if } i = n; \\ 0, & \text{if } i \neq n. \end{cases}$ We finally obtain $(\mathcal{F}^{-1}[\mathcal{F}[\mathbf{x}]])_n = x_n \ \forall n = 0, 1, \ldots, N-1$. The equality $\mathcal{F}[\mathcal{F}^{-1}[\mathbf{x}]] = \mathbf{x}$ is proved similarly.

In accordance with the definition, calculation of the vector $\mathbf{y}$ by the defined $\mathbf{x}$ requires $O(N^2)$ complex multiplications. There is a way to considerably decrease the asymptotic complexity of DFT. The algorithm of the **fast Fourier transform** (**FFT**) requires just $O(N \log_2 N)$ multiplication operations; moreover, there exist methods of parallelizing the FFT. Further, we will suppose that $N = 2^m$ for some natural number m. This limitation is not principal, since if it is necessary to realize the fast Fourier transform algorithm for $N \neq 2^m$ one of the following approaches can be applied:

(1) Fill in several cells of the array representing a signal with zeros, so that N becomes equal to the nearest power of two.
(2) Use the more complex generalizations of FFT (see, for example, [76]).

The basis of the FFT method consists in that the DFT of the vector $\mathbf{x}$ of length N can be represented as a combination of transformations of two vectors, each having the length of $\dfrac{N}{2}$, and one transformation is applied to the points $\mathbf{x}$ with even indices, and the other is applied to the points with odd indices:

$$y_n = \sum_{k=0}^{N-1} e^{\frac{2\pi i k n}{N}} x_k = \sum_{k=0}^{N/2-1} e^{\frac{2\pi i (2k)n}{N}} x_{2k} + \sum_{k=0}^{N/2-1} e^{\frac{2\pi i (2k+1)n}{N}} x_{2k+1} =$$

$$= \sum_{k=0}^{N/2-1} e^{\frac{2\pi i k n}{N/2}} x_{2k} + e^{\frac{2\pi i n}{N}} \sum_{k=0}^{N/2-1} e^{\frac{2\pi i k n}{N/2}} x_{2k+1} = y_n^{(e)} + \omega^n y_n^{(o)}.$$

In the last equality, we introduce the notations $y_n^{(e)}$ and $y_n^{(o)}$ for the nth components of DFT vectors, formed by the elements of the original vector $\mathbf{x}$, standing on even and odd positions. The obtained relation allows calculating $\mathbf{y}$ recursively.

Using the easy-to-check property of the variables ω

$$\omega^{n+N/2} = -\omega^n, \quad \text{where } n = 0, 1, \ldots, N/2 - 1,$$

calculate the sums $y_n^{(e)}$ and $y_n^{(o)}$. The sought value $\mathbf{y}$ can be obtained as follows:

(1) The first $\dfrac{N}{2}$ elements $\mathbf{y}$ are equal to $y_n^{(e)} + \omega^n y_n^{(o)}$.
(2) The remaining $\dfrac{N}{2}$ elements $\mathbf{y}$ are equal to $y_n^{(e)} - \omega^n y_n^{(o)}$.

Fig. 13.8 Computation
diagram "butterfly"

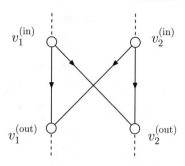

The FFT algorithm divides the calculation of the DFT vector of length N into a combination of two DFT vectors of size $\frac{N}{2}$. The results combining operation has a digraph "operations–operands," shown in Fig. 13.8. Such a sequence of computation operations is called the "butterfly" diagram.

A transition from the problem of size $\frac{N}{2}$ to the problem of size N requires $\frac{N}{2}$ complex multiplications and N assignment operations. Therefore, we can write the recurrence relation for the number of complex multiplications operations $T(N)$ in the FFT algorithm:

$$\begin{cases} T(N) = 2T\left(\frac{N}{2}\right) + \Theta(N), & N > 1, \\ T(1) = \text{const.} \end{cases}$$

The solution of the obtained recurrence relation has the form $T(N) = \Theta(N \log_2 N)$.

The recursive realization of the described version of the FFT algorithm consists of the following steps.

1. Elements of the array **x** are permuted in such a manner that the elements with even indices are located in the first half of the array and the elements with odd indices in the second half.
2. Next value of the parameter $\omega \leftarrow \omega^2$ is calculated, and FFT procedure is recursively applied to each of the two parts of the array.
3. The Fourier transforms (images) of the arrays of half-length, obtained at the previous step, form the final array **y** in accordance with "butterfly" computation diagram.

As is known, recursive algorithms are noted for high demands to the resources in comparison with the similar iterative ones. In this connection, below we give the realization of the FFT algorithm, requiring no recursive calls.

Perform permutation of the array elements, arranging them in the places required for "butterfly" diagram. Such permutation does not depend on the values x_i, where $0 \leqslant i \leqslant N - 1$, and only depends on the parameter N.

The above-mentioned transform is performed by the rule: The elements x_i and x_j change places if and only if the binary representations of the variables i and j are reversible [52]. The shuffle procedure performs the required exchanges in time $O(N)$. Note that in the solution of Exercise **13.35**, the exact number of exchanges is shown depending on the array size.

```
1    procedure shuffle(var a:ComplexArray);
2    //permutation of elements of the array a of length N
3    //for the FFT algorithm
4    var i,j,t:integer;
5    begin
6        j:=N div 2;
7        for i:=1 to N-2 do
8            begin
9                if i<j then swap(a[i],a[j]);
10               //calculation of value j,
11               //corresponding to the next i
12               t:=N div 2;
13               while (j div t = 1) do
14                   begin
15                       j:=j-t;
16                       t:=t div 2;
17                   end;
18               j:=j+t;
19           end;
20   end;
```

The text of the procedure realizing the described iterative algorithm is given in [52] and presented below in a somewhat revised form. For clarity, type of data "complex" is defined in it. The record of the arithmetic operations with the elements of this type in a real program should be naturally redefined. The auxiliary procedure get_omega(W) fills the array $W[0..N/2]$ with values $\omega_k = e^{2\pi i k/N}$, $0 \leqslant k \leqslant N/2$.

```
1    const N=64;
2    type ComplexArray = array[0..N] of complex;
3         ComplexArrayW = array[0..N div 2] of complex;
4    procedure FFT(var x:ComplexArray);
5    var W:ComplexArrayW;
6        omega,temp:complex;
7        i,j,k,l,t:integer;
8    begin
9        //generating the array of values ω
10       get_omega(W);
11       //permutation of elements of the data array
12       shuffle(x);
13       //transform "butterfly"
```

```
14 |     k:=1;
15 |     l:= N div 2;
16 |     while k<N do
17 |        begin
18 |           t:=0;
19 |           omega:=W[0];
20 |           for j:=0 to k−1 do
21 |              begin
22 |                 i:=j;
23 |                 while i<N do
24 |                    begin
25 |                       temp:=x[i];
26 |                       x[i]   := temp+omega*x[i+k];
27 |                       x[i+k]:= temp−omega*x[i+k];
28 |                       i:=i+2*k;
29 |                    end;
30 |                 t:=t+l;
31 |                 omega:=W[t];
32 |              end;
33 |           k:=2*k;
34 |           l:=l div 2;
35 |        end;
36 | end;
```

For distribution of the FFT operation by computation nodes, the following approach is used [57]: the vector $\mathbf{x} = (x_0, \ldots, x_{N-1})$, representing the readings of the input signal, is considered as a two-dimensional array of size $N_1 \times N_2$, where $N_1 = 2^{m_1}$, $N_2 = 2^{m_2}$ for some $m_1, m_2 \in \mathbb{N}$, such that $2^{m_1+m_2} = N$. Then the index i of an arbitrary element of the vector $\mathbf{x}$ can be written in the form

$$i = LN_1 + l, \text{ where } 0 \leqslant l \leqslant N_1 - 1, \ 0 \leqslant L \leqslant N_2 - 1.$$

Components of Fourier transform $\mathbf{y} = \mathcal{F}[\mathbf{x}]$ are calculated in accordance with the formula

$$y_n = \sum_{k=0}^{N_1 N_2 - 1} e^{\frac{2\pi i k n}{N_1 N_2}} x_k$$

for all $0 \leqslant n \leqslant N_1 N_2 - 1$, or

$$y_{\tilde{l} N_2 + \tilde{L}} = \sum_{L=0}^{N_2-1} \sum_{l=0}^{N_1-1} e^{\frac{2\pi i}{N_1 N_2} (\tilde{l} N_2 + \tilde{L})(L N_1 + l)} x_{L N_1 + l}.$$

After algebraic transformations, write the components of the vector **y** in the following form:

$$
y_{\tilde{l}N_2+\tilde{L}} = \sum_{l=0}^{N_1-1} e^{\frac{2\pi i \tilde{l} l}{N_1}} \left(e^{\frac{2\pi i l \tilde{L}}{N_1 N_2}} \left(\sum_{L=0}^{N_2-1} e^{\frac{2\pi i L \tilde{L}}{N_2}} x_{LN_1+l} \right) \right).
$$

Realization of a parallel program based on the obtained expression for **y** presents no problems. Using the possible parallelism over the variable l, the internal sum $\sum_{L=0}^{N_2-1} e^{\frac{2\pi i L \tilde{L}}{N_2}} x_{LN_1+l}$ is calculated for $l = 0, \ldots, N_1 - 1$. Further, the obtained values are multiplied by $e^{\frac{2\pi i l \tilde{L}}{N_1 N_2}}$, and then one more FFT is applied with the possibility of parallelism over the variable L, and $L = 0, \ldots, N_2 - 1$. The values $y_{\tilde{l}N_2+\tilde{L}}$ form the final answer $\mathbf{y} = \mathcal{F}[\mathbf{x}]$.

Let us estimate the asymptotic complexity of this algorithm. The first Fourier transform requires $\Theta(N_2 \log_2 N_2)$ arithmetic operations for each $\tilde{l} = 0, \ldots, N_1 - 1$. Then follow N_2 parallel multiplications. The second DFT will consume $\Theta(N_1 \log_2 N_1)$ operations for each $\tilde{L} = 0, \ldots, N_2 - 1$. Thus, the execution time of the described FFT version on a system with $p = \max(N_1, N_2)$ processors is equal to

$$
T(N) = \Theta(N_2 \log_2 N_2) + \Theta(1) + \Theta(N_1 \log_2 N_1) = \Theta(\max(N_1, N_2) \log_2 N).
$$

The cost $C(N)$ will be $(\max(N_1, N_2))^2 \log_2 N$, and at $N_1 = N_2 = \sqrt{N}$ it coincides with the cost of the sequential FFT version, while the speedup $S(N)$ will have the value $\sqrt{N}$.

13.8 Problems

13.1. The execution time of a sequential version of some algorithm $\mathcal{A}$ is $T_1(N) = 2N \log_2(N) \tau$, where N—the input data size, τ—the execution time of one computation operation. Supposing that the algorithm allows maximal parallelizing, i.e., the execution time on the computation system with p processors is $T_p(N) = \dfrac{T_1(N)}{p}$, calculate the execution time of the algorithm $\mathcal{A}$ in the following cases:

(1) $N = 32$, $p = 4$;
(2) $N = 32$, $p = 16$.

13.2. Solve the previous problem for the algorithm $\mathcal{B}$ with the exponential asymptotic complexity: $T_1(N) = 2^N \tau$.

13.3. Let the share of sequential computation in the program be $f = \dfrac{1}{10}$. Calculate the maximum speedup of the program $(S_p)_{\max}$ on the computation system with p processors taking into account Amdahl's law.

13.4. Let the share of sequential calculations in the program be $f = \dfrac{1}{100}$. Calculate the maximum speedup of the program S_∞ taking into account Amdahl's law.

13.5. Let the speedup for some parallel algorithm $\mathcal{A}$ during execution on the system with p processors be S_p. Taking into account Amdahl's law, calculate the algorithm's speedup when using the algorithm $\mathcal{A}$ on the system with p' processors.

13.6. Consider some computation algorithm $\mathcal{A}$, consisting of two blocks $\mathcal{A}_1$ and $\mathcal{A}_2$, where the second can only start execution after the completion of the first. Let the shares of sequential computations in $\mathcal{A}_1$ and $\mathcal{A}_2$ be f_1 and f_2, respectively, and the execution time $\mathcal{A}_2$ in the sequential mode exceeds the sequential execution time $\mathcal{A}_1$ by η times. Calculate the maximum speedup of the algorithm $\mathcal{A}$, achievable on the computation system with p processors.

13.7. Make a graph of maximum efficiency $(E_p)_{\max}$ of using the processors by the parallel algorithm depending on the number of computation nodes p taking into account Amdahl's law, if the share of sequential computations is equal to:

(1) $f = \dfrac{1}{10}$;

(2) $f = \dfrac{1}{50}$.

∗**13.8.** Prove Brent's lemma.

∗**13.9.** Prove that the algorithm parPrefix can be used for calculating partial results of any associative binary operation on the elements of some semigroup.

13.10. Demonstrate the operation of the elements sum calculation parallel algorithm by the example of the array | 20 | 12 | 18 | 16 | 24 | 10 | 22 | 14 |.

13.11. Demonstrate the operation of the partial sum calculation parallel algorithm by the example of the array from the previous exercise.

13.12. Explain what changes should be made to the algorithm parSum for calculating the sum of the elements of the array $list[1..N]$ in the case when the array size N is not equal to the integer power of two. Estimate the cost of such an algorithm.

13.13. The **continuous knapsack problem** considers N objects $a_1, \ldots, a_N$ and a knapsack of volumetric capacity V. Each object a_i, where $i = 1, \ldots, N$, has positive volume v_i and cost c_i. It is required to arrange the objects (perhaps, not wholly) in the knapsack in such a manner that the two conditions are met:

(1) Their total volume should not exceed V.

(2) Their total cost should be maximum.

Formally, the problem reduces to the following:

$$\sum_{i=1}^{N} f_i c_i \to \max$$

on conditions that $\sum_{i=1}^{N} f_i v_i \leqslant V$ and $0 \leqslant f_i \leqslant 1$ for $i = 1, \ldots, N$, where f_i is the share of the object a_i, placed into the knapsack.

Write the parallel algorithm for solution of the knapsack problem within the framework of CREW model.

13.14. The sequential algorithm for raising the real number x to the Nth power, where $N \in \mathbb{Z}_0$, is based on the relation

$$x^N = \begin{cases} \left(x^{\lfloor N/2 \rfloor}\right)^2, & \text{if } N > 0 \text{ and is even,} \\ x\left(x^{\lfloor N/2 \rfloor}\right)^2, & \text{if } N > 0 \text{ and is odd,} \\ 1, & \text{if } N = 0. \end{cases}$$

Find the asymptotic complexity of this algorithm.

∗13.15. The method of raising a real number to a nonnegative integer power, considered in the previous exercise, allows no parallelizing. Suggest a parallel algorithm for calculation of x^N, where $x \in \mathbb{R}$, $N \in \mathbb{Z}$.

∗13.16. Suggest a parallel algorithm for calculation of the elements of the set $\{x^2, x^3, \ldots, x^N\}$, where $x \in \mathbb{R}$, $N > 2$. Consider the following cases:

(1) The number of processors is $p = \dfrac{N(N+1)}{2} - 1$.

(2) The number of processors is $p = N$.

The algorithm execution time should be $2T_d + O(\log_2 N)T_a$ in the first case and $4T_d + T_m + O(\log_2 N)T_a$ in the second case, where T_a, T_m, T_d is the time spent on addition, multiplication and division of two complex numbers respectively.

13.17. Assume that in the array $list[1..N]$ the elements can be repeated. What changes should be made to the algorithm:

(1) parSequentialSearch;

(2) parBinarySearch,

so that it will return the number of the first appearance of the sought element? Consider CREW and CRCW models.

13.18. Consider the problem of searching for the target element in an ordered array of size N. Show that this problem on CREW machine with $O(N^\varepsilon)$ processors for an arbitrary $\varepsilon > 0$ can be solved in time $O(1)$.

13.19. Write an algorithm for finding the maximum among N numbers on CRCW machine with N^2 processors. The algorithm execution time should be $O(1)$.

13.20. Write an algorithm for finding the maximum among N numbers on CRCW machine with N processors. The algorithm execution time should be $O(\log_2 \log_2 N)$.

13.21. As in the bubble sort case, the odd–even sort algorithm can generate a sorted array even before the completion of cycle iterations over i. Suggest a method to reduce the number of comparisons in this case. How will it influence the algorithm's cost?

∗13.22. Prove that the number of comparisons performed by the sequential version of the Shell sort based on the increments h_s, where $h_{s+1} = 3h_s + 1, h_0 = 1$, $0 \leqslant s < \lfloor \log_3(2N+1) \rfloor - 1$, is equal to $O(N^{3/2})$.

∗13.23. Prove that the execution time of the parallel version of the Shell sort based on the increments h_s, where $h_{s+1} = 3h_s + 1, h_0 = 1, 0 \leqslant s < \lfloor \log_3(2N+1) \rfloor - 1$, is equal to $O(N \log_2 N)$.

∗13.24. Find the worst-case execution time of the parallel version of the Shell sort, if the sequence of increments suggested by Pratt is used (see Exercise **12.46**).

13.25. Suggest a parallel algorithm determining the set of elements of the array $list[1..N]$, not exceeding the kth largest element. The algorithm execution time should be $\Theta(N^{1-\varepsilon})$, where N—the array size, and the parameter $\varepsilon \in (0, 1)$ depends on the number of available computation nodes p: $\varepsilon = \dfrac{\log_2 p}{\log_2 N}$.

13.26. Let $S(f) = \mathcal{F}[s(t)]$. Check that for the continuous Fourier transform the following properties are fulfilled:

(1) $\mathcal{F}[s(at)] = \dfrac{1}{|a|} S\left(\dfrac{f}{a}\right) \ \forall a \in \mathbb{R} \setminus \{0\}$ (scaling);

(2) $\mathcal{F}[s(t - t_0)] = S(f)e^{2\pi i f t_0} \ \forall t_0 \in \mathbb{R}$ (shift in the time domain).

13.27. Prove **Parseval's theorem,**[10] connecting the full energy of the square-integrable signal in the time and frequency domains:

$$\int_{-\infty}^{\infty} |s(t)|^2 dt = \int_{-\infty}^{\infty} |S(f)|^2 df.$$

13.28. Let **y** be DFT of the vector **x** of length N. Prove the **discrete analog of Parseval's theorem:**

$$\sum_{k=0}^{N-1} |x_k|^2 = \frac{1}{N} \sum_{k=0}^{N-1} |y_k|^2.$$

13.29. Using the definition of the discrete Fourier transform, calculate $\mathcal{F}[\mathbf{x}]$, if:

(1) $\mathbf{x} = (0, 1, 0, 1)$;
(2) $\mathbf{x} = (1, 0, 1, 0)$.

13.29. Calculate DFT of the vector **x** of length N with components equal to the binomial coefficients $x_n = C(N - 1, n)$, where $0 \leqslant n \leqslant N - 1$.

13.30. Calculate DFT of the vector **x** of length $N = 2^m$ for some positive integer m, if

(1) $\mathbf{x} = (\underbrace{0, 0, \ldots, 0, 0,}_{N/2 \text{ zeros}} \underbrace{1, 1, \ldots, 1, 1}_{N/2 \text{ ones}})$;

(2) $\mathbf{x} = (\underbrace{1, 1, \ldots, 1, 1,}_{N/2 \text{ ones}} \underbrace{0, 0, \ldots, 0, 0}_{N/2 \text{ zeros}})$.

13.31. Write out the elements of the array $\boxed{0\,|\,1\,|\,2\,|\,3\,|\,4\,|\,5\,|\,6\,|\,7}$ after applying the "shuffle" procedure to it (see p. 444), which permutes the elements for subsequent application of the computation diagram "butterfly" in the fast Fourier transform algorithm.

13.32. Write out the elements of the array
$$\boxed{0\,|\,1\,|\,2\,|\,3\,|\,4\,|\,5\,|\,6\,|\,7\,|\,8\,|\,9\,|\,10\,|\,11\,|\,12\,|\,13\,|\,14\,|\,15}$$
after applying the "shuffle" procedure to it.

13.34. Show that double application of "shuffle" procedure to an arbitrary array of size $N = 2^m$, where $m = 0, 1, 2, \ldots$, leaves its elements unchanged.

13.35. Consider the array x of size $N = 2^m$ for some nonnegative integer m. Find how many exchange operations "swap" will be required by the call of "shuffle" procedure.

---

[10]Marc-Antoine Parseval des Chênes (1755–1836)—French mathematician.

13.36. Sine transform of the vector **x** of length N, and $x_0 = 0$, is defined as follows:

$$(\mathcal{F}_{\sin}[\mathbf{x}])_n = \sum_{k=0}^{N-1} x_k \sin\left(\frac{\pi kn}{N}\right).$$

Show that the sine transform can be reduced to the standard DFT.

13.37. Cosine transform of the vector **x** of length $N + 1$ is defined as follows:

$$(\mathcal{F}_{\cos}[\mathbf{x}])_n = \frac{1}{2}\left[x_0 + (-1)^n x_N\right] + \sum_{k=1}^{N-1} x_k \cos\left(\frac{\pi kn}{N}\right).$$

Show that the cosine transform can be reduced to the standard DFT.

13.38. Write the parallel algorithm:

(1) Of discrete sine transform;
(2) Of discrete cosine transform.

13.39. Demonstrate the operation of the fast Fourier transform parallel algorithm, described on pp. 445–446, by the example of the array $\boxed{0\,|\,1\,|\,2\,|\,3\,|\,4\,|\,5\,|\,6\,|\,7}$.

13.40. Write the parallel algorithm for calculating the two-dimensional discrete Fourier transform of the array $x[0..(N_1 - 1), 0..(N_2 - 1)]$

$$(\mathcal{F}[x])_{n_1, n_2} = \sum_{k_1=0}^{N_1-1} \sum_{k_2=0}^{N_2-1} \omega_1^{k_1 n_1} \omega_2^{k_2 n_2} x_{k_1, k_2},$$

where $\omega_1 = e^{2\pi i/N_1}$, $\omega_2 = e^{2\pi i/N_2}$.

13.9 Answers, Hints, Solutions

13.1. *Answer*:

(1) $T_p(N) = 80\tau$;
(2) $T_p(N) = 20\tau$.

13.2. *Answer*:

(1) $T_p(N) = 2^{30}\tau$;
(2) $T_p(N) = 2^{28}\tau$.

13.3. *Answer*: $(S_p)_{\max} = \dfrac{10p}{p + 10}$.

13.4. *Answer*: $S_\infty = 100$.

13.5. *Answer*: $S_{p'} = S_p \dfrac{p'(p-1)}{p(p'-1)+(p-p')S_p}$.

13.6. *Solution.*

Denote the execution time of the block $\mathcal{A}_1$ in the sequential mode by τ, then the execution time $\mathcal{A}_2$ in the same mode will be $\eta\tau$. Substitute these values into the formula for speedup S_p of the parallel algorithm on the computation system with p processors:

$$S_p = \frac{T_1}{T_p}.$$

According to the problem statement, the sequential execution time of the whole algorithm $\mathcal{A}$ is composed of the execution time of the first block and the execution time of the second block:

$$T_1 = \tau + \eta\tau = (1+\eta)\tau.$$

In the parallel mode, the times of the computation shares $(1 - f_1)$ in $\mathcal{A}_1$ and $(1 - f_2)$ in $\mathcal{A}_2$ will be distributed by p computation nodes. Hence, the execution time of the algorithm $\mathcal{A}$ can be estimated as

$$T_p \geqslant f_1\tau + \frac{(1-f_1)\tau}{p} + f_2\eta\tau + \frac{(1-f_2)\eta\tau}{p} =$$

$$= (f_1 + \eta f_2)\tau + \frac{1 + \eta - (f_1 + \eta f_2)}{p}\tau.$$

As a result, for the maximum achievable speedup value for execution of the algorithm $\mathcal{A}$ we obtain the expression:

$$\left(S_p\right)_{\max} = \frac{(1+\eta)\tau}{(f_1 + \eta f_2)\tau + (1 + \eta - (f_1 + \eta f_2))\tau/p} = \frac{1}{\overline{f} + (1 - \overline{f})/p},$$

where the notation $\overline{f} = \dfrac{f_1 + \eta f_2}{1 + \eta}$ is introduced, meaning the averaged share of sequential operations in this algorithm.

13.7. *Answer*: see Fig. 13.9.

13.8. *Proof.*

Let us introduce into consideration the "operations–operands" digraph of the algorithm $\mathcal{A}$. By virtue of the fact that for solution of the computation problem on a paracomputer time T_∞ is required, the number of the digraph's levels is $d = \dfrac{T_\infty}{\tau}$, where τ is the execution time of one operation. Denote by n_i, $1 \leqslant i \leqslant d$ the number of nodes on the ith level.

The algorithm $\mathcal{A}'$ will used a smaller number of processors in comparison with $\mathcal{A}$, namely p. In this case, on some level with number i each processor takes no more than $\left\lceil \dfrac{n_i}{p} \right\rceil$ operations.

We will obtain the estimate of the execution time of the algorithm $\mathcal{A}'$ having summed the time of operation on each of d levels of the "operations–operands" digraph:

$$T(p) \leqslant \sum_{i=1}^{d} \left\lceil \frac{n_i}{p} \right\rceil \tau.$$

Using the property of the "ceiling" function $\left\lceil \dfrac{k}{m} \right\rceil \leqslant \dfrac{k+m-1}{m}$, valid for all $k, m \in \mathbb{N}$, transform the inequality into the form:

$$T(p) \leqslant \sum_{i=1}^{d} \frac{n_i + p - 1}{p} \tau = \frac{1}{p} \sum_{i=1}^{d} n_i \tau + \frac{p-1}{p} \sum_{i=1}^{d} \tau.$$

In accordance with our notations, the execution time of the algorithm $\mathcal{A}$ on a one-processor system is $\sum_{i=1}^{d} n_i \tau = T_1$. We finally obtain the estimate of the execution time of the algorithm $\mathcal{A}'$ on PRAM with p processors:

$$T(p) \leqslant T_\infty + \frac{T_1 - T_\infty}{p}.$$

13.9. *Proof.*
Recall that semigroup is a pair $(A, \otimes)$, where A is some set, $\otimes$ is a semigroup operation. Show that for calculation of partial results of the operation $\otimes$, applied to

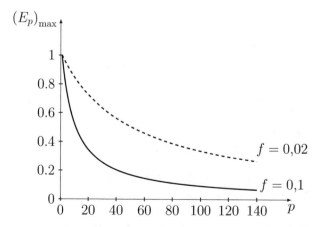

Fig. 13.9 Dependence of the values of the maximum efficiency $(E_p)_{\max}$ on the number of processors p at different f—shares of sequential computations. The solid line—$f = \dfrac{1}{10}$, the dotted line—$f = \dfrac{1}{50}$

the values $list[i]$, $1 \leqslant i \leqslant N$, the algorithm parPrefix can be used, different from the one written on p. 430 in that the row number 15 is replaced by the following:

$$prefixList[j] := temp - 2^\wedge(i-1) \otimes temp[j];$$

Instead of $\otimes$, a sign of a certain binary operation shall be placed.

Denote by $^{(i)}prefixList[j]$ the values of elements of the array $prefixList$ after execution of the ith cycle iteration in the eighth row. Using the mathematical induction method, prove the auxiliary statement:

$$^{(i)}prefixList[j] = \bigotimes_{m=\sigma(i,j)}^{j} list[m] \quad \forall j = 1, \ldots, N,$$

where $\sigma(i, j) = \max(j + 1 - 2^i, 1)$.

Denote the predicate stated in the previous sentence by $P(i)$, $i = 1, 2, \ldots,$ $\lceil \log_2 N \rceil$ (see note on p. 430).

Basis step

Consider the case $i = 1$. After execution of the first iteration, we obtain:

$$^{(1)}prefixList[j] = \begin{cases} list[j-1] \otimes list[j], & \text{if } j = 2, 3, \ldots, N, \\ list[1], & \text{if } j = 1. \end{cases}$$

Taking into account the introduced notations, this relation can be rewritten in the form: $^{(1)}prefixList[j] = \bigotimes_{m=\sigma(1,j)}^{j} list[m] \quad \forall j = 1, \ldots, N$, where $\sigma(1, j) = \max(j - 1, 1)$, with actually proves the validity of $P(1)$.

Inductive step

Assume that for $i = k$, as a result of operation of the algorithm parPrefix, the values $^{(k)}prefixList[j] = \bigotimes_{m=\sigma(k,j)}^{j} list[m]$ are obtained. Prove the validity of the predicate $P(k + 1)$, i.e., the validity of the equality $^{(k+1)}prefixList[j] = \bigotimes_{m=\sigma(k+1,j)}^{j} list[m]$ for all natural indices j, not exceeding N.

Indeed, from the algorithm analysis follows the relation (Fig. 13.10):

$$^{(k+1)}prefixList[j] = \begin{cases} ^{(k)}prefixList[j - 2^k] \otimes {}^{(k)}prefixList[j], & 2^k < j \leqslant N, \\ ^{(k)}prefixList[j], & 1 \leqslant j \leqslant 2^k. \end{cases}$$

Let us use the expression for the elements of the array $prefixList$ at the previous step:

$$^{(k+1)}prefixList[j] = \bigotimes_{m=\sigma(k,j-2^k)}^{j-2^k} list[m] \otimes \bigotimes_{m=\sigma(k,j)}^{j} list[m].$$

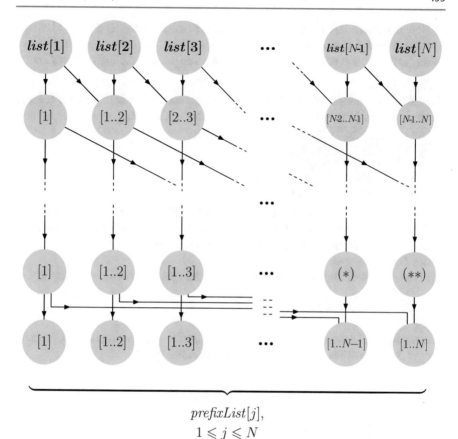

$$prefixList[j],$$
$$1 \leqslant j \leqslant N$$

by $(*)$ denote the values $[s..N-1]$,

by $(**)$ — $[s..N]$, where $s = \max(j + 1 - 2^{\lceil \log_2 N \rceil - 1}, 1)$

Fig. 13.10 Calculation of partial results of the operation $\otimes$, applied to the elements of the array *list*

Since $\sigma(k, j - 2^k) = \max(j + 1 - 2^{k+1}, 1) = \sigma(k + 1, j)$, then ${}^{(k+1)}prefixList[j] =$ $\underset{m=\sigma(k+1,j)}{\overset{j}{\otimes}} list[m]$ $\forall j = 1, \ldots, N$.

Hence, the predicate $P(i)$ takes the true value for all $i = 1, 2, \ldots, \lceil \log_2 N \rceil$. As a result, at the iteration number $i = \lceil \log_2 N \rceil$ we have $\sigma(\lceil \log_2 N \rceil, j) = 1$, and the final values of the elements are equal to $prefixList[j] = \underset{m=1}{\overset{j}{\otimes}} list[m]$.

Note that the order of operands in operations of the form $list[i] \otimes list[j]$, where $1 \leqslant i \leqslant N$ and $1 \leqslant j \leqslant N$, has not changed. It proves the correctness of the algorithm parPrefix, and as the operation $\otimes$ can act any binary associative operation.

13.10. *Solution.*
Represent the values of elements of the array after each cycle iteration over i. Since the array size is $N = 8$, the variable i takes the values $1, 2, 3$.

20	12	18	16	24	10	22	14

source data;

20	**32**	18	**34**	24	**34**	22	**36**

$i=1$;

20	32	18	**66**	24	34	22	**70**

$i=2$;

20	32	18	66	24	34	22	**136**

$i=3$.

The values of the elements changing at this iteration are highlighted in bold and underlined. After execution of the algorithm, the answer is read from the cell number $N = 8$. As a result, the algorithm returns the answer 136.

13.13. *Hint.*
Arrange the objects in nonascending order of the specific cost $\widetilde{c}_i = \dfrac{c_i}{v_i}$ for all $i = 1, \ldots, N$ and apply the parallel prefix calculation algorithm to the array $\widetilde{c}$.

13.14. *Solution.*
The asymptotic complexity of the sequential algorithm for raising to a nonnegative integer power satisfies the following recurrence relation:

$$T(N) = T\left(\left\lfloor \frac{N}{2} \right\rfloor\right) + f(N),$$

where the nonhomogeneity function $f(N)$ has the form

$$f(N) = \begin{cases} 1, & \text{if } N \text{ is even,} \\ 2, & \text{if } N \text{ is odd.} \end{cases}$$

Since $f(N) = \Theta(1)$, then, in accordance with the basic recurrence relations theorem $T(N) = \Theta(\log_2 N)$.

13.15. *Solution.*
The following computation algorithm is known x^N, suggested by Kung[11] [40]. Let the following be defined: $x \in \mathbb{R}$ and $N \in \mathbb{Z}$, and $x \neq \pm 1$.

1. Calculate ω_k and $\widetilde{\omega}_k$ for $k = 0, 1, \ldots, N - 1$, where by ω_k is denoted the kth complex root of the equation $z^N = 1, \widetilde{\omega}_k = \dfrac{\omega_k}{N}$.
2. Calculate $y_k = x - \omega_k, k = 0, 1, \ldots, N - 1$.
3. Find the relations $z_k = \dfrac{\widetilde{\omega}_k}{y_k}, k = 0, 1, \ldots, N - 1$.

---

[11] Hsiang-Tsung Kung (born 1945)—American researcher of Chinese origin, specialist in the sphere of parallel computation systems.

4. Calculate the sum $Z = \sum_{k=0}^{N-1} z_k$.

5. Invert the value Z: $\widetilde{Z} = \dfrac{1}{Z}$

6. The final answer: $x^N = \widetilde{Z} + 1$.

Prove that the value $\widetilde{Z} + 1$ is actually equal to x^N. Indeed, the obtained at the fourth step value $Z = \dfrac{1}{N} \sum_{k=0}^{N-1} \dfrac{\omega_k}{x - \omega_k}$ can be modified into the form $Z = \dfrac{1}{x^N - 1}$ (see Exercise **7.29**). Hence it follows that $\widetilde{Z} = x^N - 1$ and at the sixth step, we have $\widetilde{Z} + 1 = x^N$. Hence, Kung's algorithm is incorrect.

Estimate the algorithm execution time. The first step does not depend on the value x, which allows preparing the array with the values $\omega_k, 0 \leqslant k \leqslant N - 1$, prior to the beginning of its operation.

Let the time of the operation of summation of two complex numbers be equal to T_a, and the time of division of two complex numbers be equal to T_d. Then the execution time of the above-mentioned steps 2–6 on a multiprocessor computation system with N processors will be equal to

$$T(N) = T_a + T_d + O(\log_2 N)T_a + T_d + T_a = 2T_d + O(\log_2 N)T_a.$$

To avoid confusion, note that in the last equality the simplification $O(\log_2 N) + 2 = O(\log_2 N)$ has been performed.

13.16. *Hint.* See Exercise **13.15** and the original work [40].

13.22. *Proof.*
In order to estimate the number of comparisons, determine how many comparisons at each run of the algorithm with the increment h_s are performed, where $s = \lfloor \log_3(2N + 1) \rfloor - 1, \ldots, 1, 0$, and then sum the obtained values over s.

Separately consider the cases $h_s > \sqrt{N}$ and $h_s \leqslant \sqrt{N}$. Choice of the value $\sqrt{N}$ will become clear from the further consideration [81].

1. Let $h_s > \sqrt{N}$. Using the explicit formula for the sth increment $h_s = (3^{s+1} - 1)/2$, we conclude that $s = s_{\max}, \ldots, s^* + 1$, where the following notations are introduced: $s_{\max} = \lfloor \log_3(2N + 1) \rfloor - 1$, $s^* = \lfloor \log_3(2\sqrt{N} + 1) \rfloor - 1$.

At each iteration of the internal cycle **for**, $O\left(\dfrac{N^2}{h_s^2}\right)$ comparisons are performed; the number of algorithm runs is equal to h_s. The total number of comparisons at some stage with number s for $s = s_{\max}, \ldots, s^* + 1$ is equal to $O\left(\dfrac{N^2}{h_s}\right)$.

2. Let $h_s \leqslant \sqrt{N}$. This corresponds to the values $s = s^*, \ldots, 0$. Consider the array being sorted $a[1..N]$ after carrying out two insertion sorts that use the increments h_{s+1} and h_{s+2}. In accordance with the theorem of the h- and k-ordering for all i, divisible by h_{s+1} or h_{s+2}, we have $a[j - i] \leqslant a[j]$, where $1 \leqslant j \leqslant N$, $i < j$. Further, since for all nonnegative integers c_1 and c_2 on condition that

$j - (c_1 h_{s+1} + c_2 h_{s+2}) > 0$ the implication is valid

$$\left. \begin{array}{l} a[j - h_{s+1}] \leqslant a[j] \\ a[j - h_{s+2}] \leqslant a[j] \end{array} \right\} \Rightarrow a[j - (c_1 h_{s+1} + c_2 h_{s+2})] \leqslant a[j],$$

then the relation $a[j - i] \leqslant a[j]$ is valid for all such i, for which exist $c_1, c_2 \in \mathbb{N} \cup \{0\}: i = c_1 h_{s+1} + c_2 h_{s+2}$.

By definition of the variables h_{s+1} and h_{s+2}, their numerical values cannot have common divisors except one, at any s. From number theory it is known [36], that any natural number $d \geqslant (\alpha - 1)(\beta - 1)$ can be represented in the form $d = c_1 \alpha + c_2 \beta$, where $c_1, c_2 \in \mathbb{N} \cup \{0\}$, if the natural numbers α and β do not have common divisors other than one.

Form these considerations follows the conclusion: All numbers no smaller than

$$d = (h_{s+1} - 1)(h_{s+2} - 1) = 27 h_s^2 + 9 h_s,$$

can be represented in the form $d = c_1 h_{s+1} + c_2 h_{s+2}$. Hence, inversions in the array a at this sort stage are formed by the elements situated at a distance no more than

$$O\left(\frac{27 h_s^2 + 9 h_s}{h_s}\right) = O(h_s) \text{ from each other, and we have the estimate } O(h_s N) \text{ of}$$

the number of comparisons of the array elements.

Calculate the worst-case number of comparisons $W(N)$ after all algorithm runs with numbers $s = s_{max}, \ldots, s^* + 1, s^*, \ldots, 0$.

$$W(N) = \sum_{s=0}^{s^*} O(h_s N) + \sum_{s=s^*+1}^{s_{max}} O\left(\frac{N^2}{h_s}\right) = O\left(N \sum_{s=0}^{s^*} h_s\right) + O\left(N^2 \sum_{s=s^*+1}^{s_{max}} h_s^{-1}\right).$$

Let us use the explicit expression for the value of the sth increment $h_s = (3^{s+1} - 1)/2$ for $s = 0, \ldots, s_{max}$. Then

$$\sum_{s=0}^{s^*} h_s \leqslant \frac{1}{2} \sum_{s=0}^{s^*} 3^{s+1} = \frac{1}{2} \frac{3^{s^*+2} - 1}{2} = O(N^{1/2}),$$

$$\sum_{s=s^*+1}^{s_{max}} h_s^{-1} \leqslant 2 \sum_{s=s^*+1}^{s_{max}} \frac{1}{3^{s+2}} = \frac{2}{9} \frac{3^{-s^*} - 3^{-s_{max}}}{2} = O(N^{-1/2}).$$

Note that the formula for geometric progression has been used $\sum_{s=a}^{b} 3^{-s} = (3^{1-a} - 3^{-b})/2$. Now the choice of the variable $\sqrt{N}$ becomes clear for division of the sum $W(N)$ over the variable s into two parts, namely, with such a division we obtain an exact upper bound for the number of comparisons of the array elements.

In other words, the choice of the value h_{s^*} corresponds to the minimum of the auxiliary function $f(h) = Nh + \dfrac{N^2}{h}$, defined on the integer-valued points of the

Fig. 13.11 Minimum of the function $f(h) = Nh + \dfrac{N^2}{h}$ is located at the point with the coordinates $(\sqrt{N}, 2N^{3/2})$

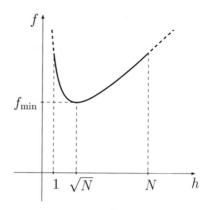

interval $(1, N)$. A graph of the function $f = f(h)$ is shown in Fig. 13.11, where, for illustration, the domain is expanded up to $\mathbb{R}^+$.

As a result, we obtain the number of comparisons $W(N) = O(N^{3/2})$, performed when using the sequential version of the Shell sort.

13.23. *Hint.*
Obtain the estimate of the value $W(N)$ of the form

$$W(N) = \sum_{s=0}^{s^*} O(N) + \sum_{s=s^*+1}^{s_{\max}} O\left(\frac{N^2}{h_s^2}\right),$$

where $s_{\max} = O(\log_2 N)$ is the number of increments used, and s^* satisfies the equation $h_{s^*}^2 = O\left(\dfrac{N}{\log_2 N}\right)$, which is equivalent to the asymptotic relation $s^* = O(\log_2 N - \log_2 \log_2 N)$.

13.24. *Solution.*
The parallel version of the Shell sort uses $p = \Theta(N)$ processors, where N—the array size. At the sth stage of the Shell sort, the increment is equal to h_s, where $s = 0, 1, \ldots, s_{\max}$, and h_s processors, independently from each other, order h_s subarrays of size $\left\lfloor \dfrac{N}{h_s} \right\rfloor$ or $\left\lceil \dfrac{N}{h_s} \right\rceil$ with the help of the insertion sort. Each subarray under consideration, according to the results of Exercise **12.48** contains $\Theta\left(\dfrac{N}{h_s}\right)$ inversions for all possible increment values. Hence, the sth sorting stage takes time $\Theta\left(\dfrac{N}{h_s}\right)$. The sum over all stages leads to the estimate of this algorithm's worst-case execution time:

$$W(N) = \sum_{s=0}^{s_{\max}} \Theta\left(\frac{N}{h_s}\right) = N\Theta\left(\sum_{s=0}^{s_{\max}} \frac{1}{h_s}\right).$$

For calculation of the obtained sum, represent h_s in the form $h_s = 2^i 3^j$, where the variables i and j have the same values as $2^i 3^j < N$, i.e., the inequalities $0 \leqslant i \leqslant i_{max}$, $0 \leqslant j \leqslant j_{max}$ are fulfilled. Now we can proceed to the estimate of the expression $\sum_{s=0}^{s_{max}} \frac{1}{h_s}$, for which purpose we will use the geometric progression formula:

$$\sum_{s=0}^{s_{max}} \frac{1}{h_s} = \sum_{i=0}^{i_{max}} \sum_{j=0}^{j_{max}} \frac{1}{2^i 3^j} = \sum_{i=0}^{i_{max}} \frac{1}{2^i} \left(\sum_{j=0}^{j_{max}} \frac{1}{3^j} \right) = \sum_{i=0}^{i_{max}} \frac{1}{2^i} \cdot \frac{3}{2} \left(1 - \frac{1}{3^{j_{max}+1}} \right) =$$

$$= 3 \left(1 - \frac{1}{2^{i_{max}+1}} \right) \left(1 - \frac{1}{3^{j_{max}+1}} \right).$$

From the inequalities $1 < h_{s_{max}} < N$ follows that $1 < 2^{i_{max}} < N$ and $1 < 3^{j_{max}} < N$. Hence,

$$3 \left(1 - \frac{1}{2} \right) \left(1 - \frac{1}{3} \right) < \sum_{s=0}^{s_{max}} \frac{1}{h_s} < 3 \left(1 - \frac{1}{2N} \right) \left(1 - \frac{1}{3N} \right),$$

or $\sum_{s=0}^{s_{max}} \frac{1}{h_s} = \Theta(1)$.

So, the execution time of the Shell sort based on the increment sequence suggested by Pratt is equal to $W(N) = \Theta(N)$.

13.29. *Answer*:

(1) $\mathcal{F}[\mathbf{x}] = (2, 0, -2, 0)$;
(2) $\mathcal{F}[\mathbf{x}] = (2, 0, 2, 0)$.

13.30. *Solution*.

In accordance with the definition of DFT

$$(\mathcal{F}[\mathbf{x}])_n = \sum_{k=0}^{N-1} e^{\frac{2\pi i}{N} kn} C(N-1, n) \text{ for } 0 \leqslant n \leqslant N-1.$$

The obtained relation for the nth component of the vector $\mathcal{F}[\mathbf{x}]$ can be transformed subject to the Newton binomial (binomial theorem) formula:

$$(\mathcal{F}[\mathbf{x}])_n = \sum_{k=0}^{N-1} C(N-1, k) \left(e^{\frac{2\pi i n}{N}} \right)^k = \left(1 + e^{\frac{2\pi i n}{N}} \right)^{N-1} = (1 + \omega^n)^{N-1}.$$

for all $n = 0, 1, \ldots, N-1$.

13.31. *Answer*:

(1) $(\mathcal{F}[\mathbf{x}])_n = \begin{cases} \dfrac{N}{2}, & \text{if } n = 0; \\ -\dfrac{1 - (-1)^n}{1 - \omega^n}, & \text{if } n = 1, 2, \ldots, N-1; \end{cases}$

$$
(2)\ (\mathcal{F}[\mathbf{x}])_n = \begin{cases} \dfrac{N}{2}, & \text{if } n = 0; \\ \dfrac{1 - (-1)^n}{1 - \omega^n}, & \text{if } n = 1, 2, \ldots, N - 1. \end{cases}
$$

13.32. *Solution.*
Denote the array elements by $x[i]$, where $0 \leqslant i \leqslant 7$. As is known, $x[i]$ and $x[j]$ change places if and only if the binary representations of the variables i and j are reversible. Write out the bit strings, corresponding to the binary notation of nonnegative integers, not exceeding seven.

$$
\begin{aligned}
0 &= (000)_2, & 4 &= (100)_2, \\
1 &= (001)_2, & 5 &= (101)_2, \\
2 &= (010)_2, & 6 &= (110)_2, \\
3 &= (011)_2, & 7 &= (111)_2.
\end{aligned}
$$

By comparison of the bit strings written out, we come to the conclusion that two pairs are reversible: $((001)_2, (100)_2)$ and $((011)_2, (110)_2)$. Hence, the "shuffle" procedure will perform changes of type swap$(x[1], x[4])$ and swap$(x[3], x[6])$, following which, the array will take the form | 0 | 4 | 2 | 6 | 1 | 5 | 3 | 7 |.

13.33. *Answer:* | 0 | 8 | 4 | 12 | 2 | 10 | 6 | 14 | 1 | 9 | 5 | 13 | 3 | 11 | 7 | 15 |

13.35. *Solution.*
"Shuffle" procedure changes places of the elements $x[i]$ and $x[j]$, where $0 \leqslant i, j \leqslant N - 1$, if and only if the binary representations of the variables i and j are reversible. Calculate the number of $\mathcal{R}$ unordered pairs $\{i, j\}$, satisfying this condition.

The bit string $(i)_2$, corresponding to the binary representation of the number i, consists of m elements. If the string $(i)_2$ is not a palindrome (see Exercise **4.2**), then there exists such index j, not exceeding the limits of the array $x[0..(N-1)]$, that $(i)_2$ and $(j)_2$ are reversible.

The number of palindromes among the bit strings of length m is found in Exercise **4.3** and is equal to $2^{\lceil m/2 \rceil}$. Therefore, the number of unordered pairs $\{i, j\}$, for which $(i)_2$ and $(j)_2$ are reversible, is equal to

$$
\mathcal{R} = \left(2^m - 2^{\lceil m/2 \rceil}\right)/2 = 2^{m-1} - 2^{\lceil m/2 \rceil - 1}.
$$

From the obtained relation follows the asymptotic estimate $\mathcal{R} = O(N)$.

13.36. *Solution.*
Arguments of the functions $\sin\left(\dfrac{\pi k}{N}\right)$ in the definition of the sine transform differ from the arguments of the exponents of the standard DFT by the multiplier $\dfrac{1}{2}$, this is why it is impossible to consider $\mathcal{F}_{\sin}[\mathbf{x}]$ as an imaginary part of $\mathcal{F}[\mathbf{x}]$. More accurate transformations are required [57].

Consider the vector $\widetilde{\mathbf{x}}$ of length $2N$, whose components satisfy the relations $\widetilde{x}[k] = x[k]$ for $0 \leqslant k \leqslant N - 1$ and $\widetilde{x}[k] = -x[2N - k]$ for $N + 1 \leqslant k \leqslant 2N - 1$, and $\widetilde{x}[N] = 0$. In other words, $\widetilde{\mathbf{x}}$ is constructed from the source vector $\mathbf{x}$ by antisymmetric reflection of the values $x[k]$ $(0 \leqslant k \leqslant N - 1)$ with respect to $k = N$.

Let us apply to the obtained tuple $\widetilde{x}[k]$, $k = 0, \ldots, 2N - 1$ the discrete Fourier transform:

$$(\mathcal{F}[\widetilde{\mathbf{x}}])_n = \sum_{k=0}^{2N-1} \widetilde{x}_k e^{2\pi i k n/(2N)}.$$

Further, consider the part of the sum, corresponding to the values $k = N, \ldots, 2N - 1$, and make a change in it $k' = 2N - k$:

$$\sum_{k=N}^{2N-1} \widetilde{x}_k e^{2\pi i k n/(2N)} = \sum_{k'=1}^{N} \widetilde{x}_{2N-k'} e^{2\pi i (2N-k')n/(2N)} =$$

$$= \sum_{k'=1}^{N} \widetilde{x}_{2N-k'} e^{2\pi i n} e^{-2\pi i k' n/(2N)} = - \sum_{k'=1}^{N} x_{k'} e^{-2\pi i k' n/(2N)}.$$

The discrete Fourier transform of the vector $\widetilde{\mathbf{x}}$ takes the form

$$(\mathcal{F}[\widetilde{\mathbf{x}}])_n = \sum_{k=0}^{N-1} x_k \left[e^{2\pi i k n/(2N)} - e^{-2\pi i k n/(2N)} \right] = 2i \sum_{k=0}^{N-1} x_k \sin\left(\frac{\pi k n}{N}\right).$$

Hence, DFT of the vector $\widetilde{\mathbf{x}}$ accurate to the multiplier $2i$ coincides with the sine transform of the vector $\mathbf{x}$.

13.39. *Solution.*
The fast Fourier transform algorithm calculates the components of the vector $\mathbf{y} = \mathcal{F}[\mathbf{x}]$, where $\mathbf{x}$ is the source data vector. Represent the input array in the form of the rowwise filled-in matrix M, whose number of columns is equal to N_1, and number of rows—N_2. Since N_1 and N_2 are numbers of the form 2^m, where $m \in \mathbb{N}$, and the number of the array elements is equal to $N_1 N_2 = 8$, then we can choose, for example, the following values: $N_1 = 4$, $N_2 = 2$. In this case, the matrix M will take the form:

$$M = \begin{bmatrix} 0 & 1 & 2 & 3 \\ 4 & 5 & 6 & 7 \end{bmatrix}.$$

Operation of the parallel version of the FFT algorithm reduces to the following steps.

1. The sequential version of FFT is applied to the columns of the matrix M. After that we obtain that $M = \begin{bmatrix} 4 & 6 & 8 & 10 \\ -4 & -4 & -4 & -4 \end{bmatrix}$.

2. Each element of the matrix M is multiplied by the value $e^{\frac{2\pi i}{N_1 N_2}(l\widetilde{L})}$, where l is the column number, $0 \leqslant l \leqslant N_1 - 1$, $\widetilde{L}$ is the row number, $0 \leqslant \widetilde{L} \leqslant N_2 - 1$ (see

p. 445). Then the matrix M takes the form:

$$M = \begin{bmatrix} 4 & 6 & 8 & 10 \\ -4 & -2\sqrt{2}(1+i) & -4i & -2\sqrt{2}(-1+i) \end{bmatrix}.$$

3. The sequential version of FFT is applied to the rows of the matrix M:

$$M = \begin{bmatrix} 28 & -4-4i & -4 & -4+4i \\ -4-4(\sqrt{2}+1)i & -4-4(\sqrt{2}-1)i & -4+4(\sqrt{2}-1)i & -4+4(\sqrt{2}+1)i \end{bmatrix}.$$

As a result, the FFT components y_i $(i = 0, \ldots, 7)$ of the original array are obtained writing out the elements of the matrix M in a columnwise manner:

$$y_0 = 28,$$
$$y_1 = -4 - 4(\sqrt{2}+1)i,$$
$$y_2 = -4 - 4i,$$
$$y_3 = -4 - 4(\sqrt{2}-1)i,$$
$$y_4 = -4,$$
$$y_5 = -4 + 4(\sqrt{2}-1)i,$$
$$y_6 = -4 + 4i,$$
$$y_7 = -4 + 4(\sqrt{2}+1)i.$$

Appendix: Reference Data

In the formulae of this section, unless otherwise specified, $a, b, c_1, c_2, C \in \mathbb{R}, k, k' \in \mathbb{Z}$.

A.1 Trigonometric Formulae

$$\sin^2 a + \cos^2 a = 1; \tag{A.1}$$

$$\tan a = \frac{\sin a}{\cos a}, \quad a \neq \frac{\pi}{2} + \pi k; \tag{A.2}$$

$$\cot a = \frac{\cos a}{\sin a}, \quad a \neq \pi k; \tag{A.3}$$

$$1 + \tan^2 a = \frac{1}{\cos^2 a}, \quad a \neq \frac{\pi}{2} + \pi k; \tag{A.4}$$

$$1 + \cot^2 a = \frac{1}{\sin^2 a}, \quad a \neq \pi k; \tag{A.5}$$

$$\sin 2a = 2 \sin a \cos a, \qquad\qquad \cos 2a = \cos^2 a - \sin^2 a; \tag{A.6}$$

$$\tan 2a = \frac{2 \tan a}{1 - \tan^2 a}, \quad a \neq \frac{\pi}{4} + \frac{\pi k}{2}, \ a \neq \frac{\pi}{2} + \pi k'; \tag{A.7}$$

$$\sin^2 \frac{a}{2} = \frac{1 - \cos a}{2}, \qquad\qquad \cos^2 \frac{a}{2} = \frac{1 + \cos a}{2}; \tag{A.8}$$

© Springer International Publishing AG, part of Springer Nature 2018
S. Kurgalin and S. Borzunov, *The Discrete Math Workbook*,
Texts in Computer Science, https://doi.org/10.1007/978-3-319-92645-2

$$\sin(a + b) = \sin a \cos b + \cos a \sin b; \tag{A.9}$$

$$\sin(a - b) = \sin a \cos b - \cos a \sin b; \tag{A.10}$$

$$\cos(a + b) = \cos a \cos b - \sin a \sin b; \tag{A.11}$$

$$\cos(a - b) = \cos a \cos b + \sin a \sin b; \tag{A.12}$$

$$\tan(a + b) = \frac{\tan a + \tan b}{1 - \tan a \tan b}, \quad a, b, a + b \neq \frac{\pi}{2} + \pi k; \tag{A.13}$$

$$\tan(a - b) = \frac{\tan a - \tan b}{1 + \tan a \tan b}, \quad a, b, a - b \neq \frac{\pi}{2} + \pi k; \tag{A.14}$$

$$\sin a + \sin b = 2 \sin\left(\frac{a + b}{2}\right) \cos\left(\frac{a - b}{2}\right); \tag{A.15}$$

$$\sin a - \sin b = 2 \cos\left(\frac{a + b}{2}\right) \sin\left(\frac{a - b}{2}\right); \tag{A.16}$$

$$\cos a + \cos b = 2 \cos\left(\frac{a + b}{2}\right) \cos\left(\frac{a - b}{2}\right); \tag{A.17}$$

$$\cos a - \cos b = -2 \sin\left(\frac{a + b}{2}\right) \sin\left(\frac{a - b}{2}\right); \tag{A.18}$$

$$\tan a \pm \tan b = \frac{\sin(a \pm b)}{\cos a \cos b}, \quad a, b \neq \frac{\pi}{2} + \pi k; \tag{A.19}$$

$$\cot a \pm \cot b = \frac{\sin(b \pm a)}{\sin a \sin b}, \quad a, b \neq \pi k; \tag{A.20}$$

$$\sin a \, \sin b = \frac{1}{2}\left(\cos(a - b) - \cos(a + b)\right); \tag{A.21}$$

$$\cos a \, \cos b = \frac{1}{2}\left(\cos(a - b) + \cos(a + b)\right); \tag{A.22}$$

$$\sin a \, \cos b = \frac{1}{2}\left(\sin(a - b) + \sin(a + b)\right). \tag{A.23}$$

A.2 Differential Calculus: The Basic Rules

Suppose that $f, g\colon \mathbb{R} \to \mathbb{R}$ have derivatives at a some point x, that is, limits $f'(x) = \lim_{\Delta x \to 0} \frac{f(x + \Delta x) - f(x)}{\Delta x}$ and $g'(x) = \lim_{\Delta x \to 0} \frac{g(x + \Delta x) - g(x)}{\Delta x}$ are exist. Then at this point

$$(c_1 f + c_2 g)' = c_1 f' + c_2 g' \qquad \text{(linearity);} \qquad \text{(A.24)}$$

$$(f(g))' = f'_g \times g'_x \qquad \text{(chain rule);} \qquad \text{(A.25)}$$

$$(f \times g)' = f'g + fg' \qquad \text{(derivative of a product);} \qquad \text{(A.26)}$$

$$\left(\frac{f}{g}\right)' = \frac{f'g - fg'}{g^2}, \ g \neq 0 \qquad \text{(derivative of a ratio).} \qquad \text{(A.27)}$$

A.3 Derivatives of the Basic Elementary Functions

$$(x^a)' = ax^{a-1}; \qquad \text{(A.28)}$$

$$(\log_a x)' = \frac{1}{x \ln a}, \quad a > 0, \ a \neq 1; \qquad \text{(A.29)}$$

$$(a^x)' = a^x \ln a, \quad a > 0; \qquad \text{(A.30)}$$

$$(\sin x)' = \cos x, \qquad (\cos x)' = -\sin x; \qquad \text{(A.31)}$$

$$(\tan x)' = \frac{1}{\cos^2 x}, \qquad (\cot x)' = -\frac{1}{\sin^2 x}; \qquad \text{(A.32)}$$

$$(\arcsin x)' = \frac{1}{\sqrt{1 - x^2}}, \qquad (\arccos x)' = -\frac{1}{\sqrt{1 - x^2}}; \qquad \text{(A.33)}$$

$$(\arctan x)' = \frac{1}{1 + x^2}, \qquad (\text{arccot}\, x)' = -\frac{1}{1 + x^2}. \qquad \text{(A.34)}$$

A.4 Integral Calculus: The Basic Rules

$$\int f'(x)\, dx = f(x) + C \qquad \text{(Integration is the inverse operation}$$

$$\text{to differentiation);} \qquad \text{(A.35)}$$

$$\int (c_1 f(x)\, dx + c_2 g(x))\, dx = c_1 \int f(x)\, dx + c_2 \int g(x)\, dx \qquad \text{(linearity);}$$

$$\text{(A.36)}$$

$$\int f(g(x))\, dg(x) = \int f(t)\, dt \bigg|_{t=g(x)} \qquad \text{(variable substitution } t = g(x)\text{);}$$

$$\text{(A.37)}$$

$$\int f\, dg = fg - \int g\, df \qquad \text{(integration by parts).} \qquad \text{(A.38)}$$

A.5   Indefinite Integrals of Some Functions

In formulae (A.42)–(A.46), we assume $a > 0$.

$$\int x^a \, dx = \frac{x^{a+1}}{a+1} + C, \quad a \neq -1; \tag{A.39}$$

$$\int \frac{dx}{x} = \ln |x| + C; \tag{A.40}$$

$$\int a^x \, dx = \frac{1}{\ln a} a^x + C, \quad a > 0, \, a \neq 1; \tag{A.41}$$

$$\int \frac{dx}{a^2 + x^2} = \frac{1}{a} \arctan \frac{x}{a} + C; \tag{A.42}$$

$$\int \frac{dx}{a^2 - x^2} = \frac{1}{2a} \ln \left| \frac{a+x}{a-x} \right| + C; \tag{A.43}$$

$$\int \frac{dx}{\sqrt{a^2 - x^2}} = \arcsin \frac{x}{a} + C; \tag{A.44}$$

$$\int \frac{dx}{\sqrt{x^2 \pm a^2}} = \ln \left| x + \sqrt{x^2 \pm a^2} \right| + C; \tag{A.45}$$

$$\int \sqrt{a^2 - x^2} \, dx = \frac{a^2}{2} \arcsin \frac{x}{a} + \frac{x}{2} \sqrt{a^2 - x^2} + C; \tag{A.46}$$

$$\int \sqrt{x^2 \pm a^2} \, dx = \frac{x}{2} \sqrt{x^2 \pm a^2} \pm \frac{a^2}{2} \ln \left| x + \sqrt{x^2 \pm a^2} \right| + C; \tag{A.47}$$

$$\int \sin x \, dx = -\cos x + C; \tag{A.48}$$

$$\int \cos x \, dx = \sin x + C; \tag{A.49}$$

$$\int \tan x \, dx = -\ln |\cos x| + C; \tag{A.50}$$

$$\int \cot x \, dx = \ln |\sin x| + C; \tag{A.51}$$

$$\int \frac{dx}{\sin^2 x} = -\cot x + C; \tag{A.52}$$

$$\int \frac{dx}{\cos^2 x} = \tan x + C. \tag{A.53}$$

A.6   Finite Summation

Let $N, s_1, s_2 \in \mathbb{N}, d > -1$. Then we have:

$$\sum_{i=1}^{N} 1 = N; \tag{A.54}$$

$$\sum_{i=s_1}^{s_2} 1 = s_2 - s_1 + 1; \tag{A.55}$$

$$\sum_{i=1}^{N} i = \frac{N(N+1)}{2}; \tag{A.56}$$

$$\sum_{i=1}^{N} i^2 = \frac{N(N+1)(2N+1)}{6}; \tag{A.57}$$

$$\sum_{i=1}^{N} i^3 = \frac{N^2(N+1)^2}{4}; \tag{A.58}$$

$$\sum_{i=1}^{N} i^d = O(N^{d+1}); \tag{A.59}$$

$$\sum_{i=1}^{N} 2^i = 2^{N+1} - 2; \tag{A.60}$$

$$\sum_{i=1}^{N} i 2^i = (N-1)2^{N+1} + 2; \tag{A.61}$$

$$\sum_{i=1}^{N} \frac{1}{i} = \ln N + \gamma + O\left(\frac{1}{N}\right), \text{ where } \gamma = 0{,}5772\ldots; \tag{A.62}$$

$$\sum_{i=1}^{N} \log_2 i = N \log_2 N + O(N). \tag{A.63}$$

A.7   The Greek Alphabet

A, α	alpha	N, ν	nu
B, β	beta	Ξ, ξ	xi
Γ, γ	gamma	O, o	omicron
Δ, δ	delta	Π, π	pi
E, ε	epsilon	P, ρ	rho
Z, ζ	zeta	Σ, σ	sigma
H, η	eta	T, τ	tau
Θ, θ	theta	Υ, υ	upsilon
I, ι	iota	Φ, φ	phi
K, $\varkappa$	kappa	X, χ	chi
Λ, λ	lambda	Ψ, ψ	psi
M, μ	mu	Ω, ω	omega

References

1. Aho, A.V., Sloane, N.J.A.: Some doubly exponential sequences. Fibonacci Q. **11**(4), 429–438 (1973)
2. Aigner, M., Ziegler, G.M.: Proofs from THE BOOK, 5th edn. Springer, Berlin (2014)
3. Akl, S.G.: The Design and Analysis of Parallel Algorithms. Prentice Hall, Englewood Cliffs (1989)
4. Anderson, J.A.: Discrete Mathematics with Combinatorics. Prentice Hall, Upper Saddle River (2003)
5. Apostol, T.M.: Mathematical Analysis, 2nd edn. Addison-Wesley, Reading (1974)
6. Arnol'd, V.I.: A mathematical trivium. Russ. Math. Surv. **46**(1), 271–278 (1991)
7. Bachmann, P.G.H.: Die Analytische Zahlentheorie. B. G. Teubner, Leipzig (1894)
8. Bang-Jensen, J., Gutin, G.: Digraphs : Theory, Algorithms and Applications, 2nd edn. Springer Monographs in Mathematics. Springer, London (2008)
9. Bentley, J.L., Haken, D., Saxe, J.B.: A general method for solving divide-and-conquer recurrences. SIGACT News **12**(3), 36–44 (1980)
10. Brent, R.P.: The parallel evaluation of general arithmetic expressions. J. Assoc. Comput. Mach. **21**(2), 201–206 (1974)
11. Breshears, C.: The Art of Concurrency. O'Reilly, Beijing (2009)
12. Chapman, B., Jost, G., van der Pas, R.: Using OpenMP: Portable Shared Memory Parallel Programming. Scientific and Engineering Computation. The MIT Press, Cambridge (2008)
13. Cormen, T.H., Leiserson, C.E., Rivest, R.L., Stein, C.: Introduction to Algorithms, 3rd edn. The MIT Press, Cambridge (2009)
14. Courant, R., Robbins, H.: What is Mathematics? An Elementary Approach to Ideas and Methods, 2nd edn. Oxford University Press, Oxford (1996)
15. Diestel, R.: Graph Theory, 5th edn. Graduate Texts in Mathematics, vol. 173. Springer, Berlin (2017)
16. Elaydi, S.: An Introduction to Difference Equations, 3rd edn. Undergraduate Texts in Mathematics. Springer, Berlin (2005)
17. Euler, L.: Solutio problematis ad geometriam situs pertinentis. Commentarii Academiae Scientiarum Imperialis Petropolitanae (Memoirs of the Imperial Academy of Sciences in St. Petersburg) **8**, 128–140 (1736)

18. Finch, S.R.: Mathematical Constants. Encyclopedia of Mathematics and Its Applications, vol. 94. Cambridge University Press, Cambridge (2003)
19. Flajolet, P., Raoult, J.C., Vuillemin, J.: The number of registers required for evaluating arithmetic expressions. Theor. Comput. Sci. **9**(1), 99–125 (1979)
20. Gamelin, T.W.: Complex Analysis. Undergraduate Texts in Mathematics. Springer, Berlin (2001)
21. Gonnet, G.H., Baeza-Yates, R.: Handbook of Algorithms and Data Structures: In Pascal and C, 2nd edn. International Computer Science Series. Addison-Wesley, Reading (1991)
22. Grafakos, L.: Classical Fourier Analysis, 2nd edn. Graduate Texts in Mathematics, vol. 249. Springer, Berlin (2008)
23. Graham, R.L., Knuth, D.E., Patashnik, O.: Concrete mathematics: A Foundation for Computer Science, 2nd edn. Addison-Wesley, Reading (1994)
24. Greene, D.H., Knuth, D.E.: Mathematics for the Analysis of Algorithms, 3rd edn. Modern Birkhäuser Classics. Birkhäuser, Boston (2008)
25. Grimaldi, R.P.: Discrete and Combinatorial Mathematics. An Applied Introduction, 5th edn. Pearson Education, New Delhi (2004)
26. Haggard, G., Schlipf, J., Whitesides, S.: Discrete Mathematics for Computer Science. Thomson Brooks/Cole, Belmont (2006)
27. Haggarty, R.: Discrete Mathematics for Computing. Addison-Wesley, Reading (2002)
28. Harary, F.: Graph Theory. Addison-Wesley, Reading (1969)
29. Hazewinkel, M. (ed.): Encyclopaedia of Mathematics, vol. 10. Springer, Berlin (1994)
30. Hoare, C.A.R.: Quicksort. Comput. J. **5**(1), 10–15 (1962)
31. Holton, D.A., Sheehan, J.: The Petersen Graph. Australian Mathematical Society Lecture Series, vol. 7. Cambridge University Press, Cambridge (1993)
32. Horowitz, E., Sahni, S., Rajasekaran, S.: Computer Algorithms, 2nd edn. Silicon Press, Summit (2008)
33. Karnaugh, M.: The map method for synthesis of combinational logic circuits. Trans. Am. Inst. Electr. Eng. Part I **72**(9), 593–599 (1953)
34. Kelley, J.L.: General Topology. Graduate Texts in Mathematics, vol. 27. Springer, New York (1995)
35. Knuth, D.E.: The Art of Computer Programming, Volume 1. Fundamental Algorithms, 3rd edn. Addison-Wesley, Reading (1997)
36. Knuth, D.E.: The Art of Computer Programming, Volume 3. Sorting and Searching, 2nd edn. Addison-Wesley, Reading (1998)
37. Koshy, T.: Discrete Mathematics with Applications. Elsevier Academic Press, Amsterdam (2004)
38. Kostrikin, A.I., Shafarevich, I.R. (eds.): Algebra I : Basic Notions of Algebra. Encyclopaedia of Mathematical Sciences, vol. 11. Springer, Berlin (2005)
39. Kruskal, C.P., Rudolph, L., Snir, M.: A complexity theory of efficient parallel algorithms. Theor. Comput. Sci. **71**(1), 95–132 (1990)
40. Kung, H.T.: New algorithms and lower bounds for the parallel evaluation of certain rational expressions. In: Proceedings of the 6th Annual ACM Symposium on Theory of Computing, April 30–May 2, 1974, Seattle, Washington, USA, ed. by R.L. Constable (1974), pp. 323–333
41. Kurosh, A.: Higher Algebra, 4th edn. Mir Publishers, Moscow (1984)
42. Lando, S.K.: Lectures on Generating Functions. Student Mathematical Library, vol. 23. American Mathematical Society, Providence (2003)
43. Levitin, A.: Introduction to the Design & Analysis of Algorithms, 3rd edn. Pearson, Boston (2012)
44. Lovász, L., Pelikán, J., Vesztergombi, K.: Discrete Mathematics: Elementary and Beyond. Undergraduate Texts in Mathematics. Springer, New York (2003)

45. Mahmoud, H.M.: Sorting: A Distribution Theory. Wiley, Wiley Series in Discrete Mathematics and Optimization. Wiley-Interscience, New York. (2011)
46. Mandelbrot, B.B.: The Fractal Geometry of Nature. W. H. Freeman and Company, New York (1983)
47. Markov, A.A.: The theory of algorithms [in russian]. Trudy Mat. Inst. Steklov **42**, 3–375 (1954)
48. Markushevich, A.I.: Rekursive Folgen, 3th edn. Kleine Ergänzungsreihe zu den Hochschulbüchern für Mathematik; Nr. 11. Verlag Wissenschaft (1973)
49. McConnell, J.J.: Analysis of Algorithms: An Active Learning Approach. Jones and Bartlett Publishers, Sudbury (2001)
50. McCool, M., Robison, A.D., Reinders, J.: Structured Parallel Programming: Patterns for Efficient Computation. Elsevier, Amsterdam (2012)
51. Miller, R., Boxer, L.: Algorithms Sequential and Parallel: A Unified Approach, 3rd edn. Cengage Learning, Boston (2013)
52. Moret, B.M.E., Shapiro, H.D.: Algorithms from P to NP, Volume. I. Design and Efficiency. Benjamin/Cummings, Redwood City (1991)
53. Olver, F.W.J., Lozier, D.W., Boisvert, R.F., Clark, C.W. (eds.): NIST Handbook of Mathematical Functions. Cambridge University Press, Cambridge (2010)
54. OpenMP (2018). http://www.openmp.org/
55. Ore, O.: Theory of Graphs. Colloquium Publications, vol. XXXVIII. American Mathematical Society, Providence (1962)
56. Perl, Y., Itai, A., Avni, H.: Interpolation search – a log log n search. Commun. ACM **21**(7), 550–553 (1978)
57. Press, W.H., Teukolsky, S.A., Vetterling, W.T., Flannery, B.P.: Numerical Recipes: The Art of Scientific Computing, 3rd edn. Cambridge University Press, Cambridge (2007)
58. Rauber, T., Rünger, G.: Parallel Programming for Multicore and Cluster Systems, 2nd edn. Springer, Berlin (2013)
59. Rosen, K.H.: Discrete Mathematics and Its Applications, 7th edn. McGraw-Hill, New York (2012)
60. Rosen, K.H., Michaels, J.G., Gross, J.L. (eds.): Handbook of Discrete and Combinatorial Mathematics. Discrete Mathematics and Its Applications. CRC Press, Boca Raton (2000)
61. Sedgewick, R.: Algorithms in C. Parts 1–4. Fundamentals, Data Structures, Sorting, Searching, 3rd edn. Addison-Wesley, Reading (1999)
62. Sedgewick, R.: Algorithms in C. Part 5. Graph Algorithms, 3rd edn. Addison-Wesley, Boston (2001)
63. Sedgewick, R., Flajolet, P.: An Introduction to the Analysis of Algorithms, 2nd edn. Addison-Wesley, Reading (2013)
64. Shell, D.L.: A high-speed sorting procedure. Commun. ACM **2**(7), 30–32 (1959)
65. Shen, A., Vereshchagin, N.K.: Basic set theory. Student Mathematical Library, vol. 17. American Mathematical Society, Providence (2002)
66. Shen, A., Vereshchagin, N.K.: Computable Functions. Student Mathematical Library, vol. 19. American Mathematical Society, Providence (2003)
67. Smith, J.R.: The Design and Analysis of Parallel Algorithms. Oxford University Press, New York (1993)
68. Snir, M.: On parallel searching. SIAM J. Comput. **14**(3), 688–708 (1985)
69. Sominskii, I.S.: The Method of Mathematical Induction. Popular Lectures in Mathematics, vol. 1. Blaisdell Publishing Company, New York (1961)
70. Stoll, R.R.: Sets, Logic, and Axiomatic Theories. A Series of Undergraduate Books in Mathematics. W. H. Freeman, San Francisco (1961)
71. Strassen, V.: Gaussian elimination is not optimal. Numer. Math. **13**(3), 354–356 (1969)
72. Tanenbaum, A.S., Austin, T.: Structured Computer Organization, 6th edn. Pearson, Boston (2013)

73. Trahtenbrot, B.A.: Algorithms and Automatic Computing Machines. Topics in Mathematics. D. C. Heath and Company, Boston (1963)
74. Uspenskii, V.A.: Pascal's Triangle. Popular Lectures in Mathematics, vol. 43. The University of Chicago Press, Chicago (1974)
75. Uspensky, V.A.: Post's Machine. Little Mathematics Library. Mir Publishers, Moscow (1983)
76. Van Loan, C.: Computational Frameworks for the Fast Fourier transform. Frontiers in Applied Mathematics. Society for Industrial and Applied Mathematics, Philadelphia (1992)
77. Vinogradov, I.M.: Elements of Number Theory. Dover Publications, Inc., New York (1954)
78. Vorob'ev, N.N., Levine, D.A., McLarnan, T.: Criteria for Divisibility. Popular Lectures in Mathematics, vol. 39. University of Chicago, Chicago (1980)
79. Vorobiev, N.N.: Fibonacci Numbers. Popular Lectures in Mathematics, vol. 2. Springer, Basel (2002)
80. Vretblad, A.: Fourier Analysis and Its Applications. Graduate Texts in Mathematics, vol. 223. Springer, Berlin (2003)
81. Weiss, M.A.: Data Structures and Algorithm Analysis in C++, 4th edn. Pearson, Boston (2014)
82. Wikipedia (2018). https://www.wikipedia.org/
83. Wilkinson, B., Allen, M.: Parallel Programming Techniques and Applications Using Networked Workstations and Parallel Computers, 2nd edn. Pearson Prentice Hall (2004)
84. Wilson, R.J.: Introduction to Graph Theory, 4th edn. Longman, Harlow (1998)
85. Wirth, N.: Algorithms and Data Structures. Prentice-Hall, Englewood Cliffs (1986)
86. Zorich, V.A.: Mathematical Analysis I, 2nd edn. Universitext. Springer, Berlin (2015)
87. Zorich, V.A.: Mathematical Analysis II, 2nd edn. Universitext. Springer, Berlin (2016)

Author Index

© Springer International Publishing AG, part of Springer Nature 2018
S. Kurgalin and S. Borzunov, *The Discrete Math Workbook*,
Texts in Computer Science, https://doi.org/10.1007/978-3-319-92645-2

Subject Index

© Springer International Publishing AG, part of Springer Nature 2018
S. Kurgalin and S. Borzunov, *The Discrete Math Workbook*,
Texts in Computer Science, https://doi.org/10.1007/978-3-319-92645-2

Printed in the United States
By Bookmasters